3 3073 00255301 2

DATE DUE

JAN 02 1997	

GAYLORD PRINTED IN U.S.A.

UCLA Symposia on Molecular and Cellular Biology
New Series

Series Editor
C. Fred Fox

Volume 1
Differentiation and Function of Hematopoietic Cell Surfaces
Vincent T. Marchesi and Robert C. Gallo, *Editors*

Volume 2
Mechanisms of Chemical Carcinogenesis
Curtis C. Harris and Peter A. Cerutti, *Editors*

Volume 3
Cellular Recognition
William A. Frazier, Luis Glaser, and David I. Gottlieb, *Editors*

Volume 4
Rational Basis for Chemotherapy
Bruce A. Chabner, *Editor*

Volume 5
Tumor Viruses and Differentiation
Edward M. Scolnick and Arnold J. Levine, *Editors*

Volume 6
Evolution of Hormone-Receptor Systems
Ralph A. Bradshaw and Gordon N. Gill, *Editors*

Volume 7
Recent Advances in Bone Marrow Transplantation
Robert Peter Gale, *Editor*

Volume 8
Gene Expression
Dean H. Hamer and Martin J. Rosenberg, *Editors*

SETON HALL UNIVERSITY
McLAUGHLIN LIBRARY
SO. ORANGE, N. J.

UCLA Symposia Published Previously

(Numbers refer to the publishers listed below.)

1972
Membrane Research (2)

1973
Membranes (1)
Virus Research (2)

1974
Molecular Mechanisms for the Repair of DNA (4)
Membranes (1)
Assembly Mechanisms (1)
The Immune System: Genes, Receptors, Signals (2)
Mechanisms of Virus Disease (3)

1975
Energy Transducing Mechanisms (1)
Cell Surface Receptors (1)
Developmental Biology (3)
DNA Synthesis and Its Regulation (3)

1976
Cellular Neurobiology (1)
Cell Shape and Surface Architecture (1)
Animal Virology (2)
Molecular Mechanisms in the Control of Gene Expression (2)

1977
Cell Surface Carbohydrates and Biological Recognition (1)
Molecular Approaches to Eucaryotic Genetic Systems (2)
Molecular Human Cytogenetics (2)
Molecular Aspects of Membrane Transport (1)
Immune System: Genetics and Regulation (2)

1978
DNA Repair Mechanism (2)
Transmembrane Signaling (1)
Hematopoietic Cell Differentiation (2)
Normal and Abnormal Red Cell Membranes (1)
Persistent Viruses (2)
Cell Reproduction: Daniel Mazia Dedicatory Volume (2)

1979
Covalent and Non-Covalent Modulation of Protein Function (2)
Eucaryotic Gene Regulation (2)
Biological Recognition and Assembly (1)
Extrachromosomal DNA (2)
Tumor Cell Surfaces and Malignancy (1)
T and B Lymphocytes: Recognition and Function (2)

1980
Biology of Bone Marrow Transplantation (2)
Membrane Transport and Neuroreceptors (1)
Control of Cellular Division and Development (1)
Animal Virus Genetics (2)
Mechanistic Studies of DNA Replication and Genetic Recombination (2)

1981
Immunoglobulin Idiotypes (2)
Initiation of DNA Replication (2)
Genetic Variation Among Influenza Viruses (2)
Developmental Biology Using Purified Genes (2)
Differentiation and Function of Hematopoietic Cell Surfaces (1)
Mechanisms of Chemical Carcinogenesis (1)
Cellular Recognition (1)

1982
B and T Cell Tumors (2)
Interferon (2)
Rational Basis for Chemotherapy (1)
Gene Regulation (2)
Tumor Viruses and Differentiation (1)
Evolution of Hormone Receptor Systems (1)

Publishers

(1) Alan R. Liss, Inc.
150 Fifth Avenue
New York, NY 10011

(2) Academic Press, Inc.
111 Fifth Avenue
New York, NY 10003

(3) W.A. Benjamin, Inc.
2725 Sand Hill Road
Menlo Park, CA 94025

(4) Plenum Publishing Corp.
227 W. 17th Street
New York, NY 10011

Symposia Board

C. Fred Fox, Director
Molecular Biology Institute
UCLA

Members

Ronald Cape, Ph.D., MBA
Chairman
Cetus Corporation

Pedro Cuatrecasas, M.D.
Vice President for Research
Burroughs Wellcome Company

Luis Glaser, Ph.D.
Professor and Chairman
of Biochemistry
Washington University School
of Medicine

Donald Steiner, M.D.
Professor of Biochemistry
University of Chicago

Ernest Jaworski, Ph.D.
Director of Molecular Biology
Monsanto

Paul Marks, M.D.
President
Sloan-Kettering Institute

William Rutter, Ph.D.
Professor and Chairman
of Biochemistry
University of California Medical
Center

Sidney Udenfriend, Ph.D.
Director
Roche Institute

The members of the board advise the director in identification of topics for future symposia.

GENE EXPRESSION

GENE EXPRESSION

Proceedings of a Cetus-UCLA Symposium held at Park City, Utah, March 26–April 1, 1983

Editors

Dean H. Hamer
National Cancer Institute
National Institutes of Health
Bethesda, Maryland

Martin J. Rosenberg
National Cancer Institute
National Institutes of Health
Bethesda, Maryland
and
Smith, Kline & French Laboratories
Philadelphia, Pennsylvania

SETON HALL UNIVERSITY
McLAUGHLIN LIBRARY
SO. ORANGE, N. J.

Alan R. Liss, Inc. • New York

QH
450
G4625
1983

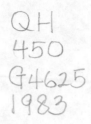

Address all Inquiries to the Publisher
Alan R. Liss, Inc., 150 Fifth Avenue, New York, NY 10011

Copyright © 1983 Alan R. Liss, Inc.

Printed in the United States of America.

Under the conditions stated below the owner of copyright for this book hereby grants permission to users to make photocopy reproductions of any part or all of its contents for personal or internal organizational use, or for personal or internal use of specific clients. This consent is given on the condition that the copier pay the stated per-copy fee through the Copyright Clearance Center, Incorporated, 21 Congress Street, Salem, MA 01970, as listed in the most current issue of "Permissions to Photocopy" (Publisher's Fee List, distributed by CCC, Inc.), for copying beyond that permitted by sections 107 or 108 of the US Copyright Law. This consent does not extend to other kinds of copying, such as copying for general distribution, for advertising or promotional purposes, for creating new collective works, or for resale.

Library of Congress Cataloging in Publication Data

Main entry under title:

Gene expression.

(UCLA symposia on molecular and cellular biology ; new ser., v. 8)

Papers presented at the 1983 Cetus-UCLA Symposium on Gene Expression held in Park City, Utah, and sponsored by the Cetus Corporation.

Bibliography: p.

Includes index.

1. Gene expression—Congresses. I. Hamer, Dean H. II. Rosenberg, Martin. III. Cetus-UCLA Symposium on Gene Expression (1983 : Park City, Utah) IV. Cetus Corporation. V. Series.

QH450.G4625 1983 575.1 83-16280

ISBN 0-8451-2607-5

Contents

Contributors

W.F. Anderson, Department of Biochemistry, University of Alberta, Edmonton, Alberta, Canada **[19]**

Charles Auffray, Department of Biochemistry and Molecular Biology, Harvard University, Cambridge, MA **[481]**

F.M. Ausubel, Department of Molecular Biology, Massachusetts General Hospital; and Department of Genetics, Harvard Medical School, Boston, MA **[175]**

J.N. Bailey, Department of Molecular Genetics, Hoffmann LaRoche, Nutley, NJ **[33]**

Carl Banner, Department of Cellular and Developmental Biology, Harvard University, Cambridge, MA **[235]**

Eric Barklls, Department of Biology, Massachusetts Institute of Technology, Cambridge, MA **[261]**

Susan Bear, Laboratory of Molecular Oncology, National Cancer Institute, National Institutes of Health, Bethesda, MD **[311]**

Jon Beckwith, Department of Microbiology and Molecular Genetics, Harvard Medical School, Boston, MA **[371]**

Vivian Berlin, Department of Biological Sciences, Stanford University, Stanford, CA **[295]**

André Bernards, Section for Medical Enzymology, Laboratory of Biochemistry, University of Amsterdam, Amsterdam, The Netherlands **[413]**

Daniel F. Bogenhagen, Department of Pharmacology, State University of New York, Stony Brook, NY **[43]**

Piet Borst, Section for Medical Enzymology, Laboratory of Biochemistry, University of Amsterdam, Amsterdam, The Netherlands **[413]**

Michael Botchan, Department of Molecular Biology, University of California, Berkeley, CA **[107]**

Michael A. Boss, Department of Molecular Immunology, Celltech, Ltd., Berks, England **[513]**

Suzanne Bourgeois, Regulatory Biology Laboratory, The Salk Institute, San Diego, CA **[571]**

Susan L. Brehm, Department of Microbiology and Immunology, University of California, Berkeley, CA **[327]**

Edith Brickman, Department of Microbiology and Molecular Genetics, Harvard Medical School, Boston, MA **[371]**

Richard R. Burgess, McArdle Laboratory for Cancer Research, University of Wisconsin, Madison, WI **[387]**

Zachary F. Burton, McArdle Laboratory for Cancer Research, University of Wisconsin, Madison, WI **[387]**

Charles R. Cantor, Department of Human Genetics and Development, College of Physicians & Surgeons, Columbia University, New York, NY **[413]**

Richard W. Carthew, Department of Biology, Massachusetts Institute of Technology, Cambridge, MA **[283]**

M. Cashel, Laboratory of Molecular Genetics, National Institute of Child Health and Human Development, National Institutes of Health, Bethesda, MD **[65]**

Thomas R. Cech, Department of Chemistry, University of Colorado, Boulder, CO **[327]**

Pierre Chambon, Laboratoire de Génétique Moléculaire des Eucaryotes, Institut de Chimie Biologique, Strasbourg, Cédex, France **[125]**

H. Cheroutre, Laboratory of Molecular Biology, State University of Ghent, Ghent, Belgium **[437]**

G. Chinali, Laboratory of Molecular Genetics, National Institute of Child Health and Human Development, National Institutes of Health, Bethesda, MD **[65]**

George E. Chisholm, Department of Biological Sciences, University of Pittsburgh, Pittsburgh, PA **[145]**

Rex L. Chisholm, Department of Biology, Massachusetts Institute of Technology, Cambridge, MA **[261]**

Su-yun Chung, Department of Biochemical Sciences, Princeton University, Princeton, NJ **[95]**

G.R. Cleaves, Department of Microbiology, University of Medicine and Dentistry of New Jersey-Rutgers Medical School, Piscataway, NJ **[33]**

S. Contente, Laboratory of Molecular Genetics, National Institute of Child Health and Human Development, National Institutes of Health, Bethesda, MD **[65]**

Terrance G. Cooper, Department of Biological Sciences, University of Pittsburgh, Pittsburgh, PA **[145]**

Donald Court, Laboratory of Molecular Oncology, National Cancer Institute, National Institutes of Health, Bethesda, MD **[311]**

B.D. Crawford, Genetics Group, Life Sciences Division, Los Alamos National Laboratory, Los Alamos, NM **[467]**

M. Cygler, Department of Biochemistry, University of Alberta, Edmonton, Alberta, Canada **[19]**

Anathbandhu Das, Department of Biological Sciences, Stanford University, Stanford, CA **[295]**

W. Degrave, Laboratory of Molecular Biology, State University of Ghent, Ghent, Belgium **[437]**

Titia De Lange, Section for Enzymology, Laboratory of Biochemistry, University of Amsterdam, Amsterdam, The Netherlands **[413]**

D.R. Dembinski, Department of Microbiology, University of Medicine and Dentistry of New Jersey-Rutgers Medical School, Piscataway, NJ **[33]**

R. Devos, Laboratory of Molecular Biology, State University of Ghent, Ghent, Belgium **[437]**

Peter DiMaria, Department of Pathology, Washington University School of Medicine, St. Louis, MO **[343]**

Deborah E. Dobson, Department of Pharmacology, Stanford University, Stanford, CA **[87]**

Jerry B. Dodgson, Departments of Microbiology and Public Health, and Biochemistry, Michigan State University, East Lansing, MI **[445]**

John J. Dunn, Biology Department, Brookhaven National Laboratory, Upton, NY **[403]**

William S. Dynan, Department of Biochemistry, University of California, Berkeley, CA **[53]**

Howard J. Eisen, Laboratory of Developmental Pharmacology, National Institute of Child Health and Human Development, National Institutes of Health, Bethesda, MD **[187]**

Spencer Emtage, Department of Molecular Genetics, Celltech, Ltd., Berks, England **[513]**

James D. Engel, Department of Biochemistry, Molecular and Cell Biology, Northwestern University, Evanston, IL **[445]**

M.D. Enger, Genetics Group, Life Sciences Division, Los Alamos National Laboratory, Los Alamos, NM **[467]**

B.F. Erlanger, Department of Microbiology, Columbia University, New York, NY **[3]**

Peggy J. Farnham, Department of Genetics, Stanford University School of Medicine, Stanford, CA **[273]**

Larry R. Feldberg, Shulton, Inc., Clifton, NJ **[159]**

Susan Ferro-Novick, Department of Microbiology and Molecular Genetics, Harvard Medical School, Boston, MA **[371]**

W. Fiers, Laboratory of Molecular Biology, State University of Ghent, Ghent, Belgium **[437]**

Richard A. Firtel, Department of Biology, University of California-San Diego, La Jolla, CA **[249]**

Robert Fisher, Department of Biological Sciences, Stanford University, Stanford, CA **[295]**

Virginia Folsom, Department of Biochemical Sciences, Princeton University, Princeton, NJ **[95]**

Marsha L. Frazier, Department of Biochemistry, The University of Texas System Cancer Center, M.D. Anderson Hospital and Tumor Institute, Houston, TX **[491]**

Theodore Friedmann, Department of Pediatrics, University of California-San Diego, La Jolla, CA **[555]**

Jeffrey Garwin, Biogen, Inc., Cambridge, MA **[371]**

Francis S. Genbauffe, Department of Biological Sciences, University of Pittsburgh, Pittsburgh, PA **[145]**

C. Gocke, Department of Microbiology, University of Medicine and Dentistry of New Jersey-Rutgers Medical School, Piscataway, NJ **[33]**

Carlos J. Goldenberg, Department of Pathology, Washington University School of Medicine, St. Louis, MO **[343]**

Paula J. Grabowski, Department of Chemistry, University of Colorado, Boulder, CO **[327]**

David K. Grandy, Departments of Microbiology and Public Health, Michigan State University, East Lansing, MI **[445]**

H.B. Gray, Jr., Department of Biophysical Sciences, University of Houston, Houston, TX **[3]**

Jack Greenblatt, Banting and Best Department of Medical Research, University of Toronto, Toronto, Ontario, Canada **[283]**

Michael Gribskov, McArdle Laboratory for Cancer Research, University of Wisconsin, Madison, WI **[387]**

B.B. Griffith, Department of Cell Biology, Cancer Research and Treatment Center, University of New Mexico School of Medicine, Alberquerque, NM **[467]**

J.K. Griffith, Department of Cell Biology, Cancer Research and Treatment Center, University of New Mexico School of Medicine, Alberquerque, NM **[467]**

Carol A. Gross, Department of Bacteriology, University of Wisconsin, Madison, WI **[387]**

Gabriel Guarneros, Department of Genetics, Centro de Investigacion y de Estudios Avanzados, Mexico City, Mexico **[311]**

Jay S. Hanas, Department of Pharmacology, State University of New York, Stony Brook, NY **[43]**

L.H. Hanau, Department of Microbiology, Columbia University, New York, NY **[3]**

J.L. Hanners, Genetics Group, Life Sciences Division, Los Alamos National Laboratory, Los Alamos, NM **[467]**

Scott D. Hauser, Department of Pathology, Washington University School of Medicine, St. Louis, MO **[343]**

Joan E. Harper, Department of Pathology, New York University Medical Center, New York, NY **[135]**

C.E. Hildebrand, Genetics Group, Life Sciences Division, Los Alamos National Laboratory, Los Alamos, NM **[467]**

N.J. Horn, Department of Microbiology, University of Medicine and Dentistry of New Jersey-Rutgers Medical School, Piscataway, NJ **[33]**

Bruce Howard, Laboratory of Molecular Biology, National Cancer Institute, National Institutes of Health, Bethesda, MD **[207]**

P.J. Jackson, Genetics Group, Life Sciences Division, Los Alamos National Laboratory, Los Alamos, NM **[467]**

W. Charles Johnson, Department of Cellular and Developmental Biology, Harvard University, Cambridge, MA **[235]**

L. Jolliffe, Department of Immunobiology, Ortho Pharmaceuticals, Raritan, NJ **[33]**

Douglas J. Jolly, Department of Pediatrics, University of California-San Diego, La Jolla, CA **[555]**

Jonathan D.G. Jones, Department of Plant Molecular Biology, Advanced Genetic Sciences, Berkeley, CA **[175]**

Jan Jongstra, Laboratoire de Génétique Moléculaire des Eucaryotes, Institut de Chimie Biologique, Strasbourg, Cédex, France **[125]**

T.M. Jovin, Max Planck Institute, Gottingen, West Germany **[3]**

David B. Kaback, Department of Microbiology, University of Medicine and Dentistry of New Jersey-New Jersey Medical School, Newark, NJ **[159]**

M.W. Kilpatrick, Department of Biochemistry, University of Alabama in Birmingham, Birmingham, AL **[3]**

J.F. Klement, Department of Microbiology, University of Medicine and Dentistry of New Jersey-Rutgers Medical School, Piscataway, NJ **[33]**

J. Klysik, Department of Biochemistry, University of Alabama in Birmingham, Birmingham, AL **[3]**

Roberto Kolter, Department of Biological Sciences, Stanford University, Stanford, CA **[295]**

Alan J. Korman, Department of Biochemistry and Molecular Biology, Harvard University, Cambridge, MA **[481]**

Laurence Jay Korn, Department of Genetics, Stanford University School of Medicine, Stanford, CA **[273]**

Michael Kriegler, Department of Molecular Biology, University of California, Berkeley, CA **[107]**

Kelly Kruger, Department of Biochemistry, College of Physicians & Surgeons, Columbia University, New York, NY **[327]**

Carol Kumamoto, Department of Microbiology and Molecular Genetics, Harvard Medical School, Boston, MA **[371]**

Ashok Kumar, Department of Anatomy, University of Texas Health Science Center, San Antonio, TX **[491]**

J.E. Larson, Department of Biochemistry, University of Alabama in Birmingham, Birmingham, AL **[3]**

Frank Lee, DNAX Ltd., Palo Alto, CA **[87]**

Michael J. Lewis, M.R.C. Laboratory of Molecular Biology, Cambridge, England **[75]**

James W. Lillie, Department of Biochemistry and Molecular Biology, Harvard University, Cambridge, MA **[481]**

John W. Little, Department of Biochemistry, University of Arizona, Tucson, AZ **[135]**

Alvin Y.C. Liu, Section for Medical Enzymology, Laboratory of Biochemistry, University of Amsterdam, Amsterdam, The Netherlands **[413]**

Harvey F. Lodish, Department of Biology, Massachusetts Institute of Technology, Cambridge, MA **[261]**

J. Longmire, Genetics Group, Life Sciences Division, Los Alamos National Laboratory, Los Alamos, NM **[467]**

Linda C. Lopez, Department of Biochemistry, The University of Texas System Cancer Center, M.D. Anderson Hospital and Tumor Institute, Houston, TX **[491]**

Richard Losick, Department of Cellular and Developmental Biology, Harvard University, Cambridge, MA **[235]**

Erika Lötscher, Institut Für Molekularbiologie II, Universität Zürich, Zürich, Switzerland **[359]**

Michael Mackett, Laboratory of Biology of Viruses, National Institute of Allergy and Infectious Diseases, National Institutes of Health, Bethesda, MD **[543]**

R.S. MacWright, Department of Microbiology, University of Medicine and Dentistry of New Jersey-Rutgers Medical School, Piscataway, NJ **[33]**

Ned Mantei, Institut für Molekularbiologie I, Universität Zürich, Zürich, Switzerland **[219]**

Bruce E. Markham, Department of Internal Medicine, University of Arizona College of Medicine, Tucson, AZ **[135]**

J.C. Martin, Department of Physics, University of Alabama in Birmingham, Birmingham, AL **[3]**

B.W. Matthews, Institute of Molecular Biology, University of Oregon, Eugene, OR **[19]**

W.T. McAllister, Department of Microbiology, University of Medicine and Dentistry of New Jersey-Rutgers Medical School, Piscataway, NJ **[33]**

Mona C. Mehdy, Department of Biology, University of California-San Diego, La Jolla, CA **[249]**

Paul A.M. Michels, Section for Medical Enzymology, Laboratory of Biochemistry, University of Amsterdam, Amsterdam, The Netherlands **[413]**

J.J. Miglietta, Department of Biochemistry, University of Wisconsin, Madison, WI **[3]**

A. Dusty Miller, Molecular Biology and Virology Laboratory, The Salk Institute, San Diego, CA **[555]**

Cecilia Montanez, Department of Genetics, Centro de Investigacion y de Estudios Avanzados, Mexico City, Mexico **[311]**

Charles Moran, Jr., Department of Cellular and Developmental Biology, Harvard University, Cambridge, MA **[235]**

Bernard Moss, Laboratory of Biology of Viruses, National Institute of Allergy and Infectious Diseases, National Institutes of Health, Bethesda, MD **[543]**

David W. Mount, Department of Molecular and Medical Microbiology, University of Arizona College of Medicine, Tucson, AZ **[135]**

A.C. Munk, Genetics Group, Life Sciences Division, Los Alamos National Laboratory, Los Alamos, NM **[467]**

Daniel W. Nebert, Laboratory of Developmental Pharmacology, National Institute of Child Health and Human Development, National Institutes of Health, Bethesda, MD **[187]**

Masahiko Negishi, Laboratory of Developmental Pharmacology, National Institute of Child Health and Human Development, National Institutes of Health, Bethesda, MD **[187]**

D.H. Ohlendorf, Institute of Molecular Biology, University of Oregon, Eugene, OR **[19]**

Donald Oliver, Department of Microbiology, State University of New York, Stony Brook, NY **[371]**

Amos B. Oppenheim, Department of Microbiological Chemistry, The Hebrew University, Jerusalem, Israel **[311]**

Stuart H. Orkin, Division of Hematology-Oncology, Children's Hospital, Dana-Farber Cancer Institute, Department of Pediatrics, Harvard Medical School, Boston, MA **[457]**

Pierre Oudet, Labortoire de Génétique Moléculaire des Eucaryotes, Institut de Chimie Biologique, Strasbourg, Cédex, France **[125]**

David W. Ow, Department of Molecular Biology, Massachusetts General Hospital; and Department of Genetics, Harvard Medical School, Boston, MA **[175]**

Raji Padmanabhan, Laboratory of Molecular Biology, National Cancer Institute, National Institutes of Health, Bethesda, MD **[207]**

Hugh R.B. Pelham, M.R.C. Laboratory of Molecular Biology, Cambridge, England **[75]**

Carl Perez, Department of Molecular Biology, University of California, Berkeley, CA **[107]**

Kenneth R. Peterson, Department of Molecular and Medical Microbiology, University of Arizona College of Medicine, Tucson, AZ **[135]**

G. Plaetinck, Laboratory of Molecular Biology, State University of Ghent, Ghent, Belgium **[437]**

Terry Platt, Department of Molecular Biophysics & Biochemistry, Yale University, New Haven, CT **[567]**

B. Pontius, Department of Biology, Massachusetts Institute of Technology, Cambridge, MA **[261]**

Jennifer Price, Department of Genetics, Stanford University School of Medicine, Stanford, CA **[273]**

Hermann Ragg, Institut für Molekularbiologie I, Universität Zürich, Zürich, Switzerland **[219]**

David Ratner, Department of Cellular Biology, Scripps Clinic and Research Foundation, La Jolla, CA **[249]**

Tim Reudelhuber, Labortoire de Génétique Moléculaire des Eucaryotes, Institut de Chimie Biologique, Strasbourg, Cédex, France **[125]**

Gordon M. Ringold, Department of Pharmacology, Stanford University, Stanford, CA **[87]**

Martin J. Rosenberg, Department of Molecular Genetics, Smith, Kline, & French Laboratories; and Laboratories of Biochemistry and Molecular Biology, National Cancer Institute, National Institutes of Health, Bethesda, MD **[311, 523]**

Magali Roux-Dosseto, Department of Biochemistry and Molecular Biology, Harvard University, Cambridge, MA **[481]**

Brian C. Rymond, Department of Biological Sciences, State University of New York at Albany, Albany, NY **[523]**

P. Sarmientos, Laboratory of Molecular Genetics, National Institute of Child Health and Human Development, National Institutes of Health, Bethesda, MD **[65]**

Grady F. Saunders, Department of Biochemistry, The University of Texas System Cancer Center, M.D. Anderson Hospital and Tumor Institute, Houston, TX **[491]**

S. Scahill, Laboratory of Molecular Biology, State University of Ghent, Ghent, Belgium **[437]**

Ursula Schmeissner, Biogen, S.A., Geneva, Switzerland **[311]**

Daniel Schümperli, Institut für Molekularbiologie II, Universität Zürich, Zürich, Switzerland; and Laboratories of Biochemistry and Molecular Biology, National Cancer Institute, National Institutes of Health, Bethesda, MD **[359, 523]**

David C. Schwartz, Department of Human Genetics and Development, College of Physicians & Surgeons, Columbia University, New York, NY **[413]**

C.K. Singleton, Department of Bacteriology, University of Wisconsin, Madison, WI **[3]**

Paul Sloof, Section for Medical Enzymology, Laboratory of Biochemistry, Unversity of Amsterdam, Amsterdam, The Netherlands **[413]**

Geoffrey L. Smith, Laboratory of Biology of Viruses, National Institute of Allergy and Infectious Diseases, National Institutes of Health, Bethesda, MD **[543]**

Mary Sopta, Banting and Best Department of Medical Research, University of Toronto, Toronto, Ontario, Canada **[283]**

R.L. Stallings, University of Texas System Cancer Center, Smithville, TX **[467]**

Claudia Stauber, Institut für Molekularbiologie II, Universität Zürich, Zürich, Switzerland **[359]**

S.M. Stirdivant, Biological Laboratories, Harvard University, Cambridge, MA **[3]**

Jeannine Strobl, Laboratory of Biochemistry, National Cancer Institute, National Institutes of Health, Bethesda, MD **[207]**

Jack L. Strominger, Department of Biochemistry and Molecular Biology, Harvard University, Cambridge, MA **[481]**

F. William Studier, Biology Department, Brookhaven National Laboratory, Upton, NY **[403]**

Chung-Jey Su, Department of Biochemistry, The University of Texas System Cancer Center, M.D. Anderson Hospital and Tumor Institute, Houston, TX **[491]**

Venkatesan Sundaresan, Department of Molecular Biology, Massachusetts General Hospital; and Department of Genetics, Harvard Medical School, Boston, MA **[175]**

Y. Takeda, Department of Chemistry, University of Maryland, Baltimore County, Catonsville, MD **[19]**

J. Tavernier, Laboratory of Molecular Biology, State University of Ghent, Ghent, Belgium **[437]**

Y. Taya, Laboratory of Molecular Biology, State University of Ghent, Ghent, Belgium **[437]**

Wayne E. Taylor, McArdle Laboratory for Cancer Research, University of Wisconsin, Madison, WI **[387]**

J.G. Tesmer, Genetics Group, Life Sciences Division, Los Alamos National Laboratory, Los Alamos, NM **[467]**

E. Brad Thompson, Laboratory of Biochemistry, National Cancer Institute, National Institutes of Health, Bethesda, MD **[207]**

Robert Tjian, Department of Biochemistry, University of California, Berkeley, CA **[53]**

Robert H. Tukey, Laboratory of Developmental Pharmacology, National Institute of Child Health and Human Development, National Institutes of Health, Bethesda, MD **[187]**

Lex H.T. Van der Ploeg, Section for Medical Enzymology, Laboratory of Biochemistry, University of Amsterdam, Amsterdam, The Netherlands **[413]**

C.M. Veneziale, Mayo Medical School, Rochester, MN **[3]**

Inder M. Verma, Department of Pediatrics, University of California-San Diego, La Jolla, CA **[555]**

R.A. Walters, Genetics Group, Life Sciences Division, Los Alamos National Laboratory, Los Alamos, NM **[467]**

R.M. Wartell, Schools of Physics and Biology, Georgia Institute of Technology, Atlanta, GA **[3]**

C.F. Wei, Department of Biophysical Sciences, University of Houston, Houston, TX **[3]**

Ulrich Weidle, Institut für Molekularbiologie I, Universität Zürich, Zürich, Switzerland **[219]**

Charles Weissmann, Institut für Molekularbiolgie I, Universität Zürich, Zürich, Switzerland **[219]**

R.D. Wells, Department of Biochemistry, University of Alabama in Birmingham, Birmingham, AL **[3]**

Kenneth F. Wertman, Department of Molecular and Medical Microbiology, University of Arizona College of Medicine, Tucson, AZ **[135]**

John Wooley, Department of Biochemical Sciences, Princeton University, Princeton, NJ **[95]**

Cheng-Wen Wu, Department of Pharmacology, State University of New York, Stony Brook, NY **[43]**

Charles Yanofsky, Department of Biological Sciences, Stanford University, Stanford, CA **[295]**

W. Zacharias, Department of Biochemistry, University of Alabama in Birmingham, Birmingham, AL **[3]**

D. Zarling, Max Planck Institute, Gottingen, West Germany **[3]**

Arthur J. Zaug, Department of Chemistry, University of Colorado, Boulder, CO **[327]**

Richard S. Zitomer, Department of Biological Sciences, State University of New York at Albany, Albany, NY **[523]**

Peter Zuber, Department of Cellular and Developmental Biology, Harvard University, Cambridge, MA **[235]**

Preface

The papers presented in this volume derive from the 1983 Cetus-UCLA Symposium on "Gene Expression" held in Park City, Utah. The success of this meeting was due in large part to the excellent organizational skills of Sandy Malone and her UCLA staff.

The Symposium dealt with gene function and structure in a wide variety of biological systems. Researchers working on the molecular biology of prokaryotic, lower and higher eukaryotic systems were brought together to compare and contrast the regulatory mechanisms operating to control gene expression in different organisms.

We wish to thank Cetus Corporation for its generous sponsorship of this meeting. We also gratefully acknowledge gifts from: Abbott Diagnostics Division, Asahi Chemical Industry Co., Ltd., Beckman Instruments, Inc., Bristol-Myers Company, Industrial Division, E.I. du Pont de Nemours & Company, Genentech, Inc., Imperial Chemical Industries, Merck Sharp & Dohme Research Laboratories, New England Biolabs, Inc., Pfizer Central Research, Schering Corporation, Searle Research and Development, and Smith, Kline & French Laboratories. The generous support of these companies helped make this scientific meeting possible.

Dean H. Hamer
Martin J. Rosenberg

I. DNA STRUCTURE AND DNA-PROTEIN INTERACTIONS

Gene Expression, pages 3-18
© 1983 Alan R. Liss, Inc., 150 Fifth Avenue, New York, NY 10011

LEFT-HANDED Z-DNA HELICES, CRUCIFORMS, AND SUPERCOILING[1]

R.D. Wells[a,b], B.F. Erlanger[c], H.B. Gray, Jr.[d], L.H. Hanau[c], T.M. Jovin[e], M.W. Kilpatrick[a,b], J. Klysik[a,b,2], J.E. Larson[a,b], J.C. Martin[f], J.J. Miglietta[b], C.K. Singleton[a,b,3], S.M. Stirdivant[b,4], C.M. Veneziale[b,5], R.M. Wartell[g], C.F. Wei[d], W. Zacharias[a,b], and D. Zarling[e]

[a]University of Alabama in Birmingham, Schools of Medicine and Dentistry, Department of Biochemistry, Birmingham, Alabama 35294
[b]University of Wisconsin, Department of Biochemistry, Madison, Wisconsin 53706
[c]Columbia University, Department of Microbiology, New York, New York 10032
[d]University of Houston, Department of Biophysical Sciences, Houston, Texas 77004
[e]Max Planck Institute, Gottingen, West Germany
[f]University of Alabama in Birmingham, Department of Physics, Birmingham, Alabama 35294
[g]Georgia Institute of Technology, Schools of Physics and Biology, Atlanta, Georgia 30332

[1]This work was supported by grants from the NIH (GM30822) and the NSF (PCM-8002622).

[2]Present address: 5/46 Napierskiego, Lodz, Poland
[3]Present address: Department of Bacteriology, University of Wisconsin, Madison, Wisconsin 53706
[4]Present address: Biological Laboratories, Harvard University, Cambridge, Massachusetts 02138
[5]Present address: Mayo Medical School, Rochester, Minnesota 55905

ABSTRACT

Biological and physical studies on recombinant plasmids and restriction fragments containing tracts of DNA that can adopt left handed conformations reveal the following: 1. Left handed Z-DNA can neighbor right handed regions of B-DNA in close proximity on the same chain; thus, DNA has conformational microheterogeniety. 2. Z-DNA exists in vivo since supercoiling causes the B→Z transition at physiological conditions. 3. B-Z junctions are conformationally pliable and differ from case to case as a function of neighboring nucleotides. 4. Single strand specific nucleases (S_1 and BAL 31) recognize and cleave aberrant structural features of the B-Z junctions. 5. $(T\text{-}G)_n \cdot (C\text{-}A)_n$ adopts a left handed DNA conformation. 6. A family of left handed DNA conformations exists. 7. The stabilizing effect of methylation on the Z form in fragments and plasmids approximately offsets the free energy contributions of the B-Z junctions. 8. Left handed DNA is very unusual in vivo (replication and recombination). 9. Cruciforms are stable structures at certain inverted repeat sequences in supercoiled DNAs. 10. The stability of a cruciform appears to be sharply dependent on stem length.

INTRODUCTION

It has been known for at least 17 years that different DNA sequences have different conformations and properties (reviewed in 1-6). However, the realization in the past 3 years that certain types of sequences can adopt either right or left handed helices has generated renewed interest in DNA structure. Also, the presence of stable cruciforms at inverted repeat sequences has been demonstrated. Thus, DNA microheterogeneity has been demonstrated in a variety of systems. The role of these structures in gene expression is currently under investigation.

(dC-dG) Segments in Restriction Fragments Adopt Left Handed Helices

This laboratory has studied the chemical, physical, and biological properties of left handed DNA using polymers, restriction fragments, and recombinant plasmids with established sequences.

We demonstrated in 1970 (7,8) that a DNA polymer might exist in a left handed conformation. After the x-ray crystallographic studies on short oligomers of (dC-dG) which revealed the very unusual Z and Z' conformations (9, 10), we constructed a family of recombinant DNAs which contained the (dC-dG) tract.

The sequences of several of our cloned molecules are shown in Fig. 1. pRW751 contains an insert 157 bp in length; the 95 bp fragment containing the Escherichia coli lac promoter-operator is flanked by segments of (dC-dG). Since there are unique BamHI sites between the insert and

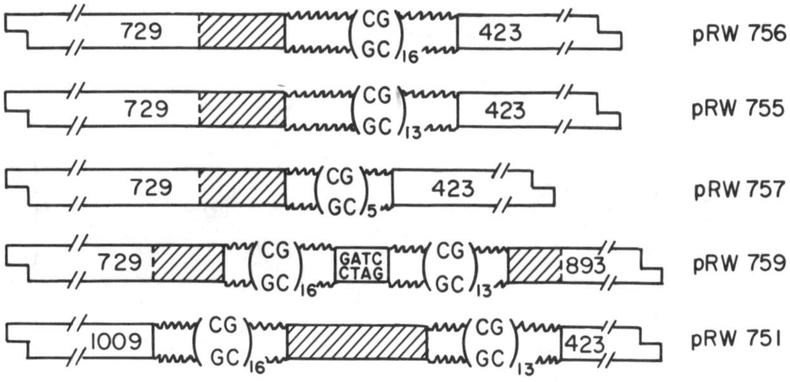

Fig. 1. (dC-dG) containing recombinant plasmids used in this study. Hatched areas denote the E. coli lac 95 bp region. From ref. 20.

the vector (pBR322), it has been possible to prepare milligram quantities of this restriction fragment for spectroscopic and physical evaluation (11-16, 18, 19). A variety of circular dichroism, 31P-NMR, and laser Raman spectroscopy studies on these types of restriction fragments (11, 16) demonstrated that the (dC-dG) regions undergo a cooperative salt induced transition from a right handed to a left handed Z type helix; the natural DNA segments in these fragments remained in a right handed B conformation. However, Raman spectroscopy studies (12) indicated that the right handed B conformation which is flanked by the Z regions was somewhat perturbed by the left handed segments.

Multiplicity of Left Handed Conformations Under Different Ionic Conditions: A Condensed Form of $(dC\text{-}dG)_n \cdot (dC\text{-}dG)_n$

A wide variety of environmental conditions cause the structural transition of a suitable sequence to a left handed helix (13). Fig. 2 shows the end point CD spectra for $(dC\text{-}dG)_n \cdot (dC\text{-}dG)_n$ under a variety of conditions.

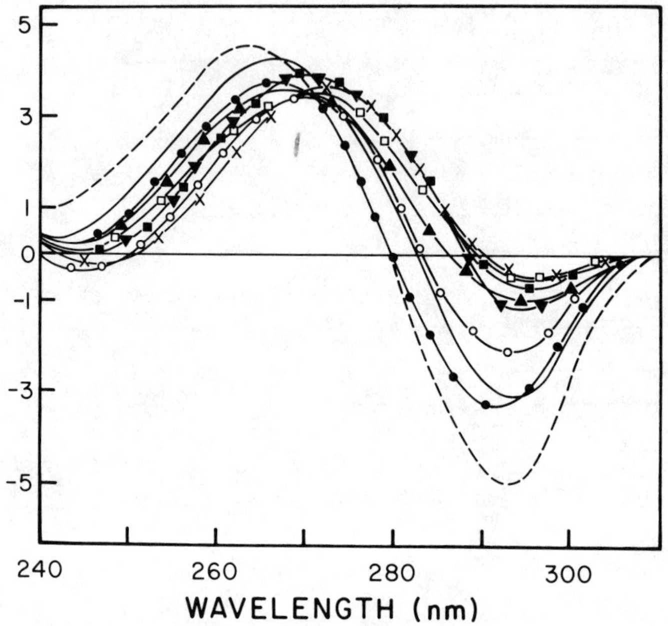

Fig. 2. CD spectra of $(dC\text{-}dG)_n$ after completion of the R to L transition induced by different metal ions in combination with a dehydrating agent. From ref. 13.

Since the spectra are substantially different quantitatively but similar in shape, we conclude that a family of left handed conformations exists.

The behavior of this DNA polymer in sodium acetate is most unusual since the cooperative transition from a B to Z form has a stable intermediate which is a condensed Ψ(+) type structure (22) (Fig. 3).

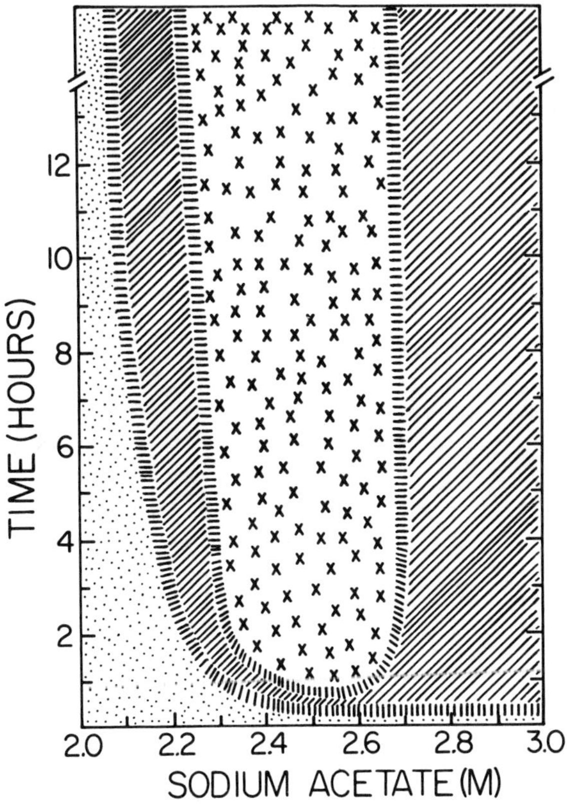

Fig. 3. Phase diagram of $(dG\text{-}dC)_n$ in aqueous sodium acetate. B-form, dotted area; Z-form, crosshatched area; Ψ-form, x area. From ref. 22.

Energy Interrelationship Between Supercoiling and the Right to Left Handed Transition

Our initial studies were performed with restriction fragments which contained a high concentration of left handed helices in order to physically and biochemically characterize the helical parameters. However, it was realized in 1981 (11) that supercoiling was a sensitive indicator of the B to Z transition, even in a tiny segment of the recombinant plasmids (Fig. 1).

Fig. 4 depicts the transition found in the supercoil when a conformational change is caused in the primary helix. When one turn of right handed primary helix is

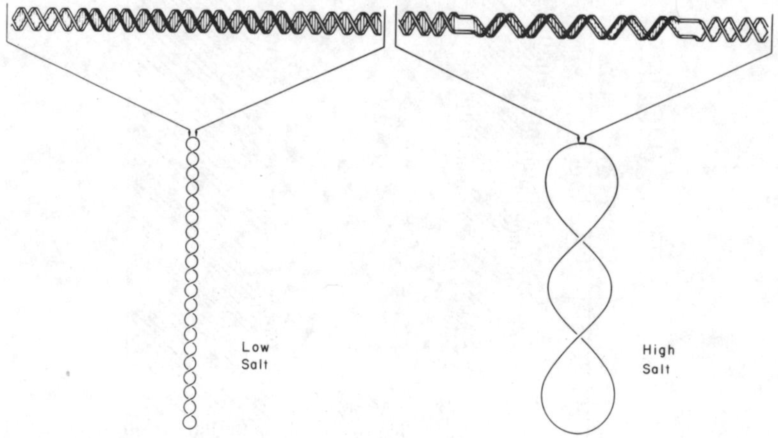

Fig. 4. Cartoon showing the relationship of B to Z transition in a small segment of a recombinant plasmid to the supercoiled state of the DNA. From ref. 19.

converted to one turn of left handed helix in a covalently closed plasmid, two supercoiled turns will be lost. Thus, this is an extremely sensitive assay for conformational transitions within supercoiled plasmids. Studies with recombinant plasmids (14, 15) which contain 58, 32, 26, and 10 bp lengths of (dC-dG) tracts revealed excellent agreement between the calculated and observed relaxation of negative supercoils as monitored by electrophoretic mobility changes of individual topoisomers. The number of supercoils relaxed were proportional to the length of the (dC-dG) segment in the plasmid. 10 bp of (dC-dG) was insufficient to allow a Z conformation under these high salt conditions. However, supercoiling is sufficient to convert 10 bp into a left handed helix (20).

Supercoiling Causes B to Z Transition: S_1 and BAL31 Nucleases Specifically Cleave Conformational Junctions

Since supercoiling is an extremely sensitive indicator of structural transitions in tiny segments of supercoiled plasmids, we decided to determine if supercoiling could cause the B to Z transition. Fig. 5 demonstrates that the $(dC\text{-}dG)_{16}$ and $(dC\text{-}dG)_{13}$ blocks in pRW751 are in a left-handed state under physiological ionic

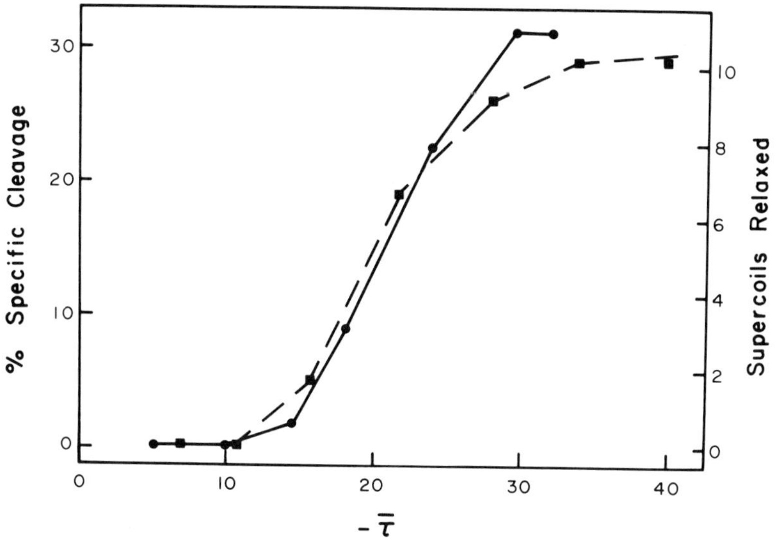

Fig. 5. Average number of negative supercoils relaxed in pRW751 versus the mean number of titratable superhelical turns. From ref. 17.

conditions (200 millimolar NaCl) when the negative superhelical density of the plasmid is greater than 0.072 (17, 19, 20). This behavior was revealed by two assays including nuclease sensitivity of the junctions. Furthermore, binding studies with antibodies raised vs. left handed DNAs agree with this conclusion.

Recent studies (17, 20, 21, 23) demonstrate that the S_1 and BAL31 nucleases, which are specific for single stranded regions, recognize and cleave the junctions between right and left handed helical segments. Restriction mapping of the sites of cleavage by the single strand specific nucleases demonstrates the presence of the B-Z junctions. Little is known concerning the structural details of these junctions. They may or may not contain non-paired nucleotides.

The range of superhelicity required to stabilize left handed structures under these conditions is well within the values of physiological superhelical densities typically found for most plasmids (24). Presuming that _in vitro_ supercoiling is a true reflection of that found _in vivo_ (25, 26), these results indicate that left handed DNA exists _in vivo_ under the torsional strain of supercoiling (17, 19-21, 23).

B-Z Junctions are Conformationally Pliable

The junctions between contiguous right and left handed helical DNA segments in supercoiled plasmids exhibit both sequence and superhelical dependent conformational flexibility (20). The effect of supercoiling on the transition from a right handed to a left handed state for the insert shown in Fig. 1 was evaluated; only moderate amounts of superhelicity (-0.03 to -0.075) are required for formation and stabilization of left handed DNA even for the (dC-dG) region of 10 bp. Fig. 6 illustrates that different junctions in pRW759 (Fig. 1) are recognized differently by S_1 nuclease. At low supercoil densities, left handedness in the (dC-dG) blocks was observed. As the titratable negative density was increased further, the center junctions corresponding to cleavage at the d(GATC) interruption became progressively less recognizable by S_1 nuclease. Also, this _BamHI_ site was no longer cleaved by this restriction enzyme (20).

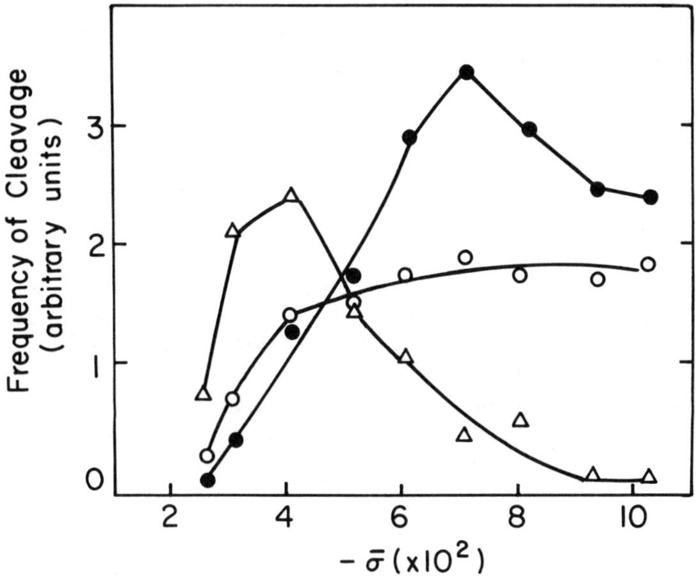

Fig. 6. S_1 nuclease cleavage of different junctions in topoisomeric pRW759 samples. Open triangles, middle junction; open circles, top junction; closed circles, bottom junction. From ref. 20.

5 Cytosine Methylation May Trigger Z-DNA Formation

Alternating (dC-dG) regions in DNA restriction fragments and recombinant plasmids were methylated at the 5 position of the cytosine residues by the HhaI methylase (18). Methylation lowers the concentration of NaCl or $MgCl_2$ necessary to cause the Z conformation when the methylated (dC-dG) tract is contiguous with regions that do not form Z structures, in contrast to the results with the DNA polymer poly (m^5dC-dG)•poly(m^5dC-dG). In supercoiled plasmids containing (dC-dG) sequences, methylation reduces the number of negative supercoils necessary to stabilize the Z-conformation. Calculations of the observed free energy contributions of the B/Z junction and cytosine methylation suggest that two junctions offset the favorable effect of methylation on the Z-conformation in (dC-dG) sequences (about 29 bp in length). The results suggest that methylation may serve as a triggering mechanism for Z-DNA formation in supercoiled DNAs (18).

d(T-G)•d(C-A) Tracts in Immunoglobin Gene Adopts a Left Handed Structure

An understanding of the biological properties of left handed DNA requires an evaluation of the types of sequences which can adopt left handed structures. d(T-G)•d(C-A) is the minimum change from the types of sequences shown in Fig. 1 which adopt left handed conformations. Recent studies (21) demonstrate that the 64 bp tract of almost perfect (T-G)•(C-A) from the mouse kappa immunoglobin gene (Fig. 7) adopts a left handed helix under the influence of

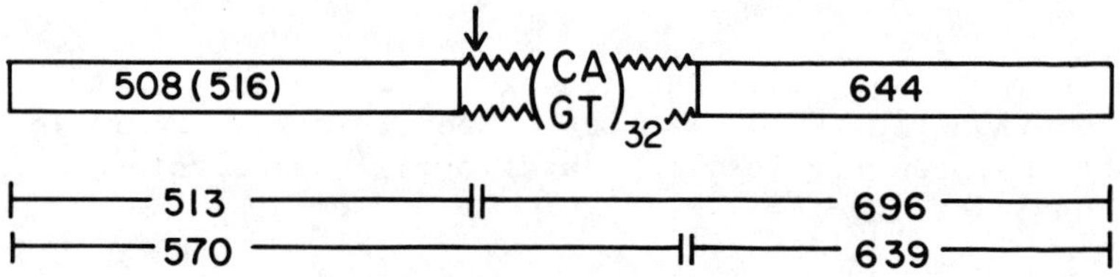

Fig. 7. Cartoon of portion of pRW777 which contains the alternating (dC-dA)•(dT-dG) region. From ref. 21.

negative supercoiling. S_1 nuclease was used as a probe for junctions between right handed and left handed segments. Furthermore, binding studies with antibodies raised vs. left handed DNAs agree with this conclusion. This result is significant since the sequence is naturally occuring (27) and since this type of sequence appears to occur widely in nature and may be involved in recombination (reviewed in 21). However, somewhat more supercoiling is required than for (dC-dG) sequences (28). Likewise, the conditions required to put the (dT-dG)•(dC-dA) polymer into a left handed conformation are somewhat more stringent than required for the C and G containing polymer (16).

Unusual Biological Properties of Plasmids Containing (dC-dG) Segments

The biological properties of the pBR322 derivative plasmids containing the (dC-dG) sequences were very unusual since deletions in the (dC-dG) regions, but not in the pBR322 nor the *lac* segments, were frequently observed. Segments of (dC-dG) longer than approximately 50 bp were not stable but suffered deletions (11, 14). Segments of approximately 30 bp or shorter were stable in most cases. The (dC-dG) tracts seemed to enhance *recA* mediated recombination when they were of suitable length and were cloned into certain sites in the recombinants (14).

Furthermore, the (dC-dG) containing plasmids were less supercoiled (by 6 - 12 turns) than expected relative to control plasmids, after isolation from *E. coli* hosts. The reason for this unusual behavior is uncertain (14), but clearly demonstrates the marked influence *in vivo* of the (dC-dG) tracts.

Possible Biological Roles of Left Handed DNA

A spectrum of biological properties can be envisaged for left handed DNA or for the structural transition to a left handed conformation. A. A unique structural site for direct protein recognition and binding; B. The transition would cause a global change in supercoiling and thus affect expression at distant locations on the genome; C. Structural perturbation of a DNA sequence neighboring the left handed conformation and thus influencing expression at a neighboring site; D. Effect on chromosome structure; E. Carcinogen binding to a segment that can undergo the transition would influence capacity of that region to adopt different conformations; F. Intermediate in DNA gyrase reaction; G. Terminator of DNA replication and/or transcription; H. DNA recombination intermediate; I. Energy sink for DNA supercoiling.

Cruciforms

Cruciforms are stable conformations found at certain inverted repeat sequences in supercoiled DNAs (29-31). The relationship between superhelical density and cruciform

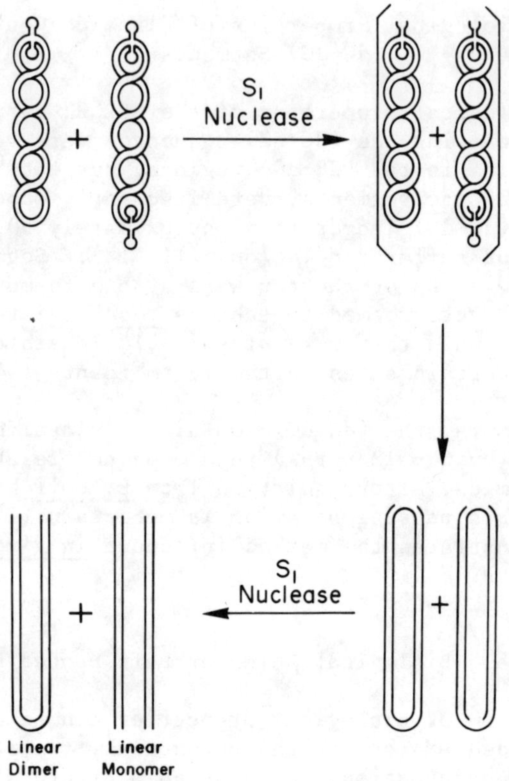

Fig. 8. Scheme of the assay used to study relative cruciform stability. From ref. 30.

formation in pVH51 has been evaluated (30) by the scheme shown in Fig. 8. At physiological superhelical densities, the cruciform state is present in a high percentage of the plasmids. A sharp transition occurs from an undetectable level to a relatively stable state at a negative superhelical density range of 0.046 to 0.066. Estimates of the free energy contribution to cruciform formation resulting from loss of negative superhelical turns suggest that about 22 kcal/mol are required to generate the cruciform structure.

The effect of salts, temperature, and stem length on supercoil induced formation of cruciforms has been evaluated recently (31). In general, conditions which stabilize duplex DNA over single stranded DNA shifted the transition to higher negative superhelical density values due to an increase in the unfavorable free energy of cruciform formation.

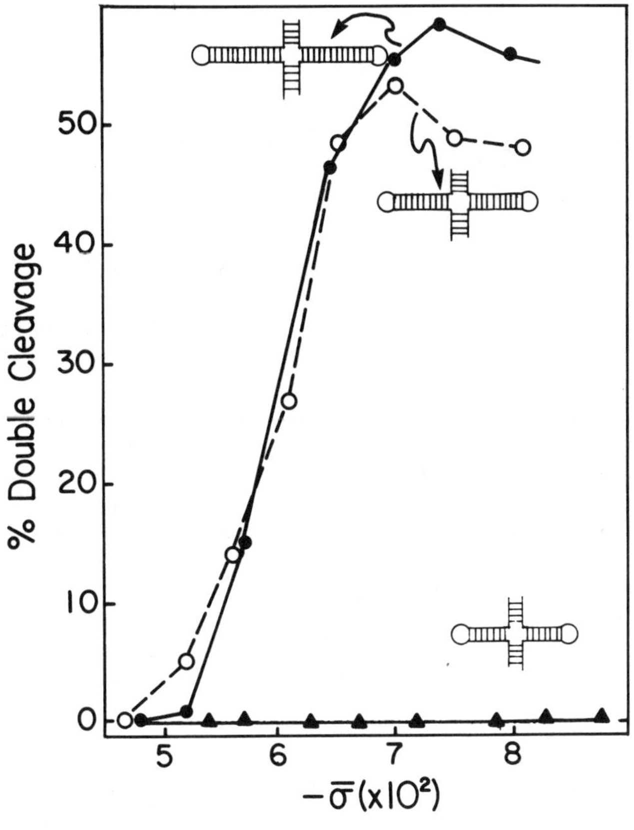

Fig. 9. Percent double cleavage of S_1 nuclease versus the mean negative superhelical density for pRW808, pRW809, and pRW810. The cartoon insets represent the relative lengths of the stems of the inverted repeats if in the cruciform state. From ref. 31.

Fig. 9 shows the influence of stem length on the supercoil induced transition to the cruciform state as studied by in vitro deletion of portions of the pVH51 major inverted repeat. Decreasing the stem length from 13 to 10 bp had no effect on the ability of this sequence to adopt the cruciform state. However, a further reduction of 3 bp to give a stem length of 7 bp completely abolished the ability of this region of DNA to exist in the cruciform state, at least up to a density of -0.15. Thus, a very sharp dependency on stem length exists for cruciform formation within an inverted repeat region possessing a potential loop of 5 nucleotides.

ACKNOWLEDGEMENTS

We thank Drs. R. M. Santella and D. Grunberger for preparation of some of the DNAs used to raise antibodies vs. left handed DNA.

REFERENCES

1. Wells RD, Goodman TC, Hillen W, Horn GT, Klein RD, Larson JE, Muller UR, Neuendorf SK, Panayotatos N, Stirdivant SM (1980). "DNA structure and gene regulation." Progress in Nucleic Acid Research and Molecular Biology 24:167.
2. Wells RD, Blakesley RW, Burd JF, Chan HW, Dodgson JB, Hardies SC, Horn GT, Jensen KF, Larson J, Nes IF, Selsing E, Wartell RM (1977). "The role of DNA structure in genetic regulation." Critical Reviews in Biochemistry 4:305.
3. Wells RD, Wartell RM (1974). "The influence of nucleotide sequence on DNA properties." Review article for Biochemistry Series One, Biochemistry of Nucleic Acids 6:41.
4. Von Hippel PH, McGhee JD (1972). "DNA-protein interactions." Annual Review of Biochemistry 41:231.
5. Zimmerman SB (1982). "The three dimensional structure of DNA." Annual Review of Biochemistry 51:395.
6. Rosenberg M, Court D (1979). Regulatory sequences involved in the promotion and termination of RNA transcription. Annual Review of Genetics 13:319.

7. Mitsui Y, Langridge R, Shortle BE, Cantor CR, Grant RC, Kodama M, Wells RD (1970). Physical and enzymatic studies on poly d(I-C)•poly d(I-C), an unusual double-helical DNA. Nature 228:1166.
8. Grant RC, Kodama M, Wells RD (1972). Enzymic and physical studies on $(dI-C)_n \cdot (dI-dC)_n$ and $(dG-dC)_n \cdot (dG-dC)_n$. Biochemistry 11:805.
9. Wang A, Quigley GJ, Kolpak FJ, Crawford JL, Van Boom JH, Van Der Marel G, Rich A (1979). Molecular structure of a left-handed double helical DNA fragment at atomic resolution. Nature 282:680.
10. Drew H, Takano T, Tanaka S, Itakura K, Dickerson RE (1980). High salt d(CpGpCpG), a left handed Z' DNA double helix. Nature 286:567.
11. Klysik J, Stirdivant SM, Larson JE, Hart PA, Wells RD (1981). Left handed DNA in restriction fragments and a recombinant plasmid. Nature 290:672.
12. Wartell RM, Klysik J, Hillen W, Wells RD (1982). The junction between Z and B conformations in a DNA restriction fragment: evaluation by Raman spectroscopy. Proc Natl Acad Sci USA 79:2549.
13. Zacharias W, Larson JE, Klysik J, Stirdivant SM, Wells RD (1982). Conditions which cause the right handed to left handed DNA conformational transitions: evidence for several types of left handed DNA structures in solution. J Biol Chem 257:2775.
14. Klysik J, Stirdivant SM, Wells RD (1982). Left handed DNA cloning, characterization and instability of inserts containing different lengths of (dC-dG) in Escherichia coli. J Biol Chem 257:10152.
15. Stirdivant SM, Klysik J, Wells RD (1982). Energetic and structural interrelationship between DNA supercoiling and the right to left handed Z-helix transitions in recombinant plasmids. J Biol Chem 257:10159.
16. Wells RD, Miglietta JJ, Klysik J, Larson JE, Stirdivant SM, Zacharias W (1982). Spectroscopic studies on acetylaminofluorene modified $(dT-dG)_n \cdot (dC-dA)_n$ suggest a left handed conformation. J Biol Chem 257:10166.
17. Singleton CK, Klysik J, Stirdivant SM, Wells RD (1982). Left handed Z-DNA is induced by supercoiling under physiological conditions. Nature 299:312.

18. Klysik J, Stirdivant SM, Singleton CK, Zacharias W, Wells RD (1983). The effects of 5 cytosine methylation on the B-Z transition in DNA restriction fragments and recombinant plasmids. J Mol Biol in the press.
19. Wells RD, Brennan R, Chapman K, Goodman TC, Hart PA, Hillen W, Kellogg DR, Kilpatrick MW, Klein RD, Klysik J, Lambert PF, Larson JE, Miglietta JJ, Neuendorf SK, O'Connor TR, Singleton CK, Stirdivant SM, Veneziale CM, Wartell RM, Zacharias W (1983). Left handed DNA helices, supercoiling, and the B-Z junction. Cold Spring Harbor Symposium 47 in the press.
20. Singleton CK, Klysik J, Wells RD (1983). Conformational flexibility of junctions between contiguous B and Z DNA in supercoiled plasmids. Proc Natl Acad Sci in the press.
21. Singleton CK, Wells RD (1983) DNA supercoiling induces left handed helices in $(dT\text{-}dG)_n \cdot (dC\text{-}dA)_n$ sequences. Submitted.
22. Zacharias W, Martin JC, Wells RD (1983). A condensed form of $(dG\text{-}dC)_n \cdot (dG\text{-}dC)_n$ as an intermediate between the B- and Z- conformations induced by sodium acetate. Biochemistry in the press.
23. Kilpatrick M, Wei C-F, Gray Jr. HB, Wells RD (1983). BAL 31 nuclease as a probe in concentrated salt for the B-Z DNA junction. Submitted.
24. Bauer WE (1978). Structure and reactions of closed duplex DNA. Ann Rev Biophys Bioeng 7:287.
25. Gellert M, (1981). DNA topoisomerases. Ann Rev Biochem 50:879.
26. Sinden RR, Carlson JO, Pettijohn DE (1980). Torsional tension in the DNA double helix measured with trimethylpsoralen in living _E. coli_ cells: analogous measurements in insect and human cells. Cell 21:773.
27. Nishioka Y, Leder P (1980). Organization and complete sequence of identical embryonic and plasmacytoma κ V-region genes. J Biol Chem 255:3691.
28. Singleton CK, Kilpatrick MW, Wells RD. unpublished work.
29. Panayotatos N, Wells RD (1981). Cruciform structures in supercoiled DNA. Nature 289:466.
30. Singleton CK, Wells RD (1982). Relationship between superhelical density and cruciform stability in plasmid pVH51. J Biol Chem 257:6292.
31. Singleton CK (1983). Effects of salts, temperature, and stem length on supercoil induced formation of cruciforms. J Biol Chem in the press.

Gene Expression, pages 19-32
© 1983 Alan R. Liss, Inc., 150 Fifth Avenue, New York, NY 10011

CRO REPRESSOR AND ITS INTERACTION WITH DNA[1]

W.F. Anderson[†], D.H. Ohlendorf[‡], M. Cygler[†], Y. Takeda[*] and B.W. Matthews[‡]

ABSTRACT The structure of the bacteriophage λ Cro repressor suggests the mode of interaction of the protein with DNA. Energy minimization and model building have been used to explore the apparent interactions between the repressor and its operators. The proposed model of the protein-DNA complex is consistent with the observed sites of anion binding on the protein and chemical modification studies of the protein. Complexes of the Cro repressor with DNA oligomers which correspond to different portions of the λ O_R3 operator have been crystallized and crystallographic analysis is in progress. The structures of these complexes will provide further information about Cro repressor-DNA interactions. Comparison of the Cro repressor with other base sequence specific DNA binding proteins suggests that a common α-helical DNA binding region occurs in many proteins which regulate gene expression.

[1]This work was supported by the Medical Research Council of Canada through the Group on Protein Structure and Function, the NIH (GM20066 to BWM; GM28138 and GM30894 to YT) and the NSF (PCM-8014311 to BWM).

†MRC Group on Protein Structure and Function, Department of Biochemistry, University of Alberta, Edmonton, Alberta, Canada T6G 2H7.

‡Institute of Molecular Biology, University of Oregon, Eugene, Oregon 97403.

*Chemistry Department, University of Maryland, Baltimore County, Catonsville, Maryland 21228.

INTRODUCTION

The interactions of proteins and nucleic acids play a central role in the control of gene expression. The determination of the structure of three proteins which carry out their biological functions through the recognition of specific base sequences of DNA has provided a firmer structural base for our understanding of protein-nucleic acid interactions. These three proteins are: the bacteriophage λ *cro* repressor (Cro) (1), the *E. coli* catabolite gene activator protein (CAP) (2) and the amino terminal domain of the bacteriophage λ *cI* repressor (λ Rep) (3).

The structure of the bacteriophage λ *cro* repressor suggests that the protein binds to specific base sequences of DNA with a pair of two-fold related α-helices placed in successive major grooves of right-handed B-form DNA (1). Making use of the known Cro structure, model building studies (4) have been carried out using computer graphics and energy minimization in an attempt to learn more about the potential interactions between this repressor and its operator DNA. The proposed model is consistent with the observed anion binding sites of the protein and chemical modification studies of the DNA (5) and the protein (6). Crystallographic analysis of co-crystals of the cro repressor with DNA will provide a further test of this model.

METHODS

The approach to the model building studies of the repressor-operator complex was to use coordinates of B-DNA derived from fibre diffraction analysis (7,8) and coordinates for Cro which were based on the reported structure determination (1) and partially refined (crystallographic residual 27% at 2.2 Å resolution) (D.H.O. and B.W.M. unpublished). The twofold axis of the protein dimer was constrained to be coincident with the pseudo-twofold axis of the operator DNA. Five parameters were then varied to obtain the best fit of the protein to the DNA: the separation between the protein and the DNA, the rotation of the protein dimer relative to the DNA about their mutual symmetry axis, the rise per base pair, the twist per base pair and the radius of curvature of the DNA. The bending of the DNA was constrained to occur about points on the twofold axis to maintain the symmetry of the complex. Examination of the models on a MMS-X computer graphics facility (9) coupled with energy minimization using the program EREF by Jack and Levitt (10) were used to evaluate the models.

The energy minimization procedure gives the calculated energy for the system under consideration and, as such, provides an indication of the relative strength of a postulated Cro-DNA complex. However, the energy minimization algorithm does not take account of solvent, and has other limitations which prevent its use for the reliable estimation of absolute interaction energies. Therefore, in attempting to optimize a given Cro-DNA complex, we used energy minimization as a guide, but supplemented this with additional criteria such as the adherence of postulated hydrogen bonds to acceptable stereochemistry.

The difference fourier analysis of anion binding sites was carried out using standard techniques (11). Data to 5 Å resolution were collected on a Cro repressor crystal soaked in 1.8 M phosphate, pH 7.3 and a crystal soaked in a solution containing 0.53 g of $(NH_4)_2SeO_4$ dissolved in 1 ml of H_2O with 15 mM phosphate buffer pH 7.3. The average difference in the amplitudes of these two data sets was 11.7% of the mean amplitude, while the variation between symmetry related reflections within the two data sets (R_{sym}) were 1.7% for the native, phosphate soaked crystal and 2.4% for the $(NH_4)_2SeO_4$ soaked crystal. Positive features in the difference electron density map that were greater than three times the r.m.s. density were tentatively assigned as anion binding sites.

DNA for co-crystallization experiments was synthesized by the solid phase triester approach (12). Crystals of complexes of Cro with the six base pair duplex and the nine base pair duplex (III and IV in Figure 4) were grown from solutions that contained approximately 0.1 M NaCl and 50 mM citrate buffer pH 4.0 at 4° C. In order to analyze the ratio of protein to DNA in the crystals they were washed briefly with 12 mM NaCl, then dissolved in 0.15 M NaCl, 10 mM Tris, pH 7.4. The amount of protein was determined by the Coomassie Blue dye binding assay (13) and the DNA by the absorbance at 260 nm.

RESULTS

Model for the Interaction of Cro with DNA.

The determination of the structure of the Cro repressor was naturally followed by the question, "How does in interact with the DNA?" Examination of the structure revealed a pair of twofold related α-helices which were 34 Å apart and inclined to the line between their centers in such a way as to fit nicely into the major grooves of right handed B-form DNA (1). Other surfaces of the protein or other forms of DNA do not exhibit such complementarity of the surfaces. In the proposed model for the complex between Cro and DNA the interactions occur primarily in two regions of the protein, residues 15-38, essentially the α_2 and α_3 helices, and residues 54-66, the last portion of the third β strand and the carboxy-terminal region (which is disordered in the crystal structure).

This preliminary model has been extended to consider the possible interactions between the atoms of the protein and the DNA. The use of computer graphics allowed the examination of many potential models and energy minimization was used to maintain acceptable geometry and relieve close contacts. Such detailed models were evaluated by the results of the energy minimization, the stereochemistry of potential hydrogen bonds and their ability to rationalize the existing biological and chemical information about Cro and its interaction with the λ operators. The resulting model is stereochemically acceptable and is consistent with the known differences in affinity of Cro for its six naturally occurring recognition sites in bacteriophage λ and for mutationally altered sites (4). The model suggests both sequence specific interactions and sequence independent ones. The sequence independent interactions are schematically shown in Figure 1 and are predominantly between amino acid side chains of the protein and the phosphate backbone of the DNA. The sequence dependent interactions are considered to be those that involve the parts of the base pairs that are exposed in the major or minor grooves of the DNA. The sequence specific interactions in the case of Cro seems to be restricted to the major groove. A schematic representation (14) of the presumed interactions with the consensus half sequence of the naturally occurring operators is shown in Figure 2. The amino acid side chains involved in the sequence specific interaction are those in the region of the third α-helix as originally proposed (1). In most cases a given amino acid side chain appears to make multiple interactions with the accessible portions of the base pairs. There is apparent bidentate-hydrogen bonding between a glutamine side chain and an adenine and an arginine side chain and a guanine, as had been suggested previously (15).

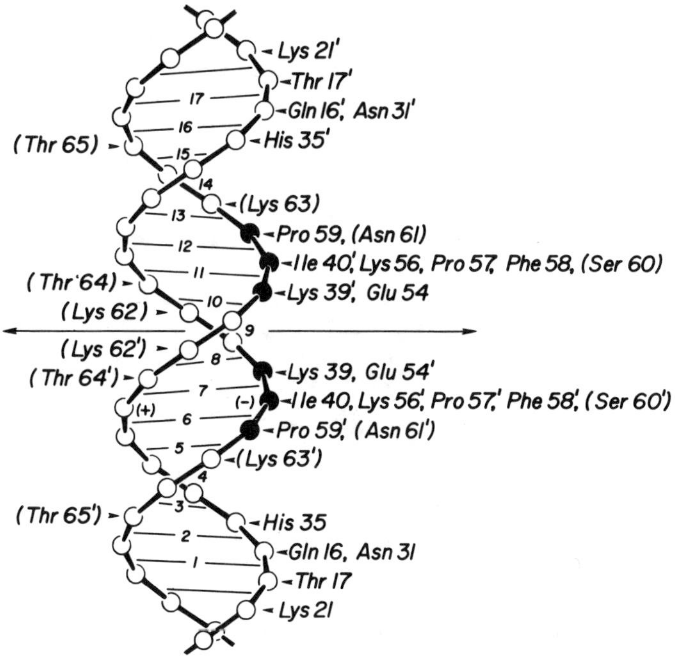

FIGURE 1. Schematic representation of the contacts and interactions which are presumed to occur between Cro and the phosphates in the DNA backbone. In the presumed Cro-DNA complex, at least one close approach occurs between a given phosphate group and the named amino acids. Those phosphates for which ethylation interferes with Cro binding (16) are drawn solid. Numbering of the base pairs and the signs (+) and (-) correspond to the identification used in the text. Amino acids belonging to 'lower' and 'upper' Cro monomers related by the horizontal dyad axis are non-primed and primed, respectively. Amino acids in parentheses are those at the C-terminus of the molecule which are disordered in the crystal (residues 62-66) or for which the backbone conformation has been adjusted (residues 60,61) to improve the contact with the DNA (Reprinted with permission from (4), copyright ©, 1982 MacMillan Journals, Ltd.).

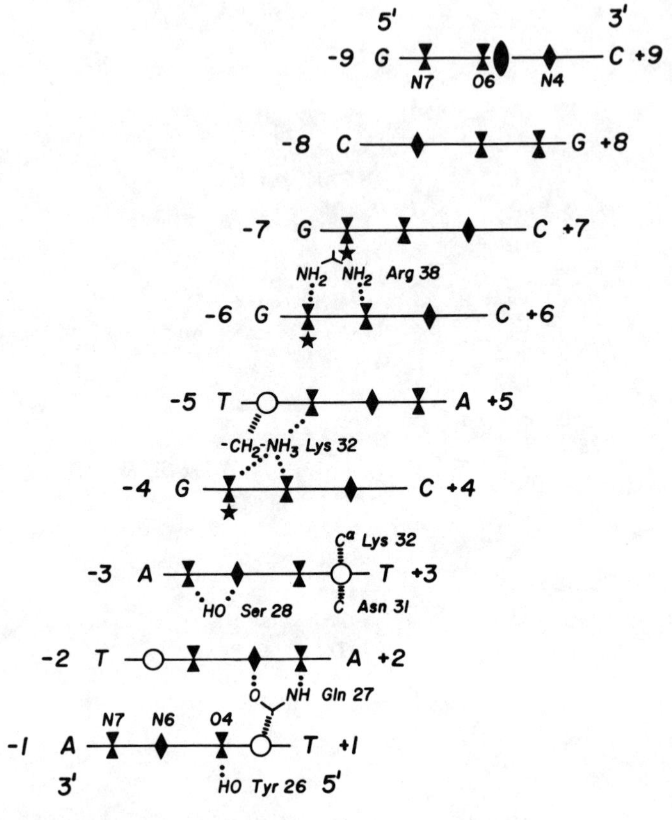

FIGURE 2. Schematic representation, following Woodbury et al. (14), of the presumed sequence-specific interactions between Cro and the parts of the base pairs exposed within the major groove of the DNA. The direction of view is imagined to be directly into the major groove of the DNA with the base pairs seen edge-on. The dyad symbol within the topmost base pair indicates the center of the overall 17-base pair binding region. The symbols are as follows: ⧗, hydrogen bond acceptor; ♦, hydrogen bond donor; ○, methyl group of thymine; ★, guanine N7 which is protected from methylation when Cro is bound (5,16). Presumed hydrogen bonds between Cro side chains and the bases are indicated (....). Apparent van der Waals contacts between Cro and the thymine methyl groups are shown (|||||||). Other van der Waals contacts are not shown (from Ohlendorf et al.(4).

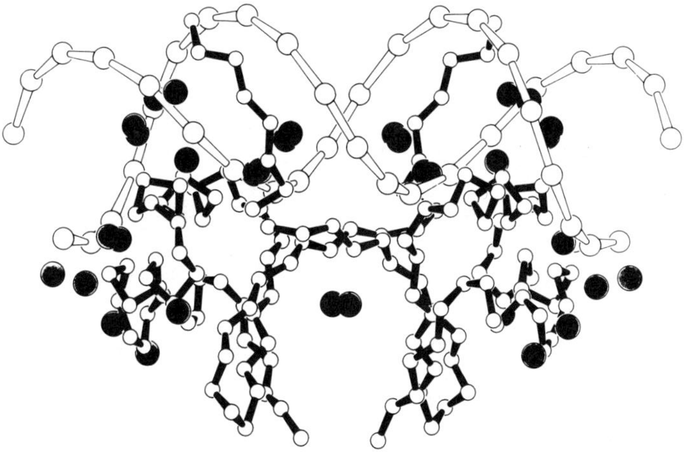

Figure 3. The locations of the positive features in the selenate-phosphate difference fourier relative to the α-carbon backbone of the Cro repressor (solid bonds) and the phosphate backbone (open bonds) of the DNA in the proposed repressor-operator complex.

Anion Binding Sites.

If the free energy of stabilization of protein-DNA complexes comes primarily from relatively non-specific interactions such as ionic interactions with the sugar-phosphate backbone of the DNA one would expect the protein to contain anion binding sites and interact with non-specific sequences of DNA. To test this hypothesis we have replaced the phosphate, which is required to obtain the repressor crystals, with ammonium selenate $[(NH_4)_2SeO_4]$. Because the selenium atom is more electron dense than the phosphate, a difference electron density map should reveal the locations to which selenate binds. It is assumed that other anions, such as phosphate, also bind to these sites and are replaced by selenate. Figure 3 shows the locations of anion binding sites relative to the Cro α-carbon backbone and the DNA phosphate backbone of the proposed complex. Because the protein in the crystal is a tetramer and each subunit occupies a different environment, the binding sites are not identical in all the subunits. To produce this figure, all of the sites were transferred onto one standard subunit and this constellation of sites is displayed on each subunit of the dimer. This array of anion binding sites is found to be similar to the locations of strong interactions with phosphates in the proposed complex (Figure 3).

The environment of two of the strongest features in the selenate-phosphate difference fourier are shown in Figure 4. The apparent anion sites generally contain a basic residue and one or more hydrogen bonding groups from the protein.

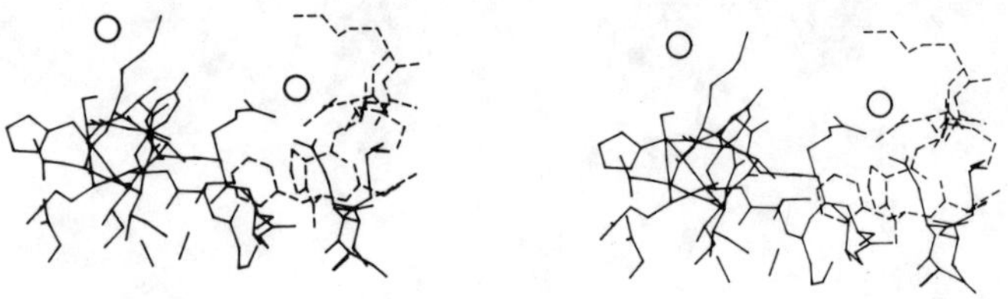

Figure 4. Stereodrawing showing the area around two of the anion binding sites (circles). The view is down α-helix three (on the right). The amino acid residues of one subunit of the Cro dimer are drawn in solid lines and the residues of the twofold related subunit are drawn in broken lines.

Crystallization of Cro-DNA Complexes.

One way to test the proposed model for the Cro-DNA complex is to study the structure of complexes of Cro with DNA oligomers. Crystals of complexes of Cro with a six base duplex (IV in Figure 5) and a nine base pair duplex (III) have been obtained (17). The crystals obtained with these two complexes are essentially isomorphous. The preliminary crystallographic data and the results of the analysis of dissolved crystals are given in Table 1. In addition, microcrystals have been obtained of the complex of Cro with the 17 base pair oligomer (I) whose sequence is that of O_R3, the operator which Cro binds most tightly (5,18).

TABLE 1

Complex	Molar Ratio Duplexes/Dimer	Space Group	Cell Dimensions		
Cro-ATCACC TAGTGG	2.3	$C222_1$	a=80.7	b=89.1*	c=80.2
Cro-ACCGCAAGG TGGCGTTCC	1.2	$C222_1$	a=81.1	b=89.2	c=80.2

*During exposure to X-rays the Cro-hexamer b cell dimension varies.

```
                 |
TATCACCGCAAGGGATA      I
ATAGTGGCGTTCCCTAT
                 |
         CAAGGGATA     II
         GTTCCCTAT
                 |
 ATCACC                III
 TAGTGG
                 |
     ACCGCAAGG         IV
     TGGCGTTCC
                 |
```

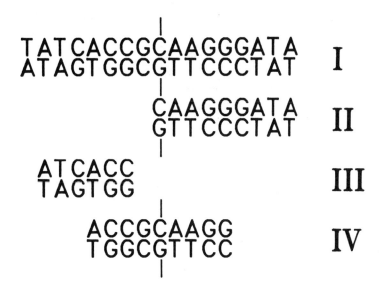

Figure 5. Base sequences of the DNA synthesized for the crystallographic studies. The vertical line indicates the position of the pseudo-dyad in the sequence.

Circular Dichroism Spectra.

To aid in the characterization of the complexes present in these co-crystals we have obtained circular dichroism spectra of the protein, the DNA oligomers, and mixtures of the protein and the oligomers. When Cro is mixed with the six base pair duplex (IV) or the nine base pair duplex (III) at either 0.15 M or 0.075 M NaCl under conditions that should give significant binding (19,20), no changes in the circular dichroism spectra between 250 nm and 320 nm are observed. However, when Cro is added to the seventeen base pair duplex (I) under the same conditions, a characteristic spectral change is produced (Figure 6). Given the conditions of these spectra the contribution of the protein is negligible and the difference spectrum in the region of 280 nm must be due to changes in the geometry or environment around the bases. Since Boschelli (20) states that no change in the CD spectrum of calf thymus DNA is produced by Cro binding, this change is not due solely to the longer DNA oligomer (seventeen base pairs vs nine or six base pairs). In addition, adding a DNA binding protein which is specific for a different base sequence (the bacteriophage λ CII protein) to the seventeen base pair duplex (I) produces no spectral change.

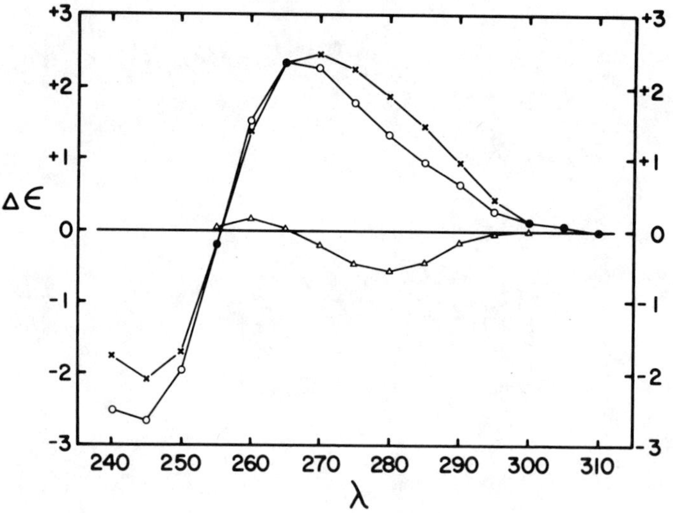

Figure 6. Circular dichroism spectra for the seventeen base pair O_R3 sequence (-x-x-), the Cro-O_R3 complex (-o-o-) and the calculated difference spectrum (-Δ-Δ-). The concentration of DNA used was 3.4×10^{-4} mole nucleotides/ℓ, the protein concentration was 4.4×10^{-6} M (in dimers) and the solutions contained 0.15 M NaCl, 5 mM Tris pH 7.2.

DISCUSSION

The model building studies of the Cro repressor-DNA complex have suggested likely interactions between the protein and the DNA. Some of these interactions, such as those with the sugar-phosphate backbone, would not seem to depend on the base sequence of the DNA and are likely to be very similar to the interactions which occur in the binding of Cro to DNA which does not contain a λ operator sequence. Other interactions would be very dependent on the sequence of bases, such as the hydrogen bonds to the portions of the base pairs exposed in the major groove. Multiple interactions between a given amino acid side chain and the base pairs appear to contribute a great deal to the specificity of the interaction of Cro with the λ operators. In particular, the model suggests bidentate hydrogen bonding interactions between Gln 27 and the N6 and N7 of adenine + 2, between Arg 38 and the O6 and N7 of guanine-6 and between Ser 28 and the N6 and N7 of adenine-3 (Figure 2). Also there appear to be multiple interactions with Lys 32 bridging two base pairs (4). In addition to these hydrogen bonds there are a number of van der Waals contacts which presumably also contribute to the specificity.

Although one of the most effective tests of this model will be the determination of the structures of complexes of the Cro repressor with DNA, only preliminary crystallographic data is available at this point. The determination of these structures is underway. In the absence of these structures, other experiments provide results which support this model.

The pattern of anion binding sites provides indirect evidence in support of the model. The sites are roughly analogous to the proposed sites occupied by the phosphate groups of the DNA. Since these anions are not constrained by the DNA structure and interact with relatively flexible amino acid side chains the sites of binding are not identical to the positions of the phosphates in the proposed complex. Also, a phosphate or selenate ion is free to optimize its location to interact both with a given protein molecule and its neighbors in the crystal lattice. To a certain extent, the anion binding simply reflects the electrostatically positive surfaces of the protein, and the location of basic and hydrophilic amino acids gives the same information. However, the fact that the same surface of the protein is complimentary to the DNA in <u>both</u> shape and polarity is consistent with the proposed model.

The change in the circular dichroism spectrum when Cro is added to duplex I, the seventeen base pair O_R3 sequence, is of interest because it appears to be due to the sequence specific interaction between the protein and the DNA. Boschelli (20) found that Cro repressor does not alter the circular dichroism spectrum of calf thymus DNA. The shorter oligomers, III and IV also do not give this spectral change at either 0.15 M or 0.075 M NaCl, even though the association constant for nonspecific DNA (20) indicates that the complex will be formed. Thus, when DNA molecules that are long enough to fully interact with both subunits of a Cro dimer are bound in a sequence specific fashion there is an alteration in the DNA which gives rise to the observed changes in the circular dichroism spectra.

Chemical modification studies of the protein are also consistent with the model presented here. As predicted from the sequence nonspecific interactions (Figure 1), removal of four or five carboxyl terminal residues with carboxypeptidase reduces DNA binding affinity (6). Also, the involvement of the lysine residues listed in Figures 1 and 2 has been confirmed by chemical protection and modification experiments of Cro (6).

The Cro repressor seems to be an excellent model system for studying protein-nucleic acid interactions. The comparison of the structures of the Cro repressor with the *E. coli* catabolite gene activator protein (21) and the λ repressor (22), together with amino acid sequence comparisons (23,24, 25,26) suggests that a common α-helical DNA binding region occurs in many proteins which regulate gene expression. On the other hand, the differences in the relative orientations of the DNA recognition helices due to differences in the interactions between protein subunits (1,2,3,6,7) suggests that there is not a simple one to one code relating the amino acid sequence of the DNA recognition helix to the DNA sequence which is recognized.

REFERENCES

1. Anderson WF, Ohlendorf DH, Takeda Y, Matthews BW (1981). Structure of the cro repressor and its interaction with DNA. Nature 290:754.
2. McKay DB, Steitz TA (1981). Structure of catabolite gene activator protein at 2.9 Å resolution suggests binding to left-handed B-DNA. Nature 290:744.
3. Pabo CO, Lewis M (1982). The operator-binding domain of λ repressor: structure and DNA recognition. Nature 298: 443.
4. Ohlendorf DH, Anderson WF, Fisher RG, Takeda Y, Matthews BW (1982). The molecular basis of DNA-protein recognition inferred from the structure of cro repressor. Nature 298:718.
5. Johnson A, Meyer BJ, Ptashne M (1978). Mechanism of action of the cro protein of bacteriophage λ. Proc Natl Acad Sci USA 75:1783.
6. Takeda Y, Cadag CG, Davis D, Stier E (1983). In preparation.
7. Watson JD, Crick FHC (1953). A structure for deoxyribose nucleic acid. Nature 171:737.
8. Arnott S, Hukins DWL (1972). Optimised parameters for A-DNA and B-DNA. Biochem. Biophys Res Commun 47:1504.
9. Molnar CE, Barry CD, Rosenberger FU (1976). Technical memo 229. Computer Systems Laboratory, Washington University, St. Louis.
10. Jack A, Levitt M (1978). Refinement of large structures by simultaneous minimization of energy and R factor. Acta crystallogr A34:931.
11. Blundell TL, Johnson LN (1976). "Protein Crystallography." New York: Academic Press.
12. Miyoshi K, Arentzen R, Huang T, Itakura K (1980). Solid-phase synthesis of polynucleotides. Nucleic Acids Res 8:5507.
13. Bradford M (1976). A rapid and sensitive method for the quantitation of microgram quantities of protein utilizing the principle of protein-dye binding. Anal Biochem 72:248.
14. Woodbury CP, von Hippel PH (1981). "Gene Amplification and Analysis." New York: Elsevier.
15. Seeman NC, Rosenberg JM, Rich A (1976). Sequence-specific recognition of double helical nucleic acids by proteins. Proc Natl Acad Sci USA 73:804.
16. Johnson AD (1980). Mechanism of action of the lambda cro protein. Ph D thesis. Harvard University.

17. Anderson WF, Cygler M, Vandonselaar M, Ohlendorf DH, Matthews BW, Kim J, Takeda Y (1983). Crystallographic data for complexes of the cro repressor with DNA. J Mol Biol (submitted).
18. Takeda Y (1980). Specific repression of in vitro transcription by the cro repressor of bacteriophage λ. J Mol Biol 127:177.
19. Boschelli F, Arndt K, Nick H, Zhang Q, Lu P (1982). Lambda phage cro repressor: DNA sequence-dependent interactions seen by tyrosine fluorescence. J Mol Biol 162:251.
20. Boschelli F (1982). Lambda phage cro repressor: non-specific DNA binding. J Mol Biol 162:267.
21. Steitz TA, Ohlendorf DH, McKay DB, Anderson WF, Matthews BW (1982). Structural similarity in the DNA-binding domains of catabolite gene activator and cro repressor proteins. Proc Natl Acad Sci USA 79:3097.
22. Anderson WF, Lewis M, Matthews BW, Ohlendorf DH, Pabo CO (1983). Comparison of the structures of cro and repressor proteins from bacteriophage λ. J Mol Biol (submitted).
23. Anderson WF, Takeda Y, Ohlendorf DH, Matthews BW (1982). Proposed α-helical super-secondary structure associated with protein-DNA recognition. J Mol Biol 159:745.
24. Matthews BW, Ohlendorf DH, Anderson WF, Takeda Y (1982). Structure of the DNA-binding region of lac repressor inferred from its homology with cro repressor. Proc Natl Acad Sci USA 79:1428.
25. Sauer RT, Yocum RR, Doolittle RF, Lewis M, Pabo CO (1982). Homology among DNA-binding proteins suggests use of a conserved super-secondary structure. Nature 298:447.
26. Weber IT, McKay DB, Steitz TA (1982). Two helix binding motif of CAP found in lac repressor and gal repressor. Nucleic Acids Res 10:5085.

Gene Expression, pages 33-41
© 1983 Alan R. Liss, Inc., 150 Fifth Avenue, New York, NY 10011

SPECIFICITY OF THE BACTERIOPHAGE T3 AND T7 RNA POLYMERASES[1]

W.T. McAllister, N.J. Horn, J.N. Bailey,[2] R.S. MacWright, L. Jolliffe,[3] C. Gocke, J.F. Klement, D.R. Dembinski, and G.R. Cleaves

Department of Microbiology, UMDNJ-Rutgers Medical School, Piscataway, New Jersey 08854

ABSTRACT To explore the basis for the template specificities of the T3 and T7 RNA polymerases we determined the nucleotide sequences of seven promoters recognized by the T3 polymerase and compared them with the previously determined T7 promoter sequences. Like the T7 promoters, the T3 promoters consist of a highly conserved 16 base pair sequence preceded by an AT rich region. The fundamental difference between the two kinds of promoters is a change in a two base pair region at -10 and -11. Whereas nearly all T7 promoters have GA at these positions, these are replaced by a single C at position -10 in the T3 promoters.

The nucleotide sequence of the gene encoding the T7 RNA polymerase (and hence the amino acid sequence of the enzyme) has been determined in other laboratories. We have determined the nucleotide sequence of the T3 RNA polymerase gene. The predicted amino acid sequence exhibits very few non-conserved changes when compared to the T7 sequence. Analysis of the data from both enzymes suggests features that may be important for polymerase function.

INTRODUCTION

The bacteriophage T3 and T7 RNA polymerases provide a uniquely attractive model system to study

1. This work was supported by NIH grant GM21783.
2. Present address: Hoffmann LaRoche, Nutley, N.J.
3. Present address: Ortho Pharmaceuticals, Raritan, N.J.

RNA polymerase:template interactions [reviewed by Chamberlin and Ryan (1)]. Compared to other RNA polymerases the phage enzymes are relatively simple in structure. Each consists of a single protein species of *ca* 100,000 daltons which is capable of accurate transcription *in vitro*. The phage polymerases are highly specific in their choice of template. Neither enzyme will transcribe *E. coli* DNA, nor will they efficiently initiate transcription on DNA from the heterologous virus.

To explore the basis for this specificity, we have begun a comparative analysis of the promoters recognized by the two RNA polymerases, and of the primary structures of the enzymes themselves.

METHODS

DNA sequence determinations were performed by the methods of Maxam and Gilbert (2). Gel readings were assembled and compared by use of the computer programs written by R. Staden (3) and adapted to run on a Hewlett Packard Model 1000 computer by GRC. Programs to analyze amino acid sequences were written in BASIC for a Radio Shack Model III computer by WTM and RSM. The latter programs are available upon request.

RESULTS

T3 Promoter Structure.

Seventeen promoters for the T7 RNA polymerase have been mapped on T7 DNA and their sequences determined (1,4). Analysis of these promoters revealed a highly conserved 16 bp sequence preceded by an uninterrupted AT region of 6-10 bp. We have cloned fragments of T3 DNA into the plasmid pBR322 and screened the recombinants for plasmids having T3 promoter activity *in vitro*. Five T3 promoters were identified and sequenced in this manner. An additional two were sequenced directly from genomic fragments (Fig. 1). The T3 promoters are highly conserved and differ from the T7 promoter consensus sequence at only two locations. Whereas nearly all T7 promoters have -GA- at positions -10 and -11, these two nucleotides are replaced by a single -C- in all T3 promoters sequenced thus far. The consensus sequences also differ at position -2, but this latter difference is probably not significant in determining the specificity of promoter recognition as five out of the seventeen T7 pro-

Figure 1. Promoter Sequences for the T3 and T7 RNA Polymerases.

Promoter	Sequence
	−10 −5 +1
a. 1.5	GTTGTCTATTTACCCTCACTAAAGGGAATAAGGTGGA
14.3	ACTTAGCATTAACCCTCACTAACGGGAGACTACTTAA
	*
16.1	TGGAAGTAATAACCCTCACTAACAGGAGAATCCTTAA
	**
22.8	AAAGCCTAATTACCCTCACTAAAGGGAACAACCCAAC
44.5	CTCTACAATTAACCCTCACTAAAGGGAAGAGGGAGCC
51.5/54.2	GTTTCTAATTAACCCTCACTAAAGGGAGAGACCATAG
51.5/54.2	TTCACCTAATTACCCTCACTAAAGGGAGACCTCATCT

b. T3 consensus (AT)$_n$AC C CTCACTAAA(+1)GGGA

T7 consensus (AT)$_n$ACGACTCACTATA(+1)GGGA

a. Sequences of T3 promoters. Identity of the promoter is by location in T3 DNA (as percent of distance from the genetic left end of the DNA) (5,6). The highly conserved region is enclosed in a box, nucleotides are numbered relative to the site of initiation (+1). Positions that differ from the conserved sequence are indicated by "*".

b. Comparison of T3 and T7 consensus sequences. Differences between the two are indicated by solid lines.

moters have a A at this position, yet are not utilized by the T3 RNA polymerase. Thus, the critical structural difference between the T3 and T7 promoters lies in the 2 bp region at -10,-11. Site directed mutagenesis experiments will be required to determine whether both of these positions are critical for promoter recognition.

T3 RNA Polymerase Gene Sequence.

Heteroduplex analysis of T3 and T7 DNA indicated that the genes which encode the phage RNA polymerases (gene 1) are 80% homologous (7). In an attempt to determine which region(s) of the structural gene is responsible for promoter specificity, Hausmann and co-workers generated a number of T3XT7 recombinant phages that produced hybrid RNA polymerases (8). The template specificity of the hybrid polymerases, in conjunction with genetic and physical mapping of the phage DNA, suggested that two regions are involved: the first extends from 25-60% of gene length, the second from 80% to the carboxy terminus (8,9).

The DNA sequence for the T7 RNA polymerase gene has been determined independently in two different laboratories (10,11). Although there are some differences in the derived sequences (see below) they are largely in agreement and predict a protein 883 amino acids in length. We have undertaken a similar analysis of the gene that encodes the T3 RNA polymerase. Over 85% of the gene 1 region of T3 DNA has been sequenced on both strands or by multiple sequencing of the same strand. Regions that have been sequenced only once are indicated by underlining in Figure 2. We would like to stress that the sequences presented for these regions are preliminary, and have not yet been confirmed.

The amino acid sequences predicted by the DNA sequences for the T3 and T7 RNA polymerases are compared in Figure 2. There are remarkably few changes in most regions of the protein, and most changes that are observed are considered to be conservative. This is particularly true for the distal portion of the sequence and less so for the amino terminal half.

An exception to the high degree of conservation is observed in the region from 380 to 424 residues (43 to 48 percent of gene length). The divergent amino acid sequence in this region can reasonably be attributed to shifts in the reading frame as a result of insertions or deletions of single base pairs, as shown in Figure 3. It is noteworthy that the two T7 DNA sequences are also out of phase in this same region, two of the frame shifts being at identical locations to those seen for T3. It is not yet known whether the disagreements between the three sequences in this narrow region are due to sequencing errors, or may represent preferred sites for potential alterations in gene structure. Resequencing of this region of T7 DNA is underway in our laboratory.

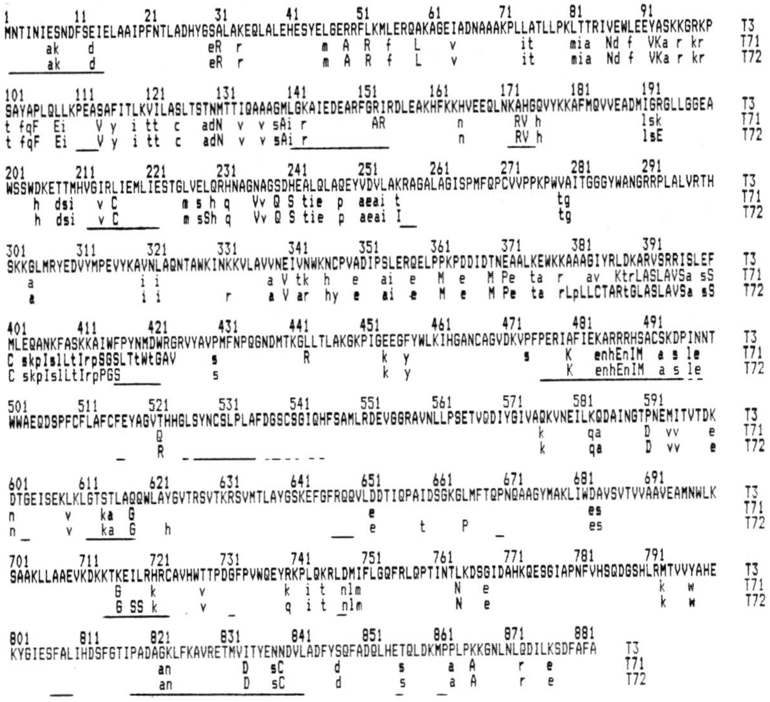

Figure 2. Comparison of the amino acid sequences of the T3 and T7 RNA polymerases.

The top line presents the amino acid sequence of the T3 RNA polymerase as predicted by the DNA sequence. The next two lines represent the sequence of the T7 RNA polymerase as predicted by Stahl and Zinn (T71) or by Grachev and Pletnev (T72) (10,11). Only positions in which each of the latter two sequences differ from that of T3 are indicated. Single letter amino acid abbreviations (12) are used throughout. A lower case letter indicates that the particular amino acid substitution is considered to be a conservative change; a capital letter indicates a non-conservative change. Non-conservative amino acid changes are those substitutions that are unlikely to occur in two related proteins based upon the accepted point mutation data (MDM_{78}) matrix of Dayhoff et al (12). A pairwise score of zero was considered to indicate non-conservation.

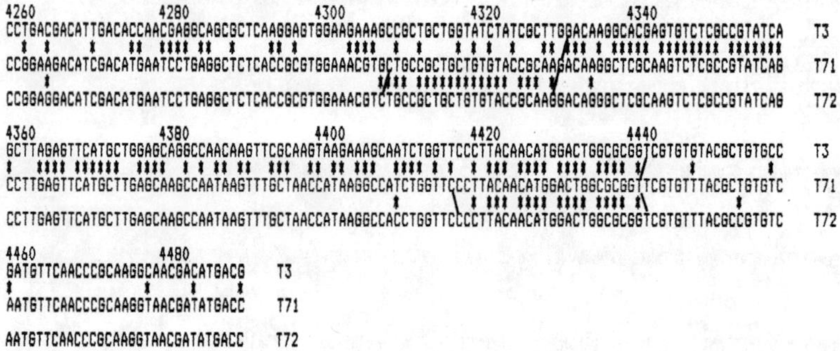

Figure 3. The sequences of T3 and T7 DNA are compared in the region from 4260 to 4490 [numbered according to Dunn and Studier (4) for T7 DNA]. The top line is that of T3 DNA (this work). The next two lines represent the sequence data of Stahl and Zinn (T71) and Grachev and Pletnev (T72). Nonmatches between each pair of sequences are indicated by *. Diagonal lines suggest regions in which frame shifts may have occurred.

A variety of modeling systems are available which facilitate the interpretation of primary structural information and which make predictions as to probable secondary structure. Two of the approaches that we have found most useful are presented in Figure 4. As is expected on the basis of their high degree of conservation, the two phage RNA polymerases are predicted to have largely similar organization. The high degree of α-helical content (ca 50%) predicted by the programs is consistent with circular dichroism spectra previously obtained for the T7 enzyme (13).

It has been reported that the T7 RNA polymerase is a membrane-associated protein (14). This would be consistent with its possible role in transferring phage DNA into the cell during the infection process (17). Although the mean hydropathicity of the protein is somewhat higher than the average for most soluble proteins, it is well below that for most membrane integrated proteins (15). Further, although there are two regions of notable hydrophobicity

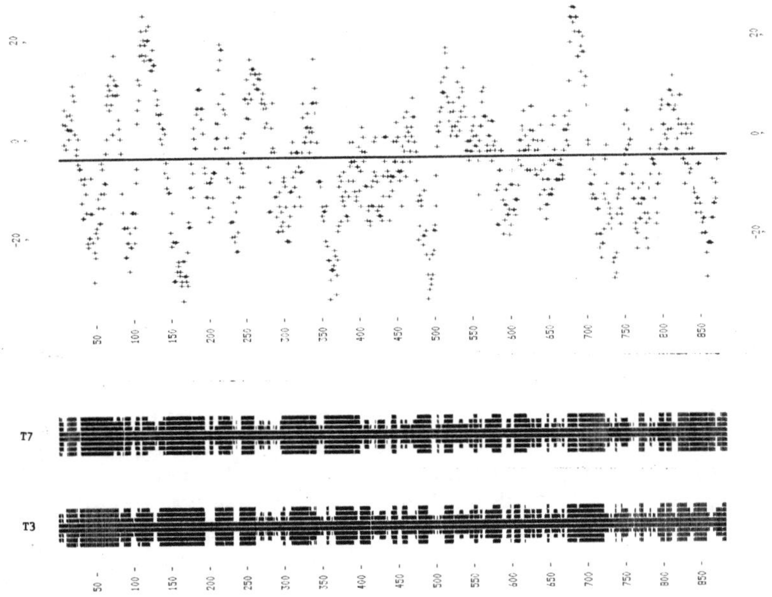

Figure 4. Structural analysis of T3 and T7 RNA polymerases.

Upper panel. The hydropathicity profile of the T3 RNA polymerase as predicted by the algorithm of Kyte and Doolittle (15). The profile for the T7 enzyme is very similar (not shown). Hydrophobic regions lie above the line of mean hydropathicity (-4), hydrophilic regions are below the line. The program considers the hydropathicity of each span of 19 amino acid residues and plots this value at the mid-point of the span length.

Lower panel. The secondary structure of the T3 and T7 RNA polymerases as predicted by the algorithm of Garnier *et al* (16). The predicted structural configuration for each residue is represented as a dashed bar. The width of the bars represent (in decreasing order) α-helix, β-sheet, turn, random coil.

near residues 110 and 690, neither of these is predicted to be a membrane spanning region (15). Additional studies will be required to characterize the nature of the membrane association of the phage RNA polymerases and the potential role of this association in regulation of transcription.

DISCUSSION

The attractiveness of the T3 and T7 RNA polymerases as model systems is due to their high degree of template specificity and their relatively simple structure. In this report we describe initial studies of the basis for this specificity at the molecular level. The promoters recognized by the two polymerases are remarkably similar and differ significantly in only a 2 bp region. Similarly, the amino acid sequences of the enzymes are highly conserved. In one region, however, shifts in the reading frame suggest a localized region in which non-conservative substitutions are rather high. Further genetic and physical experiments will be required to fully exploit and understand this system.

REFERENCES

1. Chamberlin, M, and Ryan T (1982). In Bayer CP (ed): "The Enzymes," New York: Academic Press, Vol 15, p 87.
2. Maxam A, Gilbert W (1979). In Grossman L, Moldave K (eds): "Methods in Enzymology," New York: Academic Press, Vol 65, p 499.
3. Staden R (1982). Nuc Acids Res 10;4731.
4. Dunn JJ, Studier FW. J Mol Biol (in press).
5. Bailey JN, Dembinski DR, McAllister WT (1980). J Virol 35:176.
6. Bailey JN, Klement JF, McAllister WT. Proc Nat Acad Sci (in press).
7. Davis RW, Hyman RW (1971). J Mol Biol 62:287.
8. Hausmann R, Tomkiewicz C (1976). In Losick R, Chamberlin M (eds): "RNA Polymerase," New York: Cold Spring Harbor Press.
9. Ryan T, McConnell D (1982). J Virol 43:844.
10. Stahl S, Zinn K (1981). J Mol Biol 148:481.
11. Grachev MA, Pletnev AG (1981). FEBS Lett 127:53.

12. Dayhoff MO, Schwartz RM, Orcutt BC (1978). "Atlas of Protein Sequence and Structure," p 345.
13. Oakley JL, Pascale JA, Coleman JE (1975). Biochemistry 14:4684.
14. Ennis HL, Kievitt KD (1977). J Virol 22:561.
15. Kyte J, Doolittle RF (1982). J Mol Biol 157:105.
16. Garnier J, Osguthorpe DJ, Robson B (1978). J Mol Biol 120:97.
17. Zavriev SK, Shemiakin MF (1982). Nuc Acids Res 10:1635.

Gene Expression, pages 43-52
© 1983 Alan R. Liss, Inc., 150 Fifth Avenue, New York, NY 10011

BINDING OF TRANSCRIPTION FACTOR A TO XENOPUS 5S RNA GENES

Daniel F. Bogenhagen, Jay S. Hanas and Cheng-Wen Wu

Department of Pharmacological Sciences, SUNY at Stony Brook
Stony Brook, New York 11794

ABSTRACT Accurate initiation of transcription of *Xenopus* 5S RNA genes requires the binding of a transcription factor, designated factor A, to an intragenic control region. We have studied this binding with the DNAase footprinting technique. The results indicate that two molecules of the transcription factor bind cooperatively to the 5S RNA gene with an apparent binding constant of 1 nM. An analysis of the transcriptional activities and factor binding activities of deletion mutants and linker scanning mutants suggests that sequences with dyad symmetry may be important for the dimeric binding of factor A.

INTRODUCTION

Accurate initiation of transcription of *Xenopus* 5S RNA genes is dependent on an intragenic control region bordered by gene residues 50-55 on the 5' side and by residues 80-83 on the 3' side (1,2). This control region is a binding site for a 40,000 dalton protein, transcription factor A, which is distinct from RNA polymerase III (3). When bound to the 5S RNA gene, transcription factor A protects the DNA from by DNAase I over a slightly larger region, from gene residues 46 to 97 (3,4). The intragenic control region can be roughly divided into a 5' proximal half, which is of variable sequence in somatic and oocyte type 5S RNA genes, and a 3' proximal half, which is capable of binding factor A when the 5' proximal portion is deleted (4,5). Thus, a mutant containing a 5' deletion to gene residue 64 is

[1] This work was supported by grants from the National Institutes of Health to D.F.B and to C.-W.W.

capable of binding the transcription factor (4) and of competing for transcriptional activity in vitro (5), but is, nevertheless, not transcribed. The nucleic acid binding capabilities of factor A are of particular interest since this protein is capable of binding specifically to 5S RNA as well as to 5S DNA (6,7). Indeed, Pelham and Brown (6) have proposed a feedback regulation model for the control of 5S RNA gene activity by factor A based on the unique ability of this protein to activate transcription and to bind the RNA product. Clearly, further studies are required to test this model and to determine the role of factor A in the detailed mechanism of the establishment and maintenance of active transcription complexes (8).

We have performed experiments to study both qualitative and quantitative aspects of the binding of factor A to both mutant and wild type 5S RNA genes. In this analysis we have employed the DNAase I protection, or "footprinting" method (9). The results suggest a model in which two molecules of factor A bind cooperatively to the 5S RNA gene. This 2:1 stoichiometry may result from binding of the protein to the two halves of a dyad symmetry element in the 5S RNA gene sequence.

METHODS

DNAase I protection experiments were performed with transcription factor A derived from 7S ribonucleoprotein particles as described (10). End labeled restriction fragments of cloned Xenopus borealis somatic (Xbs) 5S DNA were prepared as described (10,11). A non-radioactive 346 base pair Xbs 5S DNA fragment was prepared from milligram quantities of a plasmid containing four tandem repeats of the 5S DNA insert. This fragment was added to binding experiments as indicated in order to control the 5S DNA concentration accurately (10). The conditions for binding, DNAase I treatment, gel electrophoresis and data analysis were as described (10).

RESULTS AND DISCUSSION

Stoichiometry of Factor A binding to 5S DNA.

We have used the "footprinting" method to determine the number of factor A molecules bound to the 5S RNA gene under stoichiometric binding condition, that is, under conditions

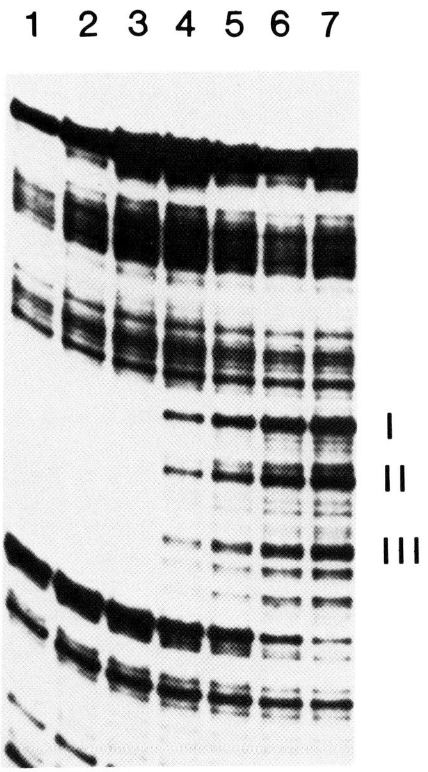

Figure 1. Footprint of the stoichiometry of binding of factor A to 5S DNA. An autoradiogram of a polyacrylamide-urea gel electrophoretic analysis of footprint reactions is shown. The factor A concentration was 40 nM in lanes 1-6, 0 in lane 7 (control). The 5S DNA concentrations were 5, 10, 20, 30, 40 and 50 nM in lanes 1-6 and 50 nM in lane 7.

in which the total factor A concentration is substantially in excess of the apparent binding constant. Figure 1 shows a series of "footprints" generated by adding increasing amounts of a non-radioactive restriction fragment of 5S DNA to a mixture of radioactive 5S DNA (at a "tracer" concentration of 1 nM) and factor A (at a concentration of 40 nM). Since factor A is present in excess in lanes 1, 2 and 3, the radioactive DNA is completely protected. However, when sufficient non-radioactive 5S DNA is added, in lanes 4 through 7, all of the factor A is bound, and there is an excess of free DNA, so that some factor A binding sites become

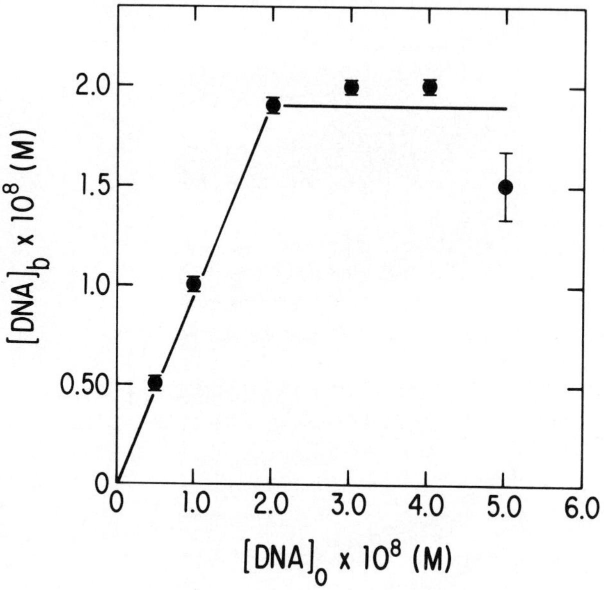

Figure 2. Stoichiometric binding plot of data from Fig. 1 The fractional saturation of factor A binding sites was determined from the data in Figure 1 by densitometry of the three bands denoted in the protected region. This data was used to calculate the amount of DNA bound to factor A, which is plotted as a function of the total input DNA concentration.

accessible to DNAase I. Data derived from this experiment are graphed in Figure 2, with the amount of DNA added plotted versus the amount of DNA bound. As shown in Figure 2, the amount of DNA bound by a 40 nM concentration of factor A saturates at a DNA concentration of approximately 19 nM. The simplest interpretation of this data is that two molecules of factor A are bound per 5S RNA gene. This experiment has been repeated several times, with various non-radioactive DNAs. The same stoichiometry is obtained with non-radioactive DNA fragments as large as 4300 base pairs. This suggests that non-specific binding to regions of the DNA outside of the 5S RNA gene control region does not contribute significantly to this analysis. Other control experiments indicated that virtually all of the factor A protein was capable of binding DNA. In addition, a fragment (or an intact plasmid) containing four 5S RNA genes appears to bind eight factor A molecules in this assay. Therefore, we consider that the transcription factor most likely binds to the 5S DNA control region with a stoichiometry of 2:1.

Determination of the affinity of factor A binding.

We have performed a titration experiment to determine the affinity of the binding of factor A to the 5S RNA gene. A series of DNAase I protection experiments was performed by adding increasing amounts of the transcription factor to a constant amount of end labeled 5S DNA. The results of an analysis of the footprinting data are graphed in Figure 3. This experiment shows that the apparent binding constant for the factor A:5S DNA complex is approximately 1 nM. Furthermore, the binding isotherm shows a distinctly sigmoid shape, indicative of cooperative binding. Since cooperativity can only result from the binding of more than one factor A molecule to the 5S RNA gene control region, this result reinforces the conclusion of the binding studies performed under stoichiometric binding conditions, that two molecules of the transcription factor bind per 5S RNA gene.

Although the data shown in Figure 3 are clearly consistent with a cooperative mode of factor A binding to the 5S RNA gene, this experiment must be interpreted cautiously. Since this experiment is performed at very low total protein concentrations, it is conceivable that inactivation of some of the protein or non-specific adsorption of the protein to surfaces could introduce artifacts into the results. We have performed a number of control experiments in an attempt to rule out such artifacts, such as including carrier protein, using siliconized plasticware, and varying the size of the DNA fragment. None of these modifications significantly affected the results shown in Figure 3. Recently, Sakonju and Brown (11) have presented a similar titration of the 5S RNA gene with factor A, with results consistent with the binding of one or two molecules of protein per gene. However, it appears that a stoichiometry of 2:1 is more consistent with our data. Two other arguments favor this interpretation. First, it is difficult to visualize how a single protein of molecular weight 40,000 could protect both strands of a 50 base pair segment of DNA extending over 170 A. Second, a 2:1 stoichiometry is also consistent with the genetic analysis presented below.

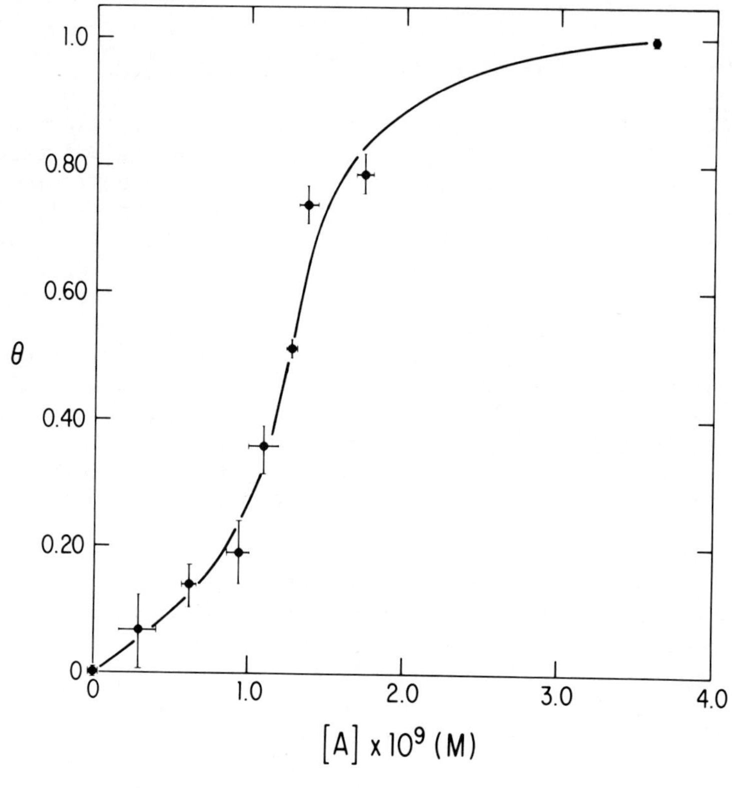

Figure 3. Analysis of footprint titration of 5S DNA with factor A. Binding reactions were performed with 0.7 nM 5S DNA and increasing amounts of factor A varied between 0 and 5 nM. The mixtures were treated with DNAase I, and footprint patterns were resolved by denaturing acrylamide gel electrophoresis. The fractional saturation, Θ, was determined by densitometry. The graph depicts as a function of the calculated concentration of free factor A.

Dyad Symmetry within the Intragenic Control Region

If factor A does bind to 5S DNA with a 2:1 stoichiometry, this dimeric binding may be reflected by symmetry within the intragenic control region. However, little is known of the fine structure of the intragenic control region. The original deletion analysis that demonstrated the intragenic control region only defined the 5' and 3' boundaries of the region (1,2). The analysis of factor A binding to deletion mutants of 5S DNA showed that the protein is capable of binding to only the 3' proximal portion of the control region (4). This is also illustrated in Figure 4, which compares the footprint titration results of the binding of factor A to a deletion mutant, Xbs 5'+67 (lanes 1-6), with the

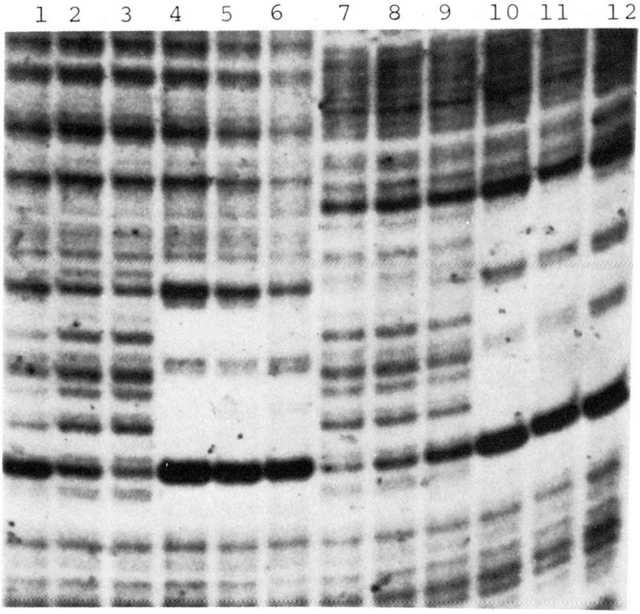

Figure 4. Comparison of footprints obtained with mutant 5S RNA genes. An autoradiogram of a polyacrylamide-urea gel is shown. Two end-labeled DNAs were used at a DNA concentration of 1 nM as described in the text. The concentrations of factor A were 0 in lanes 1 and 7; 0.5 nM in lanes 2 and 8; 1 nM in lanes 3 and 9; 5 nM in lanes 4 and 10; 10 nM in lanes 5 and 11; and 50 nM in lanes 6 and 12.

the binding of factor A to a reconstructed "linker scanning" mutant, Xbs 57/67 (lanes 7-12). This linker scanning mutant contains the first 57 residues of the 5S RNA gene joined through a synthetic Bam H1 linker to residues 67 through 120 of the 5S RNA gene. The footprint of the non-coding strand is shown for both mutants. Qualitatively, Figure 4 shows that factor A binds to and protects only the 5S DNA remnant of the deletion mutant 5'+67. However, when the 5' portion of the gene is readded in the linker scanning mutant Xbs 57/67, the entire control region is again protected. The footprint observed for Xbs 57/67 is essentially indistinguishable from the non-coding strand footprint of a wild type gene. The relatively normal footprint observed for Xbs 57/67 correlates well with the fact that the gene is capable of being transcribed in vitro (data not shown). The quantitative results of this footprint titration are somewhat surprizing: The affinity of binding of factor A to the terminal portion of the control region is only slightly less than that to the entire 5S RNA gene, as shown in Figure 3.

GENE	SEQUENCE (positions 50, 60, 70, 80, 90)	TRANSCRIBED
Xbs 1	TCGGAAGCCAAGCAGGGTCGGGCCTGGTTAGTACTTGGATGGGAG	+
Δ 5'+47	ggGGAAGCCAAGCAGGGTCGGGCCTGGTTAGTACTTGGATGGGAG	+
Δ 5'+50	cttggAGCCAAGCAGGGTCGGGCCTGGTTAGTACTTGGATGGGAG	+
Δ 5'+55	taaagcttggAGCAGGGTCGGGCCTGGTTAGTACTTGGATGGGAG	-
57/67	TCGGAAGCCAAGCccggatccgCCTGGTTAGTACTTGGATGGGAG	+
67/78	TCGGAAGCCAAGCAGGGTCGGGCccggatccggCTTGGATGGGAG	+
Δ 3'+87	TCGGAAGCCAAGCAGGGTCGGGCCTGGTTAGTACTTGGATGGccg	+
Δ 3'+83	TCGGAAGCCAAGCAGGGTCGGGCCTGGTTAGTACTTGGAccggat	+
Δ 3'+80	TCGGAAGCCAAGCAGGGTCGGGCCTGGTTAGTACTTccggatccg	-

Figure 5. Summary of sequences of deletion and linker substitution mutants. The non-coding strand sequences are shown for gene residues 45 through 90. Capital letters denote the Xbs 5S RNA gene sequence, lower case letters denote substituted sequences. Underlined sequences represent the dyad symmetry discussed in the text. The + or - signs at the right indicate qualitatively whether the gene is or is not detectably transcribed either in the oocyte germinal vesicle system (1,2) or in a factor A-supplemented high salt extract of HeLa cells (Bogenhagen, unpublished).

We are in the process of analyzing a number of other deletion and linker substitution mutants of the Xbs 5S RNA gene. The pattern of results which is emerging suggests that the 3' portion of the control region contains the information for tight binding of factor A, but that normal binding to the 5' proximal portion is also required for accurate transcription initiation. The central portion of the control region appears to be relatively unimportant for factor A binding. Several linker scanning mutants that cause replacement of portions of the center of the intragenic control region can bind factor A in a nearly normal manner and can be transcribed.

A comparison of the sequences of several deletion and linker substitution mutants of the Xbs 5S RNA gene is shown in Figure 5. This compilation suggests that two members of a dyad symmetry element located at the 5' and 3' boundaries of the minimal control region are important parts of the factor A binding site. It is also interesting to note that the 5' border of the control region contains several residues where the *Xenopus* oocyte type 5S RNA genes differ from the somatic type 5S RNA genes. These oocyte-specific base pair changes cause these genes to bind factor A less tightly and to compete poorly for transcription (5, 11).

The data presented in Figures 4 and 5 suggests that binding of factor A to the 5S RNA gene control region involves recognition of a symmetrical sequence. This dyad is only a part of the recognition sequence, as indicated by the experiments of Sakonju and Brown (11). This dyad symmetry could play any of several roles. First, it could serve as a recognition site for the binding of two molecules of factor A to the intragenic control region. Second, it remains a possibility that a single molecule of factor A could contain two similar sequence specific binding sites. Further experiments will be required to resolve this question. Third, it is conceivable that the two halves of the dyad symmetry element may form a base paired intra-strand stem, or hairpin loop, which would radically alter the local DNA structure. However, we believe that this does not occur, for two reasons. First, the hairpin is probably not sufficiently large or GC rich to be stable in naked DNA, even in a supercoiled plasmid. Second, we have performed additional experiments that indicate that factor A does not appreciably unwind the DNA helix (data not shown). Thus it appears that a hairpin structure is not formed upon binding of factor A within the control region.

In summary, we have presented data to show that two

molecules of transcription factor A are required for binding to the 5S RNA gene control region. An analysis of various mutant 5S RNA genes suggests that this binding may involve recognition of symmetrical sequences at the 5' and 3' borders of the control region.

REFERENCES

1. Sakonju, S., Bogenhagen, D.F., and Brown, D.D. (1980). Cell 19: 13.
2. Bogenhagen, D.F., Sakonju, S., and Brown, D.D. (1980). Cell 19: 27.
3. Engelke, D.R., Ng, S-Y., Shastry, B.S., and Roeder, R. (1980). Cell 19: 717.
4. Sakonju, S., Brown, D.D., Engelke, D., Ng, S-Y., Shastry, B.S., and Roeder, R. Cell 23: 665.
5. Wormington, W.M., Bogenhagen, D.F., Jordan, E., and Brown, D.D. (1981). Cell 24: 809.
6. Pelham, H.R.B., and Brown, D.D. (1980). Proc. Natl. Acad. Sci. USA 77: 4174.
7. Honda, B.M. and Roeder, R. (1980). Cell 22: 119.
8. Bogenhagen, D.F., Wormington, W.M., and Brown, D.D. (1982). Cell 28: 413.
9. Galas, D., and Schmitz, A. (1978). Nucleic Acids Res. 5: 3157.
10. Hanas, J.S., Bogenhagen, D.F., and Wu, C-W. (1983). Proc. Natl. Acad. Sci. USA, in press.
11. Sakonju, S., and Brown, D.D. (1982). Cell 31: 395.

Gene Expression, pages 53-61
© 1983 Alan R. Liss, Inc., 150 Fifth Avenue, New York, NY 10011

RECOGNITION OF UPSTREAM SEQUENCES IN THE SV40 PROMOTER REQUIRES A PROMOTER-SPECIFIC TRANSCRIPTION FACTOR[1]

William S. Dynan and Robert Tjian

Department of Biochemistry
University of California, Berkeley 94720

ABSTRACT Fractionation of a whole-cell extract capable of carrying out selective and accurate transcription in vitro reveals there are at least two factors in addition to RNA polymerase II required for transcription of the SV40 early promoter. One of these factors (Spl) is promoter-specific and is required for transcription of SV40 and BK virus early promoters but not for transcription of the polyoma virus early promoter or a variety of other viral and cellular promoters. Analysis of SV40 promoter deletion mutants suggests that the role of the Spl factor is to direct the transcription system to recognize sequences in or near the 21 base pair tandem repeats and 70 to 110 base pairs upstream from the transcription initiation sites.

INTRODUCTION

Simian virus 40 (SV40) is a circular, double-stranded, DNA virus. It has two transcription units that are transcribed divergently from a point near the origin of DNA replication (1). The early transcription unit contains the genes for large and small T antigen. The late transcription unit contains the genes for virion structural proteins.

Transcription of the SV40 early promoter can be

[1]This work was supported by the National Cancer Institute and by a National Institute of Environmental Health Sciences Center grant. W.S.D. was the recipient of a fellowship from the American Cancer Society, California Division, Inc.

carried out in vitro, using a whole-cell extract that supplies multiple factors required for promoter recognition (for references, see (2)). Transcripts synthesized in the extract begin at the same sites as RNA synthesized in vivo at early times after viral infection. Both in vivo and in vitro RNA synthesis are dependent on the presence of promoter sequences located approximately 70 to 110 base pairs upstream from the initiation sites. However, in vitro RNA synthesis proceeds equally well with or without the enhancer sequences found in the SV40 72 base pair repeats, and in this respect differs from in vivo RNA synthesis.

RESULTS

Fractionation of the Whole-Cell Extract

In order to better understand the mechanism of promoter recognition, we undertook a biochemical fractionation of the extract, with the object of separating the individual components required for transcription, purifying them, and determining how they interact to allow selective transcription in an in vitro reaction. A detailed account of the fractionation has been published elsewhere (2). A flow chart illustrating the fractionation scheme is shown in figure 1. The first step is heparin-agarose chromatography. A transcription stimulatory activity flows through heparin-agarose, but what we regard as the three important components are bound to the column and elute together in a salt step. This salt step fraction is next applied to DEAE-Sepharose. The endogenous RNA polymerase II binds to this column and is again recovered by a salt step; this fraction is used as the source of RNA polymerase in all subsequent reactions.

There are transcription factors that flow through the DEAE-Sepharose and which must be mixed with the polymerase fraction in order to reconstitute transcriptional selectivity. These factors are further purified by gel filtration chromatography using Sephacryl S-300. The peak of activity from S-300 is then pooled and applied to phosphocellulose, which resolves two activities. The first, Spl, elutes at 0.3 M KCl. It is a promoter-specific factor required for transcription of the SV40 early and late promoters, but not for transcription of other promoters we have tested. The second, Sp2, elutes at 0.55 M KCl. It is a general factor, required for transcription of all promoters tested. Fractions containing Spl and Sp2, together with the endogenous RNA polymerase faction, appear to comprise the minimum requirement for selective and accurate transcription of SV40 in an in vitro reaction.

The promoter specificity of Spl is illustrated in the experiment shown in figure 2. The presence or absence of the Spl factor makes little difference in the amount of RNA synthesized from the adenovirus 2 major late promoter (lanes 1 and 2) or from the human β-globin or avian sarcoma virus promoters (not shown). By contrast, the presence of Spl stimulates transcription from the SV40 early promoter 20 to 40-fold (lanes 3 and 4). In reactions containing a mixture of both templates, increasing amounts of Spl lead to progressive stimulation of SV40 transcription while adenovirus transcription declines slightly (lanes 5-9).

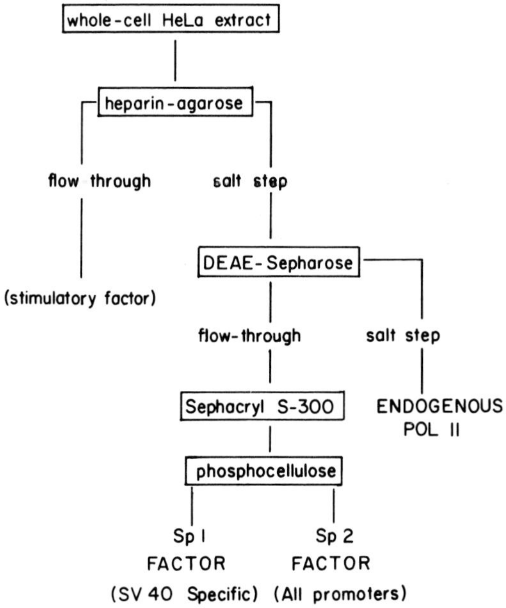

FIGURE 1. Scheme for fractionation of a whole-cell extract.

Quantitation by densitometry shows that the ratio of the two transcripts changes about 60-fold. As described in the figure legend, this experiment was done with more SV40 DNA present than adenovirus. The adenovirus promoter is stronger than that of SV40 and the ratio of DNA concentrations was chosen such that the two transcripts would be made in about equal amounts in the presence of Spl. However, the experiment has also been performed with equal amounts of the two templates, and although the levels of the two transcripts were different, a similar change in ratio was seen when Spl was added (data not shown).

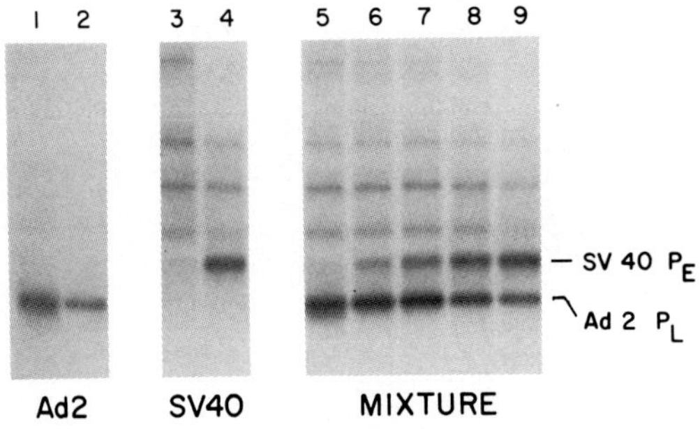

FIGURE 2. Promoter-specificity of Spl. Transcription was carried out as previously described (2), using as template the mixture of fragments generated by restriction enzyme digestion of promoter-bearing plasmids. Run-off transcripts were glyoxalated and analyzed on a 1.4% agarose gel. Template was 70 ng of adenovirus major late promoter DNA (lanes 1 and 2), 400 ng of SV40 early promoter DNA (lanes 3 and 4), or a mixture of the two (lanes 5-9). Reactions contained an endogenous RNA polymerase II fraction (24 µg protein), Sp2 (3 µg protein) and variable amounts of Spl as follows: lane 1, 0 µg; lane 2, 5 µg; lane 3, 0 µg; lane 4, 5 µg; lane 5, 0 µg; lane 6, 1 µg; lane 7, 2 µg; lane 8, 4 µg; lane 9, 6 µg protein. Data is from ref. 2.

Transcription of 5' Deletion Mutants.

The sequences responsible for the Spl effect were mapped using a series of SV40 early promoter deletion mutants constructed in vitro by standard techniques (2a). Deletion endpoints were mapped by DNA sequencing, and are shown in figure 3.

Transcription of the parental plasmid and the deletion mutants was carried out using the purified system, containing Spl, Sp2, and endogenous RNA polymerase II. Transcripts were analyzed by primer extension as previously described (2). Preliminary results were as follows: a mutant where upstream sequences were removed through -100 had essentially wild type promoter activity. The next mutant, with an endpoint at -74, had severely depressed transcriptional activity. This mutant lacks almost all of the two tandem 21 base pair repeats found in this region of the genome.

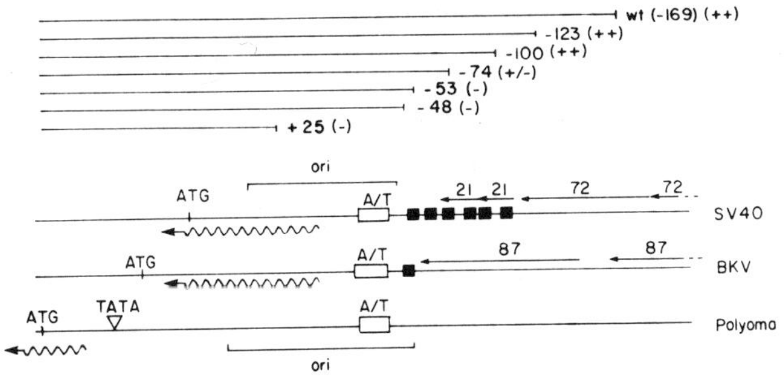

FIGURE 3. Analysis of deletion mutants. Mutants have had sequences removed from upstream of the point indicated; endpoints are given relative to the start site for early transcription, taken as nucleotide 5235 (numbering according to ref. 1). Deletion mutants were tested for their ability to direct transcription in the in vitro system, with results as indicated in parentheses. Lower portion of figure illustrates the SV40, BK (strain MM), and polyoma virus transcription control regions (1). Principal features include the start codon for T antigen synthesis (ATG), the minimal sequence required in cis for DNA replication (ori), the block of AT base pairs (A/T), the sequence CCNCCC ( ), which is repeated six times in SV40, and the 21, 72, and 87 base pair tandem repeats.

Even the severely deficient mutants showed a small amount of residual RNA synthesis. In those mutants that entirely lacked the 21 base pair repeats, such as -48, the level of RNA synthesis was the same in the presence or absence of Spl. Moreover, in the absence of Spl, the level of RNA synthesis was the same in the presence or absence of the 21 base pair repeats. These two observations together strongly suggest that the effect of Spl is to target the transcription system to recognize a sequence that lies entirely or partially within the 21 base pair repeats. This effect could be direct or indirect; Spl could bind to sequences in this region or it could influence other proteins to change their binding specificity.

Transcription of Related Papovaviruses.

Polyoma is a tumor virus similar to SV40 in size and genome organization, and with some sequence homology. The early promoter of polyoma virus has previously been shown to be transcribed in vitro (3). In our hands, cloned polyoma template gave runoff early transcripts of the expected size when cut with either of two restriction enzymes. The amount of these putative early transcripts was unaffected by the presence of Spl. This result can be rationalized by considering the differences between the control regions of polyoma and SV40, which are illustrated in figure 3. Whereas the SV40 origin and promoter are an integral, overlapping unit, the polyoma transcript starts well downstream from the origin. Moreover, polyoma does not contain extensive homology to the 21 base pair repeats of SV40. It thus appears that polyoma and SV40 may have evolved different mechanisms for producing an early transcript.

The results with BK virus were quite different. BK is very closely related to SV40. When in vitro transcripts of BK virus were analyzed by primer extension, using the same primer as was used with SV40, a single early transcript was detected. The synthesis of this transcript appeared to be completely dependent on Spl, although the maximum level of transcription was about ten-fold lower than that of SV40 under the same conditions. BK virus does not contain the 21 base pair repeats as such, although there is a stretch of about eleven base pairs with partial homology to the sequences found in the 21 base pair repeats. Although we have yet to construct the appropriate deletions of BK, the present results suggest that this region of partial homology is the target site for Spl action in BK virus.

DISCUSSION

Promoters for protein-coding genes in higher eukaryotes are typically composed of two distinct sequence elements. The downstream element, or TATA box, is an AT-rich block found 25-30 base pairs before the initiation site in many but not all promoters (reviewed in ref. 4). The upstream element, located 50 to 110 base pairs before the initiation site, is often GC-rich, and sometimes contains direct or inverted repeats. There is only limited sequence homology seen in this region when different promoters are aligned.

In some promoters, sequences in the upstream element are involved in regulating the amount of RNA that is made. In Drosophila heat shock genes, sequences in the upstream promoter element are required for induction in response to stress (5). Similarly, in the mouse metallothionein gene, sequences lying upstream of the TATA box are involved in mediating the induction by heavy metals (6,7). In several other promoters, upstream elements are required for transcription, although their regulatory role remains to be elucidated. These promoters include α- and β-globin (8-10), herpes virus thymidine kinase (11), and SV40 (12-14).

In general, in vitro transcription systems based on crude extracts respond strongly to the presence of the TATA box element, whereas the presence or absence of the upstream promoter element has little effect. SV40 is at present one of few exceptions: the early promoter requires the upstream element for transcription in vitro as well as in vivo. In vitro transcription of SV40 requires a promoter-specific factor, Spl. This factor appears to be responsible for the use of upstream sequences: without the factor, there is no dependence on the sequences, and without the sequences there is no dependence on the factor.

These findings raise several interesting questions. First, are there cellular promoters that require Spl for their expression, and does Spl play a role in biological regulation? The presence of Spl in extracts from uninfected cells suggests that it must play some role in the cell. It will be interesting, when Spl is better characterized, to see whether the amount of this factor varies from one cell type to another.

Second, are there other examples of factors like Spl that direct the recognition of other classes of upstream sequences? The existence of multiple factors, each specific for different sequences, is implicit in the idea that the upstream promoter elements have a regulatory role. One reason Spl was detected and other factors have not been may be that the AT rich block in SV40 functions very poorly as a TATA element in vitro, and thus a factor that allows recognition of the upstream sequences is necessary in order to drive transcription at an appreciable rate. In order to study the effects of upstream promoter elements in other promoters, it may be necessary to use mutants with weakened TATA elements or to use more sophisticated assays that measure the kinetics of promoter recognition rather than the overall accumulation of transcript. In addition, in some cases it may be necessary to prepare factors from cells that are actively transcribing the gene of interest.

Finally, it will be interesting to discover by what underlying mechanism Spl alters the specificity of the transcription system, for example, whether it interacts directly with either the RNA polymerase II or with sequences in the template.

ACKNOWLEDGMENTS

We would like to thank Ray Wu for the gift of BK virus DNA, R. Marc Learned for providing SV40 promoter deletion mutants and Kathy Jones for cloned polyoma virus DNA. We would also like to thank Karen Erdley for typing the manuscript.

REFERENCES

1. Tooze J (1981). Molecular biology of tumor viruses. 2nd ed., revised. DNA tumor viruses. Cold Spring Harbor, New York: Cold Spring Harbor Laboratory.

2. Dynan WS, Tjian R (1983). Isolation of transcription factors that discriminate between different promoters recognized by RNA polymerase II. Cell 32:669.

2a. Learned RM, Myers RM, Tjian R (1981). Replication in monkey cells of plasmid DNA containing the minimal SV40 origin. In Ray DS, Fox CF (eds): ICN-UCLA Symposia on Molecular and Cellular Biology, v. 21. New

York: Academic Press. p 555.

3. Jat P, Roberts, J, Cowie, A, Kamen, R (1982). Comparison of the polyoma virus early and late promoters by transcription in vitro. Nuc Acids Res 10:871.
4. Breathnach, R, Chambon, P (1981). Organization and expression of eukaryotic split genes coding for protein. Ann Rev Biochem 50:349.
5. Pelham HRB (1982). A regulatory upstream promoter element in the Drosophila Hsp70 heat-shock gene. Cell 30:517.
6. Brinster RL, Chen HY, Warren R, Sarthy A, Palmiter RD (1982). Regulation of metallothionein-thymidine kinase fusion plasmids injected into mouse eggs. Nature 296:39.
7. Hamer DH, Walling M (1982). Regulation in vivo of a cloned mammalian gene: cadmium induces the transcription of a mouse metallothionein gene in SV40 vectors. J Mol Appl Genet 1:273.
8. Dierks P, van Ooyen A, Mantei N, Weissmann C (1981). DNA sequences preceding the rabbit β-globin RNA with the correct 5' terminus. Proc Nat Acad Sci USA 78:1411.
9. Grosveld GC, deBoer E, Shewmaker CK, Flavell RA (1982). DNA sequences necessary for transcription of the rabbit β-globin gene in vivo. Nature 295:120.
10. Mellon P, Parker V, Gluzman Y, Maniatis T (1981). Identification of DNA sequences required for transcription of the human α1-globin gene in a new SV40 host-vector system. Cell 27:279.
11. McKnight SL, Gavis ER, Kingsbury R, Axel R (1981). Analysis of transcriptional regulatory signals of the HSV thymidine kinase gene: identification of an upstream control region. Cell 25:385.
12. Myers RM, Rio DC, Robbins AK, Tjian R (1981). SV40 gene expression is modulated by the cooperative binding of T antigen to DNA. Cell 25:373.
13. Benoist C, Chambon P (1981). The SV40 early promoter region: sequence requirements in vivo. Nature 290:304.
14. Fromm M, Berg, P (1982). Deletion mapping of DNA regions required for SV40 early region promoter function in vivo. J Mol Appl Genet 1:457.

II. PROMOTERS AND ENHANCERS

Gene Expression, pages 65-74
© 1983 Alan R. Liss, Inc., 150 Fifth Avenue, New York, NY 10011

RIBOSOMAL RNA OPERON PROMOTERS P1 AND P2 SHOW DIFFERENT REGULATORY RESPONSES

P. Sarmientos, S. Contente,
G. Chinali, and M. Cashel

Laboratory of Molecular Genetics,
National Institute of Child Health
and Human Development,
N.I.H., Bethesda, Maryland 20205

ABSTRACT *E. coli rrnA* operon tandem promoters P1 and P2 have been fused to the *rrnB* terminator region on multicopy plasmids. Efficient *in vivo* transcription and termination allows direct visualization of transcripts arising from each promoter by acrylamide gel analysis of whole cell RNA extracts. Previously we found that the two promoters are governed by distinctly different regulatory mechanisms during the stringent RNA control response. Here we observe that both promoter dependent transcripts are absent in stationary phase cultures. Outgrowth from stationary phase is accompanied by rapid activation of the P2 dependent transcript whereas P1 activation is initially sluggish.

INTRODUCTION

In vivo and *in vitro* studies have shown that the expression of the seven ribosomal RNA (*rrn*) operons of *E. coli* is likely to be directed by two tandem promoters [1-6]. The promoter regions of six of the seven chromosomal operons have been sequenced, revealing two tandem promoter sites whose relative positions are strongly conserved to be within 110-120 base pairs of one another [3,7-9]. Studies on fusions of these promoter regions to galactokinase or β-galactosidase genes have shown that both stringent RNA control and growth rate control are displayed by such fusions *in vivo* [10, 11], indicating that the regulatory determinants are encoded within the promoter regions. We have constructed plasmids containing fusions of the *rrn*A operon promoters to the terminator region (containing tandem terminators T1 and T2) of the *rrn*B operon [12]. Coupling the *rrn*B terminator region with the *rrn*A promoter region is unlikely to generate heterologous operon fusion artifacts in this instance at least, because both the promoter regions and the terminator regions of *rrn*A and *rrn*B are homologous [7,8,13,14]. Although metabolically stable RNA products are normally processed out of *rrn* transcripts, the *rrn* promoter-terminator fusion transcripts are labile to rifampicin addition, having half-lives of about 2 min in growing cells [12]. *In vivo*, these transcripts terminate predominately at the first *rrn* terminator (T1) with almost no termination detected at the second (T2), 160 bp further downstream. The strength of the two promoters, coupled with efficient termination, allows measurement of the cellular abundance of plasmid specified transcripts on 7M urea-acrylamide gels [12]. Similar measurements of transcript decay rates allows estimates of the activity of each promoter.

Initially, we used this approach to measure the effects of the stringent RNA control response and found that during rapid exponential growth, P1 activity predominated about 3X over P2 activity [12]. We were surprised, however, to find that P1 and P2 activities were not coordinately controlled during the stringent response. Although P1 did behave as if severely inhibited by ppGpp, P2 activity was only moderately inhibited by aminoacyl tRNA deprivation whether or not ppGpp accumulated. This mode of regulation of P1 was preserved even when the P2 promoter region as well as *in vitro* pause sites downstream [15] were deleted. Furthermore, culture manipulations resulting in ppGpp disappearance in amino acid starved stringent cells resulted in activation of the upstream P1 promoter without appreciably changing P2 activity. We have also used these plasmids to re-evaluate the effects of ppGpp on *in vitro* transcription; because a discretely terminated transcript is produced, supercoiled templates can be used, rather than runoff transcription from linear fragments [16].

We have now explored features of P2 regulation. Examination of the relative P1 and P2 activities as a function of steady state growth rates has provided further evidence that the two promoters are differentially regulated. We have found [17] that P1 activity increases as approximately the square of the growth rate whereas P2 is only weakly and linearly dependent upon the growth rate. Here we describe finding that both P1 and P2 dependent transcripts are not detected in ethidium stained gels of stationary phase cellular RNA. However, during outgrowth from stationary phase, an unusually strong and rapid burst of accumulation of the P2 specified transcript occurs that is not immediately accompanied by strong P1 activation. The P1 promoter is only sluggishly activated and takes a long time to reach levels that are characteristic of logarithmically growing cells. These results are interpreted as a new indication of the different regulatory responses of these promoters. We suggest that P2 might be necessary for recovery from stationary phase.

METHODS AND MATERIALS

Experimental methods have been summarized elsewhere [12]. The outgrowth experiments described here were performed with relaxed (CF747) and stringent (CF748) cell strains bearing the pPS1 plasmid which contains the *rrn* promoter-terminator fusion. Cultures were grown overnight with aeration in Luria Broth. Outgrowth was initiated by 20-fold dilution in fresh, prewarmed (32 $^{\circ}$) medium beginning at an A600= 0.3. Cell aliquots were lysed in hot SDS, phenol shaken and electrophoresed as described [12]. Although both host strains carry a *recA* deletion, we have frequently observed that spontaneous oligomerization of the pPS1 plasmid occurs during maintenance of the cultures. These oligomers are detected as more slowly moving supercoiled species in agarose gels that nevertheless give normal restriction patterns. Monomeric plasmids were prepared by cutting pPS1 at a unique Eco RI site and religating in dilute solution and transforming with low DNA concentrations. Monomeric transformants are usually transiently stable before oligomeric variants progressively appear. Oligomers first appear together with monomers then later predominate as the only form of plasmid present. Exact copy number changes accompanying oligomerization are uncertain. Oligomer abundance suggests a corresponding increase in unit length copy number but this has not been accurately measured. Oligomeric plasmids that are probably analogous have been noted by others to form by recombination in recA hosts [18].

RESULTS

In search of regulatory parameters affecting P2 we have screened a number of physiological conditions that might be expected to severely restrict ribosomal promoter activity. At slow steady state growth rates we find that P1 activity levels are drastically restricted despite only moderate inhibition of P2 promoter activity [17]. When *E. coli* growth is completely limited by amino acid starvation or glucose starvation, P2 activity is only moderately inhibited as compared to P1 activity [12,17]. We have found, however, that stationary phase cultures of Luria Broth grown cells show a virtual absence of both P1-T1 and P2-T1 transcripts.

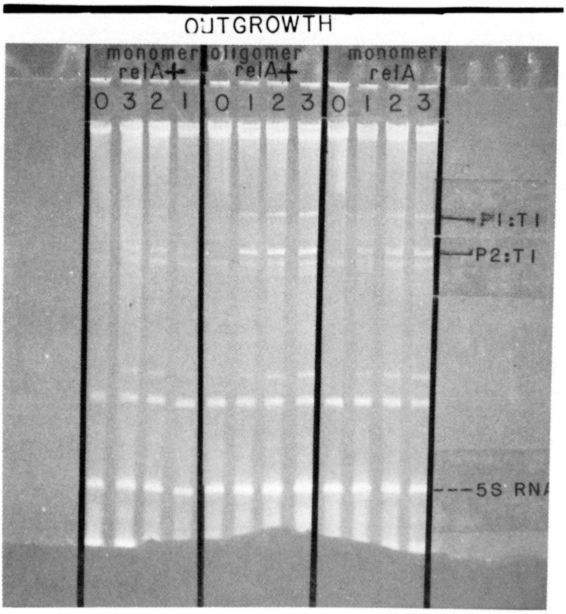

Figure 1. RNA species resolved by 5% acrylamide, 7M urea gel electrophoresis. Samples were taken from cell strains during stationary phase, labeled 0, and at ten minute intervals during outgrowth as described in the text.

Figure 1 shows an ethidium bromide stained electropherogram of RNA extracted at 0, 10, 20 and 30 min after dilution into fresh medium. The first set of lanes shows a stringent CF748 host carrying monomeric pPS1 with the times of RNA extraction marked as 0, 1, 2 and 3 respectively. The second set of lanes corresponds to the same strain bearing oligomeric pPS1 plasmid, and the third set of lanes shows a relaxed relA CF747 host bearing monomeric pPS1. During exponential growth in Luria Broth, the ratio of abundance of P1 to P2 is 3:1 [12]. During outgrowth, each culture shows a similar pattern of appearance of the two promoted RNA chains: the P2-T1 transcript is rapidly turned on while the P1-T1 transcript only gradually increases to achieve exponential phase levels.

In Figure 1, transcription in the stringent relA+ strain bearing the pPS1 plasmid is indistinguishable from that in the relaxed. With respect to P2 activity this is not surprising, for the relA allele does not affect P2 activity either during the stringent response or during growth regulation [12,17]. We ascribe the relA independence of P1 activity regulation during outgrowth to ppGpp accumulation which has been observed to be relA independent at stationary phase [19]. We presume that ppGpp slowly falls to basal levels during outgrowth in both strains. The intensity of an RNA band migrating slightly slower than 6S RNA increases markedly during outgrowth, but as this RNA species is also seen in cell controls lacking plasmid, it will not be considered further. Cells hosting the oligomeric plasmid show a striking enhancement in the abundance of both P1-T1 and P2-T1 transcripts (Fig. 1). This amplifying effect of oligomeric plasmids in comparison with monomeric plasmids is rel independent (data not shown). Figure 2 shows a densitometric tracing of the gel.

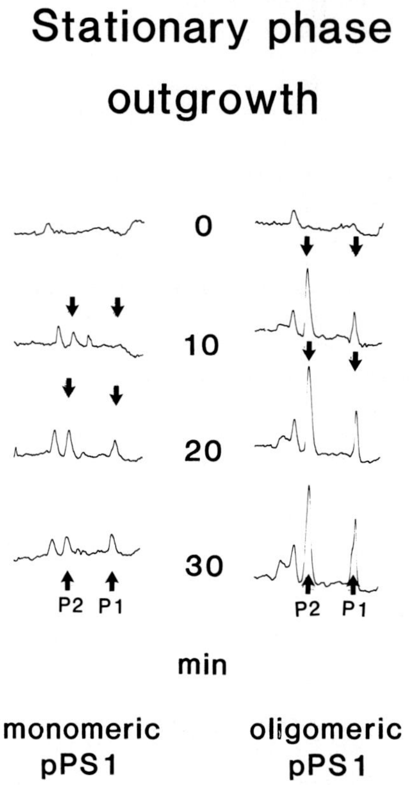

Figure 2. Densitometric tracing of P1-T1 and P2-T1 region of the ethidium bromide stained gel shown in Figure 1.

Apparently plasmid oligomers magnify the amplitude of the monomeric response without affecting the timing of differential promoter activation. Interestingly, we do not see an equivalent amplifying effect of oligomeric plasmids during steady state exponential growth (data not shown).

DISCUSSION

The regulatory features of *E. coli* tandem ribosomal operon promoters have been studied by direct analysis of transcripts arising from promoter terminator fusions. The extreme strength of these promoters and the high efficiency of termination makes this approach technically simple. We found that P2 was a weaker promoter than its upstream partner in rapidly growing cells, but that it becomes a major source of *rrn* transcripts in cells where growth is limited by amino acid or carbon source availability [12,17]. What then, is the function of P2? Conservation of P2 among all six *rrn* homology sequenced so far suggests that it is not dispensable. Previously, we suggested [16] that P2 might be a maintenance promoter that allows basal level expression of *rrn* operons when ppGpp is abundant but leaves adaptive expression to the P1 promoter. Here we have noted that both promoters are apparently turned off during stationary phase. Therefore P2 is not simply a constitutive weak cousin of P1. Perhaps the burst of P2 activity during outgrowth provides a clue to some essential function of this promoter.

The biology of stationary phase cells is imprecisely defined owing in part to uncertainty as to what actually limits further growth. Therefore, we cannot predict why P2 transcripts are not evident in stationary phase. We cannot rule out the possibility that the apparent absence of the P1 and P2 transcripts in stationary phase cells is due to either increased transcript turnover or to a failure to terminate. However, termination efficiency has not been altered by the stringent response, by slow growth rates or by glucose starvation [12,17]. Furthermore, slowly growing cells have the same turnover rates for P1 and P2 initiated RNA chains as do rapidly growing cells [17]. When cellular growth was completely stopped by chloramphenicol or amino acid deprivation, the transcripts were even more stable than they were in growing cells. We have measured transcript stability to rifampicin addition 10 min after the onset of outgrowth and find no difference in P1-T1 stability (data not shown). We therefore believe that differential promoter regulation during outgrowth occurs at the level of promoter activation.

Copy number changes as a function of growth phase are likely to introduce additional complexities. These considerations apply both to plasmid and to chromosomal copies of rrn promoters. Our rrn promoter terminator fusions are present on pBR322 plasmid derivatives containing the ColE1 replication origin in which the rop gene has been inactivated by insertion of the galK gene fragment in a PvuII site [20,21]. When XbaI linkers are inserted in the same PvuII site of pBR322, the copy number is amplified about four-fold in stationary phase as compared to exponential phase [22]. Chromosomal rrn promoter copy number, in contrast, should decrease as exponential growth slows and cells enter stationary phase because these operons are located near the replication origin of E. coli [23].

In any event ribosomal RNA promoters appear to be differentially regulated despite their close juxtaposition which creates some interesting mechanical problems. These problems have evidently been solved by the cell because the upstream and downstream promoters have independently variable activities. The addition of chloramphenicol to amino acid starved stringent cells for example, will increase P1 activity 5-10 fold without changing P2 activity. Conversely, as shown above it is clear that P2 can be strongly activated without a similar activation of P1 activity.

ACKNOWLEDGEMENTS

We would like to thank Ms Mary Thorne and EL Tron for aid in manuscript preparation.

REFERENCES

1. DeBoer H and Nomura M (1979) In vivo transcription of rRNA operons in Eschrichia coli initiates with purine nucleoside triphosphates at the first promoter and with CTP at the second promoter. J Biol Chem 254:5609.
2. Glaser G and Cashel M (1979) In vitro transcripts from the rrnB ribosomal RNA cistron originate from two tandem promoters. Cell 16:111.
3. Young R and Steitz J (1979) Tandem promoters direct E. coli ribosomal synthesis. Cell 17:225.
4. Gilbert SF, DeBoer HA and Nomura M (1979) Identification of initiation sites for the in vitro transcription of rRNA operons rrnE and rrnA in E. coli. Cell 17:211.

5. Boros IA, Kiss A and Venetianer P (1979) Physical map of the seven ribosomal RNA genes of Escherichia coli. Nucleic Acids Res 6:1817.
6. Ellwood M and Nomura M (1982) Chromosomal locations of the genes for rRNA in Escherichia coli K12. J. Bacteriol 149:458.
7. Csordas-Toth E, Boros I and Venetianer P (1979) Structure of the promoter region for the rrnB gene in Escherichia coli. Nucleic Acids Res. 7:2189.
8. De Boer HA, Gilbert SF and Nomura M (1979) DNA sequences of promoter operons of rrnE and rrnA in E. coli. Cell 17:201.
9. Shen W-F, Squires C and Squires CL (1982) Nucleotide sequence of the rrnG ribosomal RNA promoter region of E. coli. Nucleic Acids Res 10:3033.
10. Ota Y, Kikuchi A and Cashel M (1979) Gene expression of an E. coli ribosomal RNA promoter fused to the structural genes of the galactose operon. Proc Natl Acad Sci USA 76:5799.
11. Miura A, Kreuger JH, Itoh S, DeBoer HA and Nomura M (1981) Growth rate dependent regulation of ribosome synthesis in E. coli: expression of the lacZ and galK genes fused to ribosomal promoters. Cell 25:773.
12. Sarmientos P, Sylvester JE, Contente S and Cashel M (1983) Differential stringent control of the tandem E. coli ribosomal RNA promoters from rrnA expressed in vivo in multicopy plasmids. Cell in press.
13. Brosius J, Dull TJ, Sleeter DD and Noller HF (1981) Gene organization and primary structure of a ribosomal RNA operon from Escherichia coli. J Mol Biol 148:107.
14. Cashel M (1983) Sequence and termination activity of the distal region of the rrnA operon in E. coli. In preparation.
15. Kingston RE and Chamberlin MJ (1981) Pausing and attenuation of in vitro transcription in the rrnB operon of E. coli. Cell 27:523.
16. Glaser G, Sarmientos P and Cashel M (1983) Functional interrelationship between two tandem E. coli ribosomal RNA promoters. Nature 302:74.
17. Sarmientos P and Cashel M (1983) Growth rate control of the tandem ribosomal RNA (rrnA) promoters in E. coli. In preparation.
18. Fishel RA, James AA and Kolodner R (1981) recA-independent general genetic recombination of plasmids. Nature 294:184.

19. Kramer M, Keckes E and Horvath I (1981) Guanosine polyphosphate production in stringent and relaxed strains of Escherichia coli during the stationary phase of growth. Acta Microbiol Acad Sci Hung 28:165.
20. McKenney K, Shimatake H, Court D, Schmeissner U, Brady C and Rosenberg M (1981) A system to study promoter and terminator signals recognized by E. coli RNA polymerase. In Chirikjian JG and Papas TS (eds): "Gene Amplification and Analysis", New York: Elsevier North Holland, Inc: p384.
21. Cesareni G, Muesing MA and Polisky B (1982) Control of ColE1 DNA replication: the rop gene product negatively affects transcription from the replication primer promoter. Proc Natl Acad Sci USA 79:6313.
22. Steuber D and Bujard H (1982) Transcription from efficient promoters can interfere with plasmid replication and diminish expression of plasmid specified genes. EMBO Journal 11:1399.
23. Nomura M, Morgan EA and Jaskunas SR (1977) Genetics of bacterial ribosomes. Ann Rev Genet 11:297.

Gene Expression, pages 75-85
© 1983 Alan R. Liss, Inc., 150 Fifth Avenue, New York, NY 10011

ASSAY OF NATURAL AND SYNTHETIC HEAT-SHOCK PROMOTERS IN MONKEY COS CELLS: REQUIREMENTS FOR REGULATION

Hugh R.B. Pelham and Michael J. Lewis

M.R.C. Laboratory of Molecular Biology, Hills Road, Cambridge CB2 2QH, England.

ABSTRACT The promoters for the Drosophila hsp 22, 23, 26 and 27 heat shock genes have been fused to the herpesvirus thymidine kinase gene and assayed in COS cells. The hsp 22 and 26 promoters contain a good match to the previously defined heat shock consensus sequence close to their TATA boxes, and are efficiently regulated. The hsp 23 and 27 promoters, which do not fit this pattern, are inefficiently expressed. The adenovirus major late promoter can be made heat-inducible by replacing its normal upstream element with a synthetic version of the consensus sequence. None of these promoters is greatly affected by remote SV40 enhancer sequences.

INTRODUCTION

In every organism so far examined, hyperthermia and certain other stressful treatments induce the massive synthesis of a small number of so-called heat shock proteins (hsps)(see ref. 1 for review). The proteins are evolutionarily very highly conserved, and evidently serve to protect cells from the stressful conditions that induce their synthesis, although their exact functions remain unclear. The heat shock system has been studied in most detail in Drosophila; seven different proteins are induced and the genes that code for them have all been cloned and sequenced.

In almost all cases, induction of hsps involves the de novo synthesis of their mRNAs. Thus this small group of genes provides a model system for studying the coordinate regulation of transcription. The mechanism of this regulation must itself have been highly conserved during evolution, because at least some of the Drosophila heat shock genes are appropriately regulated when introduced into mammalian, Xenopus or even yeast cells (1).

It has been shown by deletion analysis that transcriptional regulation of the Drosophila hsp70 gene in monkey COS cells and Xenopus oocytes requires a short sequence just upstream from the TATA box (2,3,4). This sequence appears to be analogous to the "upstream element" which has been identified in other promoters such as those of the herpes virus thymidine kinase (tk) and rabbit beta globin genes, but it is only functional in cells that have been stressed (2). Comparison of the sequences of the different Drosophila heat shock genes suggests a symmetric consensus sequence for this element, CT-GAA--TTC-AG. Recently, we have replaced the upstream element of the tk promoter with synthetic sequences having 8 or 10 out of 10 bases matching this consensus, and have shown that the tk gene is then heat-inducible both in monkey COS cells and in Xenopus oocytes (5). Other synthetic sequences having similar symmetry but which match the consensus at only 4 or 5 positions do not confer heat-inducibility in these assays. Although the consensus sequence is close to a larger inverted repeat in most of the heat shock genes, we have been unable to demonstrate any strong requirement for such a feature in our assays (5). Our working hypothesis is that a specific protein interacts with the consensus sequence and is involved in activating the heat shock promoters in stressed cells.

In this paper we show that a synthetic heat-shock promoter element can also confer heat-inducibility on the adenovirus major late promoter, and we describe the functional assay of the promoters of the four small Drosophila heat-shock genes, which are encoded in a single cluster in the genome. The results suggest that efficient and regulated expression of the genes in monkey cells requires a good match to the consensus sequence close to the TATA box.

RESULTS AND DISCUSSION

To facilitate a direct comparison of the various promoters, we cloned them into a standard plasmid as illustrated in figure 1. The plasmid contains the coding region of the

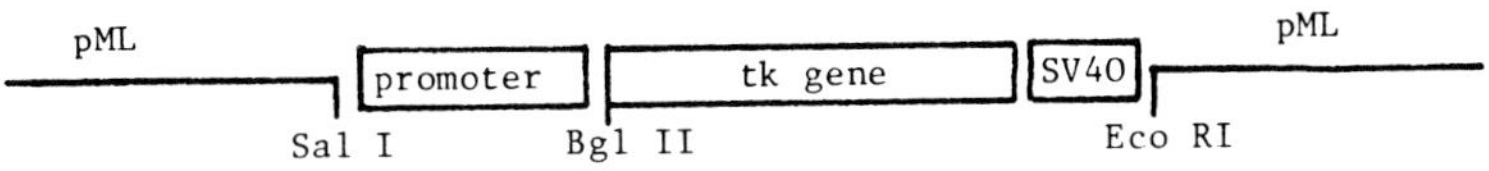

FIGURE 1. Basic plasmid structure. Each plasmid contains three fragments, ligated via appropriate linkers, between the Eco RI and Sal I sites of the vector pML. Tk sequences are from position +53 (Bgl II) to +1600. SV40 sequences from the origin region are Hind III to Eco RII (no enhancer, Eco RII on right as drawn) or Hind III to Pvu II (with enhancer, Pvu II on left or right, termed L and R in subsequent figures). Promoter fragments have the following coordinates (restriction sites): hsp 22, -404/+215 (Hind III); hsp 23, -900/+10 (Eco RI); truncated hsp 23, -119/+10 (Taq I/Eco RI); hsp 26, -2800/+11 (Eco RI); hsp 27, -1300/ +87 (Xba I); Adeno major late, -260/+33 (Xho I/Pvu II).

herpes tk gene and the SV40 origin of replication so that replication can occur in COS cells (which produce SV40 T antigen constitutively). In some cases the SV40 "enhancer" sequences (6) were also present. The promoter of the tk gene was replaced by restriction fragments containing each of the small Drosophila heat-shock gene promoters, or the adenovirus major late promoter. The plasmids were transfected into COS cells, the cells heat shocked, cytoplasmic RNA extracted and tk transcripts detected by S1 mapping as previously described (2), using radioactive probes prepared by primer extension on appropriate M13 clones.

Conversion of the Adenovirus Major Late Promoter to a Heat-inducible Promoter.

Figure 2 shows that the adenovirus major late promoter is highly active in COS cells, and despite previous suggestions to the contrary (7), it does not require the presence of an enhancer for maximal activity. However, the activity of the promoter was drastically reduced by deletion of sequences upstream of position -53 (plasmid AKSO). This deleted promoter evidently lacks its normal "upstream element". The large S1-resistant band in figure 2 represents readthrough transcripts that originate in the plasmid sequences; these are less apparent when the tk gene is flanked by long eucaryotic sequences, presumably because these contain either transcription terminators or sequences

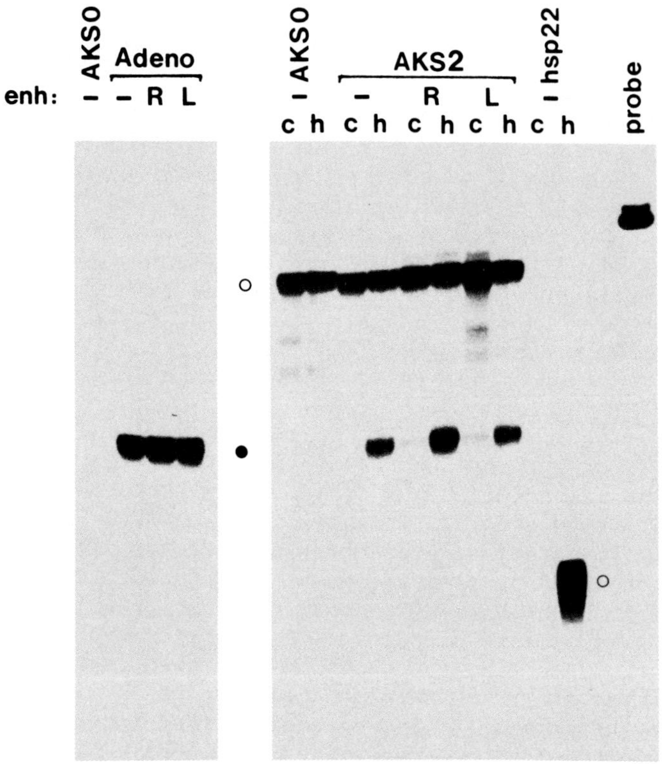

FIGURE 2. S1 mapping of transcripts from the adenovirus major late promoter constructions. An autoradiogram of a gel is shown. Plasmids contained the intact promoter (adeno), a deletion to -53 (AKSO), or this deletion with a synthetic heat shock element (AKS2), with or without enhancer sequences (enh: see legend to figure 1). The probe extended from position +153 of the tk gene through the adeno sequences. For comparison, transcripts from the hsp 22 promoter were mapped with the same probe. c = control, h = heat shocked; solid circle indicates correctly initiated transcripts, open circles indicate RNAs mapped to the point of sequence divergence between probe and template. The four lanes on the left were exposed for 1 hr, the rest for 20 hr.

that destabilize the RNA.

The upstream element of the adeno promoter was replaced with a synthetic sequence which contains two copies of the consensus heat-shock element (plasmid AKS2, figure 3). This had no effect on transcription at low temperature, but there was about a ten-fold increase in the abundance of correctly initiated transcripts after heat shock (figure 2). Other experiments (not shown) indicate that a single copy of the consensus sequence is sufficient for this response.

This result shows that the consensus sequence can substitute functionally for the upstream element of the adeno major late promoter in heat-shocked cells, and that the presence of this heat shock element is sufficient to allow regulation of the hybrid promoter. This extends the results obtained previously with the tk promoter, and shows that the regulatory element is an independent functional unit that can be combined with a variety of TATA box regions which differ markedly in their surrounding sequences. It also demonstrates that the distance of the element from the TATA box, 27 bp in this case versus 13 bp with the tk promoter, can be varied without loss of function.

Assay of the Promoters of the Small Heat Shock Genes.

Figure 3 shows the sequences of the various Drosophila heat-shock promoters, and table 1 summarizes the results we have obtained on assaying them in COS cells. From the activity of the constructs containing synthetic elements, we can predict that the hsp 26 and hsp 22 promoters, which contain good matches to the consensus sequence 15 and 28 bp from the TATA boxes respectively, should be inducible in COS cells. This is indeed the case (table 1 and figure 2). Since all the promoters are linked to the same gene, their approximate efficiencies can be estimated relative to the hsp70/tk fusions described previously (2). The hsp 26 promoter appears about as efficient as the hsp 70 one, while the hsp 22 promoter, which has the consensus sequence further away from the TATA box, is somewhat less efficient. The synthetic construct with the adeno major late promoter is relatively inefficient, producing about 5- to 10-fold fewer transcripts than the hsp 70 promoter after heat shock.

The hsp 23 and hsp 27 promoters do not fit the general

```
                       CT GAA  TTC AG
                       .. ..   ... ..
hsp 83    CCTCTAGAAGTTTCTAGAGACTTCCAGTTCGGTCGGGTTTTTCTATAAA
                       .. ...  ...
hsp 68    CTCGCAGGGAAATCTCGAATTTTCCCCTCCCGGCGACAGAGTATAAA
                       .. ...  ...  .
hsp 70    CGAGAGAGCGCGCCTCGAATGTTCGCGAAAAGAGCGCCGGAGTATAAA
                       .  ..   ... ..
hsp 26    TTTCTGTCACTTTCCGGACTCTTCTAGAAAAGCTCCAGCGGGTATAAA
                       .  . .  ... ..
hsp 22    ATTCGAGAGAGTGCCGGTATTTTCTAGATTATATGGATTTCCTCTCTGTCAAGAGTATAAA
                       .  ...  ...  .
hsp 23    GATATTTTCAGCCCGAGAAGTTTCGTGTCCCTTCTCGATG/75bp/GCGAGCGGTTGTATAAA
                        .   .    . ..
hsp 27    GTTCCGTCCTTGGTTGCCATGCACTAGTCTGTGTGAGCCCAGCGTCACTATAAA
                       .. ...  ... ..
AKS2      ACTCTAGAAGCTTCTAGAAGCTTCTAGAGGAT/CCGGCTGTTCCTCAAGGCGGGCTATATA
                        . ..   .    .
AKS0      CTCAAGGGCATCGGTCGACTCTAGAGGAT/CCGGGTGTTCCTCAAGGGGGGCTATATA
```

FIGURE 3. Upstream sequences of heat shock promoters. Dyad symmetries and TATA boxes are underlined; dots show matches to the consensus sequence. Due to the dyads, hsp 83 and AKS2 also have a second match just upstream of the one indicated. AKS0 is a deletion mutant of the adeno major late promoter, with Bam HI and Xba I linkers added; AKS2 was formed by insertion of Xba I-Hind III adapters into the Xba I site. Slashes mark the junction with adeno sequences. Note the interruption in the hsp 23 sequence.

TABLE 1
SUMMARY OF PROMOTER ACTIVITIES[a]

Promoter	Relative level of transcripts after induction	Fold induction
hsp 70	1.0	20-50
hsp 26	1.0	20-50
hsp 22	0.35	20-50
hsp 23	0.1	1
hsp 27	0.05	10
AKS2	0.15	10
AKS0	0.015	1

[a]Figures are approximate, based on an average of four independent assays of each promoter.

pattern of the other heat shock genes. There is no good match to the consensus sequence in at least the first 290 bp of 5' flanking sequence of the hsp 27 gene (8,9). The presence of a 5/10 match (figure 3) is not statistically significant, and such a weak match can be found in many constructs that are not heat-inducible (e.g. AKS0, see figure 3). We would thus not expect the hsp 27 promoter to function well in COS cells, and indeed it is 20-fold less active than the hsp 70 promoter (table 1). Nevertheless, the hsp 27 promoter is clearly heat-inducible in COS cells (figure 4) and also in Xenopus oocytes (M. Bienz and H.P., unpublished observation). It is unlikely that the low activity in COS cells is due to inefficient replication, because similar levels of transcripts were obtained on heat-shocking 24 or 48 hours after transfection.

It is conceivable that this gene is regulated by a different mechanism than that used for the others, but it seems more likely that it is recognized, albeit inefficiently, by a common heat shock factor. Sequences other than the consensus itself may have evolved to allow such recognition; for example this gene has a particularly striking dyad symmetry upstream of the TATA box. These sequences may be specifically adapted to Drosophila cells, and thus only weakly recognized in the heterologous monkey or frog cells. Whatever the explanation, the result has implications for the study of other regulatory systems: it may be impossible in some cases to identify regulatory sequences and regulated genes merely by examination of their DNA sequences. Indeed without prior knowledge we would not have predicted that the hsp 27 promoter belonged to a heat-shock gene.

The other anomalous case is the hsp 23 promoter. This differs from all the other heat-shock genes in lacking an obvious dyad symmetry upstream from the TATA box. Furthermore, although it does contain a good match to the consensus sequence, this is 99 bp away from the TATA box, much further than in the other genes. In an attempt to determine whether this distant sequence is in fact responsible for regulation, we tested the intact hsp 23 promoter and also a mutant deleted to position -119, which lacks this sequence. Surprisingly, both promoters were expressed at the same level (data not shown), and they were not reproducibly affected by heat shock. This constitutive level of expression was considerably lower than that of the hsp 70 promoter after heat shock, but higher than that of the hsp 27 promoter (figure

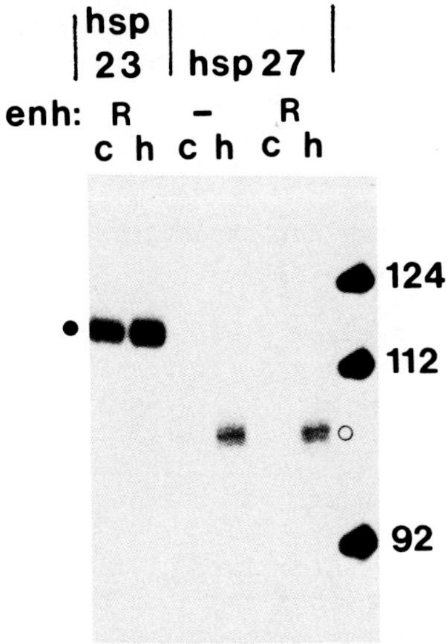

FIGURE 4. S1 mapping transcripts from the hsp 23 and hsp 27 promoters. Both were mapped with a probe that extends from position +153 of the tk gene through the hsp 23 promoter. Symbols have the same meaning as in figure 2.

4) or any of the other promoters at low temperature. S1 mapping indicates that the transcripts were correctly initiated (figure 4).

The simplest interpretation of this result is that sequences near the hsp 23 TATA box allow it to be recognized relatively efficiently by non-gene-specific factors in the heterologous monkey cells. The high copy number of the gene in COS cells would then result in a rather high level of transcripts at low temperature, and it would be hard to detect a weak heat-inducibility (such as that observed for the hsp 27 promoter) superimposed on this background. This may just be a problem with the COS cell assay: we also find that low-temperature expression from the hsp 70 promoter is relatively higher in COS cells than in stably transformed L cell lines (unpublished observations). We can conclude from the data, however, that the hsp 23 consensus sequence either does not function at all in COS cells or does so inefficiently. This is consistent with the idea that proximity of the sequence to the TATA box is important for its activity in these cells. Presumably, the functional constraints for activity are different in the Drosophila genome, and relatively long-range interactions may be possible - it might even be necessary that the TATA box region is easily recognized for such interactions to occur. Alternatively, regulation in vivo might involve some other Drosophila-specific sequence, or even some mechanism that is different from that used to regulate the other heat-shock genes.

Heat Shock Genes Are Not Greatly Affected by Remote SV40 Enhancer Sequences.

We have previously shown that the SV40 enhancer has no long-range effect on the hsp 70 gene in COS cells, although SV40 sequences can override the regulatory element and force constitutive expression when they are close (200 bp) to the promoter (2). The presence of the enhancer at the 3' end of the tk sequences (see figure 1) also has at most a small effect on transcription from the hsp 22 and hsp 23 (not shown), hsp 27 (figure 4) and synthetic heat shock/adeno promoters (figure 2). It is particularly striking that the hsp 27 promoter, which is very weakly expressed, cannot be enhanced in this way. This suggests that the enhancer does not simply increase the efficiency of all weak promoters, but rather that the ability to respond to the enhancer at long range is a specific property of certain promoter elements, and the heat shock element does not have this property. We also find that the presence of the enhancer does not greatly increase the abundance of readthrough transcripts initiated within the plasmid sequences (e.g. figure 2). These negative results are not merely an artefact of the COS cell assay: transcription of the beta globin gene can clearly be enhanced by SV40 sequences in this system (10).

CONCLUSIONS

These experiments serve to define further the requirements for heat-inducibility of a promoter in COS cells. The heat shock promoter element appears to be an independent functional entity that can regulate transcription probably from any TATA box. Since a completely symmetric synthetic sequence has activity, the regulatory element can formally be considered bidirectional. The proximity of the TATA box appears important: distances of 13 to 28 bp are tolerated, whereas the hsp 23 gene in which the distance is 99 bp is expressed only at a low constitutive level. Similarly a close (80%) match to the consensus sequence CT-GAA--TTC-AG is not only sufficient, but appears to be important: the hsp 27 gene has only a 5/10 match and is expressed very weakly, although it is still regulated. Presumably other features of these genes allow them to be expressed efficiently in Drosophila. Finally, none of the heat shock promoters tested is greatly affected by the presence of the SV40 enhancer more than 1.6 kb away.

We have yet to discover what interacts with the regulatory element, how it activates transcription, and why this only occurs in cells that have beeen subjected to stress.

ACKNOWLEDGEMENTS

We thank E. Craig, R. Southgate and P. Farrell for generously supplying clones containing the small heat shock genes and the adenovirus major late promoter.

REFERENCES

1. Schlesinger M, Ashburner M, Tissieres A (1982) (eds): "Heat Shock from Bacteria to Man", New York: Cold Spring Harbor Laboratory.
2. Pelham HRB (1982). A Regulatory Upstream Promoter Element in the Drosophila hsp 70 Heat Shock Gene. Cell 30:517.
3. Bienz M, Pelham HRB (1982). Expression of a Drosophila Heat-shock Protein in Xenopus Oocytes: Conserved and Divergent Regulatory Signals. EMBO J. 1:1583.
4. Mirault M-E, Southgate R, Delwart E (1982). Regulation of Heat-Shock Genes: a DNA Sequence Upstream of Drosophila hsp 70 Genes is Essential for their Induction in Monkey Cells. EMBO J. 1:1279.
5. Pelham HRB, Bienz M (1982). A Synthetic Heat-shock Promoter Element Confers Heat-inducibility on the Herpes Simplex Virus Thymidine Kinase Gene. EMBO J. 1:1473.
6. Banerji J, Rusconi S, Schaffner W (1981). Expression of a Beta Globin Gene is Enhanced by Remote SV40 DNA

Sequences. Cell 27:299.
7. Hen R, Sassone-Corsi P, Corden J, Gaub MP, Chambon P (1982). Sequences Upstream from the TATA box are Required in vivo and in vitro for Efficient Transcription from the Adenovirus Serotype 2 Major Late Promoter. Proc. Natl. Acad. Sci. USA 79:7132.
8. Ingolia TD, Craig E (1981). Primary Sequence of the 5' Flanking Region of the Drosophila Heat Shock Genes in Chromosome Subdivision 67B. Nucl. Acids Res. 9:1627.
9. Southgate R, Ayme A, Voellmy R (1983). Nucleotide Sequence Analysis of the Drosophila Small Heat Shock Gene Cluster at Locus 67B. J. Mol. Biol. in press.
10. Humphries RK, Ley T, Turner P, Moulton AD, Nienhuis AW (1982). Differences in Human alpha-, beta- and delta-Globin Gene Expression in Monkey Kidney Cells. Cell 30:173.

Gene Expression, pages 87-94
© 1983 Alan R. Liss, Inc., 150 Fifth Avenue, New York, NY 10011

SEPARATION OF PROMOTER AND HORMONE REGULATORY SEQUENCES IN MMTV

Deborah E. Dobson, Frank Lee*, and Gordon M. Ringold

Department of Pharmacology, Stanford University, Stanford, CA 94305 and *DNAX Ltd., Palo Alto, CA 94304

We have found the promoter and hormone regulatory regions present in the mouse mammary tumor virus (MMTV) long terminal repeat (LTR) to be distinct, dissociable domains. Chimeric plasmids were constructed by fusing a full length or deleted LTR to the E. coli XGPRT gene. These plasmids were introduced by DNA-mediated transformation into a mouse cell line. Stably transformed cells were selected and mRNA from these cells was analyzed by Northern blot hybridization and S_1 mapping techniques. Analysis of one recombinant (ΔA1, 800 bp deletion) indicates that the promoter and hormone regulatory regions are separable. The basal level of Eco-gpt mRNA was similar in cells containing either the ΔA1 or full length LTR recombinants. However, the addition of the synthetic glucocorticoid dexamethasone increased the level of Eco-gpt mRNA only in the cells harboring the full length LTR recombinants. S_1 mapping of the 5'end of these mRNAs indicates that transcription initiates at the appropriate position in both the full length and ΔA1 LTR constructions in the presence and absence of hormone.

INTRODUCTION

Mouse mammary tumor virus is unique among retroviruses in that it is known to contain promoter and hormone regulatory sequences in the long terminal repeats present at both ends of the viral genome.

Addition of glucocorticoids such as dexamethasone stimulates the transcription of MMTV RNA starting at a site in the 5' LTR in the virus (1,2) and in cloned DNA fragments containing the MMTV LTR transfected into cultured mammalian cells (3-5). Hormone-receptor protein complexes that bind to specific regions in the LTR are thought to mediate this induction (6-8). However, it is not known whether the promoter and hormone regulatory sequences act as a single functional unit or if two distinct regions exist that can be physically and functionally separated. In order to distinguish between these two possibilities, chimeric plasmids were constructed in which either a full length or deleted LTR was fused to the *E. coli* XGPRT gene. After DNA-mediated transfection of these plasmids into mouse 3T6 cells, stably transformed cells were selected and analyzed for the expression of Eco-gpt mRNA in the presence and absence of hormone. The position of the hormone regulatory sequences relative to promoter sequences and models explaining how hormone-receptor complexes stimulate transcription are discussed in the light of the data presented.

MMTV-GPT FUSIONS

The 1.4 kb Pstl fragment of the 5' LTR from MMTV (C3H strain) was fused to the *E. coli* gene encoding xanthine-guanine phosphoribosyl transferase in an SV2gpt expression vector (9); this plasmid was designated pSVMgpt (10). Deleted LTR plasmids were derived from pSVMgpt following digestion of the LTR with specific restriction endonucleases. In one plasmid, ΔAlgpt, the sequences from 67 bp to 892 bp upstream from the start of transcription are deleted. The resulting LTR is approximately 800 bp shorter, as demonstrated in Figure 1. These plasmids were used to stably transform mouse 3T6 cells using the selection outlined by Mulligan and Berg (11).

PROMOTER AND HORMONE REGULATORY SEQUENCES ARE DISTINCT

Chapman *et al.* (10) have shown that in 3T6 cells stably transformed with pSVMgpt, Eco-gpt mRNA is increased 10 to 15 fold by the addition of 10^{-6} M

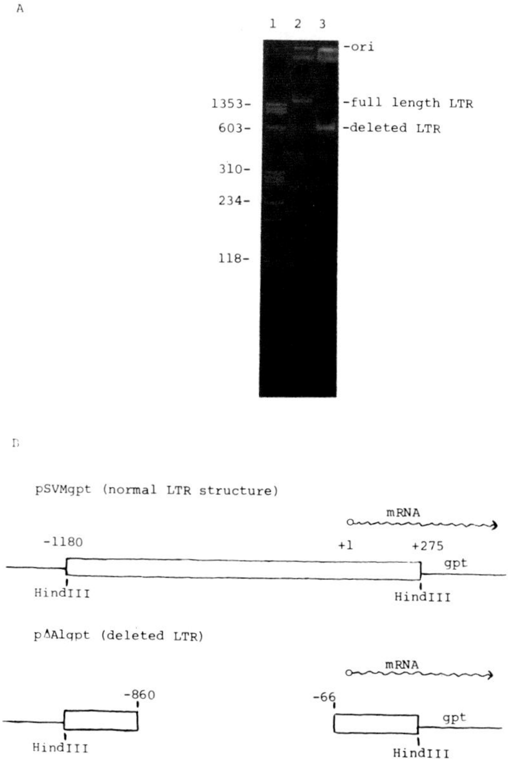

FIGURE 1. Structure of the plasmids pSVMgpt and p ΔAlgpt, with emphasis on the LTR region. Transcription from the MMTV promoter proceeds to the right, towards the Eco-gpt coding region. A more detailed description can be found in reference 10. (A) Restriction mapping of pSVMgpt and ΔAlgpt. Plasmid DNA was digested with HindIII and electrophoresed in a 1% agarose gel. HindIII cuts at two sites in each plasmid such that the entire LTR is found on a single fragment. The full length LTR is approximately 1.4 kb whereas the deleted LTR is 600 bp. Lane 1: HaeIII digest of ϕX174RF DNA, with the sizes of the fragment in base pairs; lane 2: HindIII digest of pSVMgpt; lane 3: HindIII digest of pΔAlgpt. The origin and position of the LTRs are marked. (B) A more detailed diagram showing the LTR structure. The HindIII sites in each plasmid are marked. The boxed region represents the LTR, with the transcription start site at +1 bp and the mRNA represented by a wavy line. Regions preceeding

the start site are denoted with negative numbers and regions following the start site are denoted with positive numbers, all in base pairs. The position labelled -860 in pΔAlgpt should read -893.

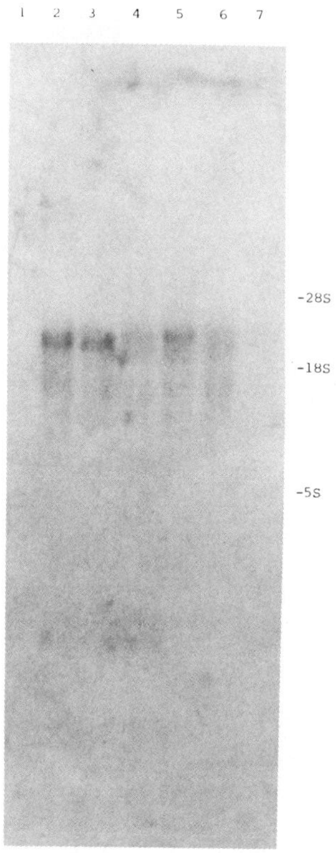

FIGURE 2. Northern blot analysis of Eco-gpt mRNA. Thirty micrograms of total cell RNA were glyoxylated and electrophoresed in a 1% agarose gel (12). The RNA was blotted onto a nitrocellulose filter (13) and then hybridized to ^{32}P-labelled pSVMgpt DNA. RNA was isolated from 3T6 cells (lane 1) or from stable transformants containing pMDSG (lanes 2,3), pSVMgpt (lanes 4,5) or pΔAlgpt (lanes 6,7). Cells were grown in the absence (lanes 2, 4, and 6) or presence (lanes 1, 3, 5, and 7) of 10^{-6} M dexamethasone added 24 hours prior to harvesting. The positions of 28S, 18S, and 5S RNAs are indicated.

dexamethasone for 24 hours. The Eco-gpt mRNA initiates at the known site of MMTV transcription in the LTR. When RNA from 3T6 cells transfected with pΔAlgpt was analyzed, the basal level of Eco-gpt mRNA was similar to that found in cells harboring pSVMgpt. However, addition of 10^{-6} M dexamethasone increased the level of Eco-gpt mRNA approximately 10 fold only in cells containing pSVMgpt, the full length LTR plasmid (Figure 2). Cells harboring pΔAlgpt actually showed a decrease in the amount of Eco-gpt mRNA after dexamethasone addition. When a promoter which is not subject to hormonal control is fused to Eco-gpt, as in the plasmid pMDSG where the SV40 early promoter is fused directly to the XGPRT gene (4), the level of Eco-gpt mRNA in transfected cells is unaffected by glucocorticoid addition.

The start site of Eco-gpt mRNA was examined by the S1 nuclease mapping procedure of Berk and Sharp (14). The BglII-EcoRI fragment containing the LTR-gpt sequences of pSVMgpt was labelled at the BglII site with ^{32}P and hybridized with RNA from cells containing either pSVMgpt or pΔAlgpt. After S1 nuclease digestion and polyacrylamide gel electrophoresis, a protected fragment of approximately 400 bp should be present if the Eco-gpt mRNA is starting at the appropriate site within the LTR. Figure 3 shows that this is indeed the case. The 5' end of the Eco-gpt mRNA maps to the same position in the deleted LTR, in the presence and absence of dexamethasone, as in the full length LTR. It is clear from these results that the minimal promoter sequences required for appropriate initiation of transcription are physically and functionally distinct from the hormone regulatory regions. In addition, the promoter functions normally in the absence of the hormone regulatory sequences i.e., inducible expression can be eliminated without interfering with basal promoter function.

The data obtained with pΔAlgpt allow us to localize the promoter and hormone regulatory sequences in the MMTV LTR. The promoter is completely contained within the 66 bp upstream from the start of transcription. Sequences that confer hormonal regulation are further upstream; additional analyses (Lee _et al_., in preparation) suggest that the hormone regulatory region is contained in the region 140 bp to 223 bp upstream of the transcription start site.

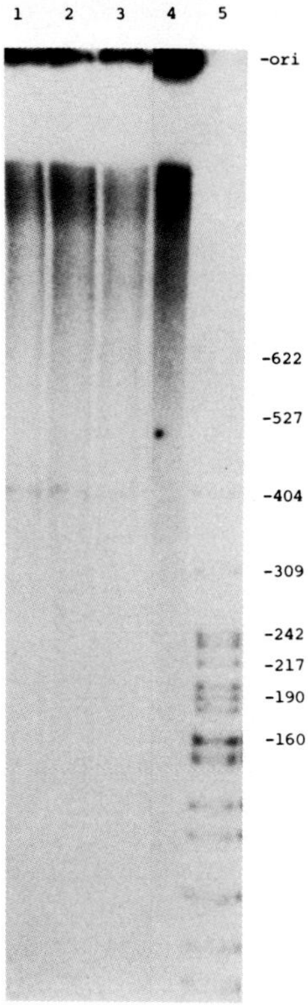

FIGURE 3. S1 nuclease mapping of Eco-gpt RNA in cells transformed with pSVMgpt or pΔAlgpt. The BglII site within Eco-gpt in pSVMgpt DNA was labelled with γ^{32}P-ATP and the 4.2 kb EcoRI-BglII fragment used as probe. The BglII site is approximately 400 bp downstream from the transcription start site in the MMTV LTR; thus a hybrid between the probe and the Eco-gpt mRNA should be 400 bp in length if the message initiates at the known start site of MMTV RNA (2-9). Twenty micrograms of total cell RNA were hybridized with end-labelled probe (30,000 cpm). After S1 treatment (1000 units, 30' at 45°C) the samples were analyzed on 5% polyacrylamide gels. Ori denotes the origin. The dark band near the top of the gel is reannealed probe. RNA was isolated from stable transformants containing pSVMgpt (lane 1) or pΔAlgpt (lanes 2,3) grown in the presence (lane 3) or absence (lanes 1,2) of 10^{-6} M dexamethasone 24 hours prior to harvesting. Lane 4: no RNA added;

lane 5: ^{32}P-labelled HpaII digest of pBR322, with the size of the fragment in base pairs.

HORMONE REGULATORY SEQUENCES IN ASSOCIATION WITH HORMONE - RECEPTOR COMPLEXES ARE POSITIVE REGULATORS OF TRANSCRIPTION

A possible model for the mechanism of glucocorticoid inducible gene expression is that the hormone regulatory region contains a negative regulatory element. In the absence of hormone, transcription would proceed at a low level, possibly due to the binding of a repressor to the hormone regulatory region. The binding of hormone-receptor complexes to this region would alleviate this inhibition, possibly by displacing the repressor. If this model were correct, one would predict that removal of the hormone regulatory region would result in a high constitutive level of transcription independent of hormone. However, as seen in Figures 2 and 3, when the hormone regulatory sequences are removed from the LTR only the low basal level of transcription is retained. It therefore seems likely that the hormone regulatory sequences in concert with steroid-receptor complexes act as positive regulators of transcription from the MMTV promoter. Additional studies will be required to understand more fully the mechanisms by which stimulation of transcription takes place.

ACKNOWLEDGMENTS

We thank Jean Luh for sequence determination of the ΔAl deletion and Steve Beverley and Tony Young for helpful comments. This work was supported by NIGMS grant GM25821 to G.R., by DNAX Research Institute (F.L.), and by a postdoctoral fellowship from the American Cancer Society to D.D.

REFERENCES

1. Ringold G, Yamamoto K, Bishop JM, Varmus,H (1977). Glucocorticoid-stimulated accumulation of mouse mammary tumor virus RNA: increased rate of synthesis of viral RNA. Proc Natl Acad Sci USA 74:2879.
2. Ucker D, Ross S, Yamamoto K (1981). Mammary tumor virus DNA contains sequences required for its hormone regulated transcription. Cell 27:257.

3. Hynes N, Kennedy N, Rahmsdorf U, Groner B (1981). Hormone responsive expression of an endogenous proviral genome of mouse mammary tumor virus after molecular cloning and gene transfer into cultured cells. Proc Natl Acad Sci USA 78:2038.
4. Lee F, Mulligan R, Berg P, Ringold G (1981). Glucocorticoids regulate expression of dihydrofolate reductase cDNA in mouse mammary tumor virus chimaeric plasmids. Nature 294:228.
5. Huang A, Ostrowski M, Berard D, Hager G (1981). Glucocorticoid regulation of the Ha-MuSV p21 gene conferred by sequences from mouse mammary tumor virus. Cell 27:245.
6. Payvar F, Firestone G, Ross S, Chandler V, Wrange O, Carlstedt-Duke J, Gustafsson J-A, Yamamoto K (1982). Multiple specific binding sites for purified glucocorticoid receptors on mammary tumor virus DNA. J Cell Biochem 19:241.
7. Govindan M, Spiess E, Majors J (1982). Purified glucocorticoid receptor-hormone complex from rat liver cytosol binds specifically to cloned mouse mammary tumor virus long terminal repeats _in vitro_. Proc Natl Acad Sci USA 79:5157.
8. Pfahl M (1982). Specific binding of the glucocorticoid receptor complex to the mouse mammary tumor proviral promoter region. Cell 31:475.
9. Mulligan R, Berg P (1980). Expression of a bacterial gene in mammalian cells. Science 209:1423.
10. Chapman A, Costello M, Lee F, Ringold G (1983). Amplification and hormone regulated expression of a MMTV-Eco gpt fusion plasmid in mouse 3T6 cells. Mol Cell Biol in press.
11. Mulligan R, Berg P (1981). Selection for animal cells that express the _Esherichia coli_ gene coding for xanthine-guanine phosphoribosyltransferase. Proc Natl Acad Sci USA 78:2072.
12. McMaster G, Carmichael G (1977). Analysis of single- and double-stranded nucleic acids on polyacrylamide and agarose gels by using glyoxal and acridine orange. Proc Natl Acad Sci USA 74:4835.
13. Thomas P (1980). Hybridization of denatured RNA and small DNA fragments transferred to nitrocellulose. Proc Natl Acad Sci USA 77:5201.
14. Berk A, Sharp P (1977). Sizing and mapping of early adenovirus mRNAs by gel electrophoresis of S1 endonuclease-digested hybrids. Cell 12:721.

Gene Expression, pages 95-105
© 1983 Alan R. Liss, Inc., 150 Fifth Avenue, New York, NY 10011

DNASE I HYPERSENSITIVE SITES IN THE J_K-C_K INTRON OF THE IMMUNOGLOBIN GENE[1]

Su-yun Chung, Virginia Folsom and John Wooley

Department of Biochemical Sciences, Princeton University
Princeton, New Jersey 08544

ABSTRACT DNase I hypersensitive sites have been mapped for the chromatin of the mouse immunoglobin C_K in rearranged and unrearranged chromosomes. Two novel, tissue-specific hypersensitive sites are observed at 0.7 kb and 1.7 kb upstream from the 5' end of the C_K gene within the J_K-C_K intron region in nuclei from myeloma cells derived from MOPC21 but not in naked DNA or in brain or liver nuclei. Inspecting the DNA sequences surrounding the C_K-proximal DNase I hypersensitive site reveals several stretches of homology with sequences within the 72 bp tandem repeat of SV40, which has been shown to function as a regulatory element and to modulate the transcriptional activity of neighboring genes. Thus, the DNase I hypersensitive site in the intron region may associate with a cellular enhancer element and play a significant role in the differential expression of the translocated V_K genes.

INTRODUCTION

Discrete localized DNase I hypersensitive sites in chromatin have been found associated with gene activity for a number of eukaryotic genes (1-5, for review see 6). The genetic role and molecular nature of these hypersensitive sites is unknown. The local structural

[1]This work was supported by NIH GM26332 and ACS CD-15

modifications might reflect different regulatory elements of specific DNA sequences. One category of DNase I hypersensitive sites is located near the 5' end of a gene and is generally believed to reflect the presence of the promoter region.

The immunoglobulin gene family which requires chromosomal rearrangement for their expression, is a unique system for examining the relationship between chromatin structure and gene expression. The kappa light chain peptide is encoded by three distinct DNA segments; the variable region by the V and J segments, and the constant region by the C segment (7-9). In germ-line cells, the V_K genes, estimated to be present at a few hundred copies per genome, are located at an unknown distance from the single copy C_K gene. For terminally differentiated, immunoglobulin secreting myeloma cells, recombination events have translocated a block of V_K genes in close proximity to the J_K-C_K region. A specific feature of the immunoglobulin expression involves this gene block which can be viewed as a multigene family. However only the V_K gene contiguous with the J_K-C_K region is preferentially expressed; the other distal translocated V_K genes are not expressed (10,11). In this communication, we report that the mapping of DNase I hypersensitive sites within the J_K-C_K intron reveals the existence of DNA sequence features that might regulate the differential expression of the translocated V_K genes.

RESULTS

DNase I hypersensitive sites have been mapped for the immunoglobulin C_K gene in the mouse myeloma cell lines and in mouse brain and liver tissues using the indirect end-labelling method of Wu (2). Fig. 1A summarizes the restriction map and DNase I hypersensitive sites of a functionally rearranged C_K allele in the X63Ag8 cell line derived from MOPC21. Fig. 1E shows the restriction map and the lack of DNase I hypersensitive sites in the germline C_K gene in liver or brain cell; the germline V_K genes are now shown in this map (7-9).

Fig. 2 shows the mapping gel for the DNase I hypersensitive sites 5' to the C_K sequence. When DNA samples from DNase I-digested X63Ag8 nuclei were restricted with

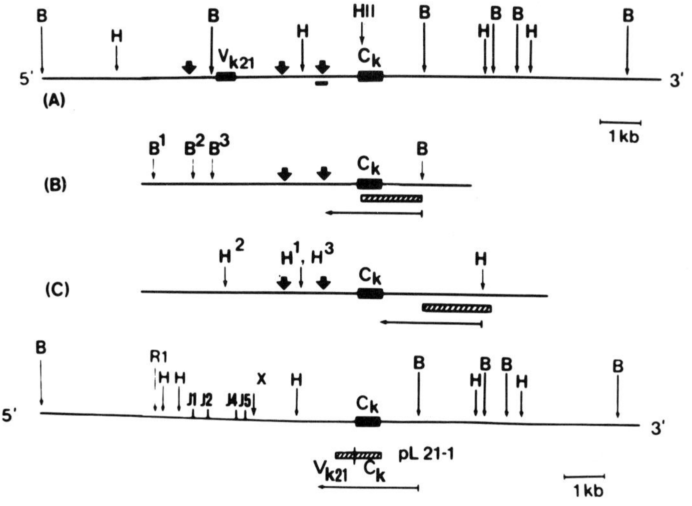

FIGURE 1. Schematic maps and DNase hypersensitive sites of immunoglobulin kappa light chain genes. (A) Restriction map of the functionally rearranged MOPC21 kappa light chain gene. V_{K21} and C_K represent sequences encoding the variable and constant region of the kappa polypeptide in MOPC21. The horizontal bar underlines the region of high homology between mouse and human identified by Hieter et al (14). HII is Hinc II. (B) Strategy for Bam HI mapping. The 6.8 kb, 5.8 kb and 5.3 kb Bam HI fragments correspond to three rearranged alleles in X63Ag8, which share a common 3' Bam HI site B and have a 5' Bam HI site designated as B^1, B^2 or B^3, respectively. B^3 represents the functionally rearranged C_K allele 3 containing V_{K21} and C_K sequences. B^1 and B^2 represent aberrantly rearranged C_K alleles 1 and 2 with unknown V_K sequences. The probe (notched bar) is a 1.5 kb Hinc II/Bam HI fragment isolated from the recombinant phage Ch4A-EC5 (15). Two DNase I hypersensitive sites (broad vertical arrows) are mapped relative to the 3' Bam HI site B according to the lengths of subfragments generated by DNase I digestion (Fig. 2A). (C) Strategy for Hind III mapping H^1, H^2 and H^3 represent the 5' Hind III sites of the three rearranged alleles which share a common 3' Hind III site H. The probe (hatched

bar) is a 1.7 kb Bam HI fragment isolated from Ch4A-EC5 (15). (D) Restriction map of germline C_K gene (7-9) and strategy for DNase I hypersensitive site mapping 5' to C_K in brain and liver nuclei (Fig. 2D, 2E). The probe (hatched bar) is a plasmid pL21-1 containing both V_{K21} and C_K sequences (16); X is XbaI.

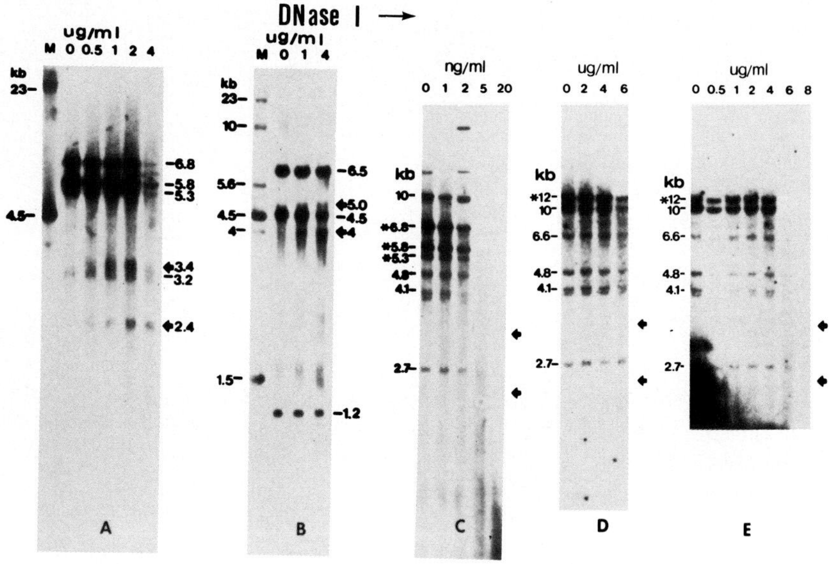

FIGURE 2. Mapping gels of DNase I hypersensitive sites in the C_K region in X63Ag8, brain, and liver nuclei and in naked DNA from X63Ag8. Broad arrows indicate subfragments generated by the hypersensitive DNase I cleavage. M is marker lane. (A) Bam HI mapping gel. (B) Hind III mapping gel. (C-E) Absence of DNase I hypersensitive sites in: (C) purified DNA from X63Ag8 nuclei, (D) mouse liver nuclei, and (E) mouse brain nuclei. Bands corresponding to rearranged C_K alleles are marked with asterisks (C) and the 12.0 kb band (marked with an asterisk) represents the germline C_K gene (C and E). The remaining bands are germline V_K genes which cross-hybridize with the V_{K21} sequence. Additional methods and details are described elsewhere (17).

Bam HI and probed with a 1.5 kb Hinc II/Bam HI fragment containing most of C_K and its 3' flanking sequence (Fig. 2A), four Bam HI fragments, 6.8 kb, 5.8 kb, 5.3 kb and 3.2 kb in length, hybridize with the probe. The 6.8 kb, 5.8 kb and 5.3 kb bands contain the rearranged C_K sequence; the origin of the 3.2 kb band is unknown. However, only the 5.3 kb band hybridizes to the V_{K21} sequence, whereas the 6.8 kb and 5.8 kb bands do not cross-hybridize to the V_{K21} sequence (data not shown). Thus, the 5.3 kb band corresponds to the functionally rearranged C_K allele 3, coding for the MOPC21 kappa polypeptide (12) and the 6.8 kb and 5.8 kb bands correspond to two aberrantly rearranged C_K alleles 1 and 2 with unknown V sequences (Fig. 1B). With increasing DNase I concentration, two broad subfragments of 2.4 kb and 3.4 kb are generated. These two subfragments map two DNase I hypersensitive sites (broad vertical arrows) at 0.7 kb and 1.7 kb from the 5' end of the C_K gene (Fig. 1B). The broadness of both subfragments reflects the breadth of the hypersensitive region which spans approximately 100-200 base pairs. When DNA samples from DNase I digested X63Ag8 nuclei were restricted with Hind III, blotted, and probed with a 1.7 kb Bam H1 fragment containing only the 3' flanking sequence of C_K (Fig. 1B), three Hind III fragments of 4.5 kb, 6.5 kb and 1.2 kb hybridize with the probe (Fig. 2B). The 4.5 kb fragment comes from the functionally rearranged C_K allele 3 and the aberrantly rearranged C_K allele 1; the 6.5 kb fragment comes from the aberrantly rearranged C_K allele 2, in which the normal Hind III site (Fig. 1A) in the J_K-C_K intron is altered (data not shown). The 1.2 kb band arises from the 3' flanking region of C_K (Fig. 1A). With increasing DNase I concentration, two subfragments of 4.0 kb and 5.0 kb are generated from the Hind III fragments. These map two DNase I hypersensitive sites located at the same positions as those identified by the Bam HI mapping (Fig. 1B and 1C). Since there are three rearranged C_K alleles in X63Ag8, the two DNase I hypersensitive sites cannot be readily assigned to each allele. The hypersensitive site generated by the 5.0 kb subfragment, which is derived from the 6.5 kb Hind III fragment, is associated with the kappa allele 2. Similar mapping procedures performed on X63Ag8 using a V_{K21} sequence specific probe or on a variant nonsecreting myeloma cell line D1/20 which has lost the functionally rearranged C_K allele 3 (13), confirm that the C_K-proximal DNase I hypersensitive site (0.7 kb from the 5' end of C_K) is definitely associated with the functionally rearranged C_K allele and at least one of the nonfunctionally rearranged C_K alleles (data not shown).

To test for the presence of DNase I hypersensitive sites in naked DNA, high molecular weight X63Ag8 DNA was digested with DNase I, restricted with Bam HI and probed with pL21-1 plasmid containing complete copies of both V_{K21} and C_K sequences. As shown in Fig. 2C, the rearranged kappa alleles of 6.8 kb, 5.8 kb and 5.3 kb are easily identified; the remaining bands correspond to germline V_K sequences that cross-hybridize with the V_{K21} portion of the probe pL21-1. Upon DNase I digestion, no discrete subfragments are detected indicating that the two regions of DNase I hypersensitivity in myeloma nuclei are not preferentially recognized by DNase I in naked DNA extracted from the same cells. Fig. 2E demonstrates that the two DNase I hypersensitive sites are specific to myeloma and are not present in brain and liver nuclei. When DNA samples from DNase I digested liver (Fig. 2D) and brain (Fig. 2E) nuclei were restricted with Bam HI and probed with pL21-1, a 12.0 kb Bam HI fragment representing the C_K gene in its germline configuration and Bam HI fragments of 10 kb, 6.6 kb, 4.8 kb, 4.1 kb, 2.7 kb representing V_K genes in germline configuration, were observed. With increasing DNase I digestion, no discrete subfragments can be detected. In addition, all the germline C_K and V_K fragments in brain or liver (Fig. 2D, 2E) are more resistant to DNase I than the rearranged kappa alleles in myeloma (Fig. 2A).

DISCUSSION

Two novel tissue specific DNase I hypersensitive sites have been identified within the gene in the J_K-C_K intron region, 0.7 kb and 1.7 kb upstream from the coding region of C_K in the myeloma cell line X63Ag8 (derived from MOPC21). At least the C_K-proximal (0.7 kb upstream from C_K) site is associated with the functional rearranged gene. These two sites are not present in naked DNA isolated from the X63Ag8 nuclei and thus reflect some aspect of chromatin structure at the kappa loci. In addition, these two sites are not present in nuclei isolated from cells (mouse brain and liver) that are not in the B-lymphocyte cell lineage and do not express immunoglobulin genes.

Although neither the nature nor the implications of the DNase I hypersensitive sites along chromatin fibers is clear, at least one class of DNase I hypersensitive sites (found for most if not all potentially active genes) is hypothesized to be associated with the promoter region for transcription (1-6). However, neither of the two sites within the intron could be associated with the promoter of the C_K transcriptional unit, since the transcription of the C_K gene in germline configuration in myeloma cells starts at approximately 8.0 kb 5' to the coding region of C_K (10). These two sites must represent different functional class(es). We have also identified a DNase I hypersensitive site 0.3 kb upstream from the V_K coding sequence that is probably associated with promoter activity (17). A similar DNase I hypersensitive site has been found in the intron region of the kappa gene in other myeloma cell lines (18) and in a mouse leukemia cell line (19), and in the switch region of heavy chain gene in T lymphocytes (20).

In searching for other functional implications, we noted that the C_K-proximal hypersensitive site coincides with a 150 bp region that is conserved between cloned mouse and human kappa genes (14). This sequence conservation in evolution might reflect functional significance. Upon inspecting the DNA sequences surrounding the C_K-proximal DNase I hypersensitive site in the J_K-C_K region (21), we found several stretches of nucleotide sequence homologous to sequences within the 72 bp tandem repeat of SV40 (shown in Fig. 3). In particular, there are three regions, namely, a seven nucleotide sequence (GGCTGCT), a
(CCGACGA)
nine nucleotide sequence (TTGAGATGC) and an eleven
(AACTCTACG)
nucleotide sequence (GGGGACTTTCC) in the DNase I hyper-
(CCCCTGAAAGG)
sensitive site that are precisely identical to sequence elements present in the 72 bp tandem repeat of SV40. Several lines of evidence indicate that genetic elements in the 72 bp repeat sequence of SV40 can act in cis, in either orientation, and at distances of several kilobases to enhance transcriptional activity of neighboring genes (22-29). Recently the same two 9 and 11 sequence elements that we identify based on their identity to elements in

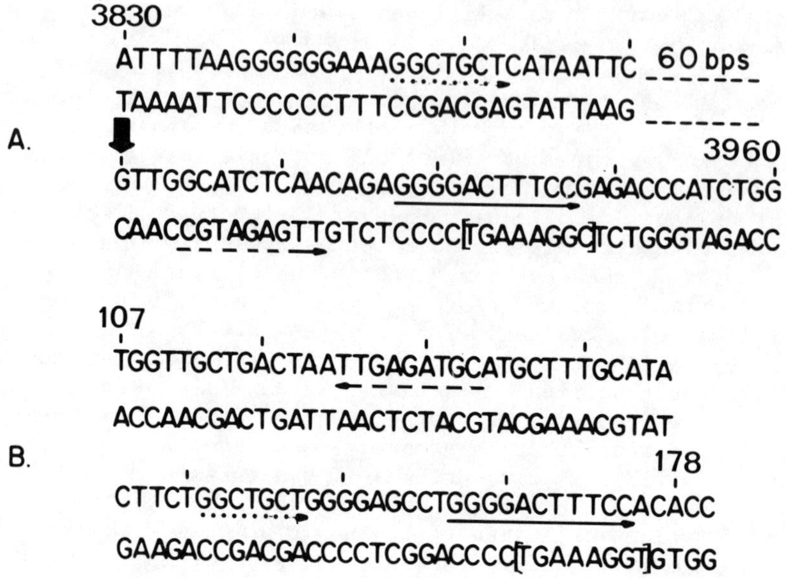

FIGURE 3. Nucleotide sequences of (a) region surrounding DNase I hypersensitive site in J_K-C_K intron (21); (b) SV40 72-bp tandem repeat (22). Broad arrow indicates center of DNase I hypersensitive site and arrows indicate homologous sequences between the DNase I hypersensitive site and SV40 72 bp tandem repeat. Brackets identify core sequence that is required for enhancing activity (29).

the SV40 72 bp repeat have been shown by genetic deletions to be essential for enhancement of the efficiency of DNA-mediated transformation (28; Botchan, personal communication). Moreover, the 11 nucleotide element contains a core sequence (TGGAAAGT) of the SV40
(ACCTCTCA)
72-bp repeat that is required for enhancing activity on the basis of transcriptional assay and point mutations (Fig. 3; 29). DNA sequences surrounding the C_K-proximal DNase I hypersensitive suggest this site may be associated with genetic elements similar to that of the

72 bp tandem repeats of SV40 and thus can act in cis to modulate neighboring gene expression. This DNase I hypersensitive site might play a significant role in immunoglobulin gene expression, namely, participating in the specific enhancement of transcriptional activity of the translocated V_K gene immediately adjacent to the J_K-C_K region.

Currently, a consistent picture has emerged, from accumulating knowledge on chromosomal abnormalities and cancer, that certain tumors derived from B-lymphocyte cell lineage are associated with specific patterns of chromosomal translocation (for review, see 30, 31). In many human Burkitt's lymphomas and mouse plasmacytomas, one of the chromosomes involved in reciprocal translocation always carries the immunoglobulin gene locus. The most excitement comes from the recent report (32-34) that the c-myc gene that normally resides in chromosome 15 is translocated adjacent to the immunoglobulin heavy-chain locus in many mouse plasmacytomas. Up to date, the transcriptional studies in these plasmacytomas revealed no consistent results on the c-myc gene transcripts qualitatively or quantitatively compared to normal tissues (32, 35). It is not known if the c-myc gene product regulates the normal growth and differentiation of B-lymphocyte cell lineage and plays a direct role in plasmacytoma induction. However, it is very plausible that the translocation event may bring the c-myc gene under the control of a cellular regulatory element, an event analogous to the activation of the translocated V_K gene immediately adjacent to the J_K-C_K intron. It is known in some mouse plasmacytomas, translocations do occur between chromosome 15 (c-myc locus) and chromosome 6 (κ-chain locus) (31). The putative cellular enhancing element identified by DNase I hypersensitivity mapping is an attractive candidate for such a regulatory element in this hypothesis. Its actual function, of course, remains to be tested.

ACKNOWLEDGEMENT

We thank Dr. N. Cowan for myeloma cell lines, Dr. P. Leder and Dr. W. Salser for recombinant clones, and Mrs. A. Bustraan for preparing the manuscript.

REFERENCES

1. Wu C, Bingham PM, Livak K, Holmgren R, Elgin SCR (1979). Cell 16:797.
2. Wu C (1980). Nature 284:856.
3. Stadler J, Larsen A, Engel JD, Dolan M, Groudine M, Weintraub H (1980). Cell 20: 451.
4. Wu C, Gilbert W (1981). Proc. Natl. Acad. Sci. USA 78:1577.
5. Weintraub H, Larsen A, Groudine M (1981). Cell 24:333.
6. Elgin SCR (1981). Cell 27:413.
7. Sakano H, Huppi K, Heinrich G, Tonegawa S (1979). Nature (London) 280:288.
8. Seidman JG, Max EE, Leder P (1979). Nature (London) 280:370.
9. Max EE, Seidman JG, Leder P (1979). Proc. Natl. Acad. Sci. USA 76:3450.
10. Van Ness GB, Weigert M, Coleclough C, Mather EL, Kelley DE, Perry RE (1981). Cell 27:593.
11. Bentley DL, Favrel PJ, Rabbits TH (1982). Nucl. Acid. Res. 10:1841.
12. Kohler G, Milstein C (1975). Nature 256:495.
13. Cowan NC, Secher DS, Milstein C (1974). J. Mol. Biol. 90:691.
14. Hieter PA, Max EE, Seidman JG, Maizel JV, Leder P. (1981). Cell 22:197.
15. Seidman JG, Leder P (1978). Nature 276:790.
16. Strathearn MD, Strathearn GE, Akopiantz P, Lui AY, Paddock GV, Salser W (1978). Nucl. Acids Res. 5: 3101.
17. Chung SY, Folsom V, Wooley JC (1983). Proc. Natl. Acad. Sci. 80, in press.
18. Weischet W, Glotov BO, Schnell H, Zachau H (1982). Nucl. Acids Res. 10:3627.
19. Parslow TG, Granner DK (1982). Nature 299:449.
20. Storb U, Arp B, Wilson R (1981). Nature 294:90.
21. Max EE, Maizel JV, Leder P (1981). J. Biol. Chem. 256:5116.
22. Gruss P, Dhor R, Khoury G (1981). Proc. Natl. Acad. Sci. USA 78:943.
23. Weiher H, Konig M, Gruss P (1983). Science 219:626.
24. Benoist C, Chambon P (1981). Nature 290:304.
25. Banerji J, Rusconi S, Schaffner W (1981). Cell 27: 299.

26. Laimins L, Khoury G, Gorman C, Howard B, Gruss P (1982). Proc. Natl. Acad. Sci. 79:6453.
27. Botchan M (1982). Personal communication.
28. Conrad SE, Botchan M (1983). Mol. Cellular Biol. 2:949.
29. Weihler H, Konig M, Gruss P (1983). Science 219:626.
30. Forman P, Rowley J (1982). Nature 300:403.
31. Klein G (1981). Nature 294:313.
32. Shen-Ong GL, Keath EJ, Piccoli SP, Cole MD (1982). Cell 31:443.
33. Taub R, Kirsch P, Morton C, Lenoir G, Swan D, Tronick S, Aaronson S, Leder P (1982). Proc. Natl. Acad. Sci. 79:7837.
34. Crews S, Barth R, Hood L (1982). Science 218:1319.
35. Marcu KB, Harris LJ, Stanton LW, Erikson J, Watt R, Croce CM (1983). Proc. Natl. Acad. Sci. 80:519.

Gene Expression, pages 107-124
© 1983 Alan R. Liss, Inc., 150 Fifth Avenue, New York, NY 10011

PROMOTER SUBSTITUTION AND ENHANCER AUGMENTATION INCREASES THE PENETRANCE OF THE SV40 A GENE TO LEVELS COMPARABLE TO THAT OF THE HARVEY MURINE SARCOMA VIRUS *ras* GENE IN MORPHOLOGIC TRANSFORMATION

Michael Kriegler, Carl Perez, and Michael Botchan

Department of Molecular Biology, University of California, Berkeley, California 94720

It is well known that certain competent cells, effectively phenotypically transformed by a DNA marker gene such as HSV-tk, will be genotypically transformed by carrier DNA fragments (e.g., ΦX174 or pBR322) mixed together with the marker (1, 2, 3). Indeed, for markers whose expression levels are equivalent, genotypic and phenotypic co-transformation is close to one. Utilizing the contact inhibited Rat-2 (tk^-) line, we have extended these observations to the co-transfection of marker DNAs containing genes capable of rendering cells morphologically transformed. When cloned Harvey murine sarcoma virus proviral DNA is co-transfected along with Herpes Simplex-2 tk DNA, 80% of the colonies selected in HAT medium are morphologically transformed. Moreover, in a 3 factor transfection (G418 resistance, tk selection and H-*ras* mediated morphological transformation), each marker independently has an 80% chance of being phenotypically co-transformed. That is, 80% of the tk^+ transformed colonies are also resistant to the antibiotic G418 and 80% of the tk^+ transformed, G418 resistant colonies are morphologically transformed. In contrast, under conditions identical to the experiments described above, SV40 mediated morphological co-transformation of these cells is strikingly low. We report that addition of "enhancer" fragments from the Harvey murine sarcoma virus, inserted either 3' to the SV40 early transcription unit or in 5' proximity to the SV40 transcriptional start sites, raises the co-transformation index of SV40 to that obtained with the retrovirus.

This non-selective analysis of SV40 transformation implies that expression of the viral DNA may often be limiting in abortively transformed cells and that transcription sequence elements adapted for function in murine cells help eliminate the apparently cryptic insertions of SV40.

INTRODUCTION

Infection of rodent cells with SV40 virus causes a large fraction of these cells to divide and transiently manifest a morphological and characteristically oncogenic transformed phenotype. However, upon suspension in soft agar, the vast majority of these cells cease to retain this phenotype and are termed abortive transformants. Abortive transformation is thought to occur as the result of the loss of the infecting viral genome. It is believed that the infecting viral genome of abortively transformed cells exists as an episome for a limited number of cell divisions after which the viral DNA is lost through dilution and degradation (4). However, a small population of infected cells, usually less than 0.1%, continue to divide in soft agar and form visible macrocolonies after a few weeks incubation. These stable transformants have been shown to express the SV40 early gene products and are known to contain at least one complete copy of the SV40 early region integrated into the genome of the host cell. It is assumed that the essential difference between an abortively transformed cell and a stable transformant is that the abortively transformed cell does not contain an integrated copy of the SV40 early region. Results presented here and elsewhere (3, 5, 6) indicate that in a fraction of these abortive transformants (10%), ineffective expression of integrated copies may occur.

The results of co-transformation studies with selectable and non-selectable marker DNAs indicate that co-integration, or genotypic co-transformation, occurs at a very high frequency (approx. 80%) (1, 3, 5). Therefore, it was surprising when we and others observed that co-expression of the tk^+ phenotype and the morphologically transformed phenotype, or phenotypic co-transformation, occurred with low frequency (approx. 5%) in cells co-transfected with both HSV-tk and SV40 DNAs (3, 5). In fact, 90-95% of the tk^+ co-transfectants fail to express the SV40 early proteins when examined via immunofluorescence analysis (3). Surprisingly, many of these T antigen negative co-transfectants have been shown to contain at least one but in several cases many intact copies of the SV40 early region (5, Kriegler and Botchan, unpublished observations).

These non-transformed tk^+ cells, termed "cryptic transformants," serve to demonstrate that stable integration of an intact SV40 early transcription unit into the genome of the host cell is insufficient to insure stable expression of the SV40 early gene products. It appears as if the transfected cell has managed to effectively silence these intact SV40 early regions. The phenomenon of abortive transformation is, at least in part, the consequence of a similar effect. Thus, in some cells effectively competent for DNA uptake and integration of the marker, the infecting viral DNA is "lost" through a different mechanism than what has been previously proposed, in that the transcriptionally active infectious viral DNA is silenced upon integration into the host genome. We suggest that efficient expression of the integrated copy of the SV40 early region is an important factor that determines whether a cell becomes an abortive transformant or a stable transformant. The results presented in this report strongly support the notion that the site of integration of the SV40 genome in the host chromosome may play a vital role in determining whether or not the early genes are expressed in that cell. In addition, our results indicate that such "position effects" can be overcome through 1) the substitution of a murine retroviral promoter for the endogenous SV40 promoter or 2) the supplementation of the endogenous SV40 enhancer/

promoter assembly with a murine retroviral enhancer element, a finding consistent with our earlier report (4).

MATERIALS AND METHODS

Cells and Viruses

Rat-2 (tk^-) cells and Rat-1 Moloney leukemia virus producer cells were grown in Dulbecco modified Eagle essential medium (DMEM) containing 10% fetal calf serum (FCS) in a 5% CO_2 containing atmosphere.

Generation of Recombinant Plasmid DNAs

Recombinant plasmid DNAs were transformed into E. coli strains HB101 or DH1. Bacteria were grown in χ-broth 50γ/ml in ampicillin. Plasmid DNA was prepared according to the procedure of Birnboim and Doly (7), banded twice in cesium chloride and dialyzed into TE buffer (10 mm Tris (pH 7.4), 1 mM EDTA). See legend to Figure 1 for construction details.

DNA Transformation of Rat-2 (tk^-) Cells

Calcium phosphate mediated gene transfer was carried out by using the modificaiton of the method of Graham and van der Eb developed by Wigler et al. (8) for preparing the DNA calcium phosphate co-precipitate. Carrier DNA extracted from Ltk^- cells (10γ per 100 mM petri dish) was diluted with sterile 1 mM Tris (pH 8.1), 0.1 mM EDTA and plasmid DNA (50 to 200 nanograms per 100 mm petri dish) was added followed by 2.5 m $CaCl_2$. After thorough mixing, an equal volume of 2x HEPES (N-2-hydroxyethylpiperazine-N'-2-ethanesulfonic acid) buffered saline (pH 7.1) was added dropwise while introducing air bubbles at the bottom of the tube with a micropipette. After 30 minutes, 1 ml of the suspension was added to cultures (2.5×10^5 cells) in 100 mm petri dishes containing 10 ml of growth medium. The cultures were then incubated at $37^\circ C$ for 16 hr. at which time the medium was replaced with fresh DMEM-10% FCS. One day later, the growth medium was replaced with selective medium (either 1x HAT or 400γ/ml of G418 (Schering)). The selective medium was changed every 3 days thereafter.

T Antigen Immunofluorescence Analysis of Co-transfected Cells

Cell colonies surviving exposure to selective media were stained *in situ* as described previously (6).

RESULTS

Absolute and relative concentrations of selectable marker DNAs have a dramatic effect on the co-transformation index of the phenotypes conferred by those DNAs.

Previous reports served to establish the anamolous behavior of HSV-tk + SV40 DNA in co-transformation studies. When one seeks to determine the frequency of phenotypic co-transformation of the two marker genes at roughly equimolar concentrations of HSV-tk and SV40 DNA, approximately 5% of the tk^+ colonies manifest the morphologically transformed phenotype.

Early in our studies of this phenomenon, we determined that variation of two experimental parameters has a dramatic effect on the phenotypic co-transformation of HSV-tk and SV40 DNA in calcium phosphate mediated gene transfer. The first parameter is the molar ratio of the two transfected marker DNAs. For example, transfection with a 20-fold molar excess of SV40 DNA relative to HSV-tk DNA will result in a high percentage of tk^+ colonies that manifest the morphologically transformed phenotype. Within a limited range, the significance of these ratios is independent of absolute concentrations. The second parameter is the absolute amount of the selectable marker DNAs transfected into the cells, even if both marker DNAs are present in equimolar amounts. For example, co-transfection of Rat-2 cells with a large amount (i.e., 10γ) of each selectable marker DNA (essentially an equimolar ratio) also results in a high percentage of the co-transfected cells manifesting both the tk^+ and the morphologically transformed phenotypes. These results imply that, in cell lines containing many independent insertions of SV40 DNA, cryptic phenotypic expression is low. We have found that the anamolous expression pattern of the SV40 genome in co-transfection experiments is highly reproducable under the following conditions. First, the studies must be conducted under conditions in which the transformation of Rat-2 (tk^-) cells to the tk^+ phenotype is linear with respect to plasmid input. This occurs within the range of 50-200 nanograms of input marker gene in the precipitates added to each plate (see Materials and Methods). In the experiments described below, both high molecular weight carrier DNA and input HSV-tk plasmid were titered to determine the conditions under which transformation was linear. Second, in all experiments, input plasmid DNAs containing selectable marker genes were transfected in equimolar amounts. Therefore, none of our studies were done under conditions in which those cells competent for efficient DNA uptake were saturated for input DNA. Under these conditions, one can measure the effect of *cis*-acting elements that effect the level of stable transformation of selected marker DNAs.

Rat-2 cells co-transfected with three marker DNAs reveals an equivalent index of co-transformation.

Rat-2 (tk^-) cells were co-transfected under the non-saturating conditions described above with various combinations of either cloned HSV-tk DNA, cloned G418 resistance DNA promoted from either the HSV-tk promoter or the HaMuSV long terminal repeat (6) and/or cloned, reconstructed Harvey murine sarcoma virus provirus (see Figure 1). In our first experiment, those plates containing cells transfected with both HSV-tk DNA and the G418 resistance marker were first placed under HAT selection to measure the frequency of transformation to the tk^+ phenotype. Once all the HAT sensitive cells had been cleared from the transfected plate, the number of tk+ colonies were counted and recorded and the surviving colonies were tested for their ability to grow in media containing 400γ/ml of G418. Ten to fourteen days later, those colonies that were both HAT resistant and G418 resistant were counted and recorded. The number of G418 + HAT resistant colonies divided by the number of HAT resistant colonies x 100 yielded the co-transformation index -- in this experiment, 75% (Table 1). In our second experiment, cells trans-

TABLE 1

CO-TRANSFORMATION ANALYSIS OF RAT-2 CELLS CO-TRANSFECTED WITH HSV-tk, G418 RESISTANCE, AND HARVEY MURINE SARCOMA VIRUS PROVIRAL DNAs

	tk^+ Colonies	Co-transformed Colonies	Co-transformation Percentage
tk^+ DNA	135	N.A.	N.A.
tk^+ DNA + $pG418^r$	124	93*	75.0
tk^+ DNA + $pG418^r$ + pBVX (*ras*)	115	83*/64†	72.17*/ 55.65†

*HAT and G418 resistant colonies

†HAT and G418 resistant and morphologically transformed colonies

fected with HSV-tk DNA, the G418 resistance marker and Harvey murine sarcoma virus proviral DNA, pBVX (*ras*) (see Figure 1), were first placed under HAT selection to measure the frequency of transformation of the tk+ phenotype. Once HAT selection was complete, the number of resistant colonies was counted and recorded after which the cells were placed under selection for G418 resistance. Ten to fourteen days later, the number of surviving colonies were counted and recorded, and 10% of the surviving colonies were picked randomly. Those colonies remaining on the dish were stained to ease visualization of those cells that were both HAT resistant, G418 resistant and morphologically transformed. Visualization of the stained colonies indicated that approximately 80% of the HAT resistant, G418 resistant cells were morphologically transformed (Table 1). This number was verified when those colonies picked at random from the transfected HAT resistant, G418 resistant colonies were tested for their ability to yield infectious, transforming Harvey murine sarcoma virus after superinfection with high titer ecotropic Moloney leukemia virus. The results of these experiments, conducted under non-saturating conditions outlined above, are consistent with experiments we have conducted with other marker genes, including cloned Moloney leukemia virus and feline leukemia virus proviral DNAs. However, this is not the case with HSV-tk + SV40 DNA co-transfection of Rat-2 cells where, under the same conditions, the index of phenotypic co-transformation is approximately 5-10%.

Promoter substitution and enhancer augmentation increases the phenotypic co-transformation index of HSV-tk + SV40 DNA to levels equivalent to that of HSV-tk + Harvey murine sarcoma virus proviral DNA.

Rat-2 (tk^-) cells were co-transfected with HSV-tk DNA and various SV40 recombinant plasmid DNAs (see Figure 1), including 1) SV40 wild type DNA cloned into the Bam HI site of pBR322 or pML (pJYI, pJYM) (9), 2) a Bgl I/Bam HI linear of SV40 in which the Bgl I site had been converted to a Bam HI site and inserted into the Bam HI site of subgenomic clones of the Harvey murine sarcoma virus, containing either one of three retroviral long terminal repeats (LTRs) (which we have previously shown to be capable of promoting the expression of various marker genes including the SV40 A gene) (pRETRO T III; pRETRO T I), and 3) a Hpa II/Bam HI fragment of SV40 DNA containing an intact SV40 early transcription unit in which the

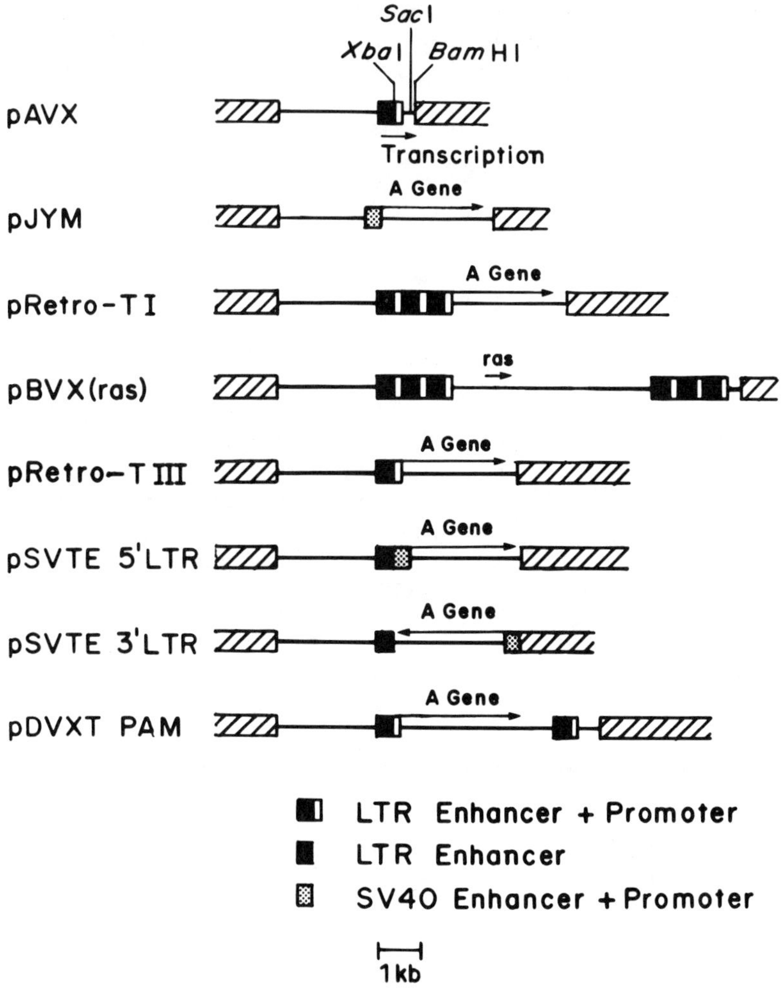

Figure 1. Physical and Functional Maps of SV40 DNA, Harvey Murine Sarcoma Virus Proviral DNA and Viral Hybrid Recombinant Plasmid DNAs

Cloned pAVX DNA is an Eco RI/Bam HI subgenomic clone of Harvey murine sarcoma virus circular proviral DNA which has been previously shown to promote the expression of a variety of well defined structural genes lacking their own promoters when those genes are inserted at the Bam HI site (6). Deletion of the DNA sequences that lie between Xba I and Sac I destroys the promoter within the retroviral long terminal repeat (LTR) but retains the DNA sequence responsible for enhancer activity (6). pJYM is a Bam HI linear of SV40 inserted into the Bam HI site of pML (9). pRETRO T I is a subgenomic clone of Harvey murine sarcoma virus proviral DNA identical to pAVX but containing three LTRs rather than one. The Bgl I/ Bam HI fragment of SV40 DNA (containing an intact structural A gene but lacking a functional SV40 promoter) has been inserted at the Bam HI site. This construction transforms murine cells and directs the synthesis of 94K SV40 T antigen (5). pBVX is a complete reconstructed Harvey murine sarcoma virus provirus flanked by six retroviral LTRs, three on each side and cloned into Bam HI/ Eco RI digested pBR322. This plasmid transforms murine cells and upon superinfection with ecotropic Moloney leukemia virus, can serve to generate authentic infectious Harvey murine sarcoma virus.

pRETRO T III is identical to pRETRO T I except that SV40 A gene expression is promoted from one retroviral LTR rather than three (6). pSVTE 5' LTR was derived from a subclone of pAVX in which the DNA fragment flanked by Xba I and Sac I has been removed thus destroying the promoter within the retroviral LTR but retaining the enhancer element. The Hpa II/Bam HI fragment of SV40 containing an intact SV40 early transcription unit was inserted into this promoter minus pAVX by converting the Hpa II site to a Bam HI site. In pSVTE 5' LTR, the retroviral LTR enhancer element is situated 5' to the SV40 early region. In pSVTE 3' LTR, the retroviral LTR enhancer element is situated 3' to the SV40 early region. pDVXT PAM is a subclone in which a Bgl I/Hpa I fragment of SV40 containing an intact structural A gene but lacking the SV40 early region polyadenylation signal has been inserted into pDVX. pDVX is a derivative of pAVX into which a Bam HI/Pst I fragment of a cloned integrated provirus of Moloney leukemia virus containing the 3' retroviral LTR has been inserted into the Bam HI/Pst I sites of pAVX. The unique Pst I site in pAVX is in the pBR322 DNA and is not shown in the figure. The construction pDVX is thus a hybrid between Harvey murine sarcoma proviral DNA 5' and Moloney leukemia virus derived proviral DNA 3' to a unique Bam HI cloning site. The Bgl I/Hpa I fragment of SV40 described above was inserted into the Bam HI site of pDVX by converting the Bgl I and Hpa I sites to Bam HI sites after which the SV40 fragment was inserted into the unique Bam HI site of pDVX. pDVXT PAM transforms murine cells in culture and such cells are positive in T antigen immunofluorescence assays.

Hpa II site had been converted into a Bam HI site. This fragment was then inserted into a Harvey murine sarcoma virus subgenomic clone in which a region known to be essential for efficient transcription from the retroviral long terminal repeat (LTR) contained within this clone had been deleted leaving behind a fragment of the LTR known to contain "enhancer" activity. This deletion joins the Xba I/Sac I sites shown in pAVX (see Figure 1) and eliminates the -30 TATA homology through the 5' cap site and extends past the coding sequences for the tRNA primer binding site outside the LTR. In these plasmid subclones, the intact SV40 transcription unit was inserted either 5' or 3' to the retroviral enhancer element (pSVTE 5' LTR; PSVTE 3' LTR). It is clear that, in a circular DNA, any point is either 3' or 5' to any other sequence. However, upon integration, the circle is linearized, yielding situations wherein the LTRs lie either 5' or 3' to the SV40 early transcription unit.

The phenotypic co-transformation index of each of these SV40 reconbinant plasmids was determined under the conditions described above. The results of five experiments are shown in Table 2. We summarize these results as follows: promotion of the SV40 early region from the promoter within the Harvey murine sarcoma virus long terminal repeat is sufficient to increase the index of co-transformation from approximately 5% to approximately 80%. Insertion of a fragment containing an intact SV40 early transcription unit (containing an intact SV40 enhancer/promoter assembly) either 5' or 3' to a retroviral enhancer element results in a similar increase in the co-transformation index (approx. 80%).

TABLE 2

CO-TRANSFORMATION ANALYSIS OF RAT-2 CELLS CO-TRANSFECTED WITH RETRO T AND HSV-tk DNAs

	tk^+ Colonies	Morphologically Transformed/ tk^+	Co-transformation Percentage
Experiment I			
SV40+tk DNA	>100	5	<5
pBVX (Harvey murine sarcoma provirus)+tk DNA	93	76	81.7
pRETRO T+tk DNA	86	73	84.8
Experiment II			
SV40+tk DNA	>100	7	<7
pBVX (Harvey murine sarcoma provirus)+tk DNA	85	75	88.2
pRETRO T+tk DNA	78	63	80.7
Experiment III			
SV40+tk DNA	>100	9	<9
pBVX (Harvey murine sarcoma provirus)+tk DNA	113	88	77.8
pRETRO T+tk DNA	107	80	74.7

TABLE 2

CO-TRANSFORMATION ANALYSIS OF RAT-2 CELLS CO-TRANSFECTED WITH RETRO T AND HSV-tk DNAs (continued)

	tk^+ Colonies	Morphologically Transformed/ tk^+	Co-transformation Percentage
Experiment IV			
SV40+tk DNA	>100	10	<10
pBVX (Harvey murine sarcoma provirus)+tk DNA	76	66	86.8
pRETRO T+tk DNA	74	57	77.0
pSVTE LTR 5' +tk DNA	67	51	76.1
pSVTE LTR 3' +tk DNA	64	47	73.4
Experiment V			
SV40+tk DNA	>100	7	<7
pBVX (Harvey murine sarcoma provirus)+tk DNA	82	70	85.3
pRETRO T+tk DNA	78	67	85.8
pSVTE LTR 5' +tk DNA	80	63	78.0
pSVTE LTR 3' +tk DNA	85	72	84.7

DISCUSSION

Our initial experiments were designed to more completely characterize calcium phosphate mediated co-transfection in an attempt to use the assay to determine why SV40 + HSV-tk co-transfection of murine cells gives a high index of genotypic co-transformation but a surprisingly low index of phenotypic co-transformation.

Initially, we observed that the variation of two experimental parameters can have a dramatic effect on the efficiency of phenotypic co-transformation of a variety of selectable marker DNAs (HSV-tk, G418 resistance, various retrovirus proviruses, etc.). Firstly, the molar ratios of the co-transfected DNAs are very important. A high molar excess of one DNA relative to the other can dramatically skew the co-transformation index in favor of the over-represented DNA. Secondly, saturation of the transfection potential of a given plate of Rat-2 (tk^-) cells with μg quantities of two selectable marker DNAs, even if they are present in equimolar amounts, will insure a co-transformation index of 80%, minimizing the effect of any *cis*-acting control element linked to the structural gene of interest. Because we set out to study such *cis*-acting control elements, all experiments in this report were designed to eliminate the effect of these variables.

Under such conditions, with all selectable marker plasmid DNAs we have studied, the absolute number of cells phenotypically transformed is linear with respect to plasmid input. Under these conditions, we found that phenotypic co-transformation of cells with two selectable marker DNAs present in equimolar amounts is roughly 80% and that phenotypic co-transformation with three selectable marker DNAs is roughly 80% of 80%, or 60%. Thus it was surprising that we and others found that, under identical conditions, while genotypic co-transformation of murine cells with HSV-tk and SV40 DNA is quite high, phenotypic co-transformation of the cells is very low with 5-10% of the tk^+ transformants manifesting the morphologically transformed phenotype. Why is SV40 DNA different than other marker DNAs? From unpublished data based upon virus rescue (J. Miller, P. Bullock, and M. Botchan) and blot data previously reported (5), a substantial fraction of SV40 tk co-transfected lines contain integrated SV40 DNA. Earlier studies from our laboratory demonstrated that we could increase the ability of wild type SV40 virus (which is a feeble transforming virus) to transform murine cells 20-fold by inserting a retroviral "enhancer" element into the late region of an infectious virus particle. In those studies, we suggested that the retroviral "enhancer" element might serve to insure that the integrated copy of the SV40 genome present in infected cells would produce T antigen in amounts sufficient to transform the cell (6). In fact, SV40 transformed cells contain substantially greater amounts of T antigen specific cytoplasmic messenger RNA when compared to the amounts of HSV-tk specific cytoplasmic messenger RNA present in tk^+ transformants. It appears that a higher level of SV40 A gene expression is required to morphologically transform a cell relative to the level of HSV-tk gene expression required to convert a (tk^-) cell to the tk^+ phenotype. However, in our studies cited above, we were unable to eliminate the possibility that the "enhancer" element indirectly served to increase the efficiency of integration of the viral DNA into the host genome by some mysterious mechanism. Because we have demonstrated that integration of co-transfected marker DNAs occurs at high efficiency, in these studies we were able to eliminate the effect of those factors that serve to stabilize integration of the viral DNA on phenotypic transformation and thus were able to focus on the roles gene expression and cell competence play. The observation of (5) that genotypic transformation occurs at much higher frequency than phenotypic transformation (as measured by Southern blot analysis and T antigen immunofluorescence) supports the notion that inefficient SV40 transformation of murine cells is the result of inefficient gene expression. The observation that substitution of the retroviral promoter for the SV40 promoter or that augmentation of the SV40 enhancer with the retroviral enhancer serves to increase the level of phenotypic co-transformation by the SV40 A gene to levels equivalent to that of other selectable marker genes further strengthens this argument. These enhancer elements may 1) alter the expression of linked genes by either providing a site for entry of factors which are required for efficient promotion of transcription or 2) provide a *cis*-acting signal that actively prevents a promoter from being shut off. In either case, markers linked to the appropriate enhancer element would be more likely to work in a given chromosomal domain than a marker lacking the enhancer element.

Our results may serve to explain, in part, the biology of SV40 abortive transformation of murine cells. It has long been assumed that infected cells appear to be abortively transformed because the unintegrated infective viral DNA is lost through dilution and degradation and therefore fails to establish itself through stable integration into the genome of the host cell (4). Such a mechanism would explain why SV40 is such a poor transforming virus. In contrast, the transforming retroviruses are more efficient transforming agents because they contain a highly efficient integration mechanism. If this notion is correct, then stable integration of an intact SV40 early region into the host genome of an infected or transfected cell should be sufficient to transform a cell. The effect of retroviral enhancer sequences upon SV40 co-transformation argues that, in at least a fraction of the abortively transformed cells, effective expression of the early antigens plays some role.

We and others have shown that genotypic co-transformation of cells occurs with high efficiency. In most cases, high genotypic co-transformation is associated with a high phenotypic transformation index. However, the high index of genotypic co-transformation of HSV-tk and SV40 DNA is not associated with a high index of phenotypic co-transformation and therefore appears to mimic the phenomenon of abortive transformation in that a high percentage of infected or transfected cells pick up the infected or transfected viral DNA but only a minority of those cells stably express the transformed phenotype. As the phenotypic co-transformation index goes up a factor of 10 upon insertion of an LTR into the SV40 constructions, we argue that only one in ten random insertions of SV40 lead to effective gene expression. While it is likely that establishment of an integrated SV40 provirus into the host genome will play a role in limiting the efficiency of SV40 transformation of murine cells by infectious virus particles, it is clear from the experiments described above that stable integration of wild type SV40 DNA into the host genome is by itself insufficient to insure that the cell carrying the integrated DNA will become transformed. Efficient gene expression is necessary as well since *cis*-acting elements that promote efficient gene expression, enhance the penetration of the A gene in these transfection studies. Further experimentation will be necessary to evaluate the relative contribution of stable integration of the infecting viral DNA during both abortive and stable transforming infection of murine cells by SV40 virus.

ACKNOWLEDGMENTS

We wish to thank Judy Kramer for her expert assistance in the preparation of this manuscript.

This work was supported in part by grants from the Public Health Service National Cancer Institute (CA 30496) and the American Cancer Society (MV-91) to M. B. M. K. was a postdoctoral fellow of the Damon Runyon-Walter Winchell Cancer Fund and the National Institutes of Health during the course of this research and is a recipient of Public Health Service National Research Service Award CA 06446 from the National Cancer Institute and Damon Runyon-Walter Winchell Cancer Fund Award 403-F.

REFERENCES

1. Wigler, M., R. Sweet, G. K. Sim, B. Wold, A. Pellicer, E. Lacy, T. Maniatis, S. Silverstein, and R. Axel (1979), Transformation of Mammalian Cells With Genes From Procaryotes. *Cell 16*, 777-785.

2. Perucho, M., D. Hanahan, and M. Wigler (1980), Genetic and Physical Linkage of Exogenous Sequences in Transformed Cells. *Cell 22*, 309-317.

3. Perucho, M. and M. Wigler (1980), Linkage and Expression of Foreign DNA in Cultured Animal Cells. Cold Spring Harbor Symp. Quant, Biol. 44, 829-838.

4. Stoker, M. (1968), Abortive Transformation by Polyoma Virus. *Nature (London) 218*, 234-238.

5. Hanahan, D., D. Lane, L. Lipsich, M. Wigler, and M. Botchan (1980), Characteristics of an SV40-Plasmid Recombinant and Its Movement Into and Out of the Genome of a Murine Cell. *Cell 21*, 127-139.

6. Kriegler, M. and M. Botchan (1983), Enhanced Transformation by a Simian Virus 40 Recombinant Virus Containing a Harvey Murine Sarcoma Virus Long Terminal Repeat. *Mol. Cell. Biol. 3*, 325-339.

7. Birnboim, H. C. and J. Doly (1979), A Rapid Alkaline Extraction Procedure for Screening Recombinant Plasmid DNA. *Nucleic Acids Res. 7*, 1513-1523.

8. Wigler, M. A., A. Pellicer, S. Silverstein, and R. Axel (1978), Biochemical Transfer of Single Copy Eukaryotic Genes Using Total Cellular DNA as a Donor. *Cell 14*, 729-731.

9. Lusky, M. and M. Botchan (1981), Inhibition of SV40 Replication in Simian Cells by Specific pBR322 DNA Sequences. *Nature (London) 293*, 253-258.

Gene Expression, pages 125-132
© 1983 Alan R. Liss, Inc., 150 Fifth Avenue, New York, NY 10011

THE SV40 ENHANCER INDUCES AN ALTERED CHROMATIN STRUCTURE[1]

Jan Jongstra, Tim Reudelhuber, Pierre Oudet and Pierre Chambon

Laboratoire de Génétique Moléculaire des Eucaryotes du CNRS, Unité 184 de Biologie Moléculaire et de Génie Génétique de l'INSERM, Faculté de Médecine, 11 Rue Humann, 67085 Strasbourg Cedex, France.

ABSTRACT We have investigated whether the SV40 72 bp repeat can induce an alteration in chromatin structure using viable SV40 mutants in which an intact or truncated enhancer element is inserted at the HpaI site. We find that the intact 72 bp sequence induces a region of increased DNaseI sensitivity on itself. However a fragment of the 72 bp sequence which has lost the ability to enhance transcription does not induce DNaseI sensitivity. It is also shown that a dimer of the 72 bp repeat is capable of inducing a nucleosome free gap.

INTRODUCTION

Studies from several groups have shown that when SV40 infected cell nuclei are digested with a variety of endonucleases, a fragment in the SV40 genome located between the unique restriction sites BglI and HpaII is preferentially cut (the Ori region, see fig. 1 "ORI A") (1-3). In particular, studies using DNAse I have revealed that the sensitivity at the ORI region is not homogeneous, but that there are two small regions of increased DNAse I sensitivity at ORI (4). Electron microscopic observations of purified minichromosomes have also shown the presence

[1]This work was supported in part by the CNRS (ATP 6182), the INSERM (CRL 801033) and the DGRST (81V1458-509179). J.J. is supported by a long term EMBO fellowship.

of a nucleosome-free region of approximately 300 bp in the same position (the nucleosome gap) (4,5). In addition, studies using viral mutants in which the ORI region is duplicated have shown that this region contains the DNA sequence information required to generate the nuclease sensitive chromatin structure (6) and a nucleosome gap (7).

The ORI region contains the viral replication origin (around the BglI site, see fig. 1) and the DNA sequences involved in the regulation of transcription of the early and late SV40 genes. The three following early gene promoter elements have been identified 1) the "TATA box" (fig. 1, dark triangle), which is not absolutely required, but dictates the proper initiation sites for transcription, 2) the 21 bp repeat region, the deletion of which decreases early gene expression (8,9) and 3) the 72 bp repeat which is indispensable for early gene transcription (10,11). A striking property of the 72 bp repeat is that it can "enhance" transcription from heterologous promoter sequences irrespective of its orientation relative to transcription and that it can exert this effect at distances longer than 4 kb. Deletion of one of the two 72 bp repeated sequences has very little effect on the "enhancing" activity of this ORI fragment, but deletion of an additional 38 bps between the EcoRII and PvuII sites (see fig. 1) reduces considerably the capacity to "enhance" transcription from the SV40 early promoter or from heterologous promoters (8,12).

Since the SV40 72 bp repeat is located in the same fragment of the SV40 genome which displays an altered chromatin structure, we decided to investigate whether this repeat might exert at least part of its "enhancing" activity through the induction of an altered chromatin structure and thereby facilitate initiation of transcription.

Our experimental approach is based on the observation that SV40 mutants with deletions in the large-T intron can accomodate extra DNA sequences in the HpaI site at position 2666 (see fig. 1). Such insertion mutants are viable and have growth characteristics similar to wild-type SV40 (13). We used SV40 dl222 which contains a 234 bp deletion around the TaqI site (see fig. 1, dotted line), and converted the HpaI site at bp 2666 to a unique XhoI restriction site by inserting a linker. This SV40 derivative, SV40 dl2122X, was then used to construct a series of viable insertion mutants (the "In" series) which carry a second complete or truncated ORI region at position 2666 (see fig. 1, ORI B). The ORI fragments were inserted in the same orientation with respect to early gene transcription as at ORI A. Both ORI regions were analyzed late in infection for the presence of DNAseI sensitive sites and a nucleosome gap. Regions of increased DNAseI sensitivity were mapped by the indirect end labelling technique of Wu (14) from the EcoRI site using probe B to map sites at ORI A and probe A to map sites at ORI B (see fig. 1).

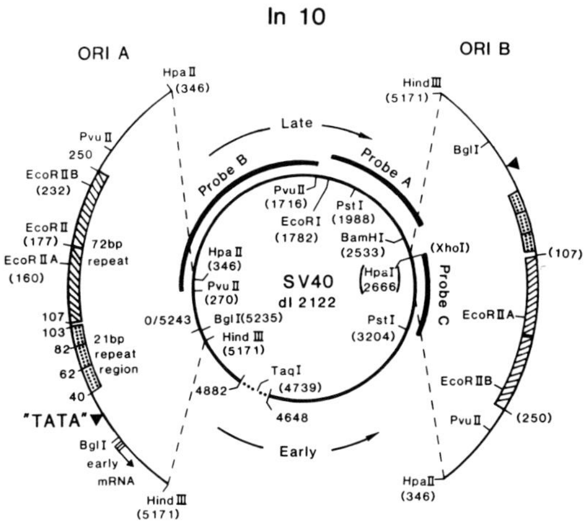

FIGURE 1. The diagram shows a schematic representation of the SV40 mutant In10. The inner circle represents the genome of SV40 dl2122 which carries a 234 bp deletion around the TaqI site (dotted line), and in which the HpaI site at bp 2666 is converted into a XhoI restriction site. Shown enlarged at the left side is the ORI region (ORI A) which contains the origin of replication around the BglI site) and the early and late gene promoter sequences. The three early gene promoter elements : TATA box, 21 bp repeat region and 72 bp repeat are indicated. The ORI region was inserted at bp 2666 (ORI B) in the same orientation with respect to early gene transcription as at ORI A. The dark bars indicate the fragments used as hybridization probes to map sensitive sites at ORI A and ORI B.

RESULTS

THE TRANSPOSED HindIII-HpaII ORI FRAGMENT IN In10 RETAINS ITS ALTERED CHROMATIN STRUCTURE

ORI A of wild-type and mutant SV40 contains two regions of increased DNAseI sensitivity. Region I contains the early mRNA cap sites and the "TATA-box" and the origin of replication. Region II extends from bp 110 ± 25 over the 72 bp repeat till approximately bp 320. A similar pattern of DNAseI sensitivity is induced at ORI B in In10 (see Fig. 2, In10, ORI B) which confirms that the 418 bp HindIII-HpaII ORI fragment contains all the DNA sequences necessary to induce this "wild-type" pattern.

THE 72 bp REPEAT INDUCES DNaseI SENSITIVITY. THE 38 bp EcoRII-PvuII ORI FRAGMENT IS ESSENTIAL FOR THE INDUCTION OF DNaseI SENSITIVITY OVER AN ISOLATED 72 bp SEQUENCE

Since region II includes the 72 bp repeat, we tested whether sequences within this repeat are capable of inducing a DNAseI sensitive chromatin structure. We therefore inserted a DNA fragment containing the 72 bp repeat and 20 bp on its late side (bp 113-270) at ORI B and found that it induces DNAseI sensitivity over the repeat (see fig. 2, In22). Deletion of exactly one 72 bp repeated sequence from ORI B in In22 reduces the size of the sensitive region at ORI B by approximately the size of the deletion (see fig. 2, In60). However further deletion of the EcoRII-PvuII sequence (bp 160-270) eliminates the DNAseI sensitivity over the remainder of the 72 bp repeated sequence (see fig. 2, In54). This same deletion has been shown to eliminate the "enhancing" activity of the 72 bp repeat (8, 12).

A MULTIMER OF THE 72 bp REPEAT CAN INDUCE A NUCLEOSOME GAP

To investigate the relation between induction of DNAseI sensitivity and the formation of a nucleosome gap the described "In" mutants were analyzed for the presence of a nucleosome gap over ORI B. Approximately 20 % of the minichromosomes extracted from In10 infected cells contains a nucleosome gap (see fig. 3a) which was mapped over ORI A using the single cut restriction enzymes EcoRI (fig. 3b) or MspI (HpaII, fig. 3c). On approximately one-third of these minichromosomes a second gap (fig. 3d) was seen which was mapped over ORI B using EcoRI (fig. 3e) or MspI (fig. 3f). Minichromosomes extracted from In22, In60 or In54 infected cells did not show a gap over ORI B. To determine whether this was due to the absence of specific ORI sequences rather than the shorter size of the DNAseI sensitive chromatin, we inserted a dimer of the 72 bp repeat at ORI B, thus inducing a long stretch of DNAseI sensitive chromatin (see fig. 2, In222). Minichromosomes extracted from In222 infected cells contained a gap over ORI B.

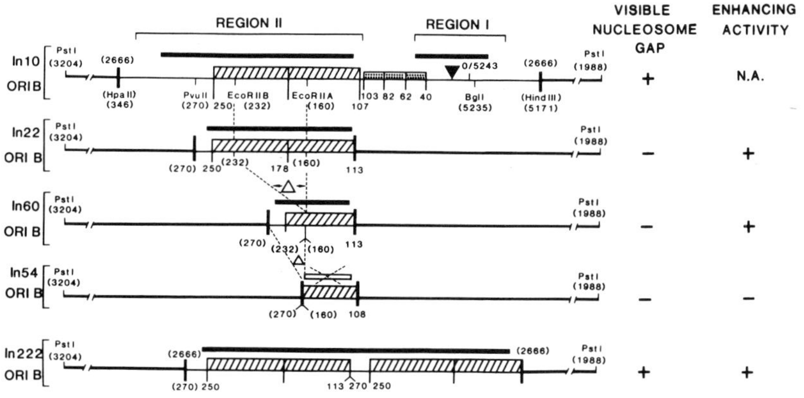

FIGURE 2. The diagrams show the results of mapping DNAseI sensitive sites at ORI B in several "In" mutants. The PstI fragments containing ORI B of these "In" mutants are represented schematically with the induced sensitive regions represented above each map by a black bar. The absence of a sensitive region in In54 is represented by an open bar. SV40 early and late gene DNA is shown as a thick line and the point of insertion of different ORI fragments as a thick vertical bar. The symbols used to denote early gene promoter sequences are as above. In22 was constructed by inserting an ORI fragment (bp 113-270) containing the 72 bp repeat at bp 2666. Deletion of exactly one 72 bp repeated sequence (from bp 160-232) gave rise to In60, while further deletion of the EcoRII-PvuII fragment (160/232-270) gave rise to In54. In222 contains a dimer of the ORI fragment used to construct In22 at ORI B.

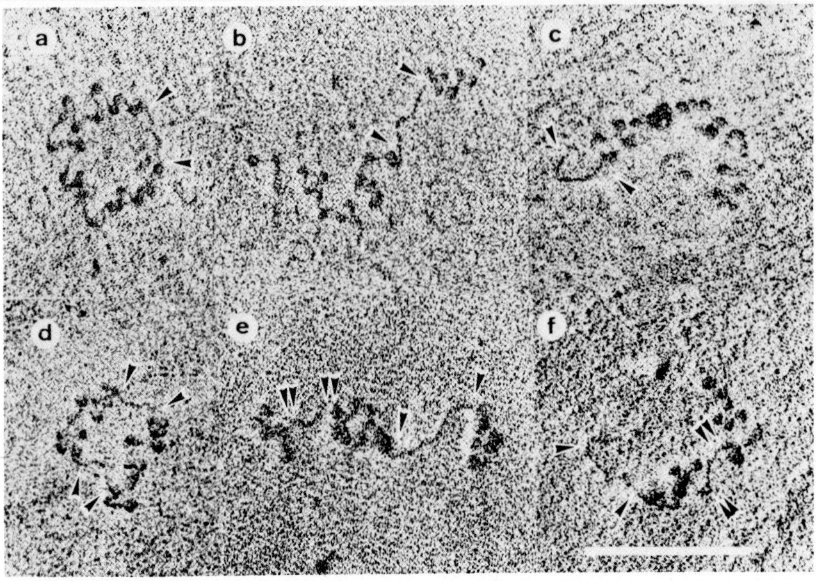

FIGURE 3. Mapping nucleosome gaps over ORI A and ORI B IN IN10. Panel a shows an In10 minichromosome with one gap which was mapped over ORI A after digestion with EcoRI (panel b) or MspI (panel c). When a second gap is present as in panel d, digestion with EcoRI (panel e) or MspI (panel f) maps the second gap over ORI B.

DISCUSSION

Our results show that sequences within the SV40 72 bp repeat can induce an altered chromatin structure over the entire enhancer element. Deletion of one of the two 72 bp repeated sequences does not prevent the assembly of the altered chromatin structure. However when we deleted from the remaining 72 bp repeated sequence a short DNA sequence which is crucial for the enhancing activity, the induction of DNAseI sensitivity was prevented. This correlation between enhancing activity and alteration in chromatin structure over the 72 bp repeat suggests that enhanced initiation of transcription may result at least in part from the 72 bp repeat being more accessible for cellular proteins involved in the transcriptional process. This proposed mechanism of action of the 72 bp repeat is particularly appealing in view of its recently postulated role as an entry site for RNA polymerase B (8, 15).

It is unclear how the induction of a visible nucleosome gap relates to the enhancing effect of the 72 bp repeat. It appears from the present results and other unpublished observations that formation of a visible gap depends on the length of the induced DNAseI sensitive chromatin. At present, we cannot rule out the existence of a nucleosome gap over the insert in In22, too short however to be visible.

ACKNOWLEDGEMENTS

We thank Edith Badzinski and Colette Kutschis for typing the manuscript, Monique Acker for the tissue culture work and Christiane Werlé and Bernard Boulay for the artwork.

REFERENCES

1. Scott WA, Wigmore DJ (1978). Sites in Simian Virus 40 Chromatin Which are Preferentially Cleaved by Endonucleases. Cell 15: 1511.
2. Varshavsky A, Sundin OH, Bohn MJ (1978). SV40 viral minichromosomes : preferential exposure of the origin of replication as probed by restriction endonucleases. Nucleic Acids Res 5: 3469.
3. Shakhov AN, Nedospasov SA, Georgiev GP (1982). Deoxyribonuclease II as a probe to sequence-specific chromatin organization : preferential cleavage in the 72 bp modulator sequence of SV40 minichromosomes. Nucleic Acids Res 10: 3951.
4. Saragosti S, Moyne G, Yaniv M (1980). Absence of Nucleosomes in a Fraction of SV40 Chromatin between the Origin of Replication and the Region Coding for the Late Leader RNA. Cell 20: 65.
5. Jakobovits EB, Bratosin S, Aloni Y (1980). A nucleosome-free region in SV40 minichromosomes. Nature 285: 263.
6. Wigmore DJ, Eaton RW, Scott WA (1980) Endonuclease-Sensitive Regions in SV40 Chromatin from Cells Infected with Duplicated Mutants. Virology 104: 462.

7. Jakobovits EB, Bratosin S, Aloni Y (1982). Formation of a nucleosome-free region in SV40 minichromosomes is dependent on a restricted segment of DNA. Virology 120: 340.
8. Moreau P, Hen R, Wasylyk B, Everett RD, Gaub MP Chambon P (1981). The SV40 72 base pair repeat has a striking effect on gene expression both in SV40 and other chimeric recombinants. Nucleic Acids Res 9: 6047.
9. Everett R, Baty D, Chambon P (1983). The repeated GC-rich motifs upstream from the TATA box are important elements in the SV40 early promoter. Nucleic Acids Res (in press).
10. Benoist C, Chambon P (1981). In vivo sequence requirements of the SV40 early promoter region. Nature 290: 305.
11. Fromm M, Berg P (1982). Deletion Mapping of DNA Regions Required for SV40 Early Region Promoter Function In Vivo. J. Molec. and Applied Genetics 1: 457.
12. Hen R, Sassone-Corsi P, Corden J, Gaub MP, Chambon P. (1982). Sequences upstream from the TATA box are required in vivo abd in vitro for efficient transcription from the Adenovirus-2 major late promoter. Proc Natl Acad Sci USA 79: 7132.
13. Shenk T (1978). Construction of a Viable SV40 Variant Containing Two Functional Origins of DNA Replication. Cell 13: 791.
14. Wu C (1980). The 5'ends of Drosophila heat shock genes in chromatin are hypersensitive to DNaseI. Nature 286: 854.
15. Wasylyk B, Wasylyk C, Augereau P, Chambon P (1983). The SV40 72 bp repeat preferentially potentiates transcription starting from proximal natural or substitute promoter elements. Cell 32: 503.

III. TRANSCRIPTIONAL REGULATION

Gene Expression, pages 135-143
© 1983 Alan R. Liss, Inc., 150 Fifth Avenue, New York, NY 10011

REGULATION OF THE SOS RESPONSE OF ESCHERICHIA COLI BY THE lexA AND recA GENES[1]

David W. Mount, Kenneth F. Wertman, Kenneth R. Peterson, John W. Little, Bruce E. Markham[2] and Joan E. Harper[3]

Departments of Molecular and Medical Microbiology and Biochemistry
University of Arizona College of Medicine
Tucson, Arizona 85724

ABSTRACT The SOS regulatory system of *E. coli* controls a cellular response to conditions which interfere with DNA replication. In this response, the SOS response, a large number of genes having different types of functions are derepressed. This derepression occurs through destruction of a repressor, the product of the *lexA* gene, by a protease activity of RecA protein. The protease is reversibly activated by cofactors produced in response to the inducing treatment. After the SOS functions have acted, the protease disappears and repression is again established. Present research is aimed at characterization of the repressor-operator and protease-repressor interactions and further analysis of the mechanism of induction.

INTRODUCTION

E. coli displays a complex physiological response, known as the SOS response, following DNA damage or inhibition of DNA synthesis (1). The regulatory system which governs the response has attracted a great deal of interest because the system regulates the expression of many genes having different biological functions and

[1]Supported by grants from the National Institutes of Health and the National Science Foundation

[2]Present address: Dept. of Internal Medicine, Univ. of Arizona Coll. of Medicine

[3]Present address: Dept. of Pathology, NYU Medical Center

because the cellular physiology changes drastically and reversibly. In addition, the expression of the SOS functions leads to induction of temperate phage, colicins and other gene activity in bacterial plasmids.

The SOS regulatory system comprises a repressor, the product of the lexA gene, and a specific protease which cleaves this repressor, inactivating it. This protease is an activity of RecA protein which results from interaction with signal molecules produced as a result of the DNA damage. A detailed review of the SOS

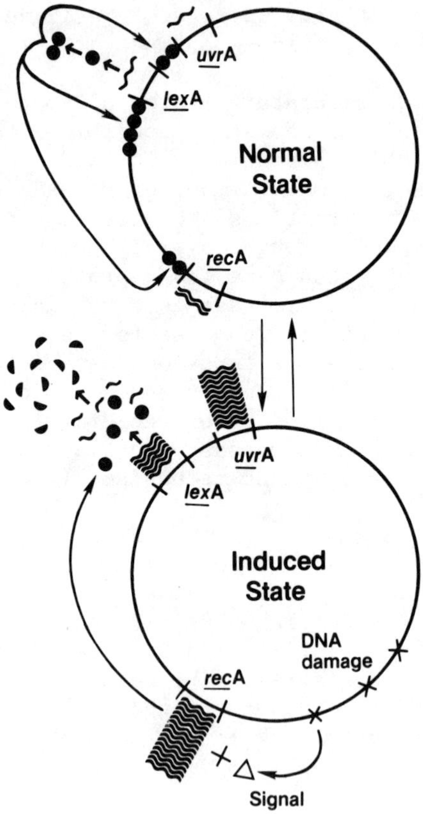

Figure 1. Model of the SOS regulatory system

regulatory system has appeared elsewhere (2), and will only be summarized here. Briefly, we now believe that in response to DNA damage, the RecA protease is activated and cleaves LexA repressor, with the result that 11 or more genes, controlling various functions of the SOS response, are all derepressed (Fig. 1). The cell then enters a new physiological state which augments DNA repair processes and it remains in this state until the damage has been repaired or bypassed in some way. Then RecA protein loses its activity, LexA repressor accumulates and shuts off its target genes, and the cell returns to a normal state.

At least 11 genes repressed by lexA have been identified (Fig. 2) and, in addition, lexA is autoregulatory.

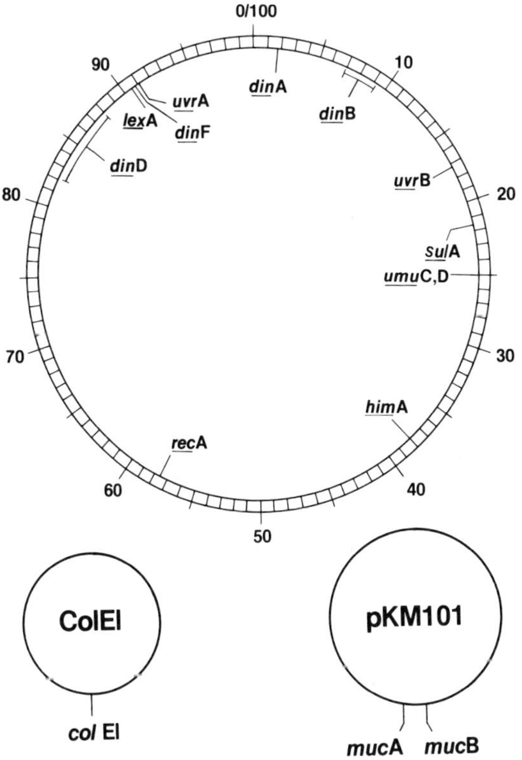

Figure 2. Partial genetic maps of E. coli and two plasmids controlled by the SOS reglatory system.

Several target genes have been located by fusion of lacZ to damage-inducible chromosomal promoters at loci denoted din. Purified LexA protein has been shown to inhibit in vitro transcription from many of these genes and, in some cases, the operator sequence has been determined. The consensus sequence for the known binding sites, including two sites which regulate lexA itself, is CTGTxTxxxxxxxCAG, where x is variable (see ref. 2). It is likely that not all SOS genes have been identified. In particular, genes essential for growth would not have been detected by the genetic screens which have been used.

In work to be published in detail elsewhere, we have measured mRNA synthesis from several lexA target genes and analyzed the structural and functional interaction among mutant LexA proteins and operators. We present here a brief overview of this work.

RESULTS AND DISCUSSION

Induction phase of the SOS response

We have detected increased rates of mRNA synthesis from lexA, recA, uvrA, uvrB, and two din loci following induction by u.v. light. Although the repressed rates of mRNA synthesis varied 35-fold, the rates for each gene increased 5 to 10-fold to a maximum within 15 to 30 min following induction (also see ref. 3) and then decreased to the repressed levels within another 60 to 120 min. This response is very similar to the induction response of the SOS functions themselves (4,5), and similar kinetics have been observed in the synthesis of RecA protein (6,7).

We have also measured the cleavage of LexA protein in similarly treated cells (8). Cleavage of LexA protein was examined by pulse labeling the protein and detecting the amount present as intact polypeptide or as fragments by gel electrophoresis of immunoprecipitates from crude cell extracts. Whereas LexA protein is stable in normally growing cells for at least 60 min at 37 degrees, greater than 80% of it is cleaved within 4 min after induction. We attribute these rapid kinetics to production of protease cofactors which activate the RecA protease very soon after the inducing treatment, and to subsequent rapid cleavage of LexA repressor by the protease.

Decline in protease activity and mRNA synthesis

Cells are able to recover from induction of their SOS genes. We predict that as the impairment to DNA synthesis is removed, the level of protease decreases, thus allowing a reestablishment of repression. We have verified this prediction directly; there was a progressive decline in the rate of LexA protein cleavage during the recovery phase. Also, the rate of mRNA synthesis gradually declined. Interestingly, induced mRNA synthesis from lexA continued approximately 30 min longer from lexA than from the other SOS genes. This result is that expected if the concentration of LexA protein in the cell was increasing and if more protein must be present to reestablish lexA autoregulation than repression of the other SOS target genes. These observations are consistent with the known repressor/operator binding affinities; LexA protein binds to its own operator with 10-fold lower affinity than to the recA operator (9). They are also reasonable in terms of the expectations that repressor should first return to high, steady state levels and only then repress its own synthesis, but that other SOS genes should be repressed at lower concentrations.

Operator Mutants

To expedite the isolation and analysis of operator mutants in SOS target genes, we have exploited the availability of phage M13mp8 for rapid DNA sequence analysis by the dideoxy nucleotide method. We prepared a derivative of phage M13mp8 (10) with the lac promoter replaced by a DNA fragment bearing a lexA or recA promoter. In each case, the fusion produced a hybrid fragment of β-galactosidase in which the first few amino acids were replaced by several N-terminal amino acids from the lexA or recA genes. The new fragments have α-complementing activity, as indicated by their ability to hydrolyze X-gal, and their synthesis can be controlled by the SOS regulatory system. For the recA-lacZ fusion, the phage produced white plaques on a $RecA^+LexA^+$ host, indicative of the presence of only very small amounts of fragment. If the host cells were induced by the addition of a small amount of Mitomycin C to the medium, the plaques became blue, indicating a response to the inducing treatment given the cell. By contrast, the phage with the lexA-lacZ fusion produced blue plaques on a RecA+LexA+ host even without an inducing treatment, indicating lack of repression. However, when LexA protein was overproduced in the host due to the presence of a multicopy plasmid bearing the lexA gene, white plaques indicative of repression were obtained, as if the higher level of LexA protein were required to repress the lexA fusion than the recA fusion. These results are consistent with the weaker binding of LexA protein to the lexA operator than to the recA operator measured in vitro (9), and demonstrate that such differences in binding may be detected by gene fusion methods.

Operator mutants were then produced by growth of these fusion phage strains on a mutD5 host or by chemical mutagenesis, and were detected by their ability to produce a blue plaque under repressed conditions, as described above. These mutations are shown in Fig. 3. Apart from a few mutations in the presumptive ribosomal binding site and several within the structural gene itself, all of them fell within the two binding sites in lexA. The quantitative effects of these mutations on repressor binding and promoter expression have not yet been measured, but their appearance in both binding sites in lexA suggests that binding to both sites is required for effective repression of lexA. In addition to the above mutations

Mutant Sequence Changes in *lex*A

mRNA	−17		1	5 8	18	21
WT DNA	CTG TATATACTCA	CAG	CATAA	CTG TATATACACC	CAG	GGGGCGGA
MUTANTS	T	GA		T	T	AAT
	oc	oc,oc		oc	oc	rib. bind. site
Repressor protection	----------------------	----------		----------------------	----------	

mRNA	29	278		419		632
WT DNA	ATG AAG	......................GCC	GGT	....................CTG	GCA	........................TGT TAA
MUTANTS			A		A	
Protein	Met Lys	...........................Ala	Gly Asp	.......................Leu	Ala Thr	.........................Leu TER
			*lex*A3 change		*lex*A41 (*tsl*-1) change	

Figure 3. Mutations obtained in the lexA operator and structural gene

in the lexA operator, we have also obtained the following changes in the recA operator, shown in parenthesis after the normal base; CT(C)GT(C)ATGAGCAT(A)A(G)C(T)AG. Thus, the method seems to have general application for obtaining operator-type mutations in regulatory sequences.

Analysis of mutant LexA proteins

Four classes of mutants in the lexA structural gene have been described; one class (Ind$^-$) synthesizes a repressor which is not cleaved by the recA protease, a second (Ts or tsl), a repressor which appears to be temperature sensitive, a third (Def$^-$ or spr), a repressor which does not produce a functional repressor, and a fourth, a repressor which acts as if it selectively represses only some of its target operators because only some of the SOS functions become induced, called the "split-phenotype" effect (1).

We have recently initiated a study of the effects of some of these mutations on repressor stability, sensitivity to protease, and repression of operator targets using gene and operon fusions. Previous analysis of one particular mutant, lexA3, revealed it to produce a repressor that is highly resistant to cleavage (11). Two other mutants whose repressors are altered in cleavage have now been identified by immunoprecipitation of LexA protein from cells irradiated with u.v. light. lexA1 protein, which is derived from a mutant as sensitive to DNA damage as the lexA3 mutant, shows no detectable cleavage in vivo, and the protein has an altered electropheretic mobility. On the other hand, the lexA2 protein, from a mutant much less sensitive to DNA damage than lexA3, is cleaved at an approximately 2-fold lower rate than lexA$^+$. These results indicate that efficient cleavage of LexA protein is important for induction, since the lexA2 mutant shows a considerable reduction in induction of SOS functions.

lexA$^+$ protein is cleaved by the protease at an ala-gly bond similar to that in phage repressors (12). The lexA3 mutation would change the sequence of the cleavage site from ala-gly to ala-asp (13).

A mutant derivative of the lexA3 strain, tsl-1 (now

called lexA41), is resistant to DNA damage but fails to grow at high temperature and shows an abnormal response to mutagenesis by u.v. light. The DNA sequence of the part of the gene encoding the C-terminal half of the protein has been determined, and a single change was detected (see Fig. 3). In addition, the presence of the original lexA3 mutation may be inferred from the absence of an MspI site caused by the mutation. If the repressor structure is similar to that of phage lambda repressor, this region should determine interaction of repressor monomers to form dimers, which is the form that binds tightly to operators (see ref. 2).

The *in vivo* stability of the mutant lexA41 protein has been examined by the antibody methods already described. lexA41 protein has an approximate half-life of 2 min. at 42° and 4 min at 32°, compared to the high stability of wildtype protein. Moreover, the mutant protein is synthesized at a 3 to 5-fold higher rate than is wildtype protein, suggesting that the instability leads to a lower steady-state level and partial derepression of lexA (also see ref. 14). Interestingly, the stability of this mutant protein is increased in a lon^- host which is deficient in certain proteolytic processes (15). These experiments suggest that mutations which affect stability of LexA protein can give rise to aforementioned "split-phenotype" effects. We are presently extending this type of analysis to a number of other lexA mutant proteins.

REFERENCES

(1) Witkin EM (1976). Ultraviolet mutagenesis and DNA repair in *Escherichia coli*. Bacteriol Rev 40:669.
(2) Little JW, Mount DW (1982). The SOS regulatory system of *Escherichia coli*. Cell 29:11.
(3) McPartland A, Green L, Echols H (1980). Control of recA gene RNA in *E. coli*: regulatory and signal genes. Cell 20:731.
(4) DeFais M, Caillet-Faquet P, Fox MS, Radman M (1976). Induction kinetics of mutagenesis DNA repair activity in *E. coli* following ultraviolet irradiation. Molec Gen Genet 148:125

(5) Darby V, Holland IB (1979). A kinetic analysis of cell division, and induction and stability of recA protein in u.v. irradiated lon+ and lon− strains of Escherichia coli. Molec Gen Genet 176:121.

(6) Quillardet P, Moreau P,Ginsberg H, Mount DW, Devoret R (1982). Cell survival, uv-reactivation and induction of phage lambda in E. coli K-12 over-producing RecA protein. Molec Gen Genet 188:37.

(7) Salles B, Paoletti C (1983). Control of uv induction of recA protein. Proc Natl Acad Sci USA 80:65

(8) Little J (1983). The SOS regulatory system: control of its state by the level of RecA protease. J Mol Biol (in press).

(9) Brent R, Ptashne M (1981). Mechanism of action of the lexA gene product. Proc Natl Acad Sci USA 78:4204

(10) Messing J, Vieira J (1982). A new pair of M13 vectors for selecting either DNA strand of double digest restriction fragments. Gene 19:269.

(11) Little JW, Edmiston SH, Pacelli LA, Mount DW (1980). Cleavage of Escherichia coli lexA protein by the recA protease. Proc Natl Acad Sci USA 77:3225.

(12) Horii T, Ogawa T, Nakatani T, Hase T, Matsubara H, Ogawa H (1981). Regulation of SOS functions: purification of E. coli LexA protein and determination of its specific site cleaved by the RecA protein. Cell 27:515.

(13) Markham BE, Little JW, Mount DW (1981). Nucleotide sequence of the lexA gene of Escherichia coli K-12. Nucl Acids Res 9:4149.

(14) Gudas LJ (176). The induction of protein X in DNA repair and cell division mutants of Escherichia coli J Mol Biol 104:567.

(15) Mount DW (1980). Genetics of protein degradation in bacteria. Ann Rev Genetics 14:279.

Gene Expression, pages 145-157
© 1983 Alan R. Liss, Inc., 150 Fifth Avenue, New York, NY 10011

GENETIC CONTROL OF GENE EXPRESSION IN S. CEREVISIAE[1]

Terrance G. Cooper, George E. Chisholm, and Francis S. Genbauffe

Department of Biological Sciences, University of Pittsburgh, Pittsburgh, PA 15260

ABSTRACT We have isolated and characterized five classes of mutations which alter expression of the genes responsible for allantoin degradation in *Saccharomyces cerevisiae*. In one class, all of the allophanate inducible genes are expressed constitutively. These same genes are rendered uninducible in the second and third classes. The remaining two groups are cis-dominant mutations which are tightly linked to the DUR1,2 gene and alter only its expression. All of the mutations alter steady state levels of allantoin system poly(A) RNAs. While analysing one of the cis-dominant mutants, we discovered a previously unreported sequence that occurs at reasonably high frequency in the yeast genome.

INTRODUCTION

The biochemistry underlying control of eucaryotic gene expression is currently a subject of intense interest. Good progress has been made

[1]This work was supported by U.S. Public Health Service Grants, GM-19386, GM-20693 and GM-24383 from the Institute of General Medical Sciences.

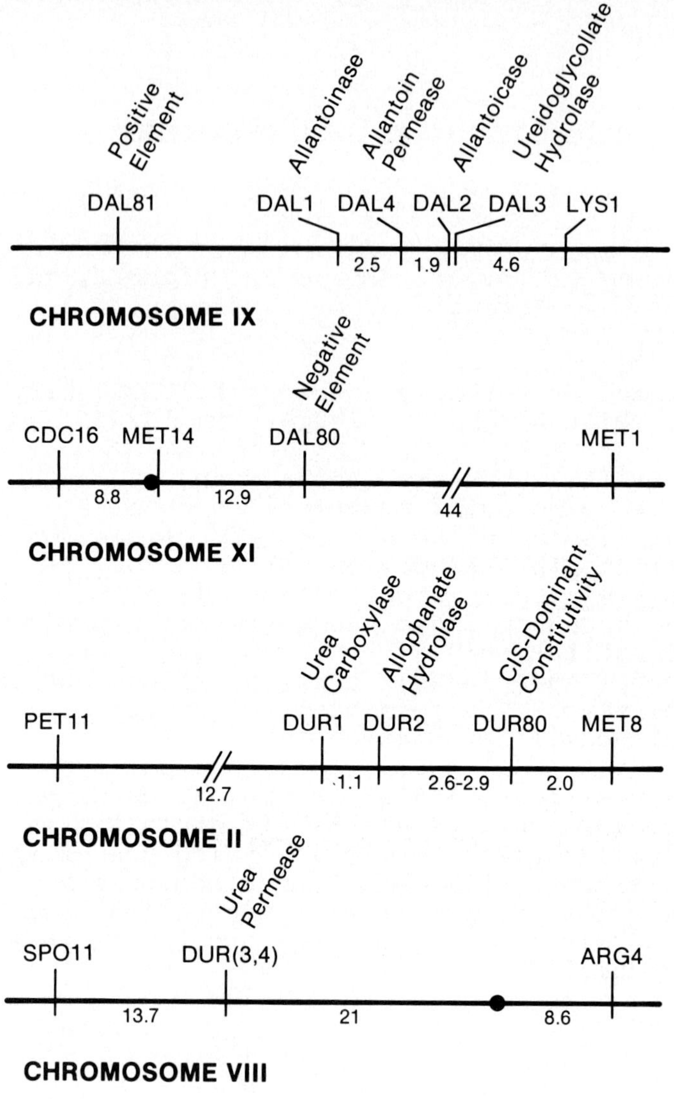

FIGURE 1. Genetic organization of the allantoin degradative system in *S. cerevisiae.*

toward identifying sequences adjacent to regulated genes that are important to their control (1-3).

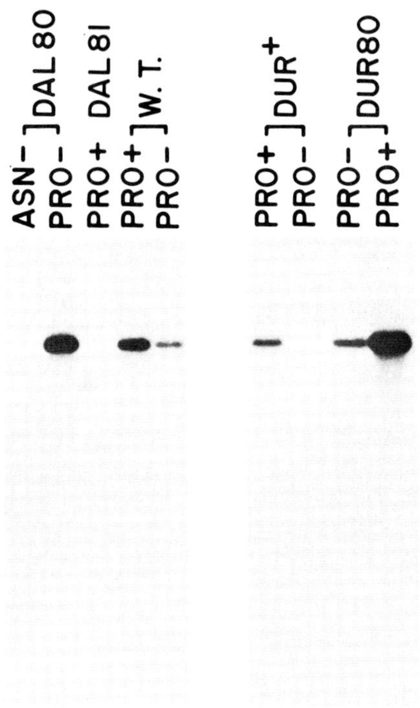

FIGURE 2. Steady state levels of DUR1,2 specific poly(A)-containing RNA in wild-type and control mutant strains of *S. cerevisiae.* + or - indicate the presence or absence of the gratuitous inducer, oxalurate, in the culture medium. Pro and Asn indicate the use of proline (non-repressive) or asparagine (repressive) as sole nitrogen source. W.T. and DUR$^+$ both indicate poly(A) RNA derived from wild-type cells.

In contrast, less is known about how these sequences participate in the regulation of gene expression or the protein elements with which they interact. We have established the allantoin degradative system in *Saccharomyces cerevisiae* as a model through which these questions may be addressed. Allantoin is a purine degradative intermediate that will serve as sole nitrogen source for this organism by virtue of being degraded to ammonia. The five degrading enzymes and at least one of the associated transport systems are inducible (see Figure 1, DAL1-3 and DUR1-3 respectively). Allophanate, the last pathway intermediate, and oxalurate (OXLU) function as native and gratuitous inducers respectively (see Figure 2, W.T. pro+ vs. W.T. pro-) (4,5).

GENETIC CONTROL OF THE INDUCTION PROCESS

The involvement of allophanate in the induction of at least five distinct enzyme activities suggested that control of the cognate genes might be regulated in parallel by common elements. This hypothesis is supported by the isolation of several classes of putative regulatory mutants which are described below.

Pleiotropic constitutive mutants

Strains with a recessive lesion in the DAL80 locus (formerly dur5) (6) were found to produce all of the allophanate-inducible gene products at high constitutive levels in the absence of inducer (6,7). Comparable constitutive enzyme levels were also observed in dur1,dal80 double mutants which are unable to synthesize the native inducer, allophanate. This and the observation that arginase activity remained at its uninduced, basal level in dal80 mutants eliminated the trivial explanation of internal induction as the basis for constitutive enzyme synthesis. The negative regulatory action of the DAL80 gene product was found to be exquisitely specific, modulating production of only those functions induced by allophanate. Other pathway activities, such as allantoate transport (DAL5) or arginase (CAR1) were shown to be immune to control by the DAL80-encoded element just as they are to the presence of allophanate.

Data, such as that shown in Figure 2, point to a strong correlation between constitutive production of enzyme activity and hybridizable mRNA. In the example shown, one can see a far greater steady-state level of DUR1,2 mRNA in a dal80 mutant grown on proline without inducer (dal80 pro-) than in wild-type cells grown even in its presence (W.T. pro+). Recessiveness of mutations in the dal80 locus argue that control is exerted by a diffusible molecule. This conclusion is also supported by the pleiotropic phenotype of dal80 mutations and the fact that the DAL80 locus, situated on the right arm of chromosome XI (Figure 1), is unlinked to any of the genes whose expression it regulates.

Pleiotropic uninducible mutants

The decision to search for constitutive mutants, such as the dal80 strains just described, was based on the quite arbitrary hypothesis that control of the pathway functions was negative. Therefore, we tested the alternative possibility of positive control by searching for mutants with a pleiotropic inability to induce the allantoin - degrading enzymes. Several mutant classes of this type were found. Strains containing mutations in a locus designated DAL81 (Figure 1), possessed the same basal levels of allantoin pathway enzymes whether or not inducer was present in the growth medium (8). Wiame and his colleagues reported isolation of a phenotypically similar mutant strain (9); however, it is not known whether the mutations harbored in these strains are allelic with dal81 mutations, because we have isolated several additional, distinct mutant classes with similar phenotypes. As a consequence of failure to induce production of the allantoin degrading enzymes, dal81 mutants grew poorly when provided with any of the allantoin-related metabolites as sole nitrogen source; they could, however, use other nitrogen sources normally. The DAL81 gene product, like that of DAL80, probably exerts its action at the level of gene expression. As shown in Figure 2, there was little if any detectable DUR1,2 mRNA in dal81 mutants grown in the presence of inducer (compare DAL81 pro+ and W.T. pro+), a result similar to those observed at the protein and enzyme activity levels (8). Inducer exclusion was eliminated as the basis for the observed phenotype by testing induction with three different sources of inducer (urea, arginine and oxalurate) which enter the cell via transport systems that are either constitutively produced or unrelated to the allantoin system. Further experiments demonstrated that the pleiotropic, positive action of this element was again exquisitely specific. A functional allele of the DAL81 locus was required for production of only those activities that respond to allophanate. Other closely related activities, such as arginase, remained unaffected by loss of the DAL81 gene function. The recessiveness of mutations in dal81 argued that its effect on control, like that of the DAL80 product, was mediated by a diffusible molecule.

These observations collectively pointed to the existence of at least two elements which appeared to specifically modulate expression of the allantoin system genes, but in an opposite manner. They prompted us to ask: (1) which elements interact with the pathway structural genes, (2) which interact with the inducer and (3) do they potentially interact with one another? Epistasis experiments with mutant alleles of the DAL80 and DAL81 genes were performed as a first attempt to gain some insight into these questions. The data obtained suggested that the two elements possessed a reasonable degree of independence (8). Unlike most of the prokaryotic control systems where epistasis is complete, here it was not. The phenotype of one mutation did not dominate over the other. Rather, the double mutant had phenotypic attributes in common with both single mutants. The level of enzyme activity observed in the dal80,dal81 double mutant was between those observed in constitutive and uninducible mutants alone, i.e. the strain was modestly constitutive. Yet it did not respond to the presence of inducer. Our current interpretation of these data is that the DAL80 gene product regulates the basal level of gene expression, perhaps as a result of interactions with a regulatory target site linked to the controlled genes. Sequence-mediated variations in the strength of these interactions could account for marked differences in observed basal levels of the various allophanate inducible enzymes. The DAL81 gene product, on the other hand, seems to be responsible for the increase in gene expression on addition of inducer. This may result from an interaction between the DAL81-gene product and allophanate.

It would be tempting to construct a molecular model of the regulatory system with the information in hand. Several observations compel us to resist for the moment. There are, for example, several additional mutant classes with phenotypic characteristics very similar to those of the dal81 mutants. However, they have not yet been characterized in detail and their epistatic relationships to dal80 mutations remain to be determined. Until the cast of players is complete, unravelling the story of their interactions with one another and the genes they control could well be difficult.

Although work on the induction process has not yet proceeded far enough to construct molecular models, the genetic data just described point with certainty to the existence of common control for the allantoin system genes. Such control predicts that homologous sequences or structures must be situated adjacent to each of the regulated stuctural genes. These sequences are expected to serve as target sites for interactions between the pathway regulatory elements and the genes they control. This hypothesis is presently being tested by comparing DNA sequences upstream from all of the allantoin system genes.

Cis-dominant constitutive mutants

Two classes of mutations which alter the level of DUR1,2 gene expression have been isolated. In both cases, the mutations have been shown to be tightly linked to the gene whose expression they regulate. The first class of mutants express allophanate hydrolase constitutively in MATa, MATα, MATa/MATa or MATα/MATα

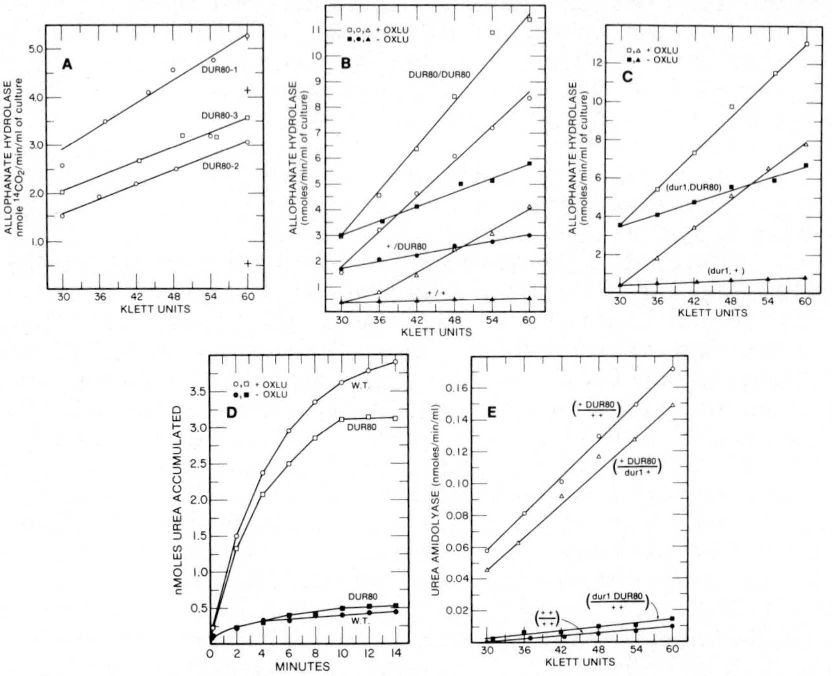

FIGURE 3. Physiological characterization of the DUR80 phenotype. Panel A. Levels of constitutivity observed for three independent DUR80 alleles. (+) signs indicate uninduced and induced levels of allophanate hydrolase activity observed in wild-type cells. Panel B. Behavior of the DUR80-1 alleles in heterozygous and homozygous condition. Panel C. Behavior of the DUR80-1 phenotype in a cell that cannot synthesize the pathway inducer due to the presence of a mutation in the DUR1,2 gene. Panel D. Assessment of the effect of the DUR80-1 allele on an unlinked gene (urea transport function, DUR3) that responds to allophanate. Panel E. Test of the cis-dominance of the DUR80-1 allele.

cells, but have essentially normal levels in MATa/MATα cells. Mutations with similar phenotypes have also been identified and characterized in several other inducible and/or repressible systems (10,11). We and others have shown that in mutants of this type, a transposable element of yeast designated Ty has inserted itself into the 5' control sequences of the affected gene. In the case of the mutant CAR1-Oh allele, we found the point of insertion to be one base 5' to the TATAT sequence upstream from the coding portion of the gene (Sumrada and Cooper, in preparation).

Three independent isolates of the second mutant class also express the DUR1,2 gene constitutively (see Figure 3A); other allophanate-inducible genes are completely unaffected (see Figure 3D). This phenotype was not affected by the mating type of the cells as mentioned above for a TY insertion mutation. In addition, the mutant locus, which has been designated DUR80, is situated 2-3 cM from the DUR1,2 gene near the MET8 locus (see Figure 1), or a distance much farther from the affected structural gene than has been previously reported for TY insertions.

Cis-dominance of the DUR80 mutations (see Figure 3B and 3E) pointed toward a role at the level of transcription. This suspicion was verified by assessing steady-state levels of DUR1,2 mRNA in wild-type and mutant cells. As shown in Figure 2, there was as much DUR1,2 mRNA in the dur80 mutant grown without inducer as there was in wild-type cells grown with it (DUR+ pro+ vs. DUR80 pro-). However, addition of inducer to a DUR80 mutant evoked a normal inductive response and, hence, a striking overproduction of DUR1,2 mRNA. DUR80 mutations did not alter the general pattern of induction, but rather increased the level of gene expression in both the presence and absence of inducer. In other words, the DUR80 phenotype is superimposed on specific control of the gene.

Given the involvement of DUR80 mutations in constitutive transcription of the DUR1,2 gene, three questions seemed appropriate: (1) were other genes situated in the vicinity of DUR80-1 and DUR1,2, (2) was the expression of such genes affected by DUR80-1, and (3) did the DUR80 mutations increase transcription in a uni- or a bi-directional manner? We addressed all three questions by cloning the entire region between the

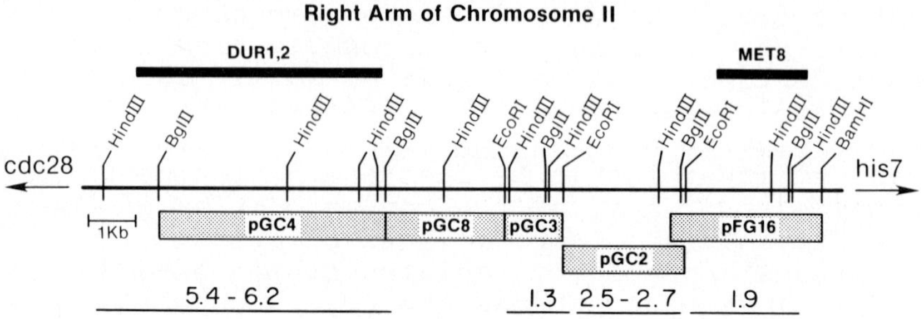

FIGURE 4. Summary of results obtained from Northern blot analysis of transcipts produced in wild-type and DUR80 strains. Diploid cultures were grown overnight in glucose-proline medium to a cell density of 50 Klett units. Total RNA was then extracted and poly(A)-enriched RNA isolated from poly(dT) cellulose columns. The RNA species were then resolved in formaldehyde-agarose gels and transferred to nitrocellulose paper. The Northern blots were hybridized with 32P-labelled YIP5 plasmids containing the indicated segments of chromosomal DNA between MET8 and DUR1,2. The sizes (in kb) of the transcripts observed are indicated in the figure. The lines cover the plasmids which hybridized to each transcript.

two loci. As summarized in Figure 4, three poly(A)-containing transcripts are encoded by the region. Interestingly, they are densely packed near the MET8 locus, resulting in a large area situated 5' to the DUR1,2 gene that is either untranscribed or transcribed below the limits of our detection. None of these transcripts, with exception of the one derived from the DUR1,2 gene, were affected by the three DUR80 mutations. This observation argues that DUR80 mutations increase gene expression in a unidirectional manner or, alternatively, are specific for the DUR1,2 gene. We subsequently observed a very small poly(A) minus transcript that hybridized to the BglII-HindIII portion of plasmid pGC8. We have not, as yet, determined whether or not the amount of this

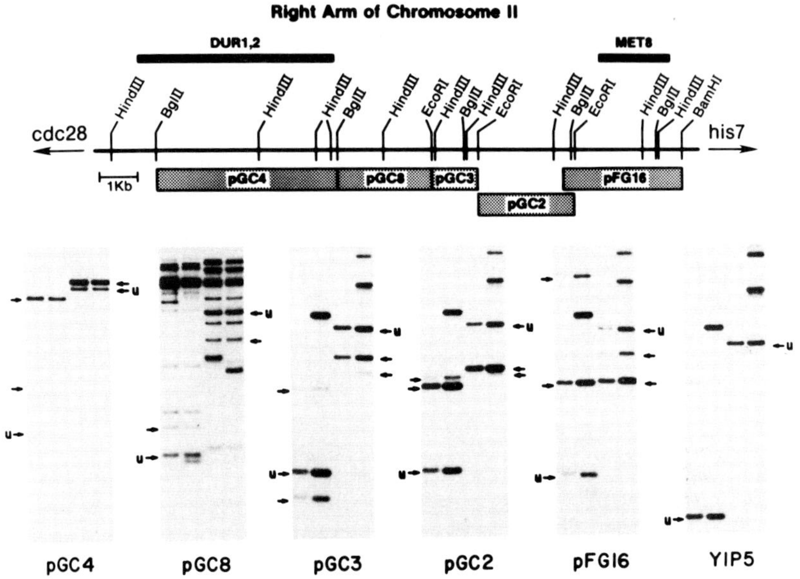

FIGURE 5. Southern blot analysis of the genomic region between MET8 and DUR1,2 in wild-type and DUR80 strains. Chromosomal DNA from both strains was digested with HindIII or BglII. The fragments were resolved in 1% agarose gels, and transferred to nitrocellulose paper. Southern blots were probed with 32P-labelled YIP5 plasmids containing the indicated segments of chromosomal DNA located between MET8 and DUR1,2. The four lanes in each set are (left to right): M1362 (wild-type) digested with HindIII; M1385 (DUR80-1) digested with HindIII; M1362 digested with BglII; and M1385 digested with BglII. A "u" indicates fragments that hybridize to the URA3 gene (a 1.1 kb fragment when DNA was digested with HindIII and a 4.7 kb species when BglII was used instead); the small arrows indicate the positions of the chromosomal fragments predicted from the restriction map of this region that is presented above the autoradiograms. The region covered by each probe is also indicated beneath the restriction map.

transcript is altered by DUR80 mutations.

In an attempt to develop a molecular explanation for the DUR80 phenotype, we used the plasmids described above to compare wild-type and mutant genomic sequences of the DUR1,2-MET8 region by Southern blotting. As shown in Figure 5, digestion of the DNA preparations with either HindIII or BglII generated the same pattern of fragments for both DNAs (arrows indicate the fragments predicted from the wild-type restriction map). Although this data is consistent with the absence of any chromosomal rearrangement, more recent cloning experiments suggest that there is, in fact, a rearrangement in the area covered by plasmid pGC8.

A second result of this experiment is seen by inspecting the genomic fragments that hybridized to plasmid pGC8. This sequence hybridized to many different wild-type genomic sequences and may be tandemly repeated in some of the larger DNA fragments (compare densities of the various bands to those indicated with arrows). We have localized the repeated sequence to the region adjacent to that coding for the small poly(A) minus RNA. There is neither sequence homology nor cross hybridization between this sequence and that of Ty elements or sigma sequences reported (12) to be associated with tRNA genes. If the poly(A) minus RNA turns out to be a tRNA, it is possible that the repeated sequence we have isolated is analogous to the sigma sequences, but distinct from them.

The mutants described in this report represent the requisite materials for a detailed study of eucaryotic gene expression by pathway specific regulatory elements. The cis-dominant mutants may provide the additonal capability of assessing perturbation of chromosome structure on gene expression and the interactions of these protein control elements with their target sequences.

REFERENCES

1. Struhl K (1982). Regulatory sites for his3 gene expression in yeast. Nature 300:284.
2. Donahue TF, Farabaugh PJ, Fink GR (1982). The nucleotide sequence of the HIS4 region of yeast. Gene 18:47.
3. Guarente L, Ptashne M (1981). Fusion of Escherichia coli lacZ to the cytochrome c gene of Saccharomyces cerevisiae. Proc Natl Acad Sci USA 78:2199.
4. Cooper TG, Lawther RP (1973). Induction of the allantoin degradative enzymes in Saccharomyces cerevisiae by the last intermediate of the pathway. Proc Natl Acad Sci USA 70:2340.
5. Sumrada R, Cooper TG (1974). Oxaluric acid: a non-metabolizable inducer of the allantoin degradative enzymes in Saccharomyces cerevisiae. J Bacteriol 117:1240.
6. Cooper TG (1980). Selective gene expression and intracellular compartmentation: two means of regulating nitrogen metabolism in yeast. Trends in Biochem Sci 5:332.
7. Chisholm G, Cooper TG (1982). Isolation and characterization of mutants that produce the allantoin-degrading enzymes constitutively in Saccharomyces cerevisiae. Mol Cell Biol 2:1088.
8. Turoscy V, Cooper TG (1982). Pleiotropic control of five eucaryotic genes by multiple regulatory elements. J Bacteriol 151:1237.
9. Lemoine Y, Dubois E, Wiame JM (1978). The regulation of urea amidolyase of Saccharomyces cerevisiae. Mol Gen Genet 166:251.
10. Errede B, Cardillo TS, Sherman F, Dubois E, Deschamps J, Wiame JM (1980). Mating signals control expression of mutations resulting from insertion of a transposable repetitive element adjacent to diverse yeast genes. Cell 25:427.
11. Roeder GS, Farabaugh PJ, Chaleff DT, Fink GR (1980). The origins of gene instability in yeast. Science 209:1375.
12. Del Rey FJ, Donahue TF, Fink GR (1982). sigma, a repetitive element found adjacent to tRNA genes of yeast. Proc Natl Acad Sci USA 79:4138.

Gene Expression, pages 159-174
© 1983 Alan R. Liss, Inc., 150 Fifth Avenue, New York, NY 10011

TRANSCRIPT ACCUMULATION IN SPORULATING YEAST[1]

David B. Kaback and Larry R. Feldberg[2]
Department of Microbiology
UMDNJ-New Jersey Medical School
Newark, NJ 07103

ABSTRACT Cultures of Baker's yeast can be induced to undergo synchronous meiosis and sporulation. We have been studying the control of gene expression in sporulating yeast cells by examining the accumulation of specific transcripts as a function of time. During the course of these studies several unexpected observations were made. The histone (H)2A and (H)2B1 transcripts appeared as predicted at their highest abundance during premeiotic DNA synthesis. However, these transcripts were also found at 1/3 to 1/2 the maximal levels at later times and in mature spores when there was little or no detectable DNA synthesis. In addition, transcripts from the GAL10 and CDC10 genes which are not essential for sporulation were found at higher abundance in sporulating cells compared to nonsporulating cells. These transcripts are not the result of a general derepression of transcript accumulation during sporulation because the transcript for at least one gene (HO) is not found in sporulating cells. In addition, all the transcripts mentioned above accumulate with different kinetics indicating their appearance is regulated in sporulating cells. The increased expression of transcripts from several genes not required for sporulation suggests that genes preferentially expressed during sporulation may not always be essential for the differentiation of a vegetative cell into an ascospore.

[1]This work was supported by a grant from the US Public Health Service (GM27712).
[2]Present address: Shulton Inc. Clifton, NJ

INTRODUCTION

Cultures of Baker's Yeast (Saccharomyces cerevisiae) that are heterozygous for the mating type (MAT a/α), undergo synchronous meiosis and spore formation (collectively called sporulation) when starved for nitrogen and supplied with a nonfermentable carbon source such as potassium acetate (for a review see ref. 1). 98% of the cells in a culture of strain SK1 (or closely related strains) are capable of sporulating with a high degree of synchrony (2, 3). Approximately 80% of the cells complete sporulation in 12 hr while the remaining cells complete it by 15-16 hrs. The majority of cells carry out premeiotic DNA replication at 2.5-5 hr. (3, D. Kaback and L. Feldberg, unpublished observation). After DNA synthesis, the first and second meiotic divisions occur. Finally, between 7 and 12 hr the spore wall forms giving mature ascospores (2).

Both DNA-RNA hybridization and two dimensional protein gel electrophoresis studies have shown that 95% of the RNA species and most of the abundant protein species synthesized during sporulation are also synthesized during mitotic growth (4-7). Only a small number of transcripts and abundant protein species appear to be found specifically during sporulation of MAT a/α cells (4-7).

Normal haploid and homozygous MAT α/α and MAT a/a diploids do not undergo premeiotic DNA synthesis, meiotic recombination or produce spores when incubated in sporulation medium (1). However, they do synthesize most but not all of the RNA and protein species found in sporulating MAT a/α and vegetative cells (4-7). Therefore, the RNA and protein species that preferentially accumulate only in sporulating MAT a/α cells have been thought to be important for the differentiation of a vegetative cell into an ascospore.

We have been studying the control of gene expression in sporulating yeast cells by examining transcript accumulation as a function of time. We have shown that several different transcripts that are thought to be required for sporulation accumulate at distinct periods during sporulation (manuscript in preparation). In this communication we report that the transcripts encoding histones (H)2A and (H)2B1 appear at their highest abundance concomitant to premeiotic DNA synthesis.

However, these transcripts were also present in high abundance at later times when there was no detectable DNA synthesis and in mature spores.

In addition, we determined that transcripts of the GAL10 and CDC10 genes, that are thought not to be required for sporulation, appear at higher relative abundance in sporulating cells compared to nonsporulating cells. These results suggest that some genes that have no obvious function during sporulation can produce higher levels of transcripts during sporulation than during vegetative growth. Accordingly, genes that appear to be preferentially expressed during sporulation may not be essential for the differentiation of a vegetative cell into an ascospore.

MATERIALS AND METHODS

Growth and sporulation of yeast.

Cells were routinely grown on 1.0% (w/v) yeast extract (Difco), 2.0% (w/v) peptone (Difco) and 2.0% (w/v) glucose (YEPD). Acetate vegetative medium was according to Roth and Halvorson (12). Sporulation was carried out as described by Petersen et al (13). Cells were harvested at the intervals noted in the text by adding 200 ug/ml cycloheximide, chilling rapidly, and centrifuging at 4000 rpm for 5 min at 2°C. The mixtures of sporulating and asporogenous cells were pools of equal volumes of cells harvested at 1 hr intervals up until 12 hr.

Preparation of RNA.

Total yeast RNA was prepared by adapting the guanidinium thiocyanate procedure (14) for use with yeast (manuscript in preparation). 5 gms of cells and 35 gms of 0.5 mm glass beads were suspended in 10 ml of 4.0 M guanidinium thiocyanate buffer (14) and homogenized for 1 min in a Braun homogenizer with CO_2 cooling. Breakage of vegetative and sporulating cells, asci and spores was routinely greater than 99%. The extract was then centrifuged at 10,000 rpm for 10 min and the supernatant treated according to the method of

Hirsh and Davidson (15) to isolate total yeast RNA.

Poly (A) containing RNA was separated by two passages over an oligo (dT) cellulose column (16).

Polysomes from spores and asci were isolated as described by Harper et al (17). Polysomal RNA was isolated by three phenol extractions and ethanol precipitation.

Analysis of RNA on Northern blots.

2.0 μg of poly (A) containing RNA was electrophoresed on a denaturing 5.0 mM CH_3HgOH, 1.0% agarose gel as described by Bailey and Davidson (18). Transfer to diazotized paper, hybridization to ^{32}P-DNA probes, washes and autoradiography were carried out as described by J. Alwine (19) in the instructions supplied by the manufacturer of the diazotyzed paper (Transa-Bind^{R} supplied by Schliecher and Schuell). Densitometric scans of the Northern blot autoradiograms were performed on a Zeineh Scanning densitometer with an automatic integrator.

TABLE 1
RECOMBINANT PLASMIDS

Plasmid	Relevant Genes	Vector	Source and Reference
TRT1	H2A1-H2A2, TRT1-1	pMB9	L. Hereford(23)
pYE98F4T	CDC10	pLC544	J. Carbon(24)
pNN 76	GAL10	pBR322	T. St. John and R. W. Davis(10)
YIp5-HO(BH2)	HO, URA3	YIp5	R. Jensen and I. Herskowitz (11)

Recombinant DNA techniques.

Plasmids described in Table 1, generously provided by the person(s) cited were amplified and isolated according to previously published methods (20).

Restriction endonuclease digestions were carried out according to the manufacturers specifications. Restriction fragments were isolated from agarose gels by the method of Tabak and Flavell (21). Labeling of DNA was done by nick translation according to published procedures (22).

RESULTS

Histone 2A and 2B transcripts are present after premeiotic DNA synthesis.

In synchronized vegetative cells synthesis of histone transcripts precedes and only continues concomitant to mitotic DNA replication (25). The appearance of histone transcripts had not previously been examined throughout all the stages of meiosis and sporulation. Northern blots of poly (A) containing RNA extracted from MATa/α , MATa and MAT α/α (not shown) cells harvested at various times from sporulation medium were hybridized to ^{32}P-labeled TRT 1 plasmid DNA. This plasmid contains the H2A1-H2B1 gene cluster and an additional nonhistone transcript, TRT1-1 (23). Due to the hybridization specificity of TRT1 we only determined the behavior of H2B1 and all the H2A transcripts (25). We did not separate the quantitation of H2A and H2B transcripts because by inspection, they appeared to be coordinately regulated. SK1 MATa/α cells harvested from stationary phase (T=0) and both 1 and 2 hr in sporulation medium contained small amounts of H2A and H2B1 transcripts. From 3-6 hrs large amounts of H2A and H2B transcripts accumulated (Fig. 1). This period is coincident with the time of maximal incorporation of precursors into DNA (3, D. Kaback and L. Feldberg, unpublished observation). Unexpectedly, the H2A and H2B1 transcripts were found at approximately one half the maximal level for more than 18 hours after the cessation of premeiotic DNA synthesis and were found to increase in relative abundance in mature asci (Fig. 1) and in ascospores (data not shown) purified by the method of Rousseau and Halvorson (26). However, the transcript level decreased during the following, 72 hr so there was very little H2A and H2B1 RNA remaining at 96 hr (not shown). Since there is little or no detectable RNA synthesis in mature ascospores (27, unpublished observation) the approximate half-life of the transcripts could be estimated from the steady state levels obtained from densitometric tracings of the autoradiographs. The half-life of the combined values for the H2A and H2B1

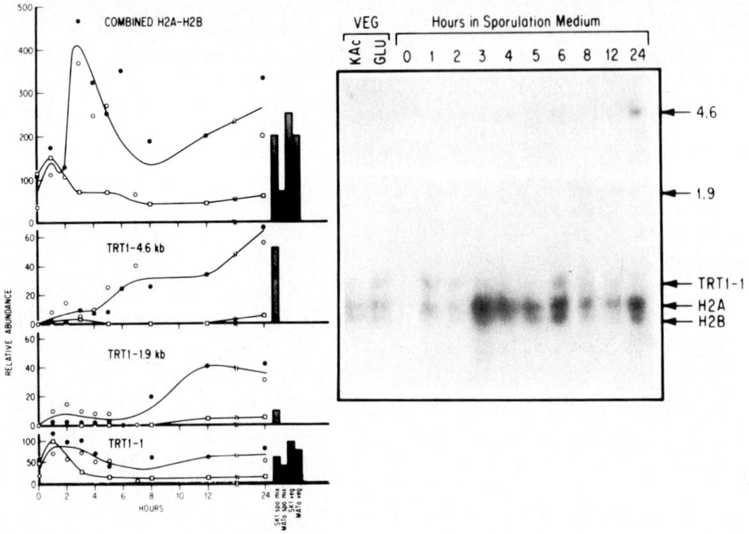

FIGURE 1. Appearance of poly (A) containing H2A, H2B1 and other transcripts complementary to TRT1 in cells incubated in vegetative and sporulation media. Northern blots and normalized densitometric tracings of time course autoradiographs are shown. o,● (2 experiments) sporulating SK1 MAT a/α ; □ asporogenous g716-5a MAT a. The bar graphs show transcript levels from mixtures of sporulating or asporogenous cells incubated in sporulation medium and from YEPD grown SK1 and g716-5a MAT a vegetative cells (see Materials and Methods). No bar indicates the transcript was not present at a measurable level.

transcripts was approximately 24 hrs (based on the average from two experiments) which is approximately 100 times longer than the 15 min half-life estimated for histone transcripts in vegetative cells (28).

Asporogenous g716-5a MATa (ho, a haploid strain closely related to SK1, courtesy of Dr. John Game; Fig. 1) and AP1 MAT α/α (a strain unrelated to SK1, courtesy of Dr. Anita Hopper, ref. 27; data not shown) cells also accumulate histone transcript but with a completely different time course. Transcripts are detected almost immediately after the cells are placed in sporulation medium. The level of histone transcript found at 1 hr in the asporogenous cells was almost equal to that found in a rapidly growing asynchronous, exponential phase, vegetative culture. This appearance of transcript is concomitant with a low level of incorporation of ^{3}H-adenine into DNA (not shown) which is probably due to replication of a small population of cells or to mitochondrial DNA synthesis. H2A and H2B1 transcript levels then decrease to approximately one-third that found in nonsynchronous populations of exponential phase vegetative cells and are detectable for the entire 26 hr period that was examined (Fig. 1). During this later period there was no detectable incorperation of ^{3}H-adenine into DNA. The prolonged presence of H2A and H2B1 transcripts in the apparent absence of DNA synthesis in both sporulating and asporogenous cells suggests histone RNA is either being synthesized or is not being degraded when cells are incubated in conditions that induce sporulation. This behavior is in contrast to vegetative cells where histone transcripts are only found in high abundance in cells that are synthesizing DNA (25). Therefore, the mechanism for regulating histone transcript appearance may be different when cells are incubated in the starvation conditions which induce sporulation.

In contrast, the adjacent TRT1-1 transcript showed different kinetics of appearance compared to H2A and H2B1. It behaved similarly in both sporulating and asporogenous cells. The apparent increased abundance of TRT1-1 RNA in sporulating cells was due to the increased level of the histone transcripts which were not completely resolvable from TRT1-1 transcripts in the autoradiogram densitometry tracings. The TRT1-1 transcript accumulated early and then decreased only slightly in abundance throughout the remaining period examined (Fig. 1). This behavior was similar for many transcripts in both sporulating and asporogenous cells (manuscript in preparation).

The CDC10 transcript is more abundant in sporulating cells than in nonsporulating cells.

It has not been previously determined if nonrequired genes are expressed during sporulation. To investigate this question we examined RNA preparations from sporulating and asporogenous cells for transcripts complementary to the cloned CDC10 gene (29).

Strains carrying a thermosensitive mutation in the CDC10 gene are unable to complete cytokinesis and bud separation at the restrictive temperature (30). In contrast to cell division, sporulation occurs at the restrictive temperature in homozygous cdc10 diploids (8). In addition, spores from cdc10/cdc10 homozygotes that have developed at the restrictive temperature are viable at the

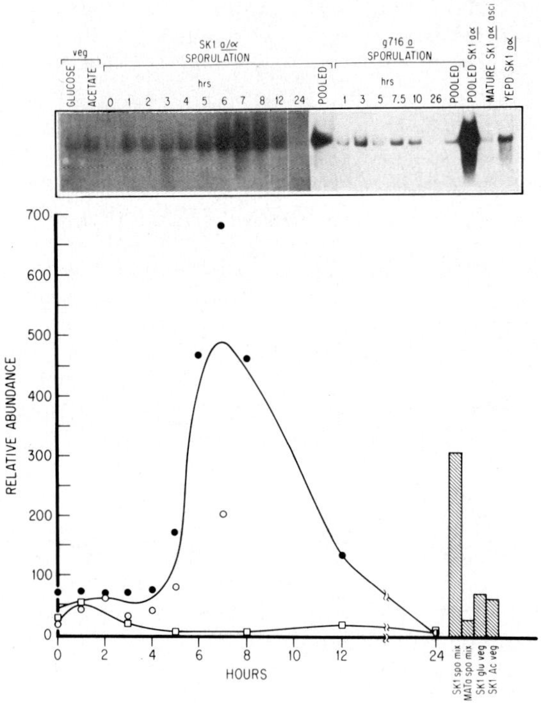

FIGURE 2. Appearance of poly (A) containing CDC 10 transcripts in cells incubated in vegetative and sporulation media. Northern blot and normalized densitometric tracings of autoradiographs using the 0.9 kb EcoR1 CDC10 DNA as the probe. o,● (2 separate experiments) sporulating SK1 MAT a/α; □ asporogenous g716-5a MAT a. The bar graphs are as explained in Figure 1 except the abundance of the transcripts in cells grown on acetate vegetative (Ac veg) and YEPD (glu veg) media were only determined for SK1.

permissive temperature (D. Kaback, unpublished observation). These results suggest the cdc10 gene product is not essential for sporulation.

Electron microscopic mapping of R loop containing DNA showed the CDC10 transcript maps directly over a 0.9 kb EcoRl fragment from pYE98F4T (manuscript in preparation). Therefore, this fragment was purified and used to probe the Northern blots described above. The CDC10 transcript was found in 6-7 fold higher relative abundance in sporulating cells than in either vegetative cells or MAT α/α and MATa cells incubated in sporulation medium (Fig. 2). Thus, the CDC10 transcript has all the qualities of a sporulation induced transcript except it appears to encode a gene not required for sporulation.

We examined the time course of the appearance of the CDC10 transcript and found it present at a low level early during sporulation. The transcript increased in relative abundance at late times (5-12 hr). Using the electron microscopic R loop quantitation (31) of the sporulation pool as a standard we estimate at 12 hr the CDC10 transcript is present at 25-50 copies per cell (manuscription in preparation). In contrast, nonsporulating MATa and MAT α/α cells incubated in sporulation medium have low levels (2-3 copies/cell) of these transcripts during the first 1-3 hr. After this time the abundance of the CDC10 transcript in the nonsporulating cells decreases even further. This behavior is typical of all the RNA species we examined in nonsporulating strains (manuscript in preparation).

The GAL10 transcript is also induced in sporulating cells.

Since homozygous galactose negative yeast are able to sporulate (D. Hawthorne, personal communication), the genes involved in galactose catabolism are also not required for sporulation. Enzymes involved in galactose catabolism are encoded by the galactose inducible GAL1,7,10 gene cluster (32). When yeast are grown on galactose, transcripts complementary to the GAL1, 7, and 10 genes accumulate to high levels. These transcripts are either not expressed in cells grown on glucose or expressed at a low or undetectable level in cells grown on acetate (9,10). We wanted to examine a transcript that is repressed during vegetative growth on both glucose and acetate to determine if it is also repressed during sporulation. GAL10 DNA (pNN 76) was used to determine if its transcript is synthesized (in the absence of exogenous galactose) during sporulation. Poly (A) containing RNA preparations from sporulating SK1 MATa/α cells, nonsporulating MATa (not shown) and MAT α/α cells incubated in sporulation medium were probed with ^{32}P-labeled pNN76 DNA as described (19). In addition, poly (A) containing RNA extracted from both glucose and acetate grown vegetative cells was also examined. Surprisingly, transcripts complementary to the GAL10 gene cluster are made in much higher relative abundance in sporulating cells than in either vegetative acetate or glucose grown cells where they were almost undetectable (Fig. 3). In addition, the GAL10 transcript was present in higher abundance and appeared with different kinetics in sporulating MATa/α cells compared with asporogenous MAT α/α and MATa (not shown) cells incubated in sporulation medium.

Therefore, the GAL10 gene is another example where a gene with no apparent sporulation related function is expressed in higher abundance during sporulation than during vegetative growth. However the presence of increased levels of the two histone transcripts, the CDC10 transcript and the GAL10 transcript does not represent a general derepression of transcript accumulation since the transcripts all appear at different times during sporulation.

The HO transcript is not detected in sporulating cells .

MATa and MATα cells that carry the HO gene switch mating types at a high frequency. Switching is repressed in normal diploid or mutant haploid cells that express both mating types (11). The HO gene has been isolated on a recombinant DNA plasmid (11). The transcript encoding the HO function is only detected in vegetative cells capable of switching and is not normally detected in diploid cells (11). Since both the GAL10 and CDC10 transcripts were unexpectedly found, we examined sporulating cells for the HO transcript. This transcript is also predicted to be repressed in sporulating cells since they do not switch (Amar Klar, personal communication). A plasmid (YIp5-HO [BH2]) containing part

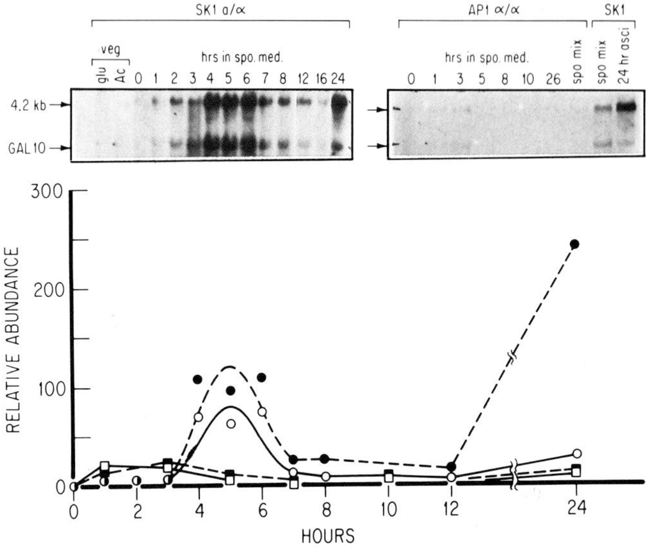

FIGURE 3. Appearance of poly (A) containing GAL 10 and 4.2 kb (see below) transcripts in cells incubated in vegetative and sporulation media. Northern blots and normalized densitometric tracings of autoradiographs using pNN76 DNA as the probe. o GAL 10, ● 4.2 kb in sporulating SK1 MATa/α cells; □ GAL 10, ■ 4.2 kb in asporogenous g716-5a MATa cells.

of the HO gene and the URA3 gene was used to probe the Northern blots used in the previous experiments. In contrast to the other genes examined, the HO (ho) transcript was only found in the asporogenous MATa and MATα/α cells incubated in sporulation medium (Fig. 4). In agreement with Jensen et al (11), the HO (ho) transcript was also detected in vegetative MATa and MAT α/α cells (not shown). In contrast, the HO transcript was not found in either poly (A) containing (Fig. 4) or total (not shown) RNA preparations from sporulating cells. The HO transcript was the first one that could not be detected in sporulating cells. Thus not all transcripts are induced by sporulation.

The URA3 transcript was found in all cells examined. The transcript accumulated with similar kinetics to the

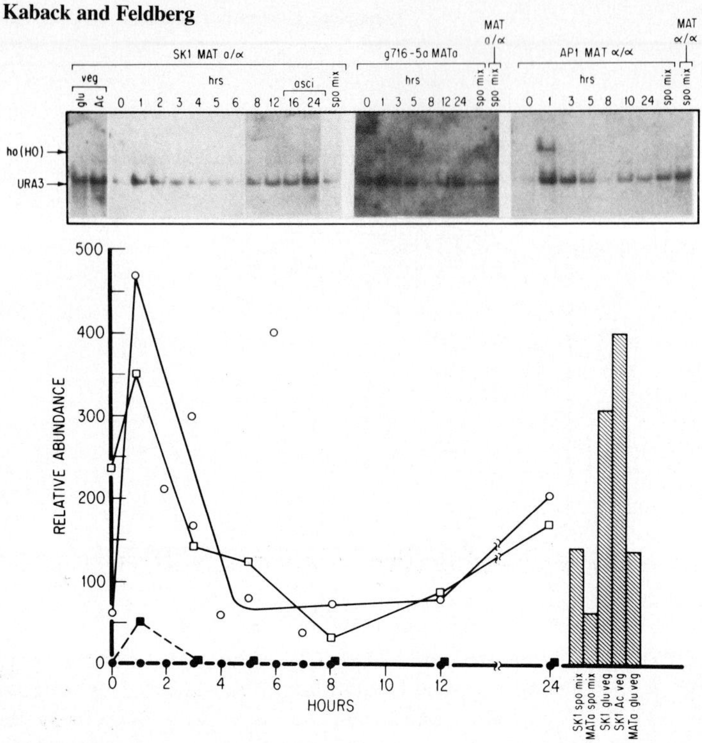

FIGURE 4. Appearance of poly (A) containing HO and URA3 transcripts in cells incubated in vegetative and sporulation media. Northern blot and normalized densitometric tracings of autoradiographs using YIp5-HO(BH2) plasmid DNA as the probe. ● HO, o URA3 in sporulating SK1 MATa/α cells; ■ HO, □ URA3 in asporogenous g716-5a MATa. The bar graphs are only for URA3 and are explained in the legends to Figures 1 and 2.

TRT1-1 transcript in MATa/α , MATa and MATα/α strains (Figure 4).

High molecular weight transcripts in sporulating cells.

In addition to the expected transcripts for the genes examined, several high molecular weight transcripts were observed. A 1.95 kb and a 4.6 kb transcript complementary to TRT1 (Fig. 1) and a 4.2 kb transcript complementary to

GAL10 were observed (Fig. 1). The transcripts were complementary to the yeast insert and not the vector sequences (data not shown). These RNA species were present in the highest abundance in sporulating MATa/α cells but were also detected in asporogenous MAT α/α and MATa cells incubated in sporulation medium. However they were not usually detected in significant quantities in vegetative cells. The transcripts also appeared with different kinetics in sporulating and asporogenous cultures. They were found to accumulate late and were in highest abundance in mature spores or asci (Fig. 1 and 3). Finally the TRT1 complementary transcripts were not detected in RNA extracted from polysomes (not shown).

DISCUSSION

We have investigated the kinetics of appearance during sporulation of transcripts from the histone (H)2A and (H)2B1 genes and several genes thought to be nonessential for sporulation. We utilized a strain of yeast that virtually sporulates to completion (2). Therefore, we assume all our observations are based on the behavior of sporulating cells. During the course of these studies several unexpected observations were made. As predicted the H2A and H2B1 transcripts appeared at their highest abundance during premeiotic DNA synthesis. However, these transcripts were also found to be relatively abundant at later times when there was little or no detectable DNA synthesis and in mature spores. In addition, the transcripts from the CDC10 and GAL10 genes, which are not required for sporulation, were found in increased abundance in sporulating cells compared to nonsporulating cells. Finally, several high molecular weight transcripts were found which were also more abundant in sporulating cells. These results suggest that there might be a general dereppression of transcription during sporulation. However, this was not the case since the transcript complementary to the HO gene which is not expressed in diploid vegetative cells (11) is also not found in sporulating cells. Indeed, all the transcripts mentioned accumulate with different kinetics indicating their appearance in sporulating cells is regulated.

It is not known if the high molecular weight transcripts have a function in sporulation. Preliminary evidence suggests the large transcripts are not translated which is consistant with them not encoding a protein with an essential role in sporulation. The large GAL10 complementary transcript may be identical to a 4.2 kb transcript seen in much lower abundance by St. John and Davis in galactose grown cells. This transcript contains both the GAL7 and GAL10 transcribed regions on a single molecule (10). In addition, large (greater than 4 kb) transcripts complementary to several other DNA sequences have been found that accumulate preferentially in sporulating cells (unpublished observation). These large transcripts which contain RNA sequences not normally detectable in vegetative cells may account for some of the additional sequence complexity seen in RNA preparations from sporulating cells.

It is possible that the CDC10 and GAL10 genes have a dispensible function for sporulation or the mutants examined provide enough activity for vegetative growth but not sporulation. However, a role for CDC10 in sporulating cells seems unlikely since CDC10 only appears to be required to complete bud separation. In contrast, sporulation can occur in the absence of budding (1). In addition, sporulation appears to be independent of other genes in the budding pathway of the cell cycle (8).

No essential role in sporulation can be assigned to the GAL10 gene which encodes UDP-galactose epimerase. Increased levels of the GAL1 gene product (galactokinase) have also been reported in sporulating cells (33). These results suggest the whole gene cluster is expressed in the absence of any exogenously added galactose. Thus, the appearance of the GAL10 transcript may be gratuitous or controlled by a different mechanism in sporulating cells (33).

In summary, the appearance of the CDC10, GAL10 and possibly the high molecular weight transcripts during sporulation suggests that some genes that have no obvious function during sporulation can produce higher levels of transcripts during sporulation than during vegetative growth. Conversely, genes that appear to be preferentially expressed during sporulation may not be essential for the differentiation of a vegetative cell into an ascospore.

ACKNOWLEDGMENTS

We are grateful to Drs. Rod Rothstein, Marjorie Brandriss, Mary Ann Osley and Jim Hopper for helpful discussions. We would also like to thank Michelle Vitale and Sharon Marotti for help in preparing the manuscript. Finally, we thank Rob Jensen and Ira Herskowitz for providing YIp5-HO(BH2) in advance of publication.

REFERENCES

1. Esposito RE and Klapholtz S (1981) (Strathern J, Jones EW and Broach J, eds) In: "The Molecular Biology of the yeast Saccharomyces: Life Cycle and Inheritance". Cold Spring Harbor, NY: Cold Spring Harbor Laboratory, p 211.
2. Kane W.S. and Roth R (1974) J Bacteriol 118:8.
3. Resnick MA, Kasimos JN, Game JC, Braun RJ and Roth RM (1980) Science 212:543.
4. Mills D (1980) Abstracts of the Annual ASM Meeting 124.
5. Trew BJ, Friesen JD and Moens PB (1979) J Bacteriol 138:60.
6. Kraig E and Haber JE (1980) J Bacteriol 144:1098.
7. Wright JF, Ajam N and Dawes IW (1981) Mol Cell Biol 1 910.
8. Simchen G (1974) Genetics 76:745.
9. St John TP and Davis RW (1979) Cell 16:443.
10. St John TP and Davis RW (1981) J Mol Biol 152:285.
11. Jensen R, Spraque GF Jr and Herskowitz I (1983) Proc Natl Acad USA 80: in press.
12. Roth B and Halvorson HO (1979) J Bacteriol 98:831.
13. Petersen JGL, Olson LW and Zickler D (1978) Carlsberg Res Commun 43:241.
14. Chirgivin JM, Przybyla AE, MacDonald RJ and Rutter WJ (1979) Biochem 18:5294.
15. Hirsh J and Davidson N (1981) Mol Cell Biol 1:475.
16. Bantle JA, Maxwell IH and Hahn WE (1976) Analyt Biochem 72:431.
17. Harper JR, Clancy M and Magee PT (1980) J Bacteriol 143:958.
18. Bailey J and Davidson N (1976) Analyt Biochem 70:75.

19. Alwine JC, Kemp DJ and Stark GR (1977) Proc Natl Acad Sci USA 74:5350.
20. Clewell DB (1972) J Bacteriol 110:667.
21. Tabak HP and Flavell RA (1978) Nucleic Acids Res 5:2321.
22. Rigby PWJ, Dieckmann M, Rhodes C and Berg P (1977) J Mol Biol 113:237.
23. Hereford L, Fahrner K, Woolford J, Rosbash M and Kaback DB (1979) Cell 18:1261.
24. Clark L and Carbon J (1980) Nature 287:504.
25. Hereford LM, Osley MA, Ludwig R and McLaughlin CS (1981) Cell 24:367.
26. Rousseau P and Halvorson HO (1969) J Bacteriol 100:1426.
27. Hopper AK, Magee PT, Welch SK, Friedman M and Hall DB (1974) J Bacteriol 119:619.
28. Osley MA and Hereford L (1981) Cell 24:377.
29. Clark L and Carbon J (1980) Proc Natl Acad Sci USA 77:2123.
30. Hartwell LH (1974) Bacteriol Rev 38:164.
31. Kaback DB, Rosbash M and Davidson N (1981) Proc Natl Acad Sci USA 78:2820.
32. Douglas HC and Hawthorne DC (1964) Genetics 49:837.
33. Ota A (1980) Microbios Letters 14:143.

Gene Expression, pages 175-185
© 1983 Alan R. Liss, Inc., 150 Fifth Avenue, New York, NY 10011

REGULATION OF NIF GENES IN KLEBSIELLA PNEUMONIAE AND RHIZOBIUM MELILOTI[1]

Venkatesan Sundaresan, Jonathan D.G. Jones[2], David W. Ow and F.M. Ausubel

Dept. of Molecular Biology, Massachusetts General Hospital and Dept. of Genetics, Harvard Medical School
Boston, MA 02114

ABSTRACT The *nif* genes of *K. pneumoniae* are under NH_4^+ regulation. Upon nitrogen starvation, the products of the *glnG* and *glnF* genes activate *nifLA*; *nifA* and *glnF* then activate the other *nif* genes. The *nif* genes of *R. meliloti* are expressed only in symbiosis with alfalfa. We present evidence that the *R. meliloti nif* genes are regulated by a *nifA* or *glnG*-like protein which is itself under symbiotic control.

INTRODUCTION

Biological nitrogen fixation involves the enzymatic reduction of N_2 to NH_4^+, and it is carried out only by prokaryotes. Within prokaryotes, the ability to fix N_2 is widespread and can be found in many widely divergent genera including the blue-greens (Anabaena), gram-positives (Clostridia) and various gram-negatives (Rhizobia, Klebsiella, Azotobacter, etc.). The nitrogen fixation (*nif*) genes encode the ability to synthesize the enzyme nitrogenase, which carries out the reduction of N_2. We have been studying the regulation of *nif* genes in two bacteria, *Klebsiella pneumoniae* and *Rhizobium meliloti*.

[1] This research was supported by N.S.F. grant PCM-8104193 awarded to F.M.A., and U.S.D.A. grant 59-2253-1-1-722-0 awarded to W. Orme-Johnson at M.I.T., with a subcontract to F.M.A.
[2] Present address: Advanced Genetic Sciences, Berkeley, California.

The free-living nitrogen-fixing bacterium K. pneumoniae reduces N_2 to NH_4^+ under conditions of NH_4^+ starvation and low O_2 tension. The reduction of N_2 is carried out by the enzyme nitrogenase which is comprised of polypeptides encoded by genes nifH, nifD, and nifK. These three genes are situated within an operon which is transcribed in the direction nifH to nifK. The nifHDK operon is itself located within a larger cluster of at least 17 contiguous nif genes which are organized into 7-8 operons. One nif operon, the nifLA operon, codes for regulatory proteins (Fig. 1; reviewed in ref. 1). The nifA product is involved in activation of all the other nif operons, while the nifL product is involved in repression of these operons under certain physiological conditions (2).

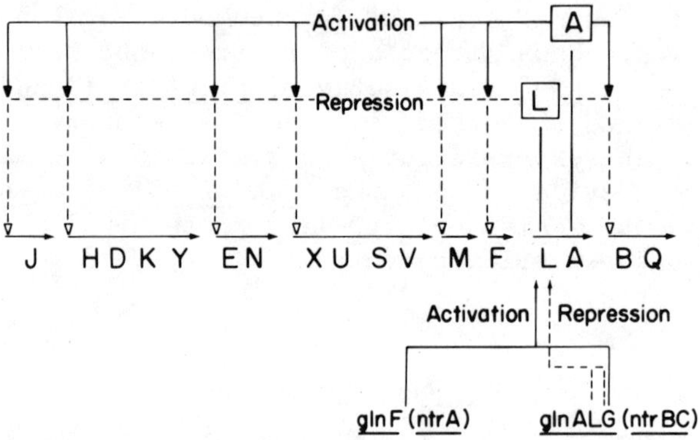

FIGURE 1. The nif genes of K. pneumoniae

Recent studies of nitrogen assimilation in enteric bacteria have shown that the process is under the control of a central regulatory system. The products of three genes, glnF (or ntrA), glnL (or ntrB), and glnG (or ntrC) have been identified as the regulatory proteins involved in this process (reviewed in ref. 3). Under conditions of nitrogen limitation, the glnG product appears to act in concert with the glnF product to activate a variety of nitrogen catabolism genes ssuch as those involved in histidine (hut) and proline (put) utilization. Under conditions of nitrogen excess, the glnG product has been postulated to act in concert with the glnL product to repress the transcription of these same genes. The nif genes of the enteric bacterium

K. pneumoniae are indirectly under the control of the gln regulatory system due to the fact that the nifLA operon is regulated by glnG and glnF. Recently, it has been shown that the nifA protein can activate the same promoters as the glnG protein, suggesting that nifA and glnG are evolutionarily related (4,5). We and others have found that glnF is required in addition to nifA, for activation of the nifHDK operon (5-7).

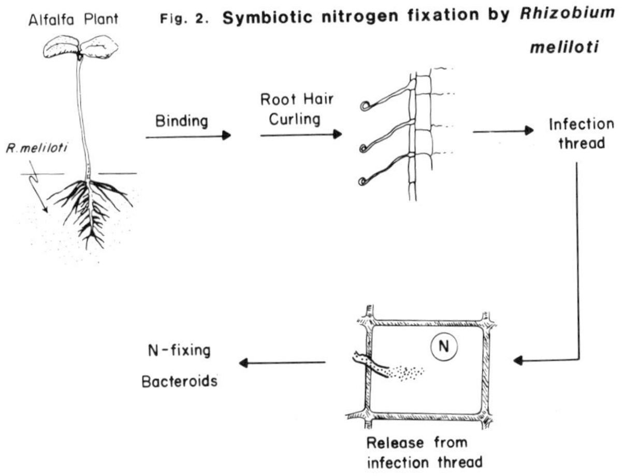

FIGURE 2. Symbiotic nitrogen fixation

In contrast to K. pneumoniae, bacteria in the genus Rhizobium normally fix nitrogen only when they interact with leguminous plants to produce on the roots a highly differentiated structure, the nodule, within which the bacteria differentiate into nitrogen-fixing bacteroids. The bacterium Rhizobium meliloti reduces nitrogen in symbiosis with alfalfa (fig. 2). Our laboratory has shown that DNA sequences in the nifHDK genes from R. meliloti are homologous to DNA sequences in the corresponding genes in K. pneumoniae (8), and that the K. pneumoniae and R. meliloti nifHDK genes are arranged similarly in an operon transcribed from nifH to nifK (fig. 3; ref. 9). We have also shown recently that the K. pneumoniae nifA product can activate the promoters for the nifHDK operons from both K. pneumoniae and R. meliloti (10), indicating that the control systems regulation nifHDK expression might also be conserved between the two genera.

We have since found that the glnG product can also activate the R. meliloti nifH promoter, although it has no effect on K. pneumoniae nifH (7). We have discussed below some recent results that may help to elucidate the mechanisms of nif regulation in Rhizobia.

MATERIALS AND METHODS

Plasmids.

We have constructed in vitro translational fusions to E. coli lacZ of the R. meliloti nifH gene (at the 29th codon) and the K. pneumoniae nifH gene (at the 1st codon). The lacZ gene was from pMC1403 (11); the structure and construction of these nifH-lacZ fusions have been described elsewhere (7, 10).

Media.

Nitrogen-free media used were as follows:

K. pneumoniae and E. coli:	NFDM medium (ref. 12)
R. meliloti:	modified LSO medium (13)

Nitrogen-excess media were the above media supplemented with 15mM NH_4^+.

Derepression studies were carried out by growing cells to saturation in nitrogen-excess medium, washing in 10mM $MgSO_4$ and resuspending in nitrogen-free or nitrogen-excess (control) media (ref. 12).

RESULTS AND DISCUSSION

The experiments described below utilize lac fusions to the nitrogenase promoters from K. pneumoniae and R. meliloti (see Material and Methods) to study the regulation of the nif genes in these species.

Regulation in K. pneumoniae.

Table 1 shows the effect of different gln (ntr) mutations on activation of the K. pneumoniae nifH promoter by the K. pneumoniae nifA product (fig. 1). The K. pneumoniae nifH-lacZ fusion was carried on a pBR322 vector. A compatible phage P4 vector by itself (control) or carrying the nifA gene under a constitutive promoter (7, 10) was then introduced, and the β-galactosidase activity was measured in nitrogen-free and nitrogen-excess conditions.

TABLE 1
ACTIVATION OF K. PNEUMONIAE NIFH-LACZ

Genotype of E. coli host	Regulatory protein supplied on multicopy plasmid	β-galactosidase activity $-NH_4^+$	$+NH_4^+$
1. gln^+	-nifA	-	-
	+nifA	++	++
2. ΔglnALG	-nifA	-	-
	+nifA	++	++
3. ΔglnF	-nifA	-	-
	+nifA	-	-
4. ΔglnALG	-glnG	-	-
	+glnG	-	-

We see that activation of the K. pneumoniae nifH promoter by nifA requires the glnF product, and the model shown in fig. 1 has to be modified accordingly, i.e., glnF not only acts at the nifLA promoter but at the nifHDK promoter, and presumably the other nif promoters as well. This result is consistent with the demonstrated similarities between glnG and nifA (4,5); nifA, like glnG, requires glnF for activation. However, only nifA and not glnG can activate K. pneumoniae nifH (Table 1, expt. 4). Since nitrogenase is O_2 sensitive, and since nitrogen fixation is energy intensive (requiring 16 or more ATPs/molecule of N_2 reduced), it is likely that evolutionary selection for specific control by nifA has occurred. The regulatory system involving nifA and nifL (which represses nifA activated promoters under conditions of high pO_2 or low to medium NH_4^+, ref. 2) allows K. pneumoniae to derepress its overall nitrogen assimilation pathways (involving gln, hut, etc.) while keeping the nif genes repressed until necessary.

Regulation in R. meliloti.

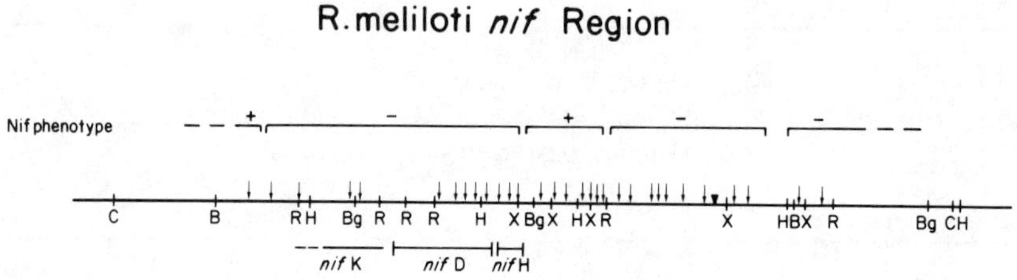

(Arrows indicate transposon Tn5 insertions

FIGURE 3. The nif genes of R. meliloti (refs. 9, 14)

In a recent study, we sequenced the R. meliloti and K. pneumoniae nifH promoters and found that they exhibited strong homologies, particularly at -32 and at -12; the significance of these homologies was demonstrated by showing that the K. pneumoniae nifA gene could activate the R. meliloti nifH promoter (10). This led us to propose that a R. meliloti "nifA" gene was responsible for the activation of the R. meliloti nitrogenase genes in the alfalfa nodule; the R. meliloti "nifA" itself would be under some form of symbiotic control.

Table 2 summarizes these and more recent results (7, 10).

TABLE 2
ACTIVATION OF R. MELILOTI NIFH-LACZ

Genotype of E. coli host	Regulatory protein supplied on multicopy plasmid	β-galactosidase activity $-NH_4^+$	$+NH_4^+$
1. gln^+	-nifA	+	-
	+nifA	++	++
2. ΔglnALG	-nifA	-	-
	+nifA	++	++
3. ΔglnF	-nifA	-	-
	+nifA	-	-
4. ΔglnALG	-glnG	-	-
	+glnG	++	++

Thus activation of R. meliloti by nifA also requires the glnF product as expected. However, we find that R. meliloti nifH, unlike K. pneumoniae nifH, can be activated by glnG (expt. 4). This accounts for the observed activation of R. meliloti nifH-lacZ under nitrogen-starvation in gln^+ E. coli (expt. 1, -nifA).

We then asked whether this derepression under NH_4^+ starvation occurred in free-living R. meliloti. The nifH-lacZ fusion plasmids were recloned into pRK290 (15) and conjugated into a R. meliloti lac^- strain. There was no activation of K. pneumoniae nifH-lacZ. However, we observed a 12-15 fold activation of R. meliloti nifH-lacZ upon NH_4^+ starvation in aerobic conditions, and a lower (4-5 fold) activation under anaerobic conditions. This is in contrast to K. pneumoniae, where activation of nifH upon nitrogen starvation takes place only under anaerobic conditions.

There was no detectable nitrogenase activity in these cultures by the acetylene reduction assay (although this might not be expected in the aerobic cultures since nitrogenase is oxygen sensitive). However, the derepression of R. meliloti nifH-lacZ in free-living R. meliloti by NH_4^+ starvation suggests that the R. meliloti glnG gene can activate its own nifHDK operon, at least to a limited extent.

These results raised the following question: Is the R. meliloti glnG gene responsible for the activation of the nif genes inside the nodule? That is, are the nif genes in R. meliloti (and perhaps in other symbiotic nitrogen-fixing organisms) under the direct control of the glnG gene, rather than under a nifA-mediated control as in K. pneumoniae? It is possible that the evolutionary selection for a nifA-type regulatory system might not be present in symbiotic nitrogen fixing systems, since both energy source and low pO_2 environment are provided by the plant. There has been previous evidence for the involvement of gln regulatory genes in regulation of nif genes in R. "cowpea" sp. 32H1 (18).

We decided to investigate the regulation of nif within the nodule using our nif-lac fusions. Alfalfa plants were inoculated with $lacZ^-$ R. meliloti harboring a plasmid that carried lacZ (control), or K. pneumoniae nifH-lacZ, or R. meliloti nifH-lacZ, and nodule extracts were assayed for β-galactosidase acitvity. We could observe strong activation of R. meliloti nifH-lacZ (50-fold over background) but no activation of K. pneumoniae nifH-lacZ could be detected. Since K. pneumonaie nifH is not activatd by glnG (Table 1, expt. 4), this result agrees with the idea that the R. meliloti regulatory protein is glnG.

Recent work in our lab by W. Szeto and L. Zimmerman (19), has led us to modify this model. They have identified a putative regulatory gene in R. meliloti, about 6 kb distal to the nifHDK operon. Tn5 insertions into this gene (rightmost insertions in fig. 3) knock out synthesis of nifHDK proteins by the bacteroids in the nodule, causing a Fix^- phenotype. This result does not support the idea that R. meliloti glnG activates nifHDK in symbiotic nitrogen fixation for the following reason: The nif genes in R. meliloti are carried on a large plasmid (16, 17), and we would expect a chromosomal location for glnG, while this regulatory gene is closely linked to nifHDK. Thus either (1) the R. meliloti glnG gene is plasmid-borne, or (2) there is a separate regulatory gene linked to nif and analogous to K. pneumoniae nifA in this respect, but functionally more like K. pneumoniae glnG since it activates R. meliloti nifH, but not K. pneumoniae nifH.

An R. meliloti Fix$^-$ strain, Rm1354, carrying a Tn5 insertion in the putative regulatory gene, was further characterized as follows. The R. meliloti nifH-lacZ fusion was conjugated into Rm1354, the exconjugants were used to nodulate alfalfa seedlings, and the nodules formed were assayed for β-galactosidase activity. There was no detectable activation of R. meliloti nifH-lacZ by Rm1354 (β-galactosidase activity remained at background levels), whereas there was strong activation of R. meliloti nifH-lacZ by a Fix$^-$ R. meliloti strain carrying a Tn5 insertion in nifH (Rm1491). Thus it seems likely that the gene inactivated in Rm1354 is a true regulatory gene and not, for example, a gene required for stability of the nitrogenase proteins. Studies of derepression of R. meliloti nifH-lacZ in free-living Rm1354 vs. wild-type R. meliloti showed identical activation of nifH in the two strains as the NH_4^+ concentration was decreased. This suggests that the insertion in Rm1354 has not affected the R. meliloti glnG gene.

The above observations can be reconciled by the following model: The nif genes in R. meliloti are under the control of a regulatory gene which is linked to the nif cluster. Further, this gene is related both to glnG and to nifA. It differs from K. pneumoniae nifA in that it is not regulated by NH_4^+ but rather by some mechanism that causes its activation in the nodule (which we have tentatively called "symbiotic control"). The absence of K. pneumoniae nifH-lacZ activation in the nodule and the other evidence discussed earlier suggests that it more closely resembles glnG. This has received some corroboration from DNA hybridization studies performed by Szeto and Zimmerman (19). They find that the regulatory gene they have identified is homologous to both E. coli glnG and K. pneumoniae nifA, but shows much better homology to E. coli glnG. For reasons discussed above, the evolutionary selection favoring divergence of nifA and glnG in K. pneumoniae may not be operative in the symbiotic nitrogen fixing R. meliloti so that the regulatory protein in the latter retains a glnG-like specificity. In this model, R. meliloti has in addition, a chromosomal glnG gene which regulates the cellular nitrogen assimilation pathways. This gene can activate the nifHDK operon under NH_4^+ starvation, but this activation is comparatively low and does not result in significant nitrogenase activity.

Structure of *nif* Promoters.

We have sequenced the *K. pneumoniae nifHDK*, *K. pneumoniae nifLA* and *R. meliloti nifHDK* promoters and have found sequence homologies that might be recognition sequences for regulatory proteins (10, 20). All three promoters can be activated by *nifA* but only the last two are activated by *glnG* (4, 7).

	-30 . . . -10	Activation *glnG*	Activation *nifA*
K. pneumoniae nifH	...ACGGCTGG...7bp...CCTGCAC...	-	+
R. meliloti nifH	...ACGGCTGG...7bp...TTTGCAC...	+	+
K. pneumoniae nifL	...GATAAGGG...7bp...TTTGCAT...	+	+

All three promoters have the sequence TGCA in the -10 region. The two *glnG* activated promoters share a longer sequence homology (TTTGCA) in this region, while *K. pneumoniae nifH*, which is not activated by *glnG*, has the sequence CCTGCA. Thus the sequence TTTGCA might be important for activation by *glnG*. The sequence TGCA appears to be involved in *nifA* recognition, since we have evidence that base-substitutions in this region abolish *nifA* binding (21). Because these promoters also require *glnF* product for activation, the *glnF* product may also be involved in sequence recognition. These putative recognition sequences lie in a region (-10) known to be involved in binding RNA polymerase in other procaryotic promoters. We note that none of the three promoters above shows good homology to the TATAAT consensus sequence at -10. The significance of the homologies at -22 to -30 is not known. Recently, Beynon *et al.* have found that the sequence TGCA in the -10 region and a sequence CTGG 9bp upstream are conserved in all the *K. pneumoniae nif* promoters that are regulated by *nifLA* (22). Thus sequences in the -26 region may also be involved in *nifA*-mediated activation.

REFERENCES

1. Ausubel FM, Brown SE, de Bruijn FJ, Ow DW, Riedel GE, Ruvkun GB, Sundaresan V (1982). In Setlow JK, Hollaender A (eds): "Genetic Engineering: Principles and Methods," New York: Plenum, p 169.
2. Merrick M, Hill S, Hennecke H, Hahn M, Dixon R, Kennedy C (1982). Molec gen Genet 185:75.
3. Magasanik B (1982). Ann Rev Genet 16:135.
4. Ow DW, Ausubel FM (1983). Nature 301:307.
5. Merrick M (1983). EMBO Journal 2:39.
6. Sibold L, Elmerich C (1982). EMBO Journal 1:1551.
7. Sundaresan V, Ow DW, Ausubel FM (1983). Proc Natl Acad Sci USA, in press.
8. Ruvkun GB, Ausubel FM (1980). Proc Natl Acad Sci USA 77:191.
9. Ruvkun GB, Sundaresan V, Ausubel FM (1982). Cell 29:551.
10. Sundaresan V, Jones JDG, Ow DW and Ausubel FM (1983). Nature 301:728.
11. Casadaban M, Chou J, Cohen S (1980). J Bacteriol 143:971.
12. Riedel GE, Brown SE, Ausubel FM (1982). J Bacteriol 153:45.
13. Elmerich C, Dreyfus BL, Reysset G, Aubert J-P (1982). EMBO Journal 1:499.
14. Buikema WB, Long SR, Brown SE, van den Bos RC, Earl CD, Ausubel FM (1983). J Molec Appl Genet, in press.
15. Ditta G, Stanfield S, Corbin D, Helinski D (1980). Proc Natl Acad Sci USA 77:7347.
16. Banfalvi Z, Sakanyan V, Koncz C, Kiss A, Dusha I, Kondorosi A (1981). Mol gen Genet 184:318.
17. Rosenberg C, Boistard P, Denarie J, Casse-Delbart F (1981). Mol gen Genet 184:326.
18. Ludwig RA (1980). Proc Natl Acad Sci USA 77:5817.
19. Szeto WW, Zimmerman JL (1983). Manuscript submitted.
20. Ow DW, Sundaresan V, Rothstein D, Brown SE, Ausubel FM (1983). Proc Natl Acad Sci USA, in press.
21. Brown SB, Ausubel FM (1983). Manuscript submitted.
22. Beynon J, Cannon MC, Buchanon-Wollaston V, Cannon FC (1983). Manuscript submitted.

Gene Expression, pages 187-206
© 1983 Alan R. Liss, Inc., 150 Fifth Avenue, New York, NY 10011

CLONED CYTOCHROME P-450 GENES REGULATED BY THE Ah RECEPTOR

Daniel W. Nebert, Robert H. Tukey,
Howard J. Eisen, and Masahiko Negishi

Laboratory of Developmental Pharmacology
National Institute of Child Health and Human Development
National Institutes of Health, Bethesda, Maryland 20205

ABSTRACT The Ah locus is an interesting model system for studying gene expression during the induction of drug-metabolizing enzymes by chemical carcinogens and other environmental pollutants. Cytochromes P1-450, P2-450, and P3-450 are controlled by the Ah receptor. "P1-450" is defined as that form of polycyclic-aromatic-induced enzyme most closely associated with polycyclic-aromatic-inducible aryl hydrocarbon hydroxylase (AHH) activity. "P2-450" is defined as that form of isosafrole-induced enzyme which metabolizes isosafrole best. "P3-450" is defined as that form of polycyclic-aromatic-induced enzyme most closely associated with acetanilide 4-hydroxylase activity. P1-450 is much more readily inducible in C57BL/6 (B6, Ah^b/Ah^b) than in DBA/2 (D2, Ah^d/Ah^d) inbred mice, due to a cytosolic Ah receptor defect in D2 mice. Using this B6-D2 difference, we have isolated and characterized clone 46 (1100 bp) and p57 (1900 bp), two cDNA clones for P1-450. In studies of AHH inducibility by benzo[a]-anthracene among 29 mouse-hamster fused hybrids, the Ah gene encoding the receptor has been assigned to mouse chromosome 17. From clone 46 we have isolated λAhP-1, a recombinant phage containing the entire P1-450 genomic gene (~4600 bp; at least five exons) in the middle of its 15.5-kbp insert. P1-450 (23 S) mRNA and P2-450

(20 S) mRNA hybridize to a common 5' fragment of the P1-450 gene. A third cDNA clone, p21, hybridizes to P3-450 (20 S) mRNA and cross-hybridizes to the 5' fragment of the P1-450 gene but is clearly associated with a different genomic gene which has been isolated and characterized by R-loop analysis. Whether p21 represents a cDNA clone for P2-450 or P3-450 must await protein and nucleotide sequence analysis. These data are consistent with multiple P-450 genes under Ah receptor control.

INTRODUCTION

We are living in a sea of foreign chemicals, most likely numbering in the hundreds of thousands. An estimated 70,000 agricultural chemicals--insecticides, pesticides, and herbicides--are dumped on United States farm land each year. Between 1,000 and 3,000 new chemicals are synthesized each year. Vertebrates, and especially the human, are at the end of the food chain. It is estimated that we consume and inhale more than 1,000--perhaps more than 10,000--foreign chemicals each day. For example, red wine alone has more than 400 polycyclic compounds (many containing oxygen in the ring) the chemical structures of which have been characterized. Spices, cosmetics, and cigarettes are other examples of agents having hundreds or thousands of distinctly different foreign chemicals.

How do organisms recognize and detoxify these foreign substances? Understanding this recognition-and-detoxication process would have widespread applications--from the design of effective new insecticides to human survival in a world filled with increasing concentrations of noxious agents. Moreover, an increasing number of these foreign chemicals are being implicated in mutagenesis, carcinogenesis, and teratogenesis.

Many of these thousands of noxious agents are very fat-soluble and would have a half-life in the organism of weeks or months, were it not for Phase I and Phase II drug-metabolizing enzymes. During Phase I metabolism, polar groups (such as alcohols) are introduced into the parent molecule, thereby presenting the Phase II enzymes with a more polar substrate. The Phase II enzymes use the polar group as a "handle" for attaching other very water-soluble moieties such as glucuronide, sulfate, or glycine. Certain Phase I products (e.g. alcohols, quinones), and especially Phase II conjugates, are sufficiently polar to be easily excreted by the cell. Living organisms often respond to foreign chemical adversity by the stimulation (or induction) of both Phase I and Phase II enzymes.

Cytochrome P-450 represents the largest class of Phase I enzymes. P-450 proteins[1] are polypeptides ranging in molecular weight between 45,000 and 60,000; they are hemoproteins (iron is present in a porphyrin ring near the enzyme active-site) and some appear to contain carbohydrate as well. This enzyme system is ubiquitous, existing in certain bacteria (Pseudomonads) and presumably all eukaryotes. Except in bacteria and certain fungi, the P-450 system is membrane-bound--principally in the smooth endoplasmic reticulum and mitochondria--and has multiple components. The cofactors NADPH and/or NADH supply electrons, which are passed by way of one or more membrane-bound flavoprotein reductases, ultimately to the P-450 enzymes. In some instances, an iron-sulfur protein, cytochrome b5, or a P-450 molecule all may be involved in transferring electrons ultimately to the P-450 protein having enzymatic activity. The substrate is oxygenated, meaning that one atom of atmospheric oxygen is incorporated into the metabolite, the other oxygen atom terminating in cellular water. The catalytic activity of the P-450 system can be a monooxygenase activity, an oxidase, a peroxidase, or a reductase, depending on the substrate and the valence state of P-450 iron.

---

[1]Cytochrome P-450 is defined as all forms of CO-binding hemoproteins having NADPH- and sometimes NADH-dependent monooxygenase activities. "P1-450" is defined as that form of polycyclic aromatic-inducible P-450 most closely associated with polycyclic aromatic-inducible aryl hydrocarbon hydroxylase activity. "P2-450" is defined as that form of isosafrole-induced enzyme which metabolizes isosafrole best. "P3-450" is defined as that form of polycyclic-aromatic-induced enzyme most closely associated with acetanilide 4-hydroxylase activity. Other abbreviations used: TCDD, 2,3,7,8-tetrachlorodibenzo-p-dioxin; AHH, aryl hydrocarbon (benzo[a]pyrene) hydroxylase (EC 1.14.14.1); B6, the C57BL/6N inbred mouse strain; and D2, the DBA/2N inbred mouse strain.

The P-450 proteins metabolize virtually everything in the Merck Index. The enzymes exhibit remarkable overlapping substrate specificities. This fact reflects the difficulty in attempting to determine the absolute number of P-450 proteins. The term "multisubstrate monooxygenases" has been suggested for describing the catalytic activities of the large P-450 multigene family (reviewed in Refs. 1-4). Other suggested names have included "miraculase" and "everything hydroxylase."

The absolute number of distinctly different P-450 isozymes is not known. The exact number of P-450 proteins biochemically proven to be different in any one tissue of a particular species is as many as 18 (Ryo Sato, personal communication), although it has been postulated (5) that most eukaryotes have the genetic capacity to synthesize hundreds or thousands of forms, depending on the stimulus (inducing chemical).

The Ah locus (Fig. 1) controls the induction of a small portion of these forms of P-450 (and their corresponding enzyme activities) by polycyclic aromatic compounds such as TCDD and 3-methylcholanthrene; P_1-450 is one of these forms. The induction process is mediated by the cytosolic Ah receptor (reviewed in Ref. 6). Numerous relatively planar foreign chemicals bind avidly (apparent $K_d \simeq 1$ nM) to the Ah receptor in direct proportion to their potency as inducers of P_1-450. No endogenous ligand for the Ah receptor has been found. Various assays for quantitating the Ah receptor include the use of [^{3}H-1,6]TCDD ($\geq$60 Ci/mmol) in combination with dextran-coated charcoal adsorption, trypsin treatment followed by isoelectric focusing, sucrose density gradient centrifugation after dextran-charcoal adsorption, a detergent-washing procedure with purified nuclei, and gel permeation and anion-exchange chromatography.

The B6 inbred mouse strain was found to express liver microsomal AHH induction following intraperitoneal 3-methylcholanthrene treatment, whereas the D2 inbred mouse strain does not express this induction process in liver (3, 6). The genetic trait was defined as aromatic hydrocarbon responsiveness, with B6 having Ah^b/Ah^b alleles and D2 having Ah^d/Ah^d alleles. The chromosomal location of the Ah locus recently has been found on the distal half of mouse chromosome 17, whereas the P_1-450 gene is located on the X chromosome. AHH (or P_1-450) induction is expressed as an autosomal dominant trait (Figs. 2, 3 and 4). When

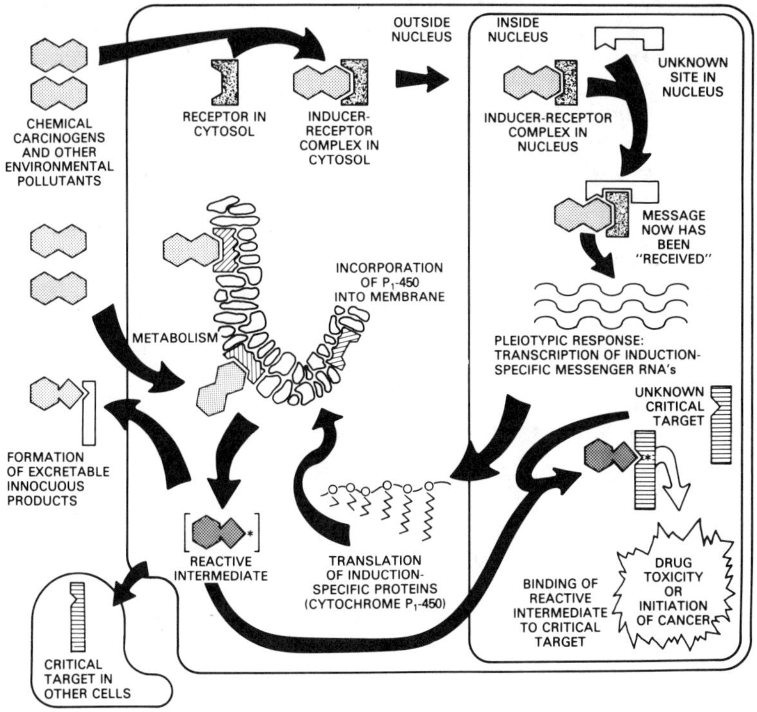

FIGURE 1. Diagram of a cell and the hypothetical scheme by which a cytosolic receptor, a product of the regulatory Ah gene, binds to inducer (5). The resultant pleiotypic response includes greater amounts of cytochrome P1-450 (and at least two other forms of P-450 still being characterized), leading to enhanced steady-state levels of reactive intermediates, which are associated with genetic increases in birth defects, drug toxicity, or chemical carcinogenesis. Depending upon the half-life of the reactive intermediate, important covalent binding may occur in the same cell in which metabolism took place, or in nearby cells. Although the "unknown critical target" is illustrated here in the nucleus, there is presently no experimental evidence demonstrating unequivocally the subcellular location of such a target or, for that matter, whether the target is nucleic acid or protein [Reproduced with permission from Dr. W. Junk Publishers].

B6 mice are crossed with D2 mice, the resulting 3-methylcholanthrene-treated heterozygote has inducible AHH (P_1-450) levels similar to those of the B6 parent. Whereas the cytosolic *Ah* receptor is read-ily observed in B6 mice, the *Ah* receptor can be detected in the nucleus but not in the cytosol of D2 mice. Of interest, the Ah^b/Ah^d heterozygote exhibits levels of receptor that are midway between those found in the B6 and D2 parents (reviewed in Ref. 6).

P_1-450 is one of many gene products controlled by the *Ah* receptor (3, 6). The definitions of three P-450 proteins regulated by the *Ah* receptor are given in Table 1.

A relatively specific antibody to P_1-450 ($M_r \cong 55{,}000$) was developed (7) and used to size P_1-450 mRNA (8). In the remainder of this report we demonstrate primarily the characterization of a P_1-450 cDNA clone and its corresponding genomic P_1-450 clone, λAhP-1.

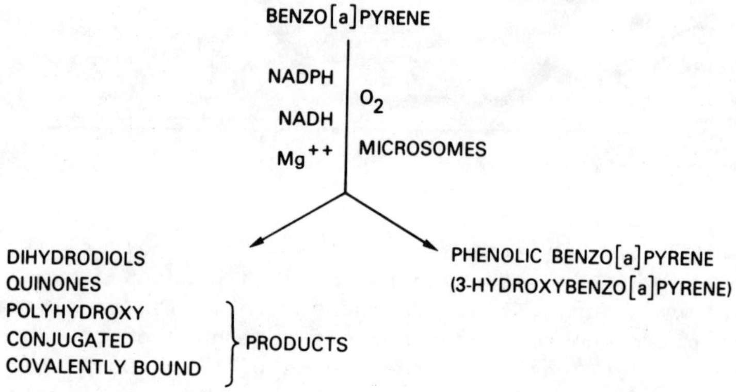

FIGURE 2. Simplified scheme for measuring "AHH activity" with benzo[a]pyrene as substrate. The AHH assay represents P_1-450. The phenolic benzo[a]pyrene metabolites are determined spectrophotofluorometrically and equated with AHH activity. This assay does not detect other oxygenative metabolism occurring concomitantly (shown at *lower left*).

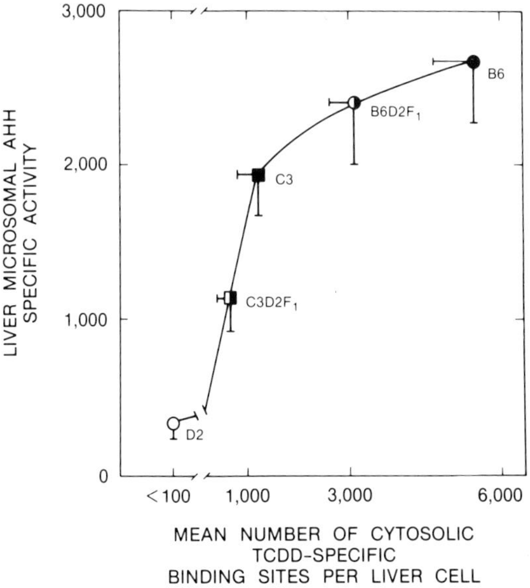

FIGURE 3. Maximal hepatic AHH inducibility as a function of number of Ah receptor molecules per liver cell among various inbred strains of mice and F1 hybrids (6). Each value is the mean of five or more individual determinations. C3, the C3H/HeN inbred mouse strain. Brackets in both directions denote standard deviations [Reproduced with permission from Academic Press, Inc.].

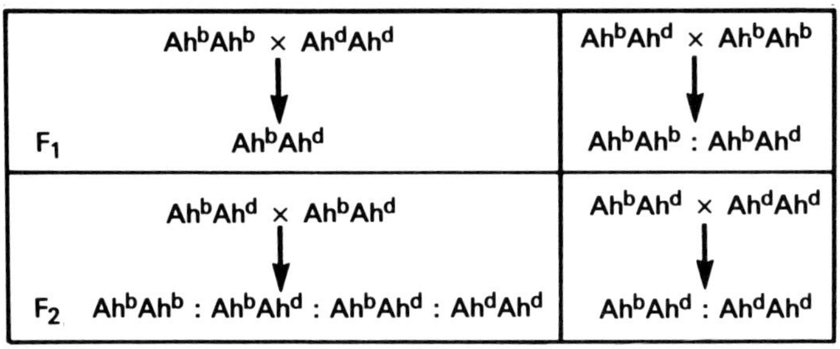

FIGURE 4. Simplified genetic scheme for aromatic hydrocarbon "responsiveness" in the mouse. Ah^b/Ah^b denotes the high-affinity receptor; Ah^d/Ah^d, the poor-affinity receptor. The Ah^b/Ah^b and Ah^b/Ah^d mouse both express similar AHH (P1-450) inducibility.

TABLE 1

Designated name of P-450 protein	One of best substrates in reconstituted enzyme	Antibody to antigen inhibits most specifically
P_1-450	Benzo[a]pyrene	Benzo[a]pyrene
P_2-450	Isosafrole	Isosafrole
P_3-450	Acetanilide	Acetanilide

RESULTS AND DISCUSSION

Colony Hybridization

Fig. 5 shows colony hybridization for the P_1-450 cDNA. From 3-methylcholanthrene-treated B6 mouse liver 23 S mRNA, double-stranded cDNA was prepared and inserted into the PstI site of pBR322. Plasmid DNA from ampicillin-sensitive tetracycline-resistant E.coli was fixed on nitrocellulose paper. The probe in this case was labeled cDNA freshly reverse-transcribed from 23 S mRNA from 3-methylcholanthrene-treated B6 or D2 mouse liver. Because there is at least 40 times more P_1-450 mRNA in 3-methylcholanthrene-treated B6 than D2 mice (8), we looked for a colony that was positive in B6 and negative in D2. Two clones, numbers 46 and 68, were found in the first 72 colonies screened (9). Because 46 contained a larger insert than 68, we chose to study clone 46 further.

Immunologic Criteria for cDNA Clone

"Positive" and "negative" translation arrest experiments provide classical immunologic proof that the cDNA clone isolated is associated with a particular antibody. The antibody precipitates the protein that has been translated from the mRNA, and this mRNA hybridizes most specifically to the cloned cDNA. There appears to be a problem with hybridization arrest experiments and P-450

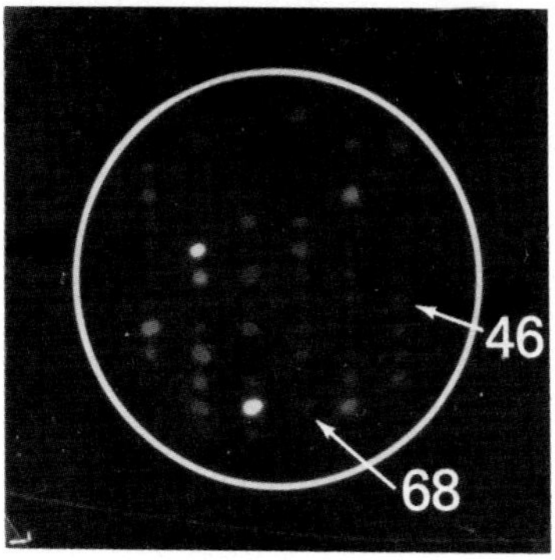

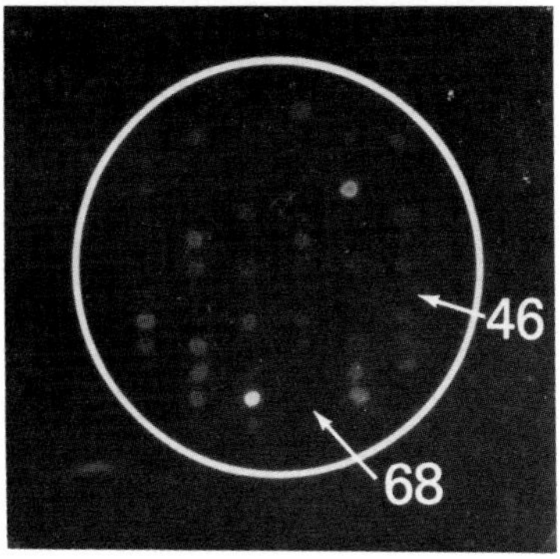

FIGURE 5. Colony hybridization of cloned mouse liver DNA probed with [^{32}P]cDNA reverse-transcribed from 3-methylcholanthrene-treated B6 and D2 liver 23 S mRNA (9).

membrane-bound proteins. Antibodies to different forms of P-450 are notoriously not monospecific (4). Even monoclonal antibodies have been reported (10) to recognize two different forms of P-450. We have evidence (R.H. Tukey and D.W. Nebert, manuscript in preparation) that several antibodies--anti-(P1-450), anti-(P2-450), and anti-(P3-450), each of which blocks one catalytic activity and not other activities--all precipitate protein translated from mRNA that hybridizes specifically to P1-450 cDNA in the translation arrest experiments. Therefore, for reasons at the protein level, translation arrest experiments involving solubilized membrane-bound proteins may lead to difficulties in the interpretation of data. Confusion about which cDNA clone represents P-450b, the major rat phenobarbital-inducible form of P-450, is an example of such problems with hybridization arrest data (reviewed in Ref. 4).

Genetic Criteria for cDNA Clone

Unfortunately, the only proof available for rabbit and rat cDNA clones is the immunologic criterion, because no genetic polymorphisms have been characterized yet in detail.

With the mouse P1-450 gene, however, we have the additional criterion of genetic differences that have been well characterized over the past decade. Hence, when the B6D2F1 heterozygote (Ah^b/Ah^d) is crossed with the D2 parent (Ah^d/Ah^d), among the progeny one finds a 50:50 distribution of Ah^b/Ah^d heterozygotes and Ah^d/Ah^d homozygotes.

mRNA was isolated from individual mouse livers (Fig. 6); samples 2 through 10 represent individual 3-methylcholanthrene-treated offspring from the B6D2F1 x D2 backcross. With a small piece of liver, their Ah phenotype was first determined, and we know that numbers 2, 4, 5, 7, and 9 are Ah^b/Ah^d heterozygotes; numbers 3, 6, 8, and 10 represent Ah^d/Ah^d homozygotes. The mRNA was electrophoresed and probed with clone 46 DNA. There is a perfect correlation between the Ah^b allele [i.e. induction of AHH activity (P1-450)] and the presence of induced P1-450 (23 S) mRNA that hybridizes to clone 46. This experiment therefore constitutes genetic proof that clone 46 is highly likely to be associated with the P1-450 cDNA. Additional proof has been obtained via Northern blots involving developmental studies (11, 12), since it is well known (7, 13) that P1-450 (AHH) induction occurs at least 1 week earlier in gestation than P2-450 or P3-450.

No Effect on DNA During Induction

Schimke et al. (14) have postulated that all forms of drug resistance might be caused by gene amplification, a process that they had found to occur for methotrexate resistance in mouse cell cultures: in response to methotrexate, the dihydrofolate reductase gene becomes amplified more than 100-fold. Because P-450 induction is a type of drug resistance, we therefore studied DNA from the livers of mice treated with various P-450 inducers (15, 16).

There was no evidence for increases in intensity of the blots or changes in the sizes of DNA fragments. We therefore conclude that, during the induction process, neither gene amplification nor some gross form of genomic rearrangement occurs in the area of the P1-450 gene represented by the clone 46 probe. In other hybridization experiments with B6 sperm, embryo and adult DNA (11), we also were unable to find any evidence for genomic rearrangement during development.

AHH ACTIVITY	LANE
1420	1 B6, Mechol
860	2 Ah^b/Ah^d MeChol
74	3 Ah^d/Ah^d MeChol
1260	4 Ah^b/Ah^d MeChol
990	5 Ah^b/Ah^d MeChol
140	6 Ah^d/Ah^d MeChol
1080	7 Ah^b/Ah^d MeChol
170	8 Ah^d/Ah^d MeChol
800	9 Ah^b/Ah^d MeChol
86	10 Ah^d/Ah^d MeChol
140	11 Control
190	12 Control
210	13 D2, Control

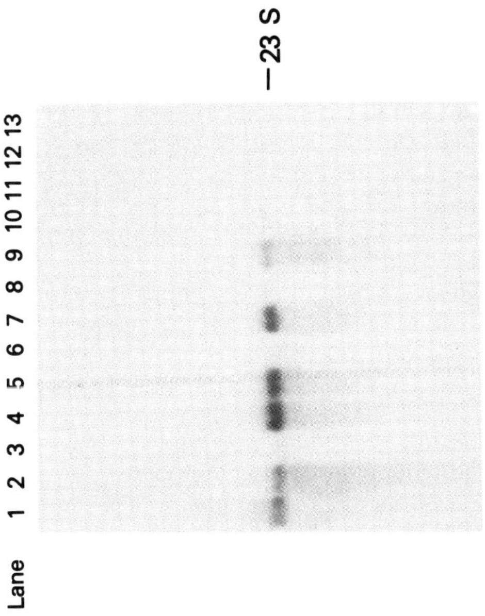

FIGURE 6. Northern blot of mRNA from individual mice probed with clone 46 [^{32}P]DNA (15). AHH, in units per mg of liver microsomal protein, was determined in a small portion of each individual's liver, and the remainder of each liver was used for the preparation of poly(A^+)-enriched RNA. MeChol, 3-methylcholanthrene treatment 12 h before killing.

Isolation and Characterization of λAhP-1

With clone 46 as the probe, a mouse plasmacytoma MOPC 41 genomic-DNA library was screened by the Benton-Davis plaque-hybridization procedure (17). Mouse tumor DNA had been digested by EcoRI and inserted into λ phage Charon 4A, and 50,000 plaques had been fixed to each filter (a generous gift of Dr. Jon Seidman, National Institute of Child Health and Human Development, Bethesda). The entire library consisted of 20 filters, so that we screened a total of about one million plaques. The first positive genomic clone was named λ3NT12 and was found to be 19 kbp in length.

By hybridization between subclones of λ3NT12 and clone 46, we found that clone 46 hybridized to the extreme 5' end of the mouse DNA insert. We therefore returned to the mouse MOPC 41 genomic-DNA library to find another recombinant phage. This time we used, as the ^{32}P-labeled probe, a 3-kbp subclone at the extreme 5' end of the mouse DNA insert (17).

By repeating the Benton-Davis plaque hybridization technique, and characterizing three more genomic clones, we have walked up the chromosome with clones λ3NT12, λ3NT13, λ3NT14, and finally λAhP-1. By R-loop analysis, the entire P_1-450 gene was found to reside in λAhP-1, spanning about 4600 bp (Fig. 7). The P_1-450 gene is in the middle of λAhP-1, a clone which has a total length of 15.5 kbp. The P_1-450 genomic gene has at least 5 exons and 4 intervening sequences. The first and last exon are remarkably large, being about 1000 and 1200 bp, respectively. Total length of all exons together is about 3000 bp (17).

Clone 46 appears to exist in the 3'-nontranslating region of the P_1-450 gene. With the clone 46 3' probe, it should be emphasized that we are unable to hybridize this probe to any other P-450 gene in mouse genomic DNA (9) or any other mouse P-450 mRNA (16). Clone 46 therefore represents a 3'-unique probe.

This laboratory has uncovered several important points of information with the use of the P_1-450 cDNA clone. An association of clone 46 with the P_1-450 protein has been demonstrated by both immunologic and genetic criteria. We know that P_1-450 induction is under transcriptional control, because increased 23 S mRNA and an intranuclear large-molecular-weight mRNA precursor occur concomitantly during the induction process (15). No evidence for gene duplication or gross form of genomic rearrangement has been found, either during induction or during development. P_1-450 mRNA is translated exclusively on membrane-bound polysomes (18). Clone 46 hybridizes with rat and rabbit DNA, and with rat but not rabbit mRNA (11, 19); these data suggest that clone 46 hybridizes to a segment of the rabbit P_1-450 gene that is not transcribed

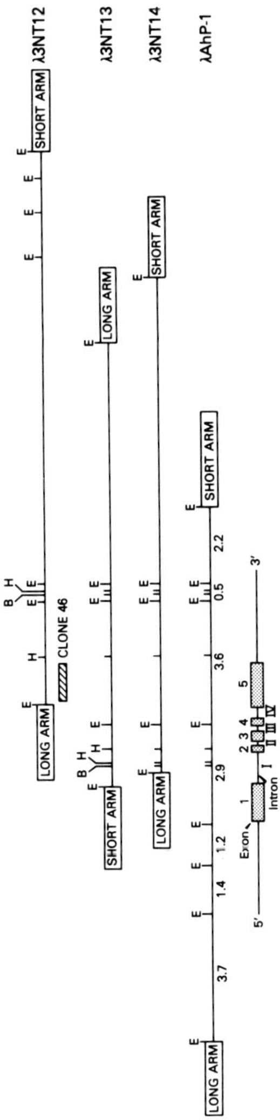

FIGURE 7. Restriction maps of four recombinant phage isolated from the mouse genomic-DNA library containing overlapping DNA regions of the P1-450 chromosomal structural gene (17). The linear DNA maps of individual phage were constructed on the basis of *EcoRI* (E), *HindIII* (H), *BamHI* (B), *PstI*, *SstI*, *XbaI*, and *XhoI* digests alone and in combination. The distances in kbp between the *EcoRI* sites of clone λAhP-1 are indicated, and the relative positions of the overlapping *EcoRI* and *XbaI* sites and clone 46 amongst the four phage are shown. In each of the four recombinant phage, the long and short arms of λ Charon 4A are indicated. By R-loop analysis between λAhP-1 and the P1-450 (23 S) mRNA from 3-methylcholanthrene-treated B6 liver, the position of the P1-450 exons and introns was determined and is depicted at *bottom* [Reproduced with permission from Springer-Verlag Inc.].

into the messenger. Clone p57, a 1900-bp cDNA clone of P1-450 that extends 800 bp further 5'-ward than clone 46, was used successfully to isolate two human genomic genes (20). We found that the P1-450 gene in adult B6 is *hypo*methylated, compared with the gene in B6 sperm, B6 embryo, or adult D2 mice; this hypomethylation pattern could be related to the increased expressivity of the P1-450 gene in B6, compared with that in D2 mice (11). Other P-450 inducers, such as benzo[a]anthracene and isosafrole, have been found to induce P1-450 mRNA, as measured by the clone 46 probe (16). Lastly, we have found an excellent correlation between the intranuclear appearance of the inducer-receptor complex and the induction of P1-450 mRNA as measured by the clone 46 probe (21).

Sequencing of the entire P1-450 gene and its flanking regions is underway. We have uncovered an interesting 104-bp sequence (Fig. 8) that includes 10-bp flanking direct repeats. This poly(dG-dT) repeat is consistent with DNA having Z-DNA-forming potential (22, 23). Of additional interest, this stretch of DNA appears to be in the noncoding region of the first exon. Most Pu-Py repeats reported to date occur either upstream from the cap site or in intervening sequences. This preliminary result is under further study.

Other Studies in Progress

p21, a 1710-bp cDNA clone that hybridizes well to 20 S mRNA (24), is associated with the Ah locus and exhibits 5' homology with the P1-450 gene. This genomic gene has been isolated and clearly differs from the P1-450 genomic gene (R.H. Tukey and D.W. Nebert, in preparation). Whether p21 represents a cDNA clone for P2-450 or P3-450 cannot be answered until we have the necessary protein and nucleotide sequence data completed. The P2-450 protein has been thoroughly characterized (25).

Benzo[a]pyrene-Resistant Mutant Clones

Benzo[a]pyrene induces high levels of AHH activity in Hepa-1 cells (26). The metabolites of benzo[a]pyrene formed by cytochrome P1-450 are extremely toxic; exposure of cells to concentrations of benzo[a]pyrene as low as 25 nM is sufficient to produce toxicity and cell death--if the cells possess sufficiently high levels of inducible

5' GAAAATAAAA TAGAGAGAGA GTGTGTGTGT GTGTGTGTGT

GTGTGTGTGT GTGTGTTGTG TGTGTGTGAA TATGATGATT

AAAATATATT GTGTGAAAAT AAAA 3'

FIGURE 8. Partial sequence near the 5' end of the P1-450 gene.

AHH activity. In the presence of 4 μM benzo[a]pyrene, however, a few Hepa-1 cells (~10^{-7} per generation) survive. These benzo[a]pyrene-resistant cells appear to represent somatic mutations in the pathways for AHH induction. Clones derived from such cells have been developed (27) with the intention of examining genetically the multiple steps during the process of AHH induction. By somatic-cell hybridization studies (28), the clones that have been developed are known to represent at least three distinct complementation groups and therefore reflect mutations in at least three different genes.

Several clones (Table 2) have essentially normal *Ah* receptor levels, compared with the wild-type Hepa-1c1c7 parent, possess normal kinetics for translocation of the inducer-receptor complex into the nucleus, yet exhibit very low or nondetectable basal or inducible AHH activity. These clones could represent a mutation in the P1-450 structural gene or other genes responsible for the induced hydroxylase activity. Other clones, c2 and c6, are receptor-deficient mutants (r^-), having no more than 10% of wild-type *Ah* receptor levels, normal kinetics of nuclear translocation of the inducer-receptor complex, and no more than 20% of wild-type AHH inducibility by either TCDD or benzo[a]anthracene. One clone, c4, has normal cytosolic levels of *Ah* receptor, is defective in nuclear translocation of the inducer-receptor complex (nt^-), and lacks any detectable basal or inducible AHH activity. These data (26) are an important prelude to planned recombinant DNA and transfection experiments designed to understand the regulatory mechanism by which the *Ah* receptor controls transcription of the P1-450 gene during induction by polycyclic aromatic compounds.

TABEL 2
Ah RECEPTOR LEVELS AND MAXIMALLY INDUCIBLE AHH ACTIVITY IN Hepa-1c1c7 PARENT LINE AND SIX MUTANT CLONES

	Ah receptor (fmol/mg protein)			Maximal AHH activity (units/mg cellular protein)	
Cell culture line	Cytosol, *in vitro* treatment	Cytosol, exposure in culture	Nuclei, exposure in culture	BzAnth as inducer	TCDD as inducer
Hepa-1c1c7	20	12	6	210	520
c1		7.6	2.0	<0.4	<0.4
c2	2.1	1.7	0.5	42	110
c3		18	3.4	22	16
c4		15	0.3	<0.4	<0.4
c5		6.0	3.4	<0.4	<0.4
c6	1.0	0.5	0.4	4	3

Complementation groups: c1 & c5; c2 & c6; c4. c3 is dominant, and therefore its complementation group cannot be determined (26). BzAnth, benzo[a]anthracene.

Concluding Remarks

With clone λAhP-1 and the surrounding regions of this mouse chromosome, we hope to understand a great deal about the regulation of P-450 induction, the evolution of the P-450 system, and perhaps the ultimate number of P-450 forms that an individual organism is genetically *capable* of expressing. With this knowledge, we hope to gain insight into the mechanism of chemical carcinogenesis, especially since P_1-450 is directly responsible for the metabolic activation of polycyclic hydrocarbons such as benzo[a]pyrene to the ultimate carcinogenic intermediate, which interacts covalently with DNA. Finally, it may be possible to develop an assay, based on recombinant DNA technology, in order to assess the human *Ah* phenotype (reviewed in Ref. 29); such an assay may predict who is at increased risk for certain types of environmentally-caused cancers.

ACKNOWLEDGMENTS

This work would not have been possible without the combined efforts of Drs. Yuan-Tsong Chen, Michitoshi Nakamura, Toshihiko Ikeda, Mario Altieri, Catherine Legraverend, Tohru Ohyama, Sirpa Kärenlampi, Allan B. Okey, Rita R. Hannah, Dana J. Kessler, Lynn W. Enquist, David C. Swan, and Oliver Hankinson. The expert secretarial assistance of Ms. Ingrid E. Jordan is greatly appreciated.

REFERENCES

1. Lu AYH, West SB (1980). Multiplicity of mammalian microsomal cytochromes P-450. Pharmacol Rev 31:277.
2. Mannering GJ (1981). Hepatic cytochrome P-450-linked drug-metabolizing systems. In Jenner P, Testa B (eds): "Concepts in Drug Metabolism," New York: Marcel-Dekker, Inc, p 53.
3. Nebert DW, Negishi M, Lang MA, Hjelmeland LM, Eisen HJ (1982). The Ah locus, a multigene family necessary for survival in a chemically adverse environment: Comparison with the immune system. Advanc Genet 21:1.
4. Nebert DW, Negishi M (1982). Multiple forms of cytochrome P-450 and the importance of molecular biology and evolution. Biochem Pharmacol 31:2311.
5. Nebert DW (1979). Multiple forms of inducible drug-metabolizing enzymes. A reasonable mechanism by which any organism can cope with adversity. Mol Cell Biochem 27:27.
6. Eisen HJ, Hannah RR, Legraverend C, Okey AB, Nebert DW (1983). The Ah receptor: Controlling factor in the induction of drug-metabolizing enzymes by certain chemical carcinogens and other environmental pollutants. In Litwack G (ed): "Biochemical Actions of Hormones," New York: Academic Press, p 227.

7. Negishi M, Nebert DW (1979). Structural gene products of the Ah locus. Genetic and immunochemical evidence for two forms of mouse liver cytochrome P-450 induced by 3-methylcholanthrene. J Biol Chem 254: 11015.
8. Negishi M, Nebert DW (1981). Structural gene products of the Ah complex. Increases in large mRNA from mouse liver associated with cytochrome P1-450 induction by 3-methylcholanthrene. J Biol Chem 256:3085.
9. Negishi M, Swan DC, Enquist LW, Nebert DW (1981). Isolation and characterization of a cloned DNA sequence associated with the murine Ah locus and a 3-methylcholanthrene-induced form of cytochrome P-450. Proc Natl Acad Sci USA 78:800.
10. Thomas PE, Reik LM, Ryan DE, Levin W (1982). Some hybridoma antibodies against rat liver cytochrome P-450c cross-react with cytochrome P-450d. Fed Proc 41:297 [Abstract].
11. Chen Y-T, Negishi M, Nebert DW (1982). Cytochrome P1-450 structural gene in mouse, rat and rabbit. Differences in DNA methylation and developmental expression of mRNA. DNA 1:231.
12. Ikeda T, Altieri M, Chen Y-T, Nakamura M, Tukey RH, Nebert DW, Negishi M (1983). Characterization of P2-450 (20 S) mRNA. Association with the P1-450 genomic gene and differential response to the inducers 3-methylcholanthrene and isosafrole. Eur J Biochem, in press.
13. Guenthner TM, Nebert DW (1978). Evidence in rat and mouse liver for temporal control of two forms of cytochrome P-450 inducible by 2,3,7,8-tetrachloro-dibenzo-p-dioxin. Eur J Biochem 91:449.
14. Schimke RT, Kaufman RJ, Alt FW, Kellems RF (1978). Gene amplification and drug resistance in cultured murine cells. Science 202:1051.
15. Tukey RH, Nebert DW, Negishi M (1981). Structural gene product of the Ah complex. Evidence for transcriptional control of cytochrome P1-450 induction by use of a cloned DNA sequence. J Biol Chem 256: 6969.
16. Tukey RH, Negishi M, Nebert DW (1982). Quantitation of hepatic cytochrome P1-450 mRNA with the use of a cloned DNA probe. Effects of various P-450 inducers in C57BL/6N and DBA/2N mice. Mol Pharmacol 22:779.

17. Nakamura M, Negishi M, Altieri M, Chen Y-T, Ikeda T, Tukey RH, Nebert DW (1983). Structure of the mouse cytochrome P_1-450 genomic gene. Eur J Biochem, in press.
18. Chen Y-T, Negishi M (1982). Expression and subcellular distribution of mouse cytochrome P_1-450 mRNA as determined by molecular hybridization with cloned P_1-450 DNA. Biochem Biophys Res Commun 104:641.
19. Chen Y-T, Lang MA, Jensen NM, Negishi M, Tukey RH, Sidransky E, Guenthner TM, Nebert DW (1982). Similarities between mouse and rat liver microsomal cytochrome P-450 induced by 3-methylcholanthrene. Evidence from catalytic, immunologic, and recombinant DNA studies. Eur J Biochem 122:361.
20. Chen Y-T, Tukey RH, Swan DC, Negishi M, Nebert DW (1983). Characterization of the human P_1-450 genomic gene. Pediat Res, in press [Abstract].
21. Tukey RH, Hannah RR, Negishi M, Nebert DW, Eisen, HJ (1982). The Ah locus. Correlation of intranuclear appearance of inducer-receptor complex with induction of cytochrome P_1-450 mRNA. Cell 31:275.
22. Wang A H-J, Quigley GJ, Kolpak FJ, Crawford JL, van Boom JH, van der Marel G, Rich A (1979). Molecular structure of a left-handed double helical DNA fragment at atomic resolution. Nature 282:680.
23. Nordheim A, Pardue ML, Lafer EM, Moller A, Stollar BD, Rich A (1981). Antibodies to left-handed Z-DNA bind to interband regions of Drosophila polytene chromosomes. Nature 294:417.
24. Tukey RH, Ohyama T, Negishi M, Nebert DW (1982). Isolation and characterization of cloned cDNA encoding cytochrome P_2-450. Pharmacologist 24:207 [Abstract].
25. Ohyama T, Nebert DW, Negishi M (1982). Genetic regulation of mouse liver microsomal P_2-450 induction by isosafrole (ISF) and 3-methylcholanthrene (MC). In Hietanen E, Laitinen M, Hänninen O (eds): "Cytochrome P-450, Biochemistry, Biophysics and Environmental Implications." Amsterdam: Elsevier/North-Holland Biomedical Press, p 177.
26. Legraverend C, Hannah RR, Eisen HJ, Owens IS, Nebert DW, Hankinson O (1982). Regulatory gene product of the Ah locus: Characterization of receptor mutants among mouse hepatoma clones. J Biol Chem 257:6402.

27. Hankinson O (1979). Single-step selection of clones of a mouse hepatoma line deficient in aryl hydrocarbon hydroxylase. Proc Natl Acad Sci USA 76:373.
28. Hankinson O (1981). Evidence that benzo[a]pyrene-resistant, aryl hydrocarbon hydroxylase-deficient variants of mouse hepatoma line, Hepa-1, are mutational in origin. Somat Cell Genet 7:373.
29. Nebert DW (1981). Genetic differences in susceptibility to chemically induced myelotoxicity and leukemia. Environ Health Perspect 39:11.

Gene Expression, pages 207-217
© 1983 Alan R. Liss, Inc., 150 Fifth Avenue, New York, NY 10011

RAT GROWTH HORMONE GENE REGULATION STUDIED BY GENE TRANSFER INTO HYBRID CELLS

Jeannine Strobl,[1] Raji Padmanabhan,[2] Bruce Howard,[2] and E. Brad Thompson[1]

[1]Laboratory of Biochemistry
and
[2]Laboratory of Molecular Biology
National Cancer Institute
National Institutes of Health
Bethesda, Maryland 20205

ABSTRACT Growth hormone is a peptide hormone synthesized exclusively in somatotroph cells of the anterior pituitary. Our goal is to understand the molecular mechanisms which allow the growth hormone gene to be expressed in pituitary cells and prevent its expression in other cell types. A rat pituitary cell line, GH_3, which expresses the growth hormone gene was fused with mouse fibroblast cells, LB82, which do not express growth hormone. The resulting somatic cell hybrids contain intact growth hormone genes derived from both the pituitary and fibroblast cell lines yet fail to produce either growth hormone RNA or protein. The hybrid cells, therefore, contain regulatory molecules which act in trans to inhibit growth hormone gene expression. Growth hormone gene transfer experiments were performed to clarify the mechanism of these gene regulators. The introduction of one or more stably integrated copies of the genomic rat growth hormone gene into the hybrid cells permits re-expression of growth hormone RNA and protein structurally indistinguishable from that produced in GH_3 pituitary cells. In contrast, others have found that transfer of the rat growth hormone gene into LB82, NIH-3T3 or C127 fibroblast cells results in transcription of aberrantly sized growth hormone RNA (1-4). Repressors of growth hormone gene expression in hybrid and fibroblast cells are therefore partially overcome when growth hormone genes are increased by gene transfer. Apparently, however, activator molecules present in GH_3 and hybrid cells are required for proper RNA transcript initiation, termination, and/or processing. We therefore propose that the tissue-specific expression of the growth hormone gene is controlled by both activating and repressive trans acting elements.

INTRODUCTION

Hormonal controls over the production of the pituitary hormone, growth hormone, have been studied extensively in the GH_3 tissue culture cell line derived from a rat pituitary adenoma (5,6). Experiments in a number of laboratories show that a highly complex regulatory scheme involving interactions of multiple steroid and peptide hormones at the transcriptional and post-transcriptional levels modulates growth hormone mRNA levels. The only specific regulatory molecules which have been identified in this control system are the hormone receptor proteins and the hormones themselves. No site or mechanism of interaction of the hormone-receptor complexes with the growth hormone gene have yet been defined.

Additional regulatory molecules are required to account for all of the observations concerning the expression of the growth hormone gene. Receptor proteins for two hormones which affect growth hormone mRNA levels most dramatically, thyroid hormone and dexamethasone, are ubiquitous; the existence of tissue-specific regulatory elements or chromatin modifications must therefore be postulated to explain tissue-specific hormone responsiveness. In GH_3 pituitary cells deprived of hormonal influences, a low level of growth hormone expression takes place which is not observed in non-pituitary cell types. The control molecules which prevent growth hormone gene expression in non-pituitary cells in the absence of hormones may comprise an additional set to those which modify the hormonal response in a tissue-specific manner. In the following experiments we define an experimental system in which the nature of these postulated non-receptor molecules controlling growth hormone gene expression may be further explored.

METHODS

Cell Culture.

GH_3 cells derived from a rat pituitary adenoma (5) and deficient in the enzyme hypoxanthine-guanosine phosphoribosyltransferase (C. Bancroft, pers. comm.) were co-cultivated with thymidine kinase deficient LB82 mouse fibroblast cells (7). Spontaneous fusion of GH_3 and LB82 cells led to the formation of somatic cell hybrids which were isolated by growth in Dulbecco's Modified Eagle's Medium (DMEM) supplemented with 5% fetal calf serum, 10^{-4} M hypoxanthine, 1.6 x 10^{-5} M thymidine, 10^{-6} M aminopterin, and 4 x 10^{-4} M glycine. Individual colonies of hybrid cells were isolated and karyotyped to confirm their hybrid nature. After cloning, the hybrid cells were grown in non-selective medium. Routine cell culture conditions for GH_3, LB82, and hybrid cells were as described previously (8,9).

DNA transformations were performed with hybrid cells plated 18 hours previously at a density of 1 x 10^4 cells/cm^2. Cells received fresh medium containing 10% fetal calf serum three hours before the addition of a fine calcium phosphate-DNA co-precipitate (20 μg DNA/10^6 cells) prepared as described in Graham and van der Eb (10). Cells were washed free of the co-precipitate 4-5 hours later, treated for two minutes with 10% glycerol in .14 M NaCl-25 mM HEPES-.75 mM sodium phosphate, pH 7.1 and returned to a non-selective medium for 48 hours. The cells were then plated at a density of 2 x 10^3 cells/cm^2 in DMEM + 5% dialyzed fetal calf serum + 5% dialyzed horse serum + 1 x 10^{-4} M hypoxanthine, 1.6 x 10^{-3} M xanthine, 4.5 x 10^{-6} M aminopterin, 4.1 x 10^{-5} M thymidine, 1.3 x 10^{-4} M glycine and 25 μg/ml mycophenolic acid. Colonies of transformed cells were isolated two weeks later at a frequency of 2 x 10^{-4}.

Plasmid pGH-XPT Construction.

The Hind III fragment containing the genomic rat growth hormone gene (11) with flanking regions extending ~ 2 kbp 5' and ~ 1.5 kbp 3' from the gene was cloned into the HindIII site of pBR327 (pBR327-GH). A selectable marker, the bacterial xanthine-guanosine phosphoribosyltransferase gene utilizing the SV40 early promotor, RNA splicing, and polyA addition signals (12,13), (SV-XGPT) was inserted into the BamHI site of pBR327-GH in the same orientation as the rat growth hormone gene (pGH-XPT). Plasmid used in the transfection experiments was isolated by the SDS lysis procedure (14) and double banding in cesium chloride-ethidium bromide gradients.

Biochemical Analyses.

DNA was isolated and analyzed by restriction enzyme digestion and Southern hybridization as described in Strobl et al. (9).

Total cell RNA was purified by the guanidine extraction method (15); the polyA$^+$ RNA fraction was isolated by a single cycle of oligo(dT) cellulose chromatography at 50°C (16). The growth hormone RNA species were examined by hybridization after denaturation in 50% formamide-2.2 M formaldehyde-20 mM MOPS (morpholinopropanesulfonic acid) pH 7.0-5 mM sodium acetate-1 mM EDTA for 5 minutes at 60°C, electrophoresis in 1.4% agarose-formaldehyde gels using the MOPS buffer system (17), and direct transfer to nitrocellulose paper (18). Northern hybridization conditions were exactly as described in Strobl et al. (9).

Secreted growth hormone protein was detected after incubating cells for four hours with serum-free, methionine-free DMEM supplemented with 10 µCi/ml of ^{35}S-methionine (specific activity ~ 1100 Ci/mmol). The medium was lyophilized, redissolved in water, dialyzed against 2 mM EDTA and a sample immunoprecipitated with baboon anti-rat growth hormone antibody using the Staph. A. procedure (19). Immunoprecipitates were analyzed in denaturing 12% polyacrylamide gels (20).

Glutamine synthetase activity and whole-cell dexamethasone receptor levels were assayed as described in Strobl et al. (9).

RESULTS

The clone of GH_3 x LB82 hybrid cells chosen as the recipient cell line for transfer of the pGH-XPT DNA was typical of many characterized clones of hybrid cells. Karyotypic analysis revealed a mean total chromosome number of 72, 51% clearly of rat pituitary origin and the remainder derived from the mouse fibroblast parent cell. For comparison, the mean chromosome number in GH_3 and LB82 cells is 68 and 51, respectively. The integrity of the mouse and rat growth hormone structural genes in the hybrid cells was supported by restriction enzyme analyses with EcoRI, HindIII, KpnI and MspI followed by Southern hybridization to identify the DNA fragments containing growth hormone coding sequences. The mouse and rat growth hormone genes were clearly distinguishable using this method and the hybridization patterns of each growth hormone gene in the parent and hybrid cells were identical (9).

Expression of the growth hormone genes in the hybrid cells could not be detected either by complement fixation analyses for secreted growth hormone peptide or by RNA hybridization analyses sensitive enough to detect <1-10 growth hormone mRNA molecules/cell. Growth hormone gene expression could not be induced by hormonal manipulations, dibutyryl cAMP, DMSO, sodium butyrate, N-methyl nicotinamide, or 5-azacytidine. Very strong and possibly multiple control factors therefore appear to block growth hormone expression in these hybrid cells. Acute expression of immunoreactive growth hormone protein by hybrid cells containing non-integrated copies of pGH-XPT following nuclear microinjection of pGH-XPT supports the hypothesis that growth hormone expression in GH_3 x LB82 hybrid cells is restricted by repressive elements acting in trans.

Hybrid cells transfected with pGH-XPT DNA were selected for expression of the bacterial xanthine-guanosine phosphoribosyltransferase gene. All ten clones which have been studied in detail appear to have stably integrated pGH-XPT since growth in non-selective medium for 3-6 months has not altered their capacity to grow in selective medium. The transfected hybrid cells generally contained 1-5 intact copies of the growth hormone gene and a variable number of additional hybridizing sequences in high molecular weight DNA. All stably transformed clones of hybrid cells produced RNA transcripts containing growth hormone coding sequences; 70% exhibited RNA transcripts consistent with correct transcript initiation, termination, and processing. The analysis of one of these clones, T10, is presented in greater detail in Figure 1 and below.

DNA isolated from GH_3 and pGH-XPT transfected hybrid clone T10 cells was digested with HindIII to measure intact copies of the endogenous genomic HindIII fragment

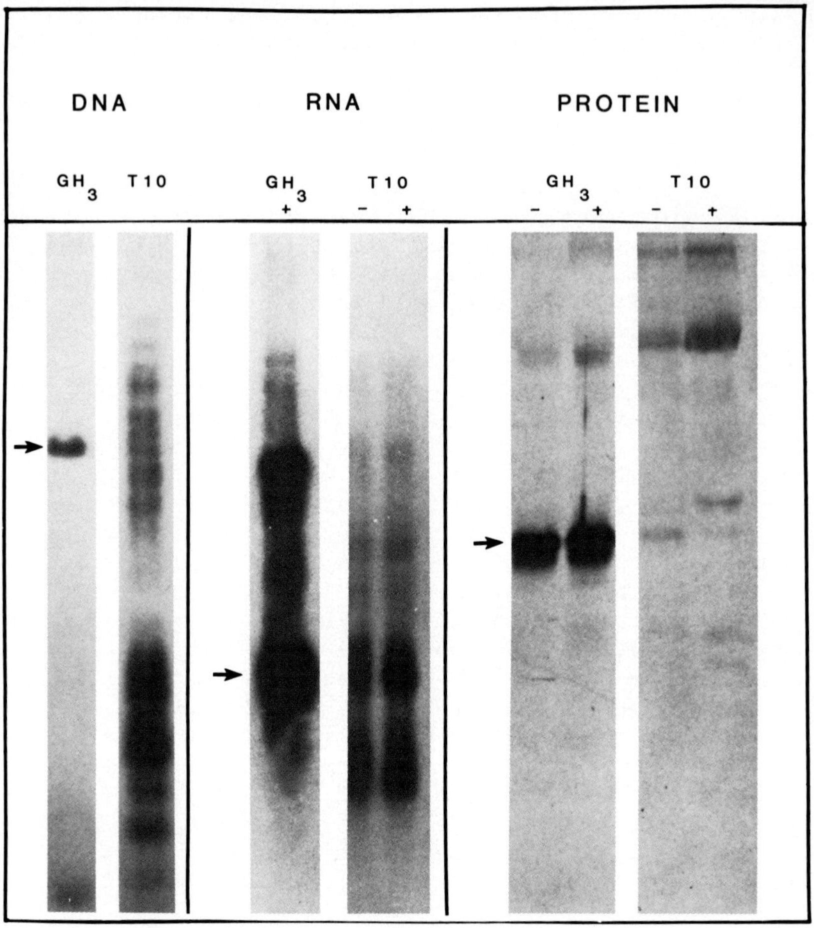

Figure 1. Growth hormone expression in GH_3 and T10 transformed hybrid cells. Cells were grown in the absence (-) or presence (+) of 5 x 10^{-7} M dexamethasone for 48 hours, harvested, and DNA, RNA or protein treated as described in Methods.

containing the growth hormone gene and the *Hind*III insert in pGH-XPT. In the first panel of Figure 1, the 5.6 kbp *Hind*III fragment containing the rat growth hormone gene in GH_3 cells is indicated with an arrow. Atypical of the transfected hybrid cells, in clone T10 the 5.6 kbp *Hind*III fragment indicative of the intact endogenous rat growth hormone gene and intact exogenous rat growth hormone genes contributed by pGH-XPT is indistinct. A 4.5 kbp fragment, the second major band in T10 migrating faster than 5.6 kbp, corresponds to the intact endogenous mouse growth hormone gene (9). The major growth hormone DNA species in T10 are clearly on HindIII fragments of ~ 2.0 kbp and ~ 1.5 kbp suggesting that a gene rearrangement occurred prior to chromosomal integration.

Clone T10 produces the highest level of GH mRNA of all 10 clones so far examined, approximately 7% of levels produced by hormone-induced GH_3 cells as compared with 0.1% or less seen in most other transformants (Fig. 1, Panel 2). The high growth hormone gene content cannot alone explain the level of growth hormone mRNA expression but may be a contributing factor. Hybrid clone T10 exhibits large growth hormone RNA species which are similar in size to the growth hormone RNA precursors produced in GH_3 cells and a prominent band of growth hormone RNA identical in size to the mature growth hormone mRNA isolated from GH_3 cells (indicated with an arrow). Some translation of the growth hormone mRNA and secretion of growth hormone peptide takes place in T10 hybrid cells. The evidence (Figure 1, Panel 3, indicated with an arrow) is the detection of a small amount of immunoreactive growth hormone protein in the culture medium which co-migrates with growth hormone produced by GH_3 cells.

Production of growth hormone mRNA by clone T10 is not regulated by the glucocorticoid hormone dexamethasone. Growth hormone peptide production is influenced by dexamethasone treatment in an unexpected manner which suggests some post-transcriptional alteration. (Dexamethasone reduced the amount of growth hormone peptide of the expected size and increased that of one larger and two smaller immunoreactive forms.) These data contrast with the intact glucocorticoid response of glutamine synthetase, for which there is also evidence of transcriptional control (21, Table 1). Both transformed and non-transformed hybrid cells have large numbers of glucocorticoid receptors with an appropriate affinity for the ligand dexamethasone. Induction of glutamine synthetase activity by dexamethasone occurs in both parental cell lines (although the effect is very small in LB82 cells) and in GH_3 x LB82 hybrid cells before and after transformation with pGH-XPT. Glucocorticoid

TABLE 1

GLUCOCORTICOID RESPONSES

	Glucocorticoid Receptors		Glutamine Synthetase Activity	
	Sites/Cell	K_d	Control	+Dexamethasone
GH_3	50,760	$6.6x10^{-9}$ M	3.6	8.0
LB82	90,670	$2.2x10^{-8}$ M	1.0	1.3
GH_3xLB82 hybrid	219,500	$1.6x10^{-8}$ M	2.0	3.5
GH_3xLB82 hybrid-T10	234,085	$3.2x10^{-8}$ M	.7	2.1

induction of glutamine synthetase activity is not limited by the additional growth hormone genes in T10 indicating that glucocorticoid receptor availability does not limit T10 responsiveness to dexamethasone. Inducibility of growth hormone is therefore restricted in T10 hybrid cells by the absence of appropriate DNA sequences, trans active gene regulators which affect growth hormone and not glutamine synthetase, or a peculiarity unique to DNA introduced by gene transfer.

DISCUSSION

Upon transfer of exogenous growth hormone genes into hybrid cells, growth hormone RNA transcript initiation, termination, processing, translation and protein secretion occur at a low level which is not detectable in hybrid cells before transformation. Acute growth hormone expression after nuclear microinjection of growth hormone genes into hybrid cells shows that expression may not depend upon chromosome integration site.

Our model of growth hormone gene re-expression in hybrid cells following gene transfer is that trans active repressor mechanisms are overloaded or by-passed. The first case predicts that expression of endogenous and exogenous genes occurs, the latter, of only the exogenous genes. Experiments to distinguish between these possibilities are in progress. If both exogenous and endogenous genes are expressed, the DNA regulatory regions may be mapped by modification of the transferred genes. If only the exogenous genes are expressed, it will be important to study how transferred genes escape normal cell regulatory mechanisms.

Transfer of rat growth hormone genes into hybrid cells is a novel approach which for the first time permits accurate rat growth hormone transcript formation; we suggest this reflects a requirement for a pituitary cell-specific positive regulatory factor. Nevertheless, the low level of growth hormone gene expression and its failure to respond to glucocorticoid regulation in the hybrid cells clearly demonstrate a requirement for additional regulatory signals. Further optimization of gene expression and regulation after gene transfer may provide a way of elucidating these additional control mechanisms.

ACKNOWLEDGMENTS

We thank Dr. John Baxter for providing the rat growth hormone cDNA, Dr. Pricilla Dannies for the antibody to rat growth hormone, Mrs. Billie Wagner for technical assistance, and Mrs. Jean Regan for secretarial assistance.

REFERENCES

1. Doehmer J, Barinaga M, Vale W, Rosenfeld MG, Verma IM, Evans RM (1982). Introduction of rat growth hormone gene into mouse fibroblasts via a retroviral DNA vector: expression and regulation. Proc Natl Acad Sci USA 79:2268.
2. Moore DD, Walker MD, Diamond DJ, Conkling MA, Goodman HM (1982). Structure, expression, and evolution of growth hormone genes. Recent Prog Horm Res 38:196.
3. Baxter JD, Eberhardt N, Selby M, Mellon-Nussman S, Spindler S, Karin M, Cooke N, Lan N, Guiterrez-Hartman

A, Cathala G (1982). Structure and expression of growth hormone related genes. J Cell Biochem Supp 6: 270.

4. Kushner PJ, Levinson BB, Goodman HM (1982). A plasmid that replicates in both mouse and E. coli cells. J Mol Appl Genet 1:527.
5. Tashjian AH Jr, Yasumura Y, Levine L, Sato GH, Parker ML (1968). Establishment of clonal strains of rat pituitary tumor cells that secrete growth hormone. Endocrinology 82:342.
6. Baxter JD, Seeburg PH, Shine J, Martial JA, Ivarie RD, Johnson LK, Fiddes JC, Goodman HM (1979). Structure of growth hormone gene sequences and their expression in bacteria and in cultured cells. In Sato GH, Ross R (eds): "Hormones and Cell Culture," Cold Spring Harbor New York: Cold Spring Harbor Press, p. 317.
7. Littlefield JW (1966). The use of drug-resistant markers to study the hybridization of mouse fibroblasts. Exp Cell Res 41:190.
8. Thompson EB, Dannies PS, Buckler CE, Tashjian AR Jr (1980). Hormonal control of tyrosine aminotransferase, prolactin and growth hormone induction in somatic cell hybrids. J Steroid Biochem 12:193.
9. Strobl JS, Dannies PS, Thompson EB (1982). Somatic cell hybridization of growth hormone-producing rat pituitary cells and mouse fibroblasts results in extinction of growth hormone expression via a defect in growth hormone RNA production. J Biol Chem 257: 6588.
10. Graham F, van der Eb A (1973). A new technique for the assay of infectivity of human adenovirus 5 DNA. Virology 52: 456.
11. Chien Y-H, Thompson EB (1980). Genomic organization of rat prolactin and growth hormone genes. Proc Natl Acad Sci USA 77:4583.
12. Mulligan RC, Berg P (1981). Selection for animal cells that express the E. coli gene coding for xanthine-guanine phosphoribosyltransferase. Proc Natl Acad Sci USA 78:2072.
13. Schümperli D, Howard BH, Rosenberg M (1982). Efficient expression of E. coli galactokinase gene in mammalian cells. Proc Natl Acad Sci USA 79:257.
14. Godson GN, Vapnek D (1973). A simple method of preparing large amounts of ØX174 RFI supercoiled DNA. Biochim Biophys Acta 299:516.

15. Strohman RC, Moss PS, Micou-Eastwood J, Spector D, Pryzbyla A, Paterson B (1977). Messenger RNA myosin polypeptide: isolation from single myogenic cell culture. Cell 10:265.
16. Aviv H, Leder P (1972). Purification of biologically active globin messenger RNA by chromatography on oligothymidylic acid-cellulose. Proc Natl Acad Sci USA 69: 1408.
17. Lehrach H, Diamond D, Wozney JM, Boedtker H (1977). RNA molecular weight determinations by gel electrophoresis under denaturing conditions, a critical reexamination. Biochemistry 16:4743.
18. Thomas PS (1980). Hybridization of denatured RNA and small DNA fragments transferred to nitrocellulose. Proc Natl Acad Sci USA 77:5201.
19. Kessler SW (1981). Use of protein-A-bearing staphylococci for the immune precipitation and isolation of antigens from cells. Methods Enzymol 72:442.
20. Laemmli UK (1970). Cleavage of structural proteins during the assembly of the head of the bacteriophage T4. Nature 227:680.
21. Harmon JM, Thompson EB (1981). Glutamine synthetase induction by glucocorticoids in the glucocorticoid-sensitive human leukemic cell line CEM-C7. J Cell Physiol 110:155.

Gene Expression, pages 219-232
© 1983 Alan R. Liss, Inc., 150 Fifth Avenue, New York, NY 10011

IDENTIFICATION OF A DNA SEGMENT REQUIRED FOR THE INDUCED TRANSCRIPTION OF A HUMAN IFN-α GENE

Ulrich Weidle[1], Hermann Ragg[2], Ned Mantei and Charles Weissmann

Institut für Molekularbiologie I
Universität Zürich
8093 Zürich, Switzerland

ABSTRACT. Mouse L cells were transformed with a hybrid gene consisting of the 700-bp 5' flanking region of the human interferon-α1 (IFN-α1) gene and the rabbit β-globin transcription unit. Correctly initiated β-globin transcripts were produced after induction by virus infection, but not otherwise. The reciprocal construction, the β-globin 5' flanking region followed by the IFN-α1 transcription unit, as well as the rabbit β-globin gene itself gave constitutive expression of β-globin and IFN-α1 RNA, respectively, and no stimulation by virus infection was observed. Thus, induction of IFN-α synthesis is due to activation of transcription, and is mediated by the 5' flanking region of the gene.

Deletion mapping showed that not more than 117 base pairs of 5' flanking sequence were required for induction of the gene.

[1]Present address: Boehringer Mannheim GmbH, Forschungszentrum Tutzing, 8132 Tutzing, FRG

[2]Present address: Hoechst AG, Pharma Biochemie H 825, 623 Frankfurt 80, FRG

INTRODUCTION

Usually only cells exposed to virus, double-stranded RNA or other inducers synthesize IFN (1, 2). IFN-mRNA appears 1-2 h after induction, peaks at 1.5-20 h and decays with a half-life of about 30 min (3-6). So far, it has not been determined whether induction of IFN is due to transient stabilization of a rapidly turning-over mRNA or to activation of transcription.

To clarify this issue we transformed mouse L cells with a hybrid gene in which the 5' flanking region of the human IFN-α1 gene was followed by the rabbit β-globin transcription unit. Correctly initiated β-globin RNA appeared only after viral induction, with the kinetics described for IFN mRNA (7). Cells transformed with the converse construction, or with the complete rabbit β-globin gene, constitutively produced correctly initiated transcripts; viral infection decreased the globin RNA level. We conclude that induction acts by activating transcription rather than by reducing turnover, and that the regulatory elements are contained in the 5' flanking region of the IFN gene.

To delineate the sequences required for induction, a set of 5' deletion mutants of the human IFN-α1 gene was constructed and the expression of the truncated genes in mouse L cells was monitored after viral or mock infection. Not more than 117 base pairs of 5' flanking sequence were required for induced expression of the gene. A purine rich sequence of 42 base pairs located immediately downstream of position -117 is highly conserved in all known human interferon-α genes.

MATERIALS AND METHODS

Plasmid construction.

pΔ425B contains the rabbit β-globin transcription unit preceded by 425 bp of 5' flanking DNA (8). An XbaI linker was inserted at position -9/-10 (numbering relative to the CAP site), to yield pΔ425B(Xba-10). pChr35-675 contains the human IFN-α1 transcription unit preceded by about 675 bp of 5' flanking region (7). XbaI linkers were introduced at positions -5/-6 and -69/-70 (the numbering is relative to the CAP site, which is 69±3 nucleotides upstream from the AUG initiator triplet; in this paper we are designating as +1 the position given as -69 in ref. 9). The β-globin 5' flanking segment between position -425 and the XbaI linker contains all elements required for expression in heterologous cells (10). It was joined to the IFN-α1 transcription unit at the XbaI site in position -5/-6, to yield pGI, positioning the IFN-α CAP site 34 bp downstream of the first A residue of the β-globin gene ATA box.

In the converse construction, pIG-I, a fragment of about 670 bp of IFN-α1 5' flanking region, extending up to the XbaI linker at position -5/-6, was joined to the β-globin gene at the XbaI linker at -9/-10, positioning the CAP site of the β-globin gene 43 bp downstream of the IFN-α1 ATA box. In pIG-II, a fragment of IFN-α1 5' flanking sequence extending to the XbaI linker at position -69/70 was linked to the β-globin DNA fragment; in this case, the hybrid gene contained no ATA box. The details of the construction are given in Weidle and Weissmann (11).

For deletion mutants, a 7.5-kb BamHI fragment containing the human IFN-α1 gene and 5.4 kb of 5' and 1.2 kb of 3' flanking sequences (11) was ligated into the PvuII site of pBR322. The resulting plasmid, pD6, in which the orientation of the IFN gene was opposite to that of the β-lactamase gene, was used to construct 5' deletion mutants, essentially as described by Frischauf et al. (12). Details of the construction are given in Ragg and Weissmann (13).

Other methods.

The plasmids described above were introduced into thymidine kinase (TK) negative mouse L cells as described (7). However, rather than analyze many individual cell lines from one transformation experiment, we pooled 500-800 TK^+ colonies and propagated the cells as mixtures. Parallel plates were either mock-induced or treated with Newcastle Disease Virus (NDV); 11 h after virus addition, poly(A) RNA was prepared and analyzed by S_1 mapping (14,15).

RESULTS AND DISCUSSION

Mouse L cells were transformed with hybrid genes in which the β-globin transcription unit was joined either to an IFN-α1 5' flanking sequence comprising the ATA box (pIG-I) or lacking it (pIG-II) (cf. Fig. 1). Transcripts of uninduced and induced cells were mapped with a 5'-^{32}P-labelled cognate minus strand probe spanning the putative cap site. The 5' end of a β-globin transcript originating 30 bp downstream (9) of the ATA box in pIG-I should map 13 bp upstream of the β-globin CAP site (Fig. 1E) and with the probe used should give an S_1 fragment of 148 bp, while authentic β-globin mRNA should give a 135-bp fragment. S_1 mapping (Fig. 2) with the probe derived from pIG-I gave the expected 135-bp signal for authentic rabbit β-globin mRNA, as well as a signal due to the renatured 219-bp probe (lanes a and b). Poly(A) RNA from the virus-treated (lane d), but not from the mock-infected cells transformed with pIG-I (lane c) gave a strong signal at about 150 bp, as well as a 219-bp band due to renatured probe and possibly to transcripts originating upstream of position -69. About 10-25 correctly initiated transcripts per cell were present 11 h after induction, as estimated by comparing the intensity of the 148-bp band with that of the 135-bp band given by known amounts of authentic β-globin mRNA, and performing the calculations described earlier (7). All samples from β-globin DNA-containing cells gave a strong 87-97 bp band due to β-globin

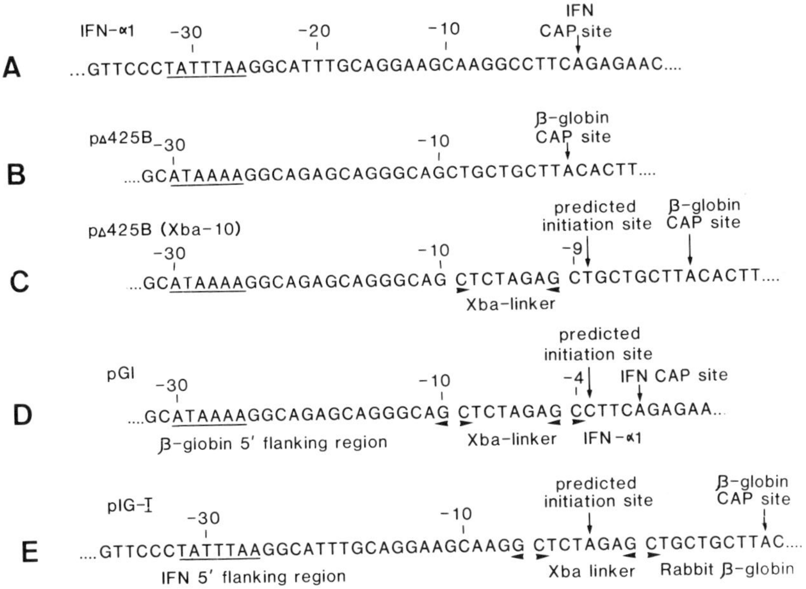

FIGURE 1. Nucleotide sequences at the junction of the 5' flanking region and the transcription unit of unmodified and hybrid genes.
(A) Human IFN-α1 (B) rabbit β-globin (pΔ425B) (9) (C) rabbit β-globin DNA pΔ425B with XbaI linker in position -9/-10 (D) β-globin-IFN-α1 hybrid (pGI) (E) IFN-α1 β-globin hybrid (pIG-I). The ATA box is underlined. The numbering is relative to the CAP site of the original sequences.

transcripts protecting the probe up to position 40. Such transcripts were previously found in cells transformed by cloned rabbit β-globin genes and are attributed to abnormal initiation or to aberrant splicing (10,16,17). As the level of these abnormal transcripts was about the same for the induced and non-induced samples, the absence of the 148-nucleotide signal in the non-induced sample was not due to degradation or losses during RNA isolation or S1 analysis.

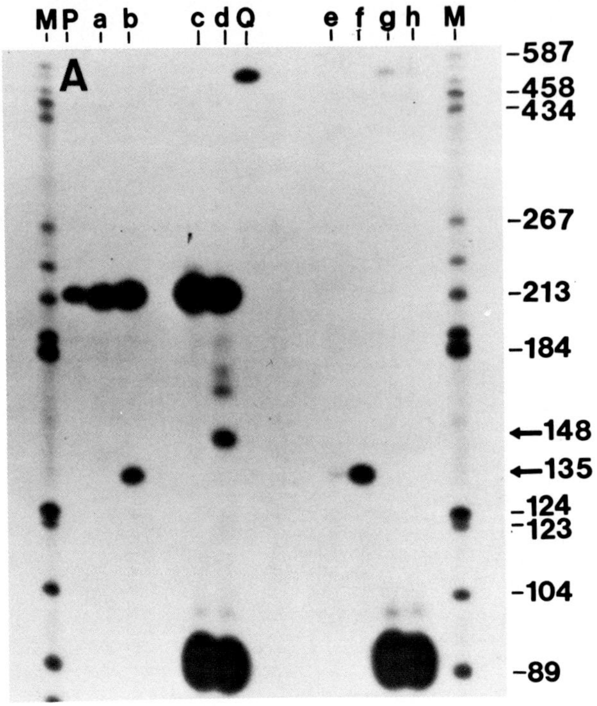

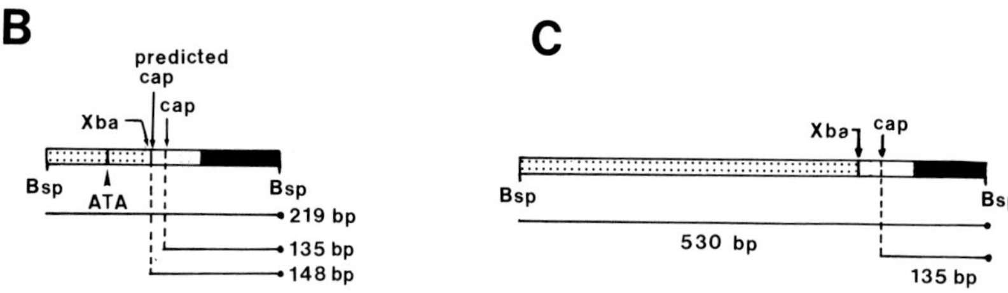

FIGURE 2. (A) S_1-mapping of rabbit β-globin transcripts in cells transformed with hybrid genes consisting of an IFN-α1 5' flanking region followed by the β-globin transcription unit (pIG). Lanes a and b, 2 pg and 100 pg β-globin mRNA; lanes c and d, RNA from cells transformed with pIG-I (c, mock infected cells: d, NDV infected cells); lanes e and f, 20 pg and 100 pg β-globin mRNA; lanes g and h, RNA from cells transformed with pIG-II (g, NDV-infected cells; h, mock-infected cells); lanes P and Q, undigested

Poly(A) RNA from induced and non-induced cells transformed with pIG-II gave no signals other than the 87 to 97-bp band due to abnormal transcripts, and a weak band in the position of completely protected probe (Fig. 2, lanes g and h).

The fact that correctly initiated β-globin RNA appeared in cells transformed with pIG-I after viral infection could be due, at least in part, not to increased transcription but to reduced turnover. To assess this possibility we examined the effect of virus infection on the expression of the rabbit β-globin and the human IFN-α1 transcription units, both under the control of the rabbit β-globin promoter. Cells were transformed with concatenates containing either the normal rabbit β-globin gene (pΔ425B) (8), the rabbit β-globin gene with an XbaI linker at position -9/-10 (pΔ425 (Xba-10)), or the IFN-α1 transcription unit joined to the β-globin promoter (pGI), and mixtures of transformed cell lines were prepared as above.

S_1 analysis of poly(A) RNA from cells transformed with pΔ425B or pΔ425B (Xba-10) was with a 5'-^{32}P-labelled 210-bp BspI fragment (-75 to 135) from pΔ425B. Correctly initiated transcripts should protect a 135-bp fragment in the case of pΔ425B. In the case of pΔ425B (Xba-10) a fragment of around 143 bp was expected because initiation should be displaced about 8 bp upstream due to the XbaI linker between the ATA box and the CAP site (9) (cf. Fig. 1). Fig. 3 (lanes d-g) shows the expected signals; the transcript levels, estimated as described above, were about 250 and 170 per cell in mock-infected cells transformed with pΔ425B and pΔ425B(Xba-10) respectively, and about

---

probes; lane M, pBR322 digested with BspI and 5'-^{32}P-labelled. (B) and (C) Probes for mapping pIG-I and pIG-II RNA, respectively. Shaded regions, 5' flanking region of the IFN-α1 gene, either with (B) or without (C) the ATA box. Open boxes, 5'-non-translated regions of transcription units. Black boxes, translated region of rabbit β-globin gene.

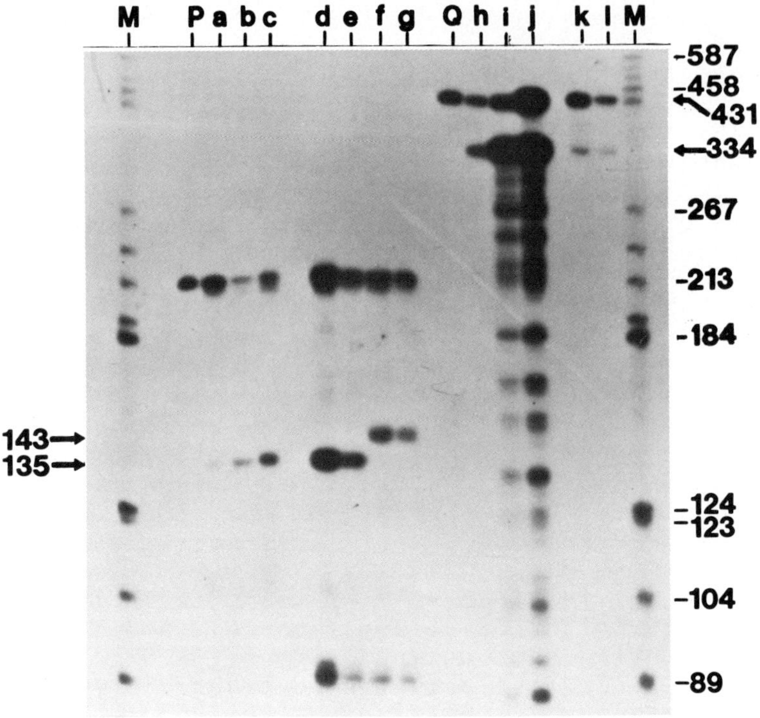

FIGURE 3. S_1-mapping of transcripts in cells transformed with a hybrid gene consisting of the rabbit β-globin 5' flanking region followed by the IFN-α transcription unit (pGI), or with β-globin genes. Lanes a, b, and c, 40 pg, 120 pg and 200 pg rabbit β-globin mRNA, respectively; lanes d and e, poly(A) RNA from 100 μg of total RNA from cells transfected with pΔ425B (d, mock-infected, e, NDV-infected); f and g, poly(A) RNA from cells transfected with pΔ425 (Xba-10), (f, mock-infected, g, NDV infected); lanes h, i and j, 10 ng, 100 ng and 300 ng, respectively, poly(A)-RNA from IFN-producing leukocytes (containing about 0.4% IFN-mRNA (7)); lanes k and l, poly(A) RNA from 400 μg of total RNA from cells transfected with pGI (k, mock-infected; l, infected with NDV). Lane P, undigested globin probe, lane Q, undigested IFN probe, lane M, pBR322 digested with BspI and 5'-^{32}P-labelled.

half that in virus-infected cells. The transcripts derived from pGI were S_1-mapped with a 5'-^{32}P-labelled minus strand probe which extended from position +330 (IFN-α1) to -99 (β-globin). RNA molecules initiating 30 bp downstream from the ATA box (i.e. 4 bp upstream from the IFN-α1 CAP site (cf. Fig. 1)) should protect a 334-bp probe fragment. As shown in Fig. 3 (lanes k and l), the expected signals appeared in all cases; mock infected cells contained about 1-2, and NDV-infected cells about 1 transcript per cell. In all cases mouse IFN-α1 mRNA was present at similar levels after virus infection and was absent after mock infection (data not given), showing that induction had been effective. We conclude that neither the IFN-α1 nor the β-globin RNA was stabilized by virus infection.

These experiments provide strong evidence that the induction of IFN-α1 mRNA is primarly due to activation of transcription, that transcriptional control is exerted via the 5' flanking region of the IFN-α1 gene and that the rate of IFN-α1 mRNA turnover is not reduced (and possibly even increased) after viral induction. We argue as follows. The level of β-globin mRNA in cells transformed with a β-globin gene decreased 11 h after induction, therefore, β-globin mRNA was not stabilized by induction. In drawing this conclusion we disregard the less likely possibility that induction stabilized β-globin transcripts but that this effect was more than offset by a reduction in β-globin transcription. The level of β-globin RNA in cells transformed with a hybrid gene consisting of the IFN-α1 promoter and the β-globin transcription unit was nil prior to, and 10-25 transcripts per cell following induction. If this rise is not attributed to stabilization of β-globin mRNA, it must be due to activation of transcription. Cells transformed with a hybrid gene consisting of the β-globin promoter and the IFN-α1 transcription unit contained more IFN-α1 transcripts prior to induction than 11 h thereafter. We conclude that IFN-α1 RNA is also not stabilized by induction, and that its accumulation is due mainly to transcriptional activation. In drawing these conclusions we assume that the transcripts of the hybrid

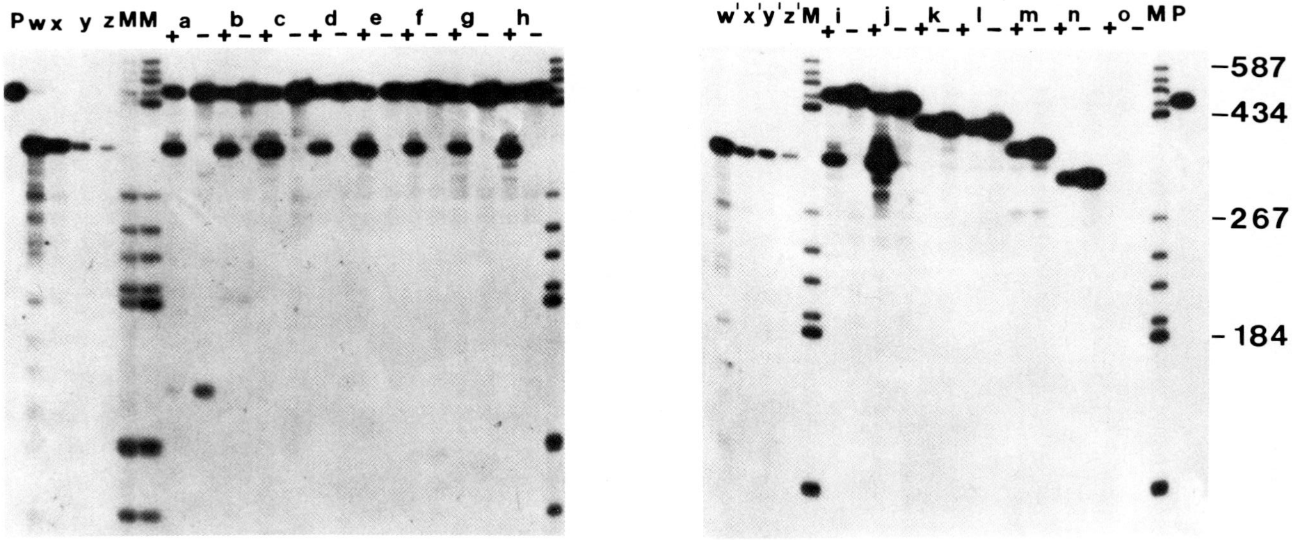

FIGURE 4. S_1-mapping of human IFN-α1 RNA from transformed mouse cells. Lanes w, x, y, z; 20, 6.6, 2 and 0.66 ng of poly(A)$^+$ RNA from IFN-producing human leukocytes; lanes w', x', y', z',; 16, 4, 2 and 0.5 ng of poly(A)$^+$ RNA from IFN-producing human leukocytes; lanes a-n; poly(A)$^+$ RNA from mouse cells transformed with the various 5' deletion mutants of the human IFN-α1 gene, either induced (+) or mock-induced (−). Deletion endpoints: (a)-5400 (b)-1100 (c)-565 (d)-385 (e)-315 (f)-235 (g)-166 (h)-145 (i)-131 (j)-117 (k)-74 (l)-68 (m)-31 (n)+18. Lanes o, RNA from mouse cells transfected with linearized pTKM-104 only; lanes M, pBR322 digested with BspI and 5'-^{32}P-labelled; lanes P, undigested probe.

genes have the same stabilities as natural β-globin and IFN-α1 mRNAs, even though their structures differ slightly at the 5' termini.

Which parts of the 5' flanking sequence of the IFN-α1 gene are required for transcription? We have shown that the ATA box is essential for β-globin transcription from the hybrid gene, and that the transcripts initiated 30$\pm$ 3 bp downstream of the ATA box and not at the CAP site of the β-globin gene. These results confirm the role generally attributed to the ATA box, namely that of determining the transcription initiation site and sustaining maximal transcription of many (but not all) genes (9,18,19).

To determine the minimal length of 5'-flanking region required for inducible transcription, a set of external 5'-deletion mutants was constructed as described in Materials and Methods. The mutants were introduced into TK^- mouse L cells and IFN-α1 transcripts from uninduced and induced cells analyzed by S_1 mapping. The probe used for S_1-mapping was a 464-bp BamHI-BglII fragment derived from a deletion mutant (end point -131), which was 5'-^{32}P-labelled at the BglII end of the minus strand and spanned the CAP site.

As shown in Fig. 4 (lanes w-z and w'-z') poly(A)$^+$ RNA from IFN-producing human leukocytes protected a probe fragment of about 330 bp. No such signal was observed with RNA from induced or mock induced mouse L cells transfected with the TK-gene alone (lane o), showing that there was no cross-hybridization of the human probe with mouse IFN RNA. RNA from induced cells transformed with the modified human IFN-α1 genes gave a strong signal indistinguishable from that of human leukocyte RNA, provided that the transcription unit was preceded by 117 bp or more of 5' flanking sequence. The steady state level of the correctly initiated transcripts after induction was estimated to be approximately 10 to 20 molecules per cell. The values for the number of strands per cell represent average values, as they were obtained from pooled clones which probably expressed the human IFN gene to variable extents (7). After mock induction no correct transcripts (less than about 0.1% per cell) were detected, except in the case of cells transformed by the gene with only 117 bp of 5' flanking region. These not only had an exceptionally high transcript level after induction, about 80 copies per cell, but also a low level (about 0.1-0.3 molecules per cell) of correctly initiated transcripts after mock induction (Fig. 4, lane j). Mutants with 74 to 31 bp of 5' flanking sequences had a very low level of correctly initiated RNA molecules, whether they were induced or mock induced.

Strong additional signals due to larger probe fragments appeared in all cases. Most likely these signals were predominantly due to transcripts originating upstream of the CAP site and to a lesser extent, if at all, to renatured probe. This is evident whenever the deletion end point of the probe DNA was located upstream of the deletion end point of the transforming DNA. In these cases (Fig. 4, lanes j-n) the transcripts originating upstream of the CAP site did not protect the entire length of the probe and generated a signal distinct from that of the renatured probe (Fig. 4, lanes P). We conclude from these experiments that infection of mouse L cells with NDV selectively stimulates correct transcription of the exogenous human IFN-α1 gene and that 117 bp of 5' flanking sequence are sufficient to mediate this induction. These findings are analogous to those made for the murine metallothionein I promoter (20) and the Drosophila heat shock (HSP70) promoter (21), where 90 and 66 bp of 5' flanking sequence, respectively, are sufficient for induction of transcription. Further insights may be obtained by testing hybrid promoters, composed of elements from the IFN-α and the β-globin 5' flanking sequence.

ACKNOWLEDGMENTS

This work was supported by the Schweizerische Nationalfonds (3.147.81), the Deutsche Forschungsgemeinschaft and the Kanton of Zürich. We thank S. Nagata and A. van Ooyen for plasmids, P. Dierks and J.-I. Fujisawa for γ^{32}P-ATP and H. Arnheiter for NDV.

REFERENCES

1. Stewart, II, W. (1979). The Interferon System (Springer, New York).
2. Berg, K. (1982). Acta Path. Micro. Immunol. Scand. Section C, Suppl. no. 279.
3. Raj, N.B.K. and Pitha, P.M. (1981). Proc. Natl. Acad. Sci. USA 78, 7426-7430.
4. Cavalieri, R.M., Havell, E., Vilček, J. and Pestka, S. (1977). Proc. Natl. Acad. Sci. USA 74, 4415-4419.
5. Pang, R.H., Hayes, T.G. and Vilček, J. (1980). Proc. Natl. Acad. Sci. USA 77, 5341-5345.
6. Morser, J., Flint, J., Meager, A., Graves, H., Baker, P., Colman, A. and Burke, D.J. (1979). J. gen. Virol. 44, 231-234.
7. Mantei, N. and Weissmann, C. (1982) Nature 297, 128-132.
8. Dierks, P., van Ooyen, A., Mantei, N. and Weissmann, C. (1981). Proc. Natl. Acad. Sci. USA 78, 1411-1415.
9. Nagata, S., Mantei, N. and Weissmann, C. (1980). Nature 287, 401-408.
10. Dierks, P., van Ooyen, A., Cochran, M.D., Dobkin, C., Reiser, J. and Weissmann, C. (1983). Cell 32, 695-706.
11. Weidle, U. and Weissmann, C. (1983). Nature, in press.
12. Frischauf, A.M., Garoff, H. and Lehrach, H. (1980). Nucl. Acids Res. 8, 5541-5549.
13. Ragg, H. and Weissmann, C. (1983). Nature, in press.
14. Berk, A.J. and Sharp, P.A. (1977). Cell 12, 721-732.
15. Weaver, R.F. and Weissmann, C. (1979). Nucl. Acids Res. 7, 1175-1193.
16. Mantei, N., Boll, W. and Weissmann, C. (1979). Nature 281, 40-46.
17. Grosveld, G., de Boer, E., Shewmaker, C.K. and Flavell, R.A. (1982). Nature 295, 120-126.
18. Grosschedl, R. and Birnstiel, M.L. (1980). Proc. Natl. Acad. Sci. USA 77, 1432-1436.
19. Benoist, C. and Chambon, P. (1981) Nature 290, 304-310.

20. Brinster, R.L., Chen, H.Y., Warren, R., Sarthy, A. and Palmiter, R.D. (1982). Nature 296, 39-42.
21. Pelham, H.R.B. (1982). Cell 30, 517-528.

IV. GENE REGULATION IN DEVELOPMENT

Gene Expression, pages 235-247
© 1983 Alan R. Liss, Inc., 150 Fifth Avenue, New York, NY 10011

ANATOMY OF A COMPLEX PROCARYOTIC PROMOTER UNDER DEVELOPMENTAL REGULATION[1]

W. Charles Johnson, Charles Moran, Jr., Carl Banner, Peter Zuber and Richard Losick

Department of Cellular and Developmental Biology, Harvard University, Cambridge, Massachusetts 02138

ABSTRACT A developmentally regulated gene (spoVG) from the spore-forming bacterium Bacillus subtilis is expressed from two overlapping promoters known as P1 and P2. Utilization of the upstream promoter P1 is determined by an RNA polymerase sigma factor of 37,000 daltons (σ^{37}) whereas a 32,000 dalton sigma species (σ^{32}) dictates transcription initiation from P2. Deletion analysis and DNase footprinting experiments show that P1 and P2 overlap extensively and that their functional boundaries extend from the vicinity of the transcription startpoints, which are 10 base pairs apart, up through an AT-rich box, which is largely composed of alternating stretches of As and Ts. Gene fusion experiments in which the spoVG promoter region was joined to the lacZ gene of E. coli show that developmental regulation is exerted at or near the region of transcription initiation.

INTRODUCTION

The formation of endospores in the gram-positive bacterium Bacillus subtilis is a favorable system for the study of cellular differentiation as this organism is subject to facile genetic and biochemical analysis. Classical systems of genetic transformation and transduction (1) and newly developed systems of genetic transposition (2) and DNA cloning (3) have made it

[1] This work was supported by NIH grant GM18568

possible to identify, isolate and manipulate with relative ease genes that control the process of spore formation. Moreover, although surprisingly complex, the B. subtilis machinery for transcribing these genes has proven to be readily accessible to biochemical reconstruction in vitro (4, 5).

Our paradigm for the study of transcriptional control during spore formation is spoVG, a developmentally regulated gene of 400 base pairs from the origin region of the B. subtilis chromosome (6, 7, 8). The transcription of spoVG is induced at the very onset of the sporulation process, although its protein product does not function until an intermediate stage of development. spoVG RNA synthesis is subject to regulation by a class of genes known as the spo0 loci; the products of the spo0 genes control the initiation phase of sporulation and a mutation at any one of at least five of these regulatory loci severely restricts the induction of spoVG. Here we report on the structure of spoVG promoter region and present evidence suggesting that developmental regulation of this sporulation gene is exerted at or near the region of transcription initiation.

STRUCTURE OF THE spoVG PROMOTER REGION: TWO TRANSCRIPTION STARTPOINTS

Fig. 1 shows the nucleotide sequence of the non-coding strand of the 5' region of the spoVG gene (9). The position of bases in this sequence is shown by distance in base pairs from an EcoR1* site (not shown) located within the spoVG protein coding sequence (7). S1 nuclease mapping experiments with RNA from cells undergoing sporulation have revealed a curious feature of spoVG: namely that its transcription apparently originates from two startpoints, which we call P1 and P2 (9). These are ten base pairs apart and are located at positions 120 and 110, respectively (that is, 120 and 110 base pairs upstream from the EcoR1* site). P1 and P2 immediately precede the ribosome binding site (AAAGGTGGTGA), initiation codon (GTG) and open reading frame for spoVG mRNA (10). The ribosome binding site displays extensive complementarity to the 3' terminal region of B. subtilis 16S rRNA, (the predicted stability of interaction with 16S

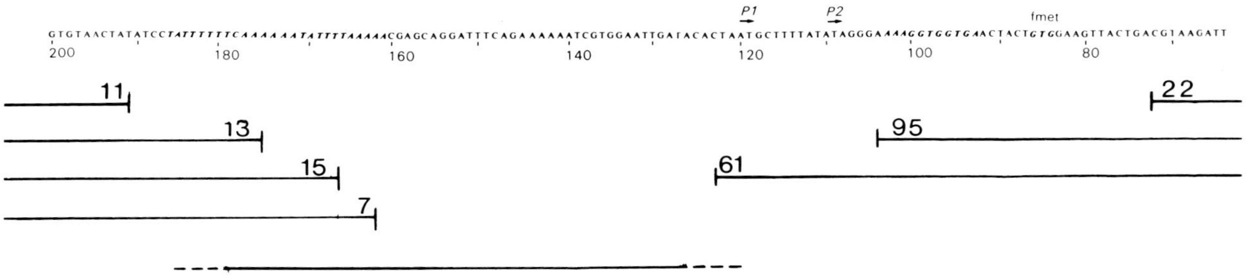

FIGURE 1. Structure of the spoVG promoter region. The nucleotide sequence of the non-coding strand is shown. The AT-rich box, the ribosome binding site and initiation codon are shown in italics. The position of bases is shown in distance from a downstream EcoR1* site (7). The numbered lines below the sequence represent the end points of upstream and downstream deletions (see the text). The line at the bottom of the figure represents the region of DNase protection observed in the footprinting experiment of Fig. 2. The dotted lines at the extremes of the footprint represent regions whose protection was uncertain.

rRNA is -19 Kcal), a characteristic which is emerging as a general and possibly necessary feature of B. subtilis mRNAs (11, 12). A fusion in frame at the nineth codon of the open reading frame to the lacZ gene of E. coli causes active β-galactosidase synthesis in B. subtilis cells, a finding which supports the notion that these translation initiation signals are physiologically significant (P. Zuber, unpublished results).

Another curious feature of spoVG is the presence upstream of the P1 and P2 start sites of a region that is strikingly rich in AT base pairs (25 out of 26 AT base pairs). This AT-rich box, which extends from position 161 to 186, is largely composed of alternating stretches of As and Ts. The significance of this unusual upstream sequence will be considered below.

TWO RNA POLYMERASE SIGMA FACTORS DETERMINE THE UTILIZATION OF P1 AND P2

In contrast to the apparent homogeneity of RNA polymerase in E. coli and other Gram-negative bacteria, B. subtilis RNA polymerase exists in multiple holoenzyme forms (13). Each form of holoenzyme contains a particular species of sigma factor, which confers on a common core RNA polymerase the recognition of a characteristic class of promoters. Five such B. subtilis sigma species have now been described in addition to two regulatory sigmas encoded by the virulent B. subtilis phage SP01. These are listed in Table 1. In at least five cases (σ^{55}, σ^{37}, σ^{29} and the phage SP01-coded sigmas) these species are known to be products of separate genes (14, 15).

Neither of the spoVG promoters is used by σ^{55}, the principal form of B. subtilis holoenzyme and the form most nearly homologous to E. coli RNA polymerase (11, 16). We have previously described a 37,000 dalton species of sigma factor (σ^{37}), which dictates transcription initiation from the upstream promoter P1, but was completely unable to promote RNA synthesis from P2 (9, 17, 18). Recently, however, we have succeeded in isolating a 32,000 dalton sigma (σ^{32}), which exclusively dictates transcription from the downstream promoter (19). Thus, spoVG is emerging as a highly complex transcription initiation region whose utilization is determined by two distinct forms of RNA polymerase holoenzyme.

TABLE 1
B. SUBTILIS SIGMA FACTORS

Sigma	Reference
B. subtilis σ^{55}	(16)
B. subtilis σ^{37}	(17,18)
B. subtilis σ^{32}	(19)
B. subtilis σ^{29}	(20)
B. subtilis σ^{28}	(21)
Phage SP01 σ^{gp28}	(22)
Phage SP01 $\sigma^{gp33-34}$	(23)

P1 AND P2 OVERLAP EXTENSIVELY

To map the functional boundaries of spoVG, we (24) constructed a series of deletion mutations that extend into the transcription initiation region from the upstream and downstream directions (Fig. 1). These deletion-mutated DNAs were cleaved at a downstream endonuclease restriction site and then used as templates in a "run off" transcription assay. In the case of deletions from the downstream direction, the elimination of DNA to position 103 (deletions 22 and 95 in Fig. 1) had no measurable effect on the ability of spoVG gene to support run off RNA synthesis by $E\sigma^{37}$ and $E\sigma^{32}$. A deletion to position 122 (deletion 61), however, completely abolished the utilization of both P1 and P2. We conclude that the right hand boundaries of the P1 and P2 promoters are in the general vicinity of the RNA synthesis start sites (between positions 103 and 122).

A more complicated picture of promoter structure emerges from an examination of the transcription of templates bearing deletions from the upstream direction. Deletion 11, which extends close to the AT-rich box (see above), had no detectable effect on promoter utilization. Deletions extending into the AT-rich box, however, severely impaired transcription from both P1 and P2 and this effect was progressive: the degree of transcription inhibition being partial for deletion 13, more severe for deletion 15 and essentially total for deletion 7. We conclude that the upstream AT-rich box strongly influences the efficiency of utilization of both promoters. Moreover, since in our "pre-binding" protocol, polymerase was pre-incubated at 37° with excess template and then allowed to initiate RNA synthesis by the addition of NTPs, the AT-rich box is presumably not simply a "promoter flag," which facilitates the ability of polymerase to find the spoVG promoters. Rather, this sequence must act, we believe, in the binding or initiation phase of promoter utilization.

Another important conclusion from this deletion analysis is that the P1 and P2 promoters must overlap extensively. This was not unexpected as the dual transcription startpoints are separated by only ten base pairs, a small distance compared with the size of promoters (approximately sixty base pairs). To attempt to visualize this directly, we employed the "footprinting" procedure of Galas and Schmitz (25) to scan spoVG DNA for

---

FIGURE 2. $E\sigma^{37}$ and $E\sigma^{32}$ binding to spoVG DNA. An EcoR1-HaeIII fragment of 300 bp that had been labeled with ^{32}P-phosphate at the EcoR1* terminus downstream from the spoVG promoters was isolated from plasmid pCB13011 (9). End-labeled DNA (0.3 pmol) was then incubated for 10 min at 37° with $E\sigma^{37}$(1.8 μg), ATP (0.1 mM) and 0.1 mM UpA ("+"; track b), a dinucleotide which primes initiation at P1 (9), or with $E\sigma^{32}$ (0.1 μg), ATP (0.1 mM) and 0.1 mM ApU ("+"; track f), a dinucleotide which primes initiation exclusively at P2 (9), or in the absence of polymerase (tracks a, c, e and g). After incubation, the radioactive DNAs were treated with DNase I as modified from the "footprinting" procedure of Galas and Schmitz (25) by Talkington and Pero (27) denatured at 90° and subjected to electrophoresis either in a 7% polyacrylamide /8.3 M urea slab gel (tracks a-e) or in 8% acrylamide /8.3 M urea slab gel (tracks e-g). Correlation of fragments with the nucleotide sequence of spoVG as represented by their distance from the EcoR1* terminus (indicated by numbers in the figure) was determined by electrophoresis on the slab gels of the products of the base specific chemical cleavage reaction "G greater than A" of Maxam and Gilbert (28) of the end-labeled DNA (for example, track d).

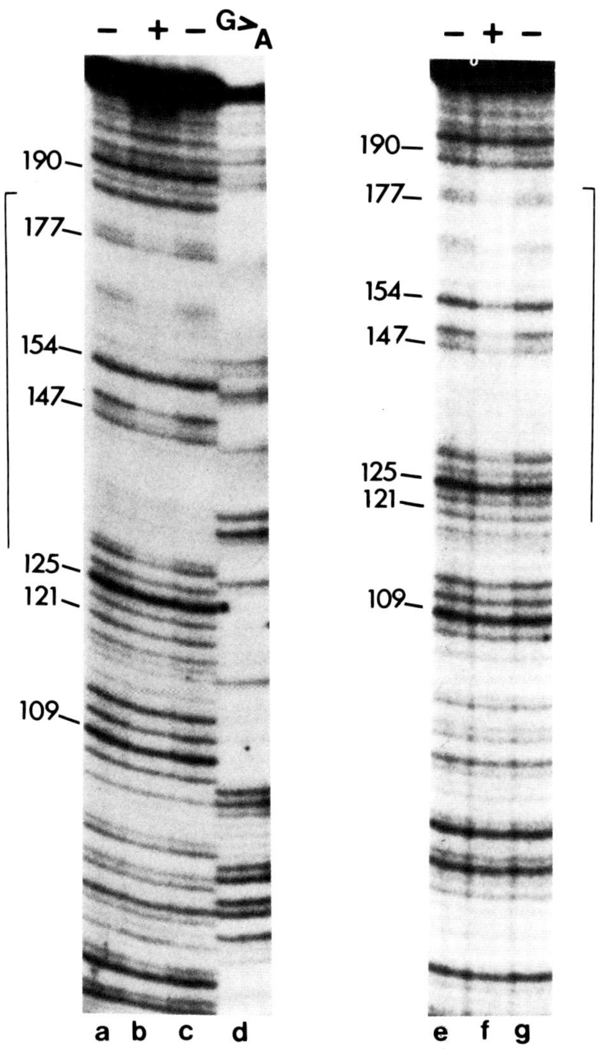

- + - G>A
190
177
154
147
125
121
109
a b c d
- + -
190
177
154
147
125
121
109
e f g

nucleotides whose susceptibility to DNase I action was reduced subsequent to binding by $E\sigma^{37}$ and $E\sigma^{32}$. To accomplish this, initiation complexes were formed between each holoenzyme and promoter-bearing DNA that had been uniquely labeled with ^{32}P-phosphate at the EcoR1* terminus downstream from P1 and P2. End-labeled DNAs containing (or lacking) prebound $E\sigma^{37}$ or $E\sigma^{32}$ were partially digested with DNase I and the products of digestion were then subjected to electrophoresis in a high resolution nucleotide sequencing gel. Although the boundaries between protected an unprotected DNA were not definite, the autoradiograph of Fig. 2 (tracks b and f) shows that each holoenzyme partially protected a stretch of fifty to seventy base pairs from the action DNase I and that the location (upstream from the transcription startpoints) of these regions of partial protection were similar in $E\sigma^{37}$ and $E\sigma^{32}$ initiation complexes. In the case of P1, a similar pattern of DNase I protection could also be observed (albeit weakly), in simple binary complexes of $E\sigma^{37}$ and spoVG DNA (not shown).

As is evident from Fig. 2, the spoVG promoters are very weak binding sites for RNA polymerase. [This is not, however, a general characteristic of novel B. subtilis promoters as we (26) have shown that another sporulation promoter known as spoVC (ctc) is a very strong binding site for $E\sigma^{37}$.] Nevertheless, both $E\sigma^{37}$ and $E\sigma^{32}$ were clearly able to protect nucleotides in the AT-rich box, a finding which supports the view that this upstream region functions in the binding or initiation phase of promoter utilization.

PROMOTER RECOGNITION SIGNALS

What specific sequences signal promoter recognition by $E\sigma^{37}$ and $E\sigma^{32}$? E. coli promoters approximately conform to each other in hexanucleotide sequences centered at about 35 and 10 base pairs upstream from the start site of transcription. These are known as the "-35" (TTGACA) and "-10" (TATAAT) sequences. In B. subtilis, where RNA polymerase is heterogenous, promoters recognized by the principal holoenzyme form $E\sigma^{55}$ also strongly conform to the canonical sequences observed in E. coli. Interestingly, this pattern of conserved -35 and -10 sequences is now emerging as a general feature of each of the promoter classes recognized by the various forms of holoenzyme in *B. subtilis*, although in each instance the particular class of conserved sequences differs to a greater or lesser extent from the *E. coli* paradigm (13). In the best documented case of *B. subtilis* phage SPO1 *middle* gene promoters, which are controlled by the phage-coded sigma factor σ^{gp28}, the canonical -35 and -10 sequences are AGGAGA and TTTNTTT, respectively (29).

Nucleotide sequence homologies among a variety of $E\sigma^{37}$ utilized promoters (9, 26, 30) and purine methylation protection experiments (31) with the strongly utilized *spoVC* (*ctc*) promoter suggest that, like other forms of holoenzyme, $E\sigma^{37}$ interacts with approximately conserved sequences centered at about 35 and 10 base pairs preceding the transcription start site. In the P1 promoter of *spoVG* these putative recognition sequences are centered at positions 152 (AGGATT) and 131 (GGAATTGAT) or, in other words at 32 and 11 base pairs upstream from the P1 start site (Fig. 1). Similarly, although our evidence in this case rests simply on homology to one other σ32-controlled promoter (19), $E\sigma^{32}$ interacts, we hypothesize, with sequences centered at position 140 (AAATC) and 117 (TAATGCTTTA), which are 30 and 7 base pairs, respectively, upstream from the P2 start site. Thus, if our proposal is correct, the *spoVG* transcription initiation region is a mosaic in which recognition sequences for $E\sigma^{37}$ and $E\sigma^{32}$ are arranged in the alternating order $E\sigma^{37}$ "-35", $E\sigma^{32}$ "-35", $E\sigma^{37}$ "-10" and $E\sigma^{32}$ "-10". We plan to test this hypothesis by creating *in vitro* point mutations that separately impair P1 and P2.

DEVELOPMENTAL REGULATION

Superimposed on the transcriptional complexity of the *spoVG* promoter region is a regulatory system that controls the expression of this developmental gene during the initiation phase of sporulation. A mutation in any one of at least eight genes known as the *spo0* loci prevent formation of the asymmetric septum, which is the first morphological manifestation of the sporulation process. A mutation in any one of at least five of these regulatory genes blocks or substantially restricts transcription of *spoVG* (7). Two lines of evidence suggest that the regulatory effect of the spo0 gene products is exerted at or near the promoter region of the spoVG gene.

We (P. Zuber, and R. Losick, unpublished work) have constructed a gene fusion in which a 157 base pair DNA segment (extending from position 60 to position 217) containing the promoter region of spoVG and the first nine codons of its protein coding sequence was joined in frame to the lacZ gene of E. coli. This gene fusion was then inserted into the genome of the temperate B. subtilis phage SPβ, thereby creating a specialized transducing phage which permitted us to transfer the spoVG-lacZ fusion into a variety of genetic backgrounds. The synthesis of β-galactosidase in lysogens of the SPβ transducing phage was induced by nutrient depletion and this induction was substantially reduced in cells harboring mutations in spo0A, spo0B, spo0E, spo0F or spo0H. We conclude from this that the spo0 gene products are transcriptional control proteins and that one or more spo0 gene products or a protein under spo0 control acts at or very near the spoVG promoters to control their utilization during the initiation phase of sporulation.

As further evidence that the spoVG promoters are the site of action of a sporulation-specific regulatory protein(s), we (24) have observed that amplification of spoVG on a multicopy plasmid inhibits sporulation at an early stage. Deletion analysis using the set of upstream and downstream deletions described in Fig. 1 has demonstrated a strong correlation between the sporulation inhibitory effect and the functional boundaries of the overlapping spoVG promoters. We interpret this to indicate that amplification of the spoVG promoter region is titrating a sporulation-specific regulatory protein that is present in very low concentrations. This protein could be the product of a spo0 gene or σ^{32}, the determinant of P2 transcription, as this sigma species is present in only about 20 molecules per cell (19).

In summary, these observations bring into sharp focus the general problem of the way in which the spo0 gene products activate the transcription of developmental genes in the initiation phase of the sporulation process. In our paradigm case of spoVG, the pathway(s) of spo0 regulatory proteins exerts its effect at or near the spoVG start sites, a complex transcription control region whose utilization is determined by two novel forms of B. subtilis holoenzyme. We now seek to understand the molecular details of the way in which the spoO gene products stimulate the utilization of P1 and P2 and the way in which transcription from these overlapping promoters is triggered by nutrient depletion.

REFERENCES

1. Henner DJ, Hoch JA (1980). The Bacillus subtilis chromosome. Microbial Rev 44:57.
2. Youngman PY, Perkins JB (1983). Genetic transposition and insertional mutagenesis in Bacillus subtilis with Streptococcus faecalis transposon Tn917. Proc Natl Acad Sci USA (April, in press).
3. Ferrari E, Henner DJ, Hoch J. (1981). Isolation of Bacillus subtilis genes from a charon 4A library. J Bacteriol 146:430.
4. Losick R (1982). Sporulation genes and their regulation. In Dubnau D (ed): " Molecular Biology of the Bacilli, Vol I: Bacillus subtilis," New York: Academic Press, p 179.
5. Doi R (1982). RNA polymerase of Bacillus subtilis. In Dubnau D, (ed): " Molecular Biology of the Bacilli, Vol I: Bacillus subtilis," New York: Academic Press, p 72.
6. Segall J, Losick R (1977). Cloned Bacillus subtilis DNA containing a gene that is activated early during sporulation. Cell 11:751.
7. Ollington JF, Haldenwang WG, Huynh TV, Losick R (1981). Developmentally-regulated transcription in a cloned segment of the Bacillus subtilis chromosome. J Bacteriol 147:432.
8. Rosenbluh A, Banner CDB, Losick R, Fitz-James PC (1981). Identification of a new developmental locus in Bacillus subtilis by construction of a deletion mutation in a cloned gene under sporulation control. J Bacteriol 148:341.
9. Moran CP Jr, Lang N, Banner CBD, Haldenwang WG, Losick R (1981). Promoter for a developmentally-regulated gene in Bacillus subtilis. Cell 25:783.
10. Moran CP Jr, Lang N, Losick R (1982). Anatomy of a sporulation gene: Nucleotide sequences that signal the initiation of transcription and translation. In Ganesan AT, Chang S, Hoch JA (eds): "Molecular

Cloning and Gene Regulation in Bacilli," New York: Academic Press, p 325.
11. Moran CP Jr, Lang N, LeGrice S, Lee G, Stephens M, Sonenshein AL, Pero J, Losick R (1982). Nucleotide sequences that signal the initiation of transcription and translation in Bacillus subtilis. Mol Gen Genet 186:339.
12. McLaughlin J, Murray C, Rabinowitz J (1981). Unique features in the ribosome binding site sequence of the Gram-positive Staphylococcus aureus β-lactamase gene. J Biol Chem 256:11283.
13. Losick R, Pero J (1981). Cascades of sigma factors. Cell 25:582.
14. Haldenwang WG, Truitt CL (1982). Peptide maps of regulatory subunits of Bacillus subtilis RNA polymerase. J Bacteriol 151:1624.
15. Wong S-L, Doi RH (1982). Peptide mapping of Bacillus subtilis RNA polymerase σ factors and core-associated polypeptides. J Bio Chem 257:11932.
16. Shorenstein RG, Losick R (1973). Comparative size and properties of the sigma subunits of RNA polymerase from Bacillus subtilis and Escherichia coli. J Biol Chem 248:6170.
17. Haldenwang WG, Losick R (1979). A modified RNA polymerase transcribes a cloned gene under sporulation control in Bacillus subtilis. Nature 282:256.
18. Haldenwang WG, Losick R (1980). A Novel RNA polymerase sigma factor from Bacillus subtilis. Proc Natl Acad Sci USA 77:7000.
19. Johnson WC, Moran CP Jr, Losick R (1983). Two RNA polymerase sigma factors from Bacillus subtilis discriminate between overlapping promoters for a developmentally regulated gene. Nature 302: in press.
20. Haldenwang WG, Lang N, Losick R (1981). A sporulation-induced sigma-like regulatory protein from Bacillus subtilis. Cell 23:615.
21. Wiggs J, Gilman M, Chamberlin M (1981). Heterogeneity of RNA polymerase in Bacillus subtilis: Evidence for an additional sigma factor in vegetative cells. Proc Natl Acad Sci USA 78:2762.
22. Fox T, Losick R, Pero J (1976). Regulatory gene 28 codes for a phage-induced subunit of RNA polymerase. J Mol Biol 101:427.
23. Fox T (1976) Identification of phage SP01 proteins

coded by regulatory genes 33 and 34. Nature 262:748.

24. Banner CDB, Moran CP Jr, Losick R (1983). Deletion analysis of a complex promoter for a developmentally regulated gene from Bacillus subtilis. J Mol Biol (submitted).

25. Galas D, Schmitz A (1978). DNase footprinting: a simple method for the detection of protein-DNA binding specificity. Nucleic Acids Res 5:3157.

26. Moran CP Jr, Lang N, Losick R (1981). Nucleotide sequence of a Bacillus subtilis promoter recognized by Bacillus subtilis RNA polymerase containing σ^{37}. Nucleic Acids Res 9:5979.

27. Talkington C, Pero J (1979). Distinctive nucleotide sequence of promoters recognized by RNA polymerase containing a phage-coded "sigma-like" protein. Proc Nat Acad Sci USA 76:5465.

28. Maxam AM, Gilbert W (1980). Sequencing end-labeled DNA with base specific chemical cleavages. Meth Enzymol 65:499.

29. Lee G, Pero J (1981). Conserved nucleotide sequences in temporally-controlled phage promoters. J Mol Biol 152:247.

30. Lang N (1982). Harvard University Ph.D. thesis.

31. Moran CP Jr, Johnson WC, Losick R (1982). Close contacts between σ^{37}-RNA polymerase and a Bacillus subtilis chromosomal promoter. J Mol Biol 162:709.

Gene Expression, pages 249-260
© 1983 Alan R. Liss, Inc., 150 Fifth Avenue, New York, NY 10011

cAMP AND CELL CONTACT REGULATION OF CELL-TYPE-SPECIFIC GENE EXPRESSION IN DICTYOSTELIUM[1]

Mona C. Mehdy,[2] David Ratner[3] and Richard A. Firtel[2]

[2]Department of Biology, B-022
University of California, San Diego
La Jolla, CA 92093

[3]Department of Cellular Biology
Scripps Clinic and Research Foundation
La Jolla, CA 92037

ABSTRACT We have identified genes that are expressed preferentially in either prestalk or prespore cells in *Dictyostelium* aggregates. The prestalk RNAs appear considerably earlier in development than the prespore RNAs. Exogenous cAMP in the absence of sustained cell contact is sufficient to induce prestalk-specific gene expression while multicellularity is required for the induction of prespore-specific gene expression. Under conditions where both sets of genes are expressed without exogenous cAMP, the addition of cAMP results in the precocious induction of their mRNAs. A gene expressed equally in both cell types, which has the same developmental kinetics as the prestalk genes, is induced in shaking culture in the absence of either cAMP or stable cell associations. Dissociation of cell aggregates results in the rapid loss of mRNA complementary to both prespore- and prestalk-specific genes, and these mRNAs can be induced to reaccumulate by the addition of cAMP to the cultures. Cycloheximide inhibits the reaccumulation of the prespore but not the prestalk mRNAs. We conclude that there are substantial differences in the timing and requirements for tissue-specific gene expression in *Dictyostelium*.

[1]This work was supported by grants from the National Institutes of Health and the National Science Foundation.
[3]Present address: Department of Biology, Amherst College, Amherst, Massachusetts 01002.

INTRODUCTION

Dictyostelium discoideum provides a simple system for studies on cell differentiation. During the life cycle, a homogenous population of vegetative cells differentiates into a fruiting body containing two different cell types, stalk and spore cells. Approximately 9 hours after starvation, the amoebae aggregate and form loose mounds. The mounds acquire EDTA resistant cell-cell contacts at 10-11 hours. Differentiation into the precursors to spore and stalk cells can be distinguished at the tip formation stage (12-13 hours; 1,2).

We are interested in understanding the mechanisms causing a uniform cell population to differentiate along two different developmental pathways. Previous studies implicate the involvement of intracellular cAMP levels and cell surface interactions in regulating differentiation (3) In this work, we examine the effects of cAMP and sustained cell-cell contacts on tissue-specific gene expression. We have isolated a set of cDNA clones complementary to developmentally regulated mRNAs that preferentially accumulate in either prespore or prestalk cells. We find that the two sets of genes differ in: 1) timing of expression; 2) requirements for cAMP and cell-cell interactions for their induction, and 3) cAMP regulation of expression later in development.

RESULTS

Cell Type Specificity and Developmental Expression of Cloned Genes

We constructed a library of cDNA clones complementary to mRNA from the culmination stage in development (20 hour). Twelve clones encoding developmentally regulated RNAs were identified by colony hybridization of labeled cDNA complementary to vegetative and 20 hour RNA. The hybridization patterns of developmental RNA blots (see Fig. 2) and genomic DNA blots indicate that the cloned genes are different and unique. To ascertain whether the genes were differentially expressed in prespore or prestalk cells, the two cell types from migrating slugs were separated by Percoll gradients (4). RNA was extracted from the two cell fractions and RNA blots were hybridized with the labeled cDNA clones.

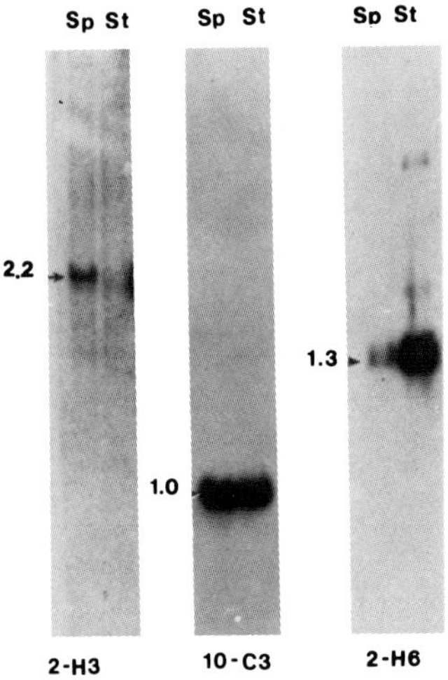

Figure 1. Cell-type specificity of mRNA expression. Prespore and prestalk mRNAs from migrating slugs were prepared as described (5) and analyzed by blot hybridization. Sizes in kilobases are indicated. Sp: prespore RNA St: prestalk RNA.

Figure 1 shows that mRNA complementary to clone 2-H3 accumulates preferentially in prespore cells while 2-H6 mRNA is found mainly in prestalk cells. 10-C3 mRNA is present approximately equally in both cell types. In all, we have identified 7 genes that are "prespore-specific", 4 genes that are "prestalk-specific", and 1 that is cell type nonspecific.

The developmental expression of the cloned genes was determined to see whether temporal differences exist. As shown in Fig. 2, mRNAs complementary to the prestalk-specific probe, 16-G1, and the cell-type nonspecific probe, 10-C3, show similar kinetics of expression. They are first detected at 7.5 hours in development, prior to the completion of cell aggregation. In contrast, the prespore-specific (2-H3) mRNA accumulates much later and is undetectable prior to the finger stage (~15 hours). The other prestalk and prespore genes show similar patterns of early and late expression respectively.

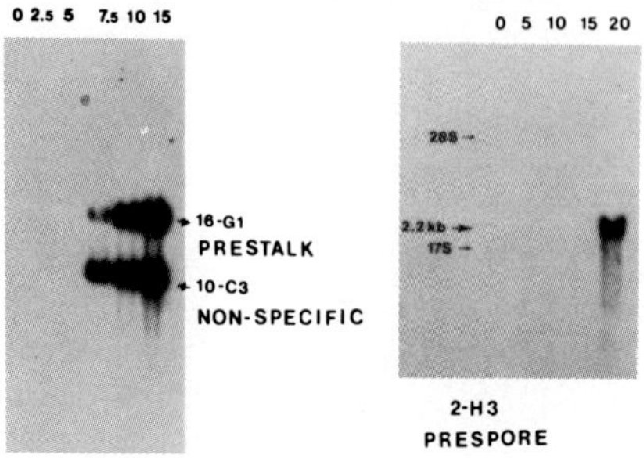

Figure 2. Time course of expression of cloned genes. Cells were developed on filter pads as described (5) and analyzed by blot hybridization. Hours in development are indicated. Some cross-hybridization to rRNAs is observed.

Induction of Late Gene Expression in Suspension

To dissect the possible regulatory roles of cAMP and sustained cell-cell contact on tissue-specific gene expression, we starved vegetative amoebae in a fast shaking suspension culture for 6 hours (5). The cells were then shaken either slowly (70 rpm) or fast (230 rpm) with or without the addition of cAMP. The fast shaken cells remained as largely single cells while most of the slowly shaken cells were in large clumps by 11 hours. Figure 3A shows the expression of tissue-specific genes in these cells. The prestalk-specific gene, 2-H6, is induced in fast shaking cells only when cAMP is present at 200μM. It is expressed in slowly shaking cells with or without added cAMP. On the other hand, the prespore-specific gene, 2-H3, is expressed only in slowly shaking cells and not in fast shaking cells whether or not cAMP is added. Also different is 10-C3, the cell-type nonspecific gene, which is expressed in fast and slowly shaking culture with (data not shown) or without (Fig. 3B) exogenous cAMP. The other probes complementary to prespore- and prestalk-specific mRNAs give patterns of accumulation similar to 2-H3 and 2-H6 respectively. The same results were obtained with 40-500μM cAMP concentrations. Thus, there are substantial differences in the requirements for prespore and prestalk gene induction. Prestalk genes can be induced in fast shaking cultures with cAMP. In slowly shaking cultures, cAMP may be generated

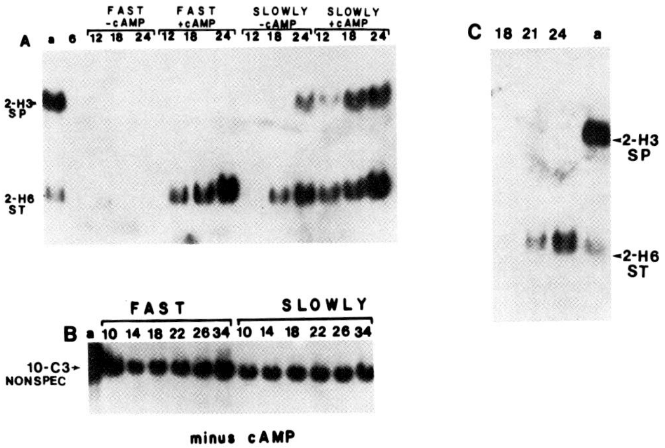

Figure 3. Effect of cAMP and cell contact on the induction of late gene expression. Cells were suspended in a buffer and shaken at 230 rpm for 6 hours (5). The suspension was divided into fast and slowly shaking cultures, with and without 200μM cAMP. cAMP was added to 200μM every 2 hours to maintain its level. Blot hybridization was performed as before. Hours after the cells were resuspended in buffer are indicated. Lane a is RNA from aggregates as a marker. (c) cells were fast shaken as described above except that cAMP was added to 200μM at 18 hours.

endogenously in the aggregates. Stable cell contacts are not required for prestalk gene expression but are necessary for prespore gene expression.

Since the prespore mRNAs start to accumulate several hours later than the prestalk mRNAs, the absence of prespore gene expression in our rapidly shaking cultures could be due to mechanical damage of the cells after the longer periods of time under these conditions. To test whether the cells could still accumulate new RNAs, we added cAMP to fast shaken cultures at 12 hours (data not shown) and 18 hours (Fig. 3C). In both cases, prestalk but not prespore genes are expressed indicating that the cells are still capable of expressing new RNAs under our culture conditions. At the end of the experiments described above, aliquots of the cells were plated on filter pads to determine if the cells were developmentally competent. In all cases, development

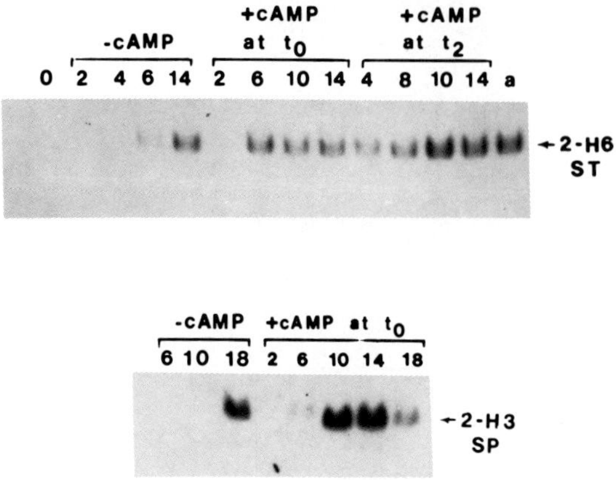

Figure 4. The precocious induction of tissue-specific mRNAs by cAMP. cAMP was added to 200μM as in Figure 3 either immediately (t_0) or 2 hours after the onset of starvation (t_2). The cells were shaken slowly and large aggregates formed in all cultures. Blot hybridization was performed as before.

was accelerated several hours relative to control cells plated from vegetative growth, and the cells formed fruiting bodies.

Exogenous cAMP caused earlier accumulation and higher levels of both prespore and prestalk mRNAs in slowly shaking cultures suggesting an inductive role for both sets of genes. To determine whether cAMP added earlier than 6 hours would cause even earlier accumulation of the tissue-specific mRNAs, we added cAMP to slowly shaking suspensions immediately upon starvation (t_0) or 2 hours later (t_2). As shown in Figure 4, both prestalk and prespore mRNAs are precociously induced relative to cells not treated with cAMP. Large aggregates formed in all cultures by 9 hours. We also examined genes whose mRNAs peak during aggregation during normal development, the discoidin I genes and I_{42} (6,7). Interestingly, the high exogenous cAMP causes markedly reduced mRNA levels relative to untreated cells (data not shown), exactly opposite to cAMP's stimulation of the tissue-specific genes expressed later in development.

The Effect of Cell Contact and cAMP on Prespore and Prestalk Gene Expression Later in Development

Earlier results have shown that the expression of some late genes is inhibited by dissociation of late aggregates and that these genes are re-expressed if the cells are allowed to reaggregate or if cAMP is added to the disaggregated cells (8,9,10). To examine the effects of cAMP and the loss of cell contact on prestalk and prespore gene expression, the following experiments were performed (5). Cells were developed to 15-hour morphology (mid-finger), shaken off the filter pads, dissociated, and shaken (230 rpm). After 3 hours, cAMP was added to half the cells, and the cells were shaken for another 5 hours. As shown in

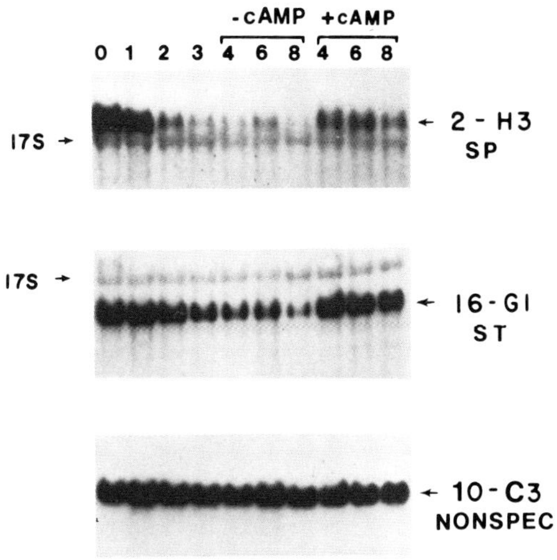

Figure 5. The effect of dissociation and cAMP on late gene expression. At the finger stage in development (~15 hours), aggregates were shaken off the filters, dissociated and fast shaken in buffer for 3 hours (5). The suspension was divided, one half received 1 mM cAMP, and the cultures continued to shake an additional 5 hours. cAMP was added to 100μM every hour. RNA blots were hybridized as before. Hours after dissociation are indicated. The rRNAs which sometimes cross-hybridize are also indicated.

Figure 5, the levels of both prestalk (16-G1) and prespore (2-H3) mRNAs decrease rapidly upon disaggregation, and their levels remain low unless cAMP is added. Within 1 hour after the addition of cAMP, the levels of both mRNAs rise appreciably and by 2-3 hours have been significantly restored. Other members of both sets of tissue-specific genes show similar patterns of expression. When cAMP was added at the time of disaggregation, there was only a slight decrease in prespore- and prestalk-specific mRNA levels (data not shown). At this time in development, gene expression in both cell types appears to be modulated by cAMP levels. The prespore genes no longer require multicellularity for expression. Neither dissociation nor the subsequent addition of cAMP affected the level of expression of 10-C3, the cell-type nonspecific developmentally regulated gene (see Figure 5).

The effect of cycloheximide on the disaggregation-induced loss and cAMP-induced restoration of mRNA levels

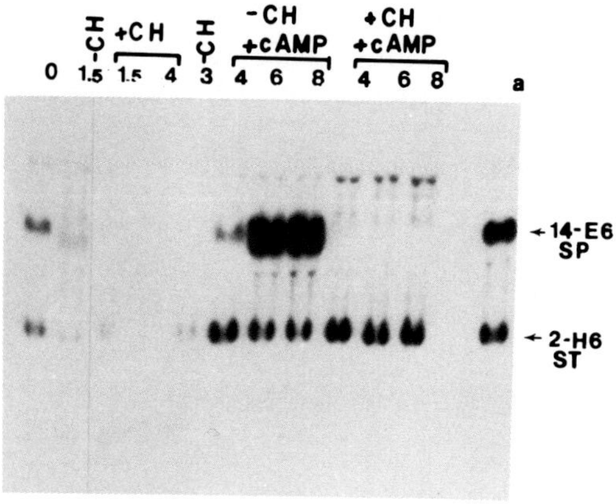

Figure 6. Effect of cycloheximide on the disaggregation-induced loss and cAMP-induced reaccumulation of prespore and prestalk mRNAs. 15-hour aggregates were dissociated and fast shaken as above (5). The suspension was divided as follows: 1) untreated control; 2) cycloheximide (500μg/ml) added immediately after disaggregation; 3) cAMP added at 3 hours, and 4) cAMP and cycloheximide added simultaneously at 3 hours. Hours after disaggregation are indicated. Lane a is RNA from 20 hour aggregates as a marker.

was also investigated (Fig. 6). Cycloheximide added at the time of disaggregation slightly accelerated the loss of 14-E6 and 2-H6 mRNA relative to disaggregated cells lacking the drug. When cycloheximide was added to dissociated cells at the same time as cAMP, there was no reaccumulation of the prespore (14-E6) mRNA while the prestalk (2-H6) mRNA reaccumulation was not affected. Corresponding results were obtained with the other cell-type specific genes. These results suggest that the reaccumulation of prespore but not prestalk gene transcripts requires protein synthesis and indicate possible differences in the mechanism by which cAMP modulates the expression of the two classes of genes. 10-C3 levels are unaffected by the disaggregation, cAMP addition, or cycloheximide addition (data not shown).

DISCUSSION

Our approach to the study of cell differentiation in Dictyostelium centers on genes that are preferentially expressed in either prespore or prestalk cells. We have found that the two sets of tissue-specific genes differ in their patterns of expression while the members of each set (7 prespore genes and 4 prestalk genes) show similar patterns of expression. In addition, a developmentally regulated gene expressed in both cell types shows a third pattern of expression.

The prestalk and prespore mRNAs we have examined differ in their kinetics during development. The prestalk mRNAs and cell-type nonspecific mRNA start to accumulate at 7.5 hours into development, prior to the completion of aggregation and previously identified cellular differentiation (1, 2). In contrast, mRNAs complementary to our prespore clones cannot be detected prior to ~15 hours, several hours after tight cell contacts are formed. Of course, there may be prestalk-specific genes which are induced later or prespore genes which are induced earlier in development. We do not know whether some or all of the cells at 7.5 hours initially express the prestalk genes, perhaps influenced by high cAMP at the aggregation center.

The prespore and prestalk sets of genes differ in their requirements for cAMP and stable cell-cell interactions for their induction. cAMP is necessary for expression of prestalk genes while sustained cell contacts are not. This agrees with Bonner's work (11) showing that isolated amoebae can differentiate into stalk cells on cAMP-containing agar.

The appearance of prestalk mRNAs in slowly shaking cells not treated with cAMP might be explained by endogenous cAMP production by aggregates. In contrast, sustained cell-cell interactions are necessary for the induction of the prespore genes. It is not known whether cell surface interactions or a diffusible molecule is the critical parameter. The cell-type nonspecific gene does not require either cAMP or cell contact for its expression.

An inductive role for cAMP for prespore as well as prestalk genes is suggested by the precocious onset of their mRNAs in cells treated with cAMP (at 0, 2, and 6 hours after the initiation of starvation) relative to untreated cells. While the levels of the late mRNAs are increased by the high cAMP, discoidin I and I_{42} mRNAs, which are normally expressed earlier in development, are substantially inhibited. Under these conditions, it appears that late gene expression may not be dependent on prior expression of at least some early genes. In suspension, the early mRNAs are induced by nanomolar levels of cAMP (Brandis, J., Mann, S., and Firtel, R.A., manuscript in preparation). These data agree with the work of Williams and co-workers (12) suggesting that increasing levels of cAMP over the course of development coordinates early and late gene expression.

We have also compared the requirements for prespore and prestalk gene expression later in development. When developing aggregates are dissociated, the mRNA levels of both prestalk- and prespore-specific genes are rapidly reduced. Addition of cAMP restores the levels of the tissue-specific mRNAs substantially. Addition of cAMP immediately upon disaggregation maintained the mRNA levels of the cloned genes examined. The observation that cAMP is sufficient for continued synthesis suggests that the effect of dissociation is the lowering of the intercellular concentration of cAMP. The loss of tissue-specific mRNAs upon disaggregation is in essential agreement with the effect of disaggregation on specific developmentally regulated enzymes and mRNAs (5,6,7). In contrast, the cell-type nonspecific mRNA is not affected by disaggregation or cAMP addition.

The differences in requirements for prestalk and prespore gene expression could be related to the time of initial induction of these genes rather than to the cell type in which they are preferentially expressed. However, gene 10-C3, which has similar developmental kinetics as the prestalk genes, is expressed in both cell types and has different requirements for its expression. We suggest that there are substantial differences in the regulation of gene activity in the two cell types.

ACKNOWLEDGMENTS

We thank Wayne Borth for technical assistance, and Stephen Poole, Cynthia Edwards, Alan Kimmel and William Loomis for helpful suggestions.

REFERENCES

1. Muller W, Hohl HR (1973). Pattern formation in Dictyostelium discoideum: temporal and spatial distribution of prespore vesicles. Differentiation 1:267.
2. Hayashi M, Takeuchi I (1976). Quantitative studies on cell differentiation during morphogenesis of the cellular slime mold Dictyostelium discoideum. Dev Biol 50:302.
3. Loomis WF (1982). The spatial pattern of cell-type differentiation in Dictyostelium. Dev Biol 93:279.
4. Ratner D, Borth W (1983). Comparison of differentiating Dictyostelium discoideum cell types separated by an improved method of density gradient centrifugation. Exp Cell Res, in press.
5. Mehdy MC, Ratner D, Firtel RA (1983). Induction and modulation of cell-type-specific gene expression in Dictyostelium. Cell 32:763.
6. Rowekamp W, Poole S, Firtel RA (1980). Analysis of the multigene family coding the developmentally regulated carbohydrate-binding protein Discoidin I in D. discoideum. Cell 20:495.
7. Rowekamp W, Firtel RA (1980). Isolation of developmentally regulated genes from Dictyostelium Dev Biol 79:409.
8. Sussman M, Newell PC (1972). Quantal control. In Sussman M (ed): "Molecular Genetics and Developmental Biology," Englewood Cliffs: Prentice-Hall, p 275
9. Firtel RA, Bonner JT (1972). Developmental control of alpha 1-4 glucan phosphorylase in the cellular slime mold Dictyostelium discoideum. Dev Biol 29:85.
10. Chung S, Landfear SM, Blumberg DD, Cohen NS, Lodish HF (1981). Synthesis and stability of developmentally regulated Dictyostelium mRNAs are affected by cell-cell contact and cAMP. Cell 24:785.

11. Bonner JT (1970). Induction of stalk cell differentiation by cyclic AMP in the cellular slime mold Dictyostelium discoideum. Proc Nat Acad Sci USA 65:110.
12. Williams JG, Tsang AS, Mahbubani H (1980). A change in the rate of transcription of a eucaryotic gene in response to cAMP. Proc Nat Acad Sci USA 77:7171.

Gene Expression, pages 261-271
© 1983 Alan R. Liss, Inc., 150 Fifth Avenue, New York, NY 10011

REGULATION OF *DICTYOSTELIUM DISCOIDEUM* mRNAS SPECIFIC FOR PRESPORE OR PRESTALK CELLS

Rex L. Chisholm, Eric Barklis, B. Pontius, and Harvey F. Lodish

Department of Biology, Massachusetts Institute of Technology
Cambridge, MA 02139

INTRODUCTION

Dictyostelium discoideum is an eukaryotic organism which offers many advantages for studies of the mechanisms and factors regulating a developmental program (see 1 for review). *Dictyostelium* grows as free living amoebae, which upon starvation begin a program of development consisting of three easily defined phases. During the *pre-aggregation* period the cells conduct a complex process of cellular signalling and chemotaxis. Certain cells spontaneously begin secreting pulses of cAMP. Upon sensing a cAMP pulse, neighboring cells move a few micra up the cAMP concentration gradient toward the original source of the cAMP. In addition, they secrete a pulse of cAMP, following which they become unable to respond to further cAMP pulses for a period of several minutes. In this way there is an outward propagation of the cAMP pulses, and an inward movement of cells. Eventually this chemotactic signalling establishes an "aggregation field" consisting of roughly 100,000 cells. Eventually, at about 10-12 hours after the initiation of development the cells in an aggregation field form a cellular aggregate. At this point, the second phase of development begins. The amoebae exhibit evidence of differentiation into prespore and prestalk cells, and the prestalk cells cluster at the tip of the aggregate. The final stage, *culmination*, consists of formation of the fruiting body in which spores are hoisted up on a vacuolated stalk.

We have concentrated our investigations on the factors and mechanisms by which regulation of cell type specific gene expression is achieved. To this end we have identified cloned cDNAs representing mRNAs differentially expressed in prespore and prestalk cells. These cloned sequences have then been used as probes to follow the expression of cell type specific genes, to identify the factors which induce their expression, and to elucidate the levels at which regulation of their expression occurs.

RESULTS

Identification of Prespore And Prestalk Specific Genes.

We have used Percoll density gradients to separate prespore from prestalk cells. Prestalk cells are enriched in low density fractions and prespore cells migrate preferentially to the higher density regions of the gradient. Figure 1 shows a typical gradient profile obtained from

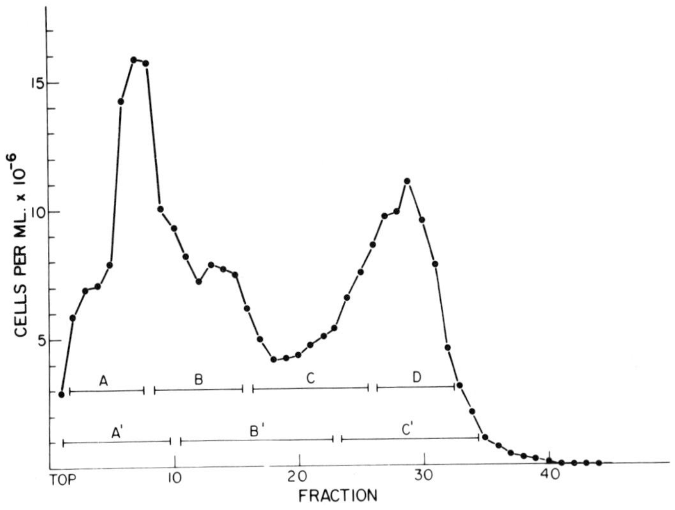

Figure 1: Percoll density gradient profile. *Dictyostelium* cells harvested after 16-18 hrs of development were fractionated by Percoll density centifugation as previously described (13). Four fractions, labeled A, B, C and D were taken for further analysis. Prespore cells are enriched at the bottom of the gradient, and prestalk cells at the top.

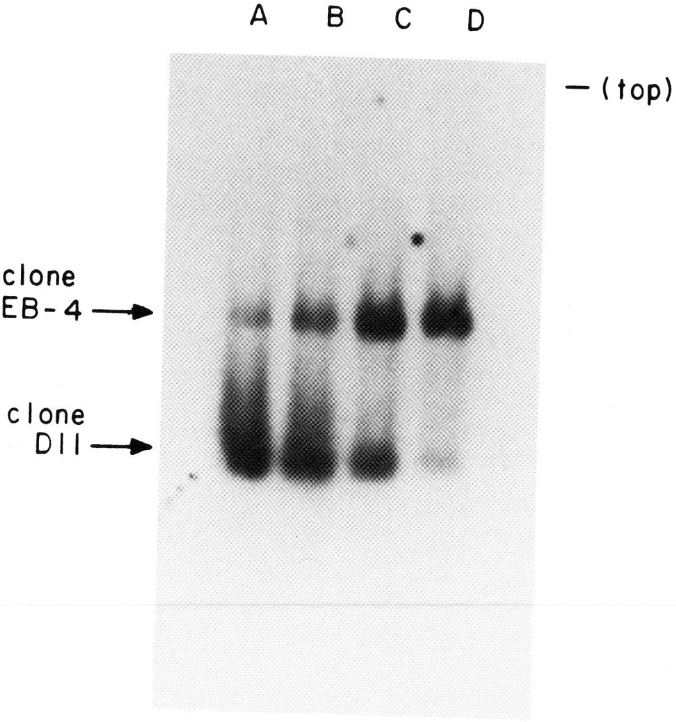

Figure 2: Identification of cell type specific Dictyostelium mRNAs. Total polyadenylated RNA (1 µg/lane) from Dictyostelium cells developed for 16 hrs and fractionated on a Percoll density gradient was run through a formaldehyde agarose gel, blotted to nitrocellulose, and hybridized to [^{32}P]labeled DNA probes. Tracks A through D contain RNA prepared from the corresponding fraction of the Percoll gradient shown in Figure 1. Clone EB-4 encodes a prespore Class II mRNA and D11 is a prestalk Class I gene.

cells which have been developed for 16 hours. Cells were pooled into four fractions, A, B, C and D and RNA prepared from them. The RNA obtained from fraction A is thus enriched for mRNAs expressed in prestalk cells, and that from fraction D is enriched in those expressed in prespore cells. Replicate filters containing samples of DNA prepared from a bank of cDNA clones made using mRNA obtained from either 15 hour or 22 hour developed cells were screened by hybridization with kinase-labeled cytoplasmic polyadenylated RNA from the four Percoll fractions. This differential screening procedure identified a series of clones representing mRNAs differentially expressed in prespore or prestalk cells. In addition we used this approach to screen a set of cloned cDNAs and genomic DNAs we had previously characterized as encoding transcripts which first appear at the time of aggregation. Most of these genes were also preferentially expressed in prespore or prestalk cells. Figure 2 shows an autoradiogram of a filter onto which gel fractionated poly (A^+) RNA has been blotted and then probed with a prespore clone (EB-4) and a prestalk clone (D11). The four gel lanes represent RNA prepared from the four gradient fractions A through D as shown in Figure 1. The RNAs encoded by the two cDNA clones are indeed enriched in different cell types.

Patterns of Expression of Prespore and Prestalk Genes

Each of the 20 cloned genes obtained has been characterized with respect to its pattern of accumulation during development. RNA prepared from cells developed for 4, 8, 12, 16, 20 and 24 hours, as well as from growing cells, was size-fractionated on formaldehyde agarose gels, transferred to nitrocellulose or Genescreen filters, and hybridized with [^{32}P]labeled cloned DNAs. Based on the time of accumulation of the RNAs, four patterns of expression were observed. Prestalk genes fall into two distinct classes--those expressed at very low levels during growth, but which increase significantly in abundance between 8 and 12 hours of development (Class I) and Class II, which begin to accumulate only about 12 hours of development. Prespore genes also fall into two classes based upon time of appearance. Both classes of mRNA are absent in growing and 4 hr cells. Class I mRNAs accumulate in cells 8 hours into development, while the Class II prespore genes begin to accumulate only at about 12 hours. The relative abundances of several examples of these cell type specific mRNAs are shown in Table 1.

TABLE 1

Clone	Hours of Development								
	Veg	4	8	12	16	20	24	Dis	Dis +cAMP
D11 (pSt I)	3	4	19	77	100	6	8	81	85
D14 (pSt I)	1	3	16	66	58	74	100	38	52
B1 (pSt II)	-	-	-	20	90	100	50	-	10
SC253 (?)	-	-	-	4	27	41	100	-	-
D18 (pSp I)	-	-	21	100	87	92	79	8	99
D19 (pSp II)	-	-	-	20	31	27	100	-	40
EB4 (pSp II)	-	-	-	7	32	28	100	-	23

Table 1: Relative concentrations of prespore and prestalk mRNAs during normal development. The relative concentrations of messenge corresponding to a given cloned sequence was determined by densitometric scanning of an autoradiogram of a Northern blot. All data are normalized to the highest concentration seen (100%) for a given clone. (-) = undetected; pSt I, II = prestalk Class I or II mRNAs; pSp I, II = prespore mRNAs Class I or II; Dis = disaggregated cells.

We have shown previously that disaggregation of the multicellular aggregates has a profound effect on gene expression (2-4). Many of the mRNAs which begin to accumulate at the time of aggregation are specifically lost following disaggregation. Transcription of these mRNAs is specifically reduced at least five-fold and there is a specific degradation of many developmentally regulated mRNAs (5, 6). Moreover, the addition of cAMP to the disaggregated cells both restores transcription of these mRNAs and causes them to be stabilized (6, 7).

To identify the behavior of the different classes of prespore and prestalk genes upon disaggregation and subsequent exposure to cAMP, RNA was prepared from cells disaggregated after 15 hours of development and shaken for an additional four hours, either in the presence or absence of 100 μm cAMP, under conditions which prevented reformation of multicellular aggregates. Quantitation of mRNA levels in disaggregated cells, and disaggregated cells treated with cAMP, is shown in the last two columns of Table 1. The concentration of both classes of prespore mRNAs is drammatically reduced in disaggregated cells, but is restored by the addition of cAMP. The abundance of the prestalk messages that are expressed at low levels in growing cells (Class I) are not greatly affected by disaggregation or by cAMP (see clones D11 and D14). The second class of prestalk genes (B1, for example) is greatly reduced in abundance following disaggregation, but is not restored to normal levels by the addition of cAMP.

What Factors Induce Accumulation of The Prespore and Prestalk mRNAs?

By transferring growing cells into starvation medium, but shaking them in suspension instead of plating them onto a solid surface, it is possible to prevent formation of cellular aggregates and to disrupt the cAMP signalling which is a feature of the preaggregation phase of development. Thus, suspension cultures can be used to assess the contribution of multicellular aggregate formation and the role of cAMP pulses in the induction of prespore and prestalk gene expression. cAMP signalling can be mimicked in suspension culture by the periodic addition of cAMP to the shaken suspension culture. The pattern of expression of the four classes of cell type specific genes was determined by preparing RNA from cells harvested at four hour intervals from

TABLE 2

	Hours of Culture									
Clone	Suspension starved					Suspension starved and cAMP pulsed				
	4	8	12	16	20	4	8	12	16	20
D11	6	2	17	24	1	2	6	1	5	30
D14	3	3	15	20	23	3	6	86	160	180
B1	-	-	-	-	-	-	-	20	82	79
SC253	-	-	-	-	-	-	-	-	8	50
D18	-	-	-	-	-	-	-	80	75	80
D19	-	-	-	-	-	-	-	-	-	-
EB4	-	-	-	-	-	-	-	-	-	-

Table 2: Relative concentration of prespore and prestalk mRNAs in suspension starvation cultures and suspension starved cAMP pulsed cultures. The values shown are normalized to the highest concentration seen in normal development (100% in Table 1). (-) = undetected.

starved suspension cultures and from similar cultures which were pulsed every 5 to 6 minutes with 50 nM cAMP. As before this RNA was subjected to gel fractionation, transfer, and hybridization with labeled, cloned prespore and prestalk specific genes. The relative concentrations of the various mRNAs is shown in Table 2. Each of the four cell type specific gene classes has a specific pattern of accumulation in suspension cultures. None of the prespore mRNAs accumulated to appreciable levels in suspension starvation culture (see, for example, D18 and D19). Clone D18 mRNA (prespore Class I), however, accumulates to essentially normal levels by 12 hours in a starved suspension culture that is pulsed with cAMP. Thus, pulsatile signalling by cAMP is sufficient for induction of Class I prespore genes. In contrast, clone D19 mRNA never accumulates in suspension culture, regardless of whether or not cAMP is present. Class II, the class of prespore genes represented by clone D19, appears to require formation of multicellular aggregates for induction.

Prestalk genes such as D11 and D14 (Class I) accumulate to normal levels in suspension culture whether or not cAMP is added. Consequently, starvation appears sufficient for induction of Class I prestalk genes. Messenger RNAs corresponding to prestalk Class II, such as clone B1, do not accumulate in suspension starved cultures without cAMP, but do accumulate to normal levels, and at a normal time, in cAMP-pulsed suspension cultures. It would appear, then, that Class II prestalk genes are induced by cAMP.

DISCUSSION

We have isolated and identified cDNA clones representing mRNAs which are specifically expressed in prespore or prestalk cells. Based upon the temporal pattern of their accumulation, they can be divided into four discrete classes, two for genes expressed in prespore cells and two for those found in prestalk cells. Table 3 shows the features which describe these classes of genes.

Class I prestalk genes are expressed at low levels in growing cells and accumulate to higher concentrations around 12 hours of development. The mRNAs encoded by Class I prestalk genes also accumulate in cells shaken in starvation buffer, with a time course similar to that normally seen during development. These genes appear to require only

TABLE 3

Gene Class	Time of Accumulation	Disag	Disag +cAMP	Suspension starved	Suspension starved and cAMP pulsed
Prestalk I	Veg	+	+	+	+
Prestalk II	8-12 h	-	-	-	+
Prespore I	8 h	-	+	-	+
Prespore II	12 h	-	+	-	-

Table 3. Characteristics of prespore and prestalk specific gene classes. (+) indicates accumulation of mRNA. (-) indicates no accumulation of mRNA.

starvation for their induction. Class II prestalk mRNAs, on the other hand, are absent in growing cells, at the level of detection by Northern blot analysis, which we estimate to be 1 to 5 copies per cell. Messages encoded by these genes accumulate between 8 and 12 hours of development. They are not induced in cells shaken in starvation buffer, but do accumulate if cellular cAMP signalling is mimicked by the introduction of cAMP pulses into the culture. This suggests that cAMP induces Class II prestalk genes.

Based on the time of accumulation during development, the two prespore gene classes differ in that mRNAs from Class I genes are expressed in cells 8 hours after the initiation of development, at a time when cell-cell contacts are just beginning to form. mRNAs from Class I prespore genes do not appear until later--8 to 12 hours. Neither of these gene classes is expressed in cells shaken in suspension culture. However, addition of pulses of cAMP induced Class I prespore genes to levels comparable to those found in filter developed cells, and with a similar time of appearance. Class II prespore genes are not expressed under these conditions; transcripts of them accumulate only after the cells have formed tight cellular aggregates. Thus, while cAMP appears able to induce the expression of prespore Class I genes, and may, indeed play some role in maintaining expression of Class II genes (see below), primary induction of Class II prespore genes requires the formation of cellular aggregates.

We (2-6) have described the effects of disaggregation on the pattern of gene expression in the post-aggregation phase of development. Disaggregated cells cease synthesis of most of the proteins (resolved by 2-D gels) that are induced between 8 and 12 hours of differentiation. Further, there is selective loss of mRNAs which were induced during the 8 to 14 hour period of development. These results suggested a role for cell contact in expression of these genes. Moreover, addition of cAMP to the disaggregated cells restored the accumulation of most of these gene products. This led to the conclusion that cAMP was also involved in gene regulation.

Our current studies show that three of the four classes of regulated mRNAs are selectively lost upon disruption of multicellular aggregates. However only two of these gene classes--prespore I and prespore II--show restored accumulation upon addition of cAMP to the disaggregated cells. It is interesting to note that, despite the fact that prestalk II genes can be induced in suspension starved cultures by cAMP pulses, addition of cAMP to disaggregated cells restores accumulation of these messages only marginally, if at all. Perhaps accumulation of these mRNAs requires the puslatile application of cAMP, and the single large dose given in the disaggregation experiments is inadequate to re-induce accumulation of these messages. In contrast, prespore II genes, although not inducible by application of cAMP pulses to cells in suspension, do respond to cAMP following their primary induction in cellular aggregates. These observations suggest that signals in addition to cAMP must also be controlling the expression of these two gene classes during normal differentiation.

Gross _et al._ (8) and others (9), have reported that cAMP alone was sufficient to induce "post-aggregation" genes, while Chung _et al._ (3) and Blumberg _et al._ (4) argued that the formation of EDTA resistant cell contacts was required to induce "post-aggregation" genes. One reason for this controversy now seems apparent: Three of our four classes of developmentally regulated cell type specific genes can be induced to accumulate in the absence of stable cell-cell contact. Cyclic AMP is required for induction of two of these classes (prespore I and prestalk II). Induction of the fourth class of regulated mRNAs, prespore II, however, does require cell-cell contact. Based on the number of cloned genes we have identified which fall into each class, prespore Class II genes appear to be the most abundant. Thus the conclusions reached by others may be a result of the particular gene or gene product they chose to follow. We note that our present conclusions are liable to the same limitation.

These results also provide an explanation for the observation that stalk cells can be induced in the absence of cell contact (10-12), but that generally, spore cells cannot. Because both types of genes specifically expressed in prestalk cell are induced in suspension starved, cAMP pulsed cultures, it is possible that stalk cell differentiation could progress normally in the absence of the formation of multicellular aggregates. Since only one of the two gene classes expressed in prespore cells is induced in suspension culture, conversion of prespore cells into spores may be precluded.

The induction of cell type specific genes in suspension cultures raises the question of differentiation in the absence of morphogenesis. One can easily envision two models by which the appropriate expression is achieved in a multicellular structure consisting of two principal cell types. In the first, cells differentiate into one cell type or another, then migrate to their proper position in the forming structure. According to the second model the cells sense their position in the structure and then differentiate to become the cell type appropriate for that position. The induction of both prestalk gene classes and one of the two prespore gene classes in the absence of morphogenesis would tend to support the first model. It is possible, however, that the cell-type specific gene induction we observe is not occuring in specific cells, but rather that our experimental conditions are inducing expression of both prespore and prestalk genes in the same cells. We are currently addressing this question.

As Dictyostelium differentiation involves principally two cells types, the classification system just described provides a useful framework upon which we can base a detailed molecular analysis of the regulation of a simple program of differentiation and development. How does addition of cAMP to the culture medium alter the state of gene expression in these cells? What is the signal transduction mechanism? How does the formation of multicellular aggregates regulate gene expression? Is modulation of the cell surface involved, or does this occur through specific receptor-ligand interactions? Do common sequences on co-regulated genes mediate their induction, or the stability of their mRNAs? These and other questions can now be studied in detail.

REFERENCES

1. Loomis WL (1982). "The Development of Dictyostelium discoideum." New York: Academic Press, Inc.
2. Landfear SM, Lodish HF (1980). A role for cAMP in expression of developmentally regulated genes in Dictyostelium discoideum. Proc Natl Acad Sci USA 77:1044.
3. Chung S, Landfear SM, Blumberg DD, Cohen NS, Lodish HF (1981). Synthesis and stability of developmentally regulated Dictyostelium mRNAs are affected by cell-cell contact and cAMP. Cell 24: 785.
4. Blumberg DD, Margolskee JP, Chung S, Barklis E, Cohen NS, Lodish HF (1982). Specific cell-cell contacts are essential for induction of gene expression during differentiation of Dictyostelium discoideum. Proc Natl Acad Sci USA 79:137.
5. Mangiarotti G, Lefebvre P, Lodish HF (1982). Differences in the stability of developmentally regulated mRNAs in aggregated and disaggregated Dictyostelium discoideum cells. Devel Biol 89:82.
6. Landfear SM, Lefebvre P, Chung S, Lodish HF (1982). Transcriptional control of gene expression during development of Dictyostelium discoideum. Mol Cell Biol 2:147.
7. Mangiarotti G, Ceccarelli A, Lodish HF (1983). Cyclic AMP stabilizes a class of developmentally regulated Dictyostelium discoideum mRNAs Nature 301:616.
8. Gross JG, Town CD, Brookman JJ, Jeermyn KA, Peacey MJ, Kay RR (1981). Cell patterning in Dictyostelium. Phil Trans R Soc Lond B295:497.
9. Kay RR, Town CD, Gross J (1979). Cell differentiation in Dictyostelium discoideum. Differentiation 13:7.
10. Bonner JT (1970). Induction of stalk cell differentiation by cAMP in the cellular slime mold Dictyostelium discoideum. Proc Natl Acad Sci USA 65:110.
11. Chia WK (1975). Induction of stalk cell differentiation by cAMP in a susceptible variant of Dictyostelium discoideum. Devel Biol 44:239.
12. Town CD, Gross JD, Kay RR (1976). Cell differentiation without morphogenesis in Dictyostelium discoideum. Nature 262:717.
13. Barklis E, Lodish HF (1983). Regulation of Dictyostelium discoideum mRNAs specific for prespore or prestalk cells. Cell in press.

Gene Expression, pages 273-281
© 1983 Alan R. Liss, Inc., 150 Fifth Avenue, New York, NY 10011

A PROTEIN FACTOR REGULATES TRANSCRIPTION OF THE *XENOPUS* 5S RIBOSOMAL RNA GENES IN A POSITIVE FASHION[1]

Jennifer Price, Peggy J. Farnham, and Laurence Jay Korn

Department of Genetics
Stanford University School of Medicine
Stanford, CA 94305 U.S.A.

ABSTRACT We have begun to characterize protein factors involved in the developmental regulation of the dual 5S RNA gene system in *Xenopus*. Although both types of 5S genes (oocyte and somatic) are expressed in oocytes, in somatic cells only somatic-type genes are expressed. We have taken three approaches to study the differential expression of the two types of 5S genes. First, we have shown that the oocyte- and somatic-type genes have a similar chromosomal location, making it unlikely that differences in chromosomal position are responsible for the regulation of 5S genes. Second, we have found that oocyte-type genes retain their regulated state when chromatin is digested with micrococcal nuclease into individual repeat units, indicating that the mechanism responsible for maintaining the inactivity does not depend on the cooperative effect of adjacent genes. Third, we have injected somatic nuclei into *Xenopus* oocytes and identified two classes of female frogs. Oocytes

[1]This work was supported by NSF Grant No. PCM 8104522 and March of Dimes Birth Defects Foundation Grant No. 5-323 to L.J.K. P.J.F. was a postdoctoral fellow of the Damon Runyon-Walter Winchell Cancer Fund.

from one class specifically reactivate the developmentally inactive oocyte-type 5S genes in somatic nuclei (activators) and oocytes from the other class do not (non-activators). However, oocyte-type genes in somatic nuclei are expressed in non-activating oocytes if such nuclei are coinjected with extracts from activating oocytes. The activating factor is heat labile, protease sensitive, and nondialyzable, indicating that a protein component is involved. Purification and analysis of the factor is in progress.

INTRODUCTION

A major question in contemporary biology concerns the control of gene expression during development. The *Xenopus* 5S gene system (for reviews see Brown, 1981; Korn, 1982; Korn and Bogenhagen, 1982) provides an excellent opportunity to study this question. There are two types of 5S RNA genes (Brown and Sugimoto, 1973), oocyte-type (Wegnez *et al.*, 1972; Ford and Southern, 1973) and somatic-type (Ford and Southern, 1973). In *Xenopus laevis*, there are 20,000 copies of the oocyte-type 5S RNA genes and 400 copies of the somatic-type 5S RNA genes per haploid genome (Peterson *et al.*, 1980). In oocytes (immature, unfertilized eggs), both oocyte- and somatic-type 5S genes are expressed (Wegnez *et al*. 1972; Ford and Southern, 1973). In somatic cells, however, only the somatic-type 5S RNA genes are expressed.

Two major categories of models have been advanced to explain the differential expression of the two types of 5S RNA genes: (1) models which invoke large scale differences in the structure of the chromosomal regions containing the oocyte-type and somatic-type gene clusters (Ford and Mathieson, 1976), and (2) models which postulate direct protein-DNA interactions to activate or inactivate each gene individually (Pelham and Brown, 1980; Korn *et al*., 1982; Bogenhagen *et al*., 1982). In this article, we review some of our work on the developmental control of the *Xenopus* 5S ribosomal RNA genes and describe the partial purification of a transcription factor which acts in a positive fashion to regulate transcription.

CHROMOSOMAL LOCATION AND CHROMATIN STRUCTURE

We have taken three approaches in the study of 5S gene regulation. First, we have localized the oocyte- and somatic-type genes by in situ hybridization to Xenopus chromosomes (Harper et al., 1983). Models for the differential expression of the two types of 5S genes based on chromosomal location (Ford and Mathieson, 1976) suggested that the oocyte-type genes, located at the distal ends of most chromosomes (Pardue et al., 1973), might be in a chromosomal structure unavailable for transcription in somatic cells. These models predicted that the somatic-type genes have a separate location from the oocyte-type genes. To test this prediction, we determined the chromosomal location of both types of genes. We found that the somatic-type genes are located at a single major site, the distal end of the long arm of chromosome 9. Since the oocyte-type genes are also at this site, it seems unlikely that large scale differences in chromosomal position are responsible for the differential expression of the 5S genes (Harper et al., 1983).

As a second approach, we have determined whether the differential expression of the oocyte- and somatic-type genes can be maintained when chromosomes are cleaved into small pieces. When intact somatic chromosomes are injected into certain oocytes, oocyte-type 5S genes remain inactive whereas the somatic-type 5S genes continue to be expressed (Korn and Gurdon, 1981). We found that when chromosomes are cleaved to a length less than one repeat unit (gene plus spacer), the oocyte-type genes still remain inactive and the somatic-type genes are transcribed as efficiently as in a somatic cell (Gurdon et al., 1982). Therefore, the mechanism responsible for maintaining the inactivity does not depend on the cooperative effect of adjacent genes or on distal DNA sequences. Nor can it depend on higher order chromatin structures, which would have been destroyed by this treatment. Rather, each repeat unit must contain the regulatory elements sufficient to maintain the inactive state of the oocyte-type 5S genes.

FACTORS THAT REGULATE TRANSCRIPTION IN A POSITIVE FASHION

A third approach has involved the coinjection of crude extracts or protein factors together with somatic nuclei into Xenopus oocytes. When purified genomic DNA isolated from somatic cells (in which the oocyte-type genes are inactive), or cloned oocyte- or somatic-type 5S DNA is injected into Xenopus oocytes, both oocyte- and somatic-type 5S genes are transcribed efficiently (Brown and Gurdon, 1977; 1978; Gurdon and Brown, 1978). Similarly, when DNA from either of these sources is added to in vitro transcription systems derived either from oocytes or from somatic cells, no differential expression of 5S RNA is observed (Birkenmeier et al., 1978; Korn et al., 1979; Brown et al., 1979; Ng et al., 1979; Weil et al., 1979). Although these transcription systems have helped identify the DNA sequences (Korn et al., 1979; Sakonju et al., 1980; Bogenhagen et al., 1980) and other macromolecular components (Engelke et al., 1980) required for faithful synthesis of 5S RNA, it has not been possible to demonstrate any developmental regulation of transcription using purified DNA. Therefore, we are studying the 5S RNA genes when they are complexed with proteins in their normal chromosomal arrangement as part of intact nuclei. Our approach is to analyze the transcription of the 5S RNA genes when somatic nuclei are injected into Xenopus oocytes (Gurdon, 1976; Korn and Gurdon, 1981). Figure 1 shows the results of injecting aliquots of a suspension of somatic nuclei into oocytes from two classes of female Xenopus. In the oocytes of one class, oocyte-type 5S genes in the injected somatic nuclei were activated (activating oocytes, Figure 1, lanes 3-5); in the other, they remained quiescent (non-activating oocytes, lanes 1,2). The synthesis of somatic-type 5S RNA from injected

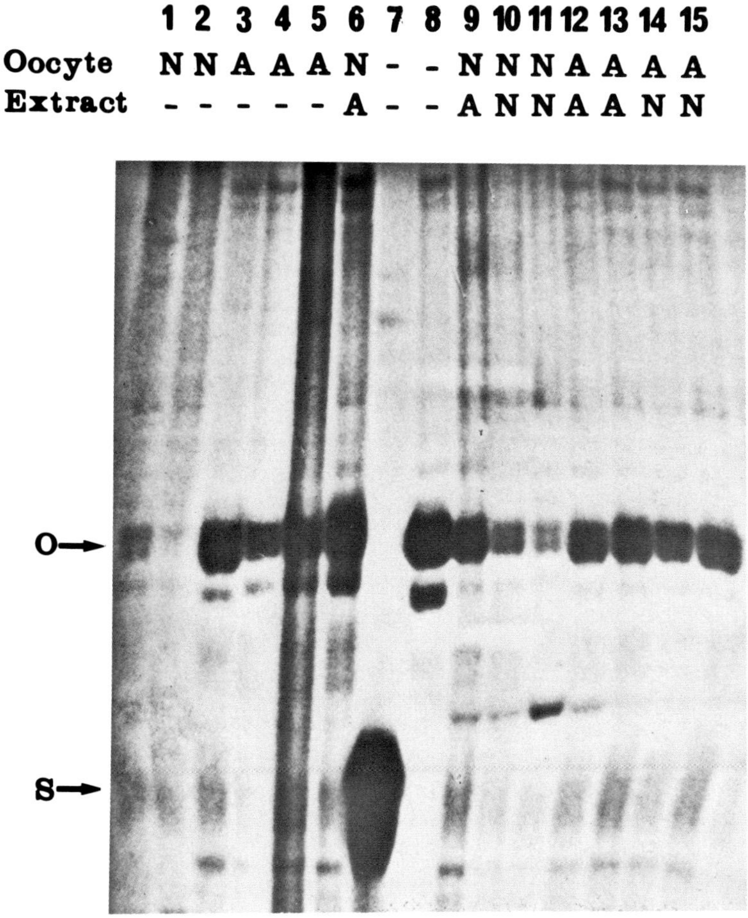

Figure 1. Transcription of 5S RNA genes with activating and non-activating oocyte extracts. Each lane contains half the RNA from one oocyte. All the oocytes shown received an injection of erythrocyte nuclei. For each lane, the type of oocyte used and the source of extract injected (if any) is indicated; oocytes or extracts are from non-activating (N) or activating females (A). Lanes 7 and 8 are markers showing somatic- (S) and oocyte- (O) type 5S RNA, respectively. (Fig. reprinted by permission from Nature 300, 354-355, Copyright (c) 1982 MacMillan Journals Limited.)

nuclei is approximately the same, whether or not the oocyte-type 5S RNA genes are transcribed. Somatic-type 5S RNA therefore provides an essential control, eliminating the possibility that the inactivity of oocyte-type 5S RNA genes could be due to the inviability of the injected nuclei in these oocytes. We conclude that the developmental inactivity of oocyte-type 5S genes takes place by a process that is readily reversible by components of the oocytes of some, but not all, females. The existence of these two types of oocytes, activating and non-activating, provides a system for studying the differential expression of the oocyte- and somatic-type 5S RNA genes.

We have further analyzed the properties of these oocytes. Somatic nuclei were injected, either alone or mixed with extracts prepared from either activating or non-activating oocytes, into the germinal vesicles of both types of oocytes. Labelled transcripts were recovered and electrophoresed on polyacrylamide gels to differentiate between oocyte- and somatic-type 5S RNA (Korn and Gurdon, 1981). When activating extract was added to the somatic nuclei, the oocyte-type 5S genes were reactivated in both the activating (Figure 1, lanes 12,13) and non-activating oocytes (lanes 6,9). Extracts from non-activating oocytes did not diminish the expression of the oocyte-type 5S genes when coinjected with somatic nuclei into activating oocytes (lanes 14,15). However, such extracts produced a slight increase in the activation of the oocyte-type 5S genes in the non-activating oocytes (lanes 10,11). This partial activation was much less than that observed when activating extracts were used (compare lanes 10,11 to lanes 6,9). One way to interpret this finding is that there is a variation in the concentration of a specific factor in the Xenopus population, some oocytes having more of this factor than others. Since 500 to 1000 somatic nuclei are injected into each oocyte, the oocyte is being challenged to regulate the expression of a large excess of 5S genes. Oocytes containing only enough of the factor to regulate their own oocyte-type 5S genes would be scored as non-activating. Concentrated extracts from the non-activating oocytes could be adding back enough of the factor to increase slightly the level of activation. In contrast, a large excess of the factor would result in an activating oocyte. When we screened a series of female Xenopus, we found some which were strongly activating, some which were non-activating, and others which gave intermediate levels of activation. Oocytes taken from the same female at different times over a three year period consistently showed the same activating or non-activating pattern as initially determined. The existence of a continuum in the extent to which oocytes can reactivate the previously inert oocyte-type genes again suggests a variation in the concentration of a specific factor. Since the activating extracts exert their influence in non-activating oocytes, and not vice versa, it appears that at least one factor controlling the reactivation of oocyte-type 5S RNA genes regulates transcription in a positive fashion. We have shown the activating factor to be heat labile, protease sensitive, and non-dialyzable, indicating that a protein component is involved in activation (Korn et al., 1982).

PURIFICATION OF ACTIVATING FACTORS

By injecting non-activating oocytes with somatic nuclei plus fractionated extracts from activating oocytes, we have begun to purify factors from the extract which specifically reactivate the oocyte-type 5S genes. Extracts were prepared as described by Korn et al. (1982). The extract was then centrifuged in an airfuge at 100,000 x g and the supernatant used for further purification. Our preliminary experiments suggest that the activity can be precipitated from the supernatant by 35% saturated ammonium sulfate. Furthermore, the activity binds to DE52 and can be eluted between 0.25 and 0.45 M NaCl. When the supernatant is chromatographed on an Amicon blue B dye-ligand column, the vast majority of proteins do not bind to the gel matrix. However, the protein(s) necessary to reactivate the oocyte-type 5S genes are retained. The activity remains bound at 0.8 M NaCl, but can be eluted at 1.6 M NaCl. A combination of these procedures results in a substantial purification of the activating factor.

We have shown that the differential expression of the oocyte- and somatic-type 5S RNA genes is not due to large scale differences in their chromosomal location. Furthermore, the developmentally regulated state of transcription is maintained when chromatin containing the 5S genes is cut into individual repeat units. Finally, transcription is regulated in a positive fashion, by specific, although so far uncharacterized, protein factors. We are currently involved in the further purification and analysis of these factors.

REFERENCES

Birkenmeier EH, Brown DD, Jordan E (1978). Cell 15:1077.

Bogenhagen DF, Wormington WM, Brown DD (1982). Cell 28:413.

Bogenhagen DF, Sakonju S, Brown DD (1980). Cell 19:27.

Brown DD (1981). Science 211:667.

Brown DD, Gurdon JB (1977). Proc. Natl. Acad. Sci. U.S.A. 74:2064.

Brown DD, Gurdon JB (1978). Proc. Natl. Acad. Sci. U.S.A. 75:2849.

Brown DD, Sugimoto K (1973). Cold Spring Harb. Symp. Quant. Biol. 38:501.

Brown DD, Korn LJ, Birkenmeier E, Peterson R, Sakonju S (1979). ICN-UCLA Symp. 14:511.

Engelke DR, Ng SY, Shastry BS, Roeder RG (1980). Cell 19:717.

Ford PJ, Mathieson T (1976). Nature 261:433.

Ford PJ, Southern EM (1973). Nature new Biol. 241:7.

Gurdon JB (1976). J. Embryol. exp. Morph. 36:523.

Gurdon JB, Brown DD (1978). Devel. Biol. 67:346.

Gurdon JB, Dingwall C, Laskey RA, Korn LJ (1982). Nature 299:652.

Harper ME, Price J, Korn LJ (1983). Nucleic Acids Res., in press.

Korn LJ (1982). Nature 295:101.

Korn LJ, Birkenmeier EH, Brown DD (1979). Nucleic

Acids Res. 7:947.
Korn LJ, Bogenhagen DF (1982). Organization and Transcription of the Xenopus 5S RNA Genes. In Rothblum L, Busch H (eds.): 'The Cell Nucleus', New York, Academic Press 12:1
Korn LJ, Gurdon JB (1981). Nature 289:461.
Korn LJ, Gurdon JB, Price J (1982). Nature 300:354.
Ng SY, Parker CS, Roeder RG (1979). Proc. Natl. Acad. Sci. U.S.A. 76:136.
Pardue ML, Brown DD, Birnstiel ML (1973). Chromosoma 42:191.
Pelham HRB, Brown DD (1980). Proc. Natl. Acad. Sci. U.S.A. 77:4170.
Peterson RC, Doering JL, Brown DD (1980). Cell 20:131.
Sakonju S, Bogenhagen DF, Brown DD (1980). Cell 19:13.
Wegnez M, Monier R, Denis H (1972). FEBS Lett 25:13.
Weil PA, Segall J, Harris B, Ng SY, Roeder RG (1979). J. Biol. Chem. 254:6163.

Gene Expression, pages 283-292
© 1983 Alan R. Liss, Inc., 150 Fifth Avenue, New York, NY 10011

A MODEL FOR THE REGULATION OF TRANSCRIPTIONAL EVENTS IN TERMINAL DIFFERENTIATION AND ONCOGENIC TRANSFORMATION BY PHOSPHOPROTEINS THAT BIND TO RNA POLYMERASE II[1]

Jack Greenblatt, Richard W. Carthew[2] and Mary Sopta

Banting and Best Department of Medical Research,
University of Toronto, Toronto, Ontario M5G 1L6

ABSTRACT There are three phosphoproteins (polIIB72, polIIB38, and polIIB30) in cultured murine and human cells which bind to micro-affinity columns containing immobilized calf thymus RNA polymerase II. PolIIB38 and polIIB30 are altered when murine erythroleukemia cells are treated with DMSO or with phorbol ester or when NIH/3T3 cells are transformed with the Kirsten sarcoma virus. PolIIB38 and polIIB30 may therefore have a role in regulating terminal erythroid differentiation and oncogenic transformation. We propose that polIIB38 and polIIB30 enable one or more of the protein kinase-mediated cellular regulatory mechanisms to regulate transcription by RNA polymerase II. A model for this type of transcriptional regulation is presented. We discuss the progress we have made in the purification of polIIB38 and polIIB30 and the possible involvement of polIIB38 and polIIB30 in oncogenic transformation and in the initiation of transcription by RNA polymerase II.

[1]This work was supported by the National Cancer Institute of Canada.
[2]Present Address: Department of Biology, Massachusetts Institute of Technology, Cambridge, Mass. 02139

INTRODUCTION

Murine erythroleukemia cells (MELC) can be induced with DMSO to undergo terminal erythroid differentiation in culture (1,2). The differentiation process takes several days and culminates in the production of highly differentiated cells resembling erythrocytes. DMSO induction of MELC increases the synthesis of some proteins, including α- and β-globins and heme biosynthetic enzymes, and decreases the synthesis of others (3). Increased production of β^{maj}-globin is partly a result of an increased rate of transcription of the β^{maj}-globin gene (4) and partly a result of increased stability of β^{maj}-globin mRNA (5). The increased transcription rate seems to result partly from an increased rate of chain initiation and partly from an increased probability of chain completion (4). A site essential for DMSO-induced transcription of the β^{maj}-globin gene is closely linked to the gene, within 78 base pairs from the site where transcription initiates (6,7).

We have been exploring the possibility that some transcriptional events during DMSO induction of MELC may be controlled by one or more proteins that associate directly with RNA polymerase II (8,9). We have detected three phosphoproteins in MELC (polIIB72, polIIB38, polIIB30) that bind to an affinity column containing covalently bound calf thymus RNA polymerase II (8,9) (see Figure 1). These proteins are not unique to MELC since all three are also present in murine NIH/3T3 cells and Y1 adrenocortical tumor cells and in human HeLa cells (9). When MELC are treated with DMSO or with a phorbol ester tumor promoter (TPA) an alkali-stable phosphate in polIIB30 is eliminated (9). PolIIB38 has increased phosphorylation in MELC treated with TPA (9). The alteration of polIIB30 induced by treatment of MELC with DMSO occurs during the first day of DMSO treatment, prior to and during the time of commitment to terminal differentiation (2). This suggests that polIIB30 may control alterations of transcription that occur early in the differentiation of MELC.

We have also observed a major alteration of polIIB38 in NIH/3T3 cells transformed by the Kirsten murine sarcoma virus (8,9). PolIIB38 forms a doublet band during SDS-PAGE, and transformation by KiMSV reduces the RNA polymerase II-binding concentration of the more abundant form of polIIB38 which has higher mobility during electrophoresis (9). It is therefore possible that polIIB38 has a role in oncogenic transformation.

We describe here the progress we have made in the purification of polIIB38 and polIIB30 on a preparative scale. We also describe some of their properties and discuss their possible relationships with other proteins thought to regulate transcription by RNA polymerase II. Finally, we shall propose models for the regulation of transcription by RNA polymerase II-binding phosphoproteins and discuss the possibility that polIIB30 and polIIB38 may be involved in controlling oncogenic transformation.

RESULTS AND DISCUSSION

Purification of PolIIB38 and PolIIB30

Chromatography on an RNA polymerase II affinity column is suitable for the study of polIIB38 and polIIB30 on an analytical scale, but not suitable for their purification on a preparative scale. We have therefore tried other approaches for the purification of polIIB38 and polIIB30 from the nuclei of MELC.

PolIIB72, polIIB38, and polIIB30 all bind to coenzyme A-agarose (Figure 1a-f). PolIIB38 eluted from coenzyme A-agarose with buffer containing coenzyme A is still able to bind to RNA polymerase II (Figure 1h,i). PolIIB30 can also be recovered from coenzyme A-agarose by washing the column with buffer containing a higher salt concentration (Figure 2). We have therefore used coenzyme A-agarose to achieve a considerable purification of polIIB38 and polIIB30 and their partial separation from one another (Figure 2). It was helpful in this regard that neither polIIB38 nor polIIB30 was precipitated when a nuclear extract was brought to 60% saturation with ammonium sulfate. In Figure 2 is shown an MELC nuclear extract precipitated with ammonium sulfate between 60% and 90% saturation and chromatographed on coenzyme A-agarose. PolIIB38 can be obtained nearly pure at this stage and is at least 90% pure after rechromatography on coenzyme A-agarose.

PolIIB30 chromatographs on coenzyme A-agarose with an activity that stimulates eightfold the transcription of calf thymus DNA by RNA polymerase II in the presence

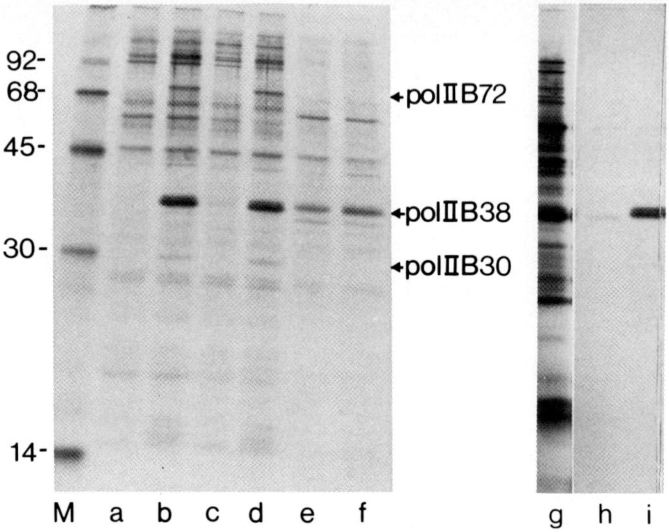

FIGURE 1. Binding of RNA polymerase II-binding proteins to coenzyme A-agarose. Extract from ^{35}S-methionine labelled MELC (9) was treated with one half volume of (a,b) agarose-hexane-NAD^{+}, (c,d) agarose-hexane-CoASH, or (e,f) agarose-hexane-S-CoA. Unbound protein was chromatographed on a control column (a,c,e) or on an AffiGel 10-RNA polymerase II column (b,d,f) (9). PolIIB72, polIIB30, and most of the polIIB38 are adsorbed by agarose-hexane-S-CoA. MELC proteins that bound to agarose-hexane-S-CoA, and were eluted by 10 mM CoA (g), were applied to a control column (h) and to an RNA polymerase II column (i) in combination with unlabeled carrier cell lysate. Eluates from control and affinity columns were analyzed by SDS-PAGE followed by fluorography (9). (M) molecular weight standards.

of Mn^{2+} (Figure 2). This activity may be intrinsic to polIIB30 because it copurifies with polIIB30 during salt gradient elution from phosphocellulose and carboxymethylcellulose, cosediments with polIIB30 in a sucrose density gradient, and flows through DEAE cellulose with polIIB30. None of these procedures is very efficient for the separation of polIIB30 from its major contaminants, and so we have not yet succeeded in purifying polIIB30 to homogeneity.

Is PolIIB38 or PolIIB30 an Initiation Factor for RNA Polymerase II?

Several groups have reported the partial purification from Hela cells of factors utilized by RNA poly-

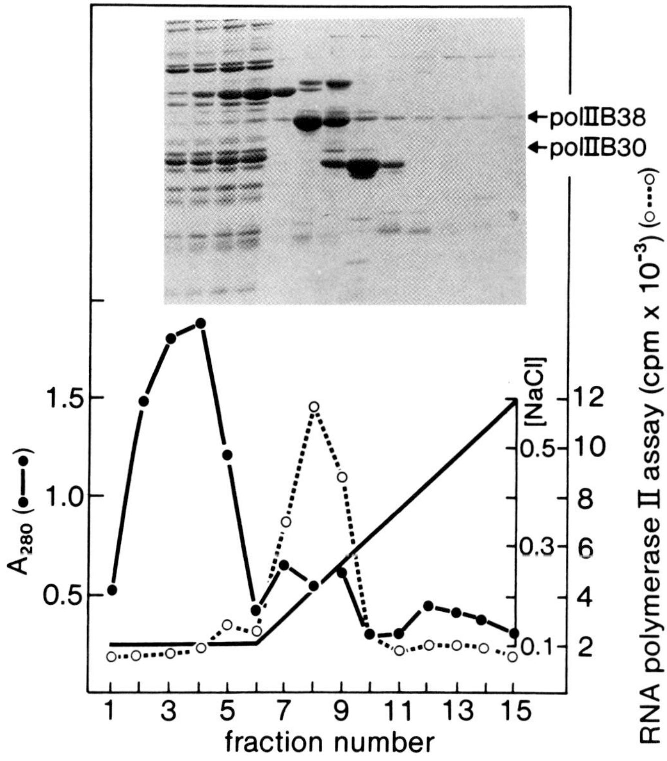

FIGURE 2. Purification of polIIB38 and polIIB30 on coenzyme A-agarose. Nuclear extract (10) from $2x10^{10}$ MELC was fractionated by ammonium sulfate precipitation (60-90% saturation) and chromatographed on a 3 ml CoA-agarose column (PL Biochemicals #5504). A 5 ul aliquot of each fraction was tested in an assay containing 0.2 ug RNA polymerase II and 0.5 ug native calf thymus DNA with Mn^{2+} as divalent cation. A 25ul (fr.1-5) or 50 ul (fr.6-14) aliquot of each fraction was analyzed by SDS-PAGE (inset). PolIIB30 elutes with a peak of RNA polymerase II-stimulating activity.

merase II for the accurate initiation of transcription at the major late promoter of adenovirus 2 (11-14). An antibody against an RNA polymerase II stimulatory factor, called SII, isolated by Natori and his coworkers (15), blocks initiation of transcription at this promoter (16). The stimulatory activity that copurifies with polIIB30 (Figure 2) is strikingly similar in its properties to those of SII (15), and they may be related proteins.

Matsui et al (11) and Samuels et al (12) have purified initiation factors initially on phosphocellulose and subsequently on DEAE cellulose (11) or by other methods. PolIIB30 and polIIB38 behave the same way during purification as an initiation factor that elutes from phosphocellulose with 0.6 M KCl and fails to bind to DEAE cellulose (11,12).

Dynan and Tjian (13) and Davison et al (14) have purified initiation factors initially on heparin-Sepharose and subsequently by gel filtration and other methods. They have obtained evidence that the RNA polymerase II initiation factor is a high molecular weight complex that can be dissociated by chromatography on phosphocellulose (13,14). We have noted that the activity associated with polIIB30 (Figure 2) sediments heterogeneously after coenzyme A-agarose chromatography, but homogeneously at 3.2S when the coenzyme A-agarose fraction has been subjected to phosphocellulose chromatography (not shown).

It is therefore possible that polIIB38 and polIIB30 form part of a multi-protein complex required for initiation of transcription by RNA polymerase II. This question will only be answered rigorously when the initiation factors are purified to homogeneity or when polIIB38 and polIIB30 are tested in specific initiation assays.

Protein Phosphorylation as a Transcriptional Regulatory Mechanism

All three RNA polymerase II-binding proteins that we have identified in murine and human cells are phosphoproteins (8,9). Furthermore, the phosphorylation of polIIB38 and polIIB30 changes when MELC are treated with DMSO or with TPA. This suggests that the state of phosphorylation of these proteins can be used as a signal for the regulation of transcription.

However, transcriptional regulation depends for its

specificity on the recognition of particular sequences in nucleic acids, and it is not obvious how phosphorylation of particular sites in an RNA polymerase II-binding protein could alter its ability to sense many different sequences in DNA or RNA.

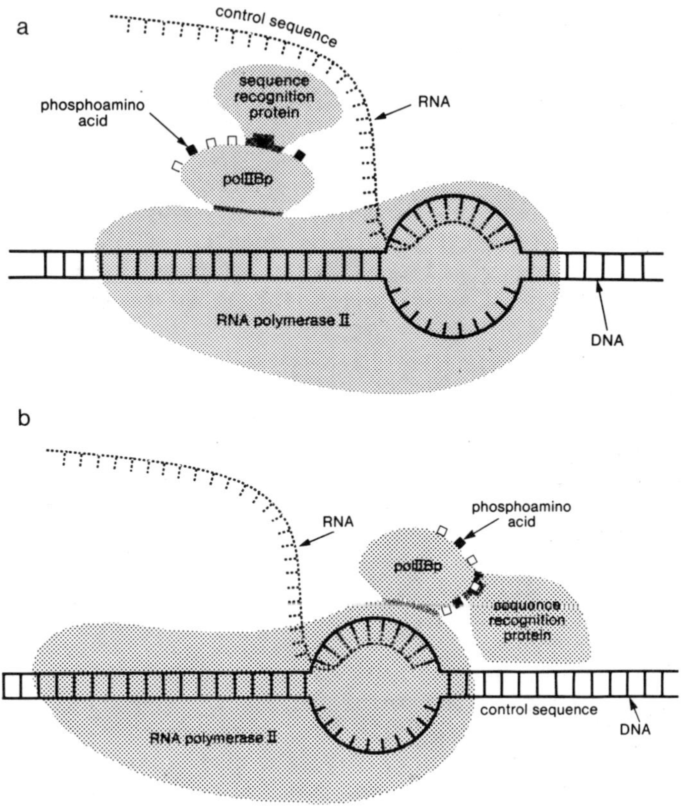

FIGURE 3. Schemes for the regulation of transcription by phosphoproteins that bind to RNA polymerase II. We postulate the existence of sequence recognition proteins (srp) that are also capable of recognizing sites on polIIB38 or polIIB30 that can be altered by phosphorylation. In (a) the srp binds to a control sequence in nascent RNA and may regulate chain elongation in a manner analogous to the N gene protein of bacteriophage λ (18). In (b) the srp binds to a control sequence in DNA which could be located in the promoter or elsewhere.

We therefore propose (see Figure 3) that there are sequence recognition proteins which bind specifically to DNA or nascent RNA and are also able to recognize whether a particular amino acid in polIIB38 or polIIB30 is phosphorylated. Such a regulatory mechanism would potentially allow as many protein kinase systems to regulate transcription by RNA polymerase II as there are phosphorylation sites in polIIB38 and polIIB30 (at least four) (9).

It is noteworthy in this regard that at least two of the core subunits of RNA polymerase II are also phosphorylated in mammalian cells (17). RNA polymerase II and its associated proteins could therefore reflect as many protein kinase-mediated regulatory mechanisms as there are phosphorylation sites. The transcriptional flexibility this could provide is substantial = if there are n phosphorylation sites, there could be as many as 2^n possible ways of regulating a gene.

Are PolIIB38 and PolIIB30 Involved in Oncogenic Transformation?

PolIIB38 and polIIB30 are modified in various ways when murine cells are treated with the tumor promoter TPA or are transformed by the Kirsten sarcoma virus (8,9). This raises the possibility that the modification of polIIB38 and polIIB30 is an important pathway for the alterations of gene expression that accompany an oncogenic phenotype.

Cooper and Hunter (19) have identified two proteins which are probably the main targets for tyrosine phosphorylation in murine cells transformed by the retroviral oncogenes src, fes, and abl. One has a molecular weight of 39kD and a pI of 7.5 and is similar in its molecular parameters to polIIB38. The other has a molecular weight of 29kD and a pI of 7.38 and is similar in its molecular parameters to polIIB30. It is tempting to speculate that polIIB38 and polIIB30 may be phosphorylated on tyrosine by the protein kinases encoded by various oncogenes. The major targets of $pp60^{src}$ are believed to be located mainly in the cytoplasm (20). However, an S100 lysate prepared from human KB cells grown in tissue culture contains all the factors necessary for the specific initiation of transcription by RNA polymerase II (21). Perhaps polIIB38 and polIIB30 diffuse freely between nucleus and cytoplasm in order to carry information to RNA polymerase II concerning the status of protein kinase regulatory systems in the cytoplasm. If polIIB38 and polIIB30 are modified by $pp60^{src}$, their modification may represent an important common pathway for the modification of gene expression in oncogenic transformation.

REFERENCES

1. Friend C, Scher W, Holland JG, Sato T (1971). Hemoglobin synthesis in murine virus-induced leukemic cells in vitro = stimulation of erythroid differentiation by dimethyl sulfoxide. Proc Natl Acad Sci USA 68:378.
2. Gusella J, Geller R, Clarke B, Weeks V, Housman, D (1976). Commitment to erythroid differentiation by Friend erythroleukemia cells: a stochastic analysis. Cell 9:221.
3. Marks P, Rifkind R (1978). Erythroleukemic differentiation. Ann Rev Biochem 47:419.
4. Hofer E, Hofer-Warbinek R, Darnell Jr JE (1982). Globin RNA transcription: a possible termination site and demonstration of transcriptional control correlated with altered chromatin structure. Cell 29:887.
5. Volloch V, Housman D (1981). Stability of globin mRNA in terminally differentiating murine erythroleukemia cells. Cell 23:509.
6. Chao MV, Mellon P, Charnay P, Maniatis T, Axel R (1983). The regulated expression of β-globin genes introduced into mouse erythroleukemia cells. Cell 32:483.
7. Mellon P, Maniatis T (1983). Regulation of globin genes after DNA transformation into mouse erythroleukemia cells. J Cell Biochem Supp 7A (in press).
8. Greenblatt J, Carthew RW (1983). Identification and characterization of proteins that bind to mammalian RNA polymerase II. J Cell Biochem Supp 7A:0312.
9. Carthew RW, Greenblatt J (1983). RNA polymerase II-binding proteins in cultured mammalian cells. Submitted.
10. Dignam JD, Martin PM, Shastry BS, Roeder RG (1982). Methods Enzymol (in press).
11. Matsui T, Segall J, Weil PA, Roeder RG (1980).

Multiple factors required for accurate initiation of transcription by RNA polymerase II. J Biol Chem 255:11992.
12. Samuels M, Fire A, Sharp PA (1982). Separation and characterization of factors mediating accurate transcription by RNA polymerase II. J Biol Chem 257:14419.
13. Dynan WS, Tjian R (1983). Isolation of factors needed to direct site-specific initiation of transcription by RNA polymerase II at SV40 and adenovirus promoters. In press.
14. Davison BL, Egly J-M, Mulvihill ER, Chambon P (1983). Formation of stable preinitiation complexes between eukaryotic class B transcription factors and promoter sequences. Nature 301:680.
15. Nakanishi Y, Mitsuhashi Y, Sekimizu K, Yokoi H, Tanaka Y, Horikoshi M, Natori S (1981). Characterization of three proteins stimulating RNA polymerase II. FEBS Lett 130:69.
16. Sekimizu K, Yokoi H, Natori S (1982). Evidence that stimulatory factor(s) of RNA polymerase II participates in accurate transcription in a HeLa cell lysate. J Biol Chem 257:2719.
17. Dahmus ME (1981). Phosphorylation of eukaryotic DNA-dependent RNA polymerase. Identification of calf thymus RNA polymerase subunits phosphorylated by two purified protein kinases, correlation with in vivo sites of phosphorylation in HeLa cell RNA polymerase II. J Biol Chem 256:3332.
18. Greenblatt J (1981). Regulation of transcription termination by the N gene protein of bacteriophage lambda. Cell 24:8.
19. Cooper JA, Hunter T (1981). Four different classes of retroviruses induce phosphorylation of tyrosines present in similar cellular proteins. Mol Cell Biol 1:394.
20. Cooper JA, Hunter T (1982). Discrete primary locations of a tyrosine protein kinase and of three proteins that contain phosphotyrosine in virally transformed chick fibroblasts. J Cell Biol 94:287.
21. Weil PA, Luse DS, Segall J, Roeder RG (1979). Selective and accurate initiation of transcription at the Ad2 major late promoter in a soluble system dependent on purified RNA polymerase II and DNA. Cell 18:469.

V. TERMINATION AND POST-TRANSCRIPTIONAL CONTROLS

Gene Expression, pages 295-310
© 1983 Alan R. Liss, Inc., 150 Fifth Avenue, New York, NY 10011

ATTENUATION CONTROL OF trp OPERON EXPRESSION[1]

Charles Yanofsky, Anathbandhu Das, Robert Fisher, Roberto Kolter and Vivian Berlin

Department of Biological Sciences, Stanford University
Stanford, California 94305

ABSTRACT Our current model of regulation of transcription termination at the attenuator of the tryptophan (trp) operon of *Escherichia coli* is presented. This model invokes transcription pausing as the event that synchronizes transcription of the trp leader region with translation of the trp leader transcript; it explains the extent of transcription termination at the attenuator in cultures growing with excess tryptophan; it proposes that leader peptide synthesis is shut down once the transcription termination decision has been made.

INTRODUCTION

Transcription of the structural genes of many bacterial operons concerned with amino acid biosynthesis and utilization is regulated by transcription termination at a site, the attenuator, located in the leader region

---

[1]The studies described in this summary were supported by grants from the U.S. Public Health Service, the National Science Foundation and the American Heart Association. Charles Yanofsky is a Career Investigator of the American Heart Association, Anath Das, Robert Fisher, and Roberto Kolter are postdoctoral fellows of the American Heart Association, the Damon Runyon - Walter Winchell Foundation and the Helen Hay Whitney Foundation, respectively, and Vivian Berlin is a predoctoral trainee of the U.S. Public Health Service.

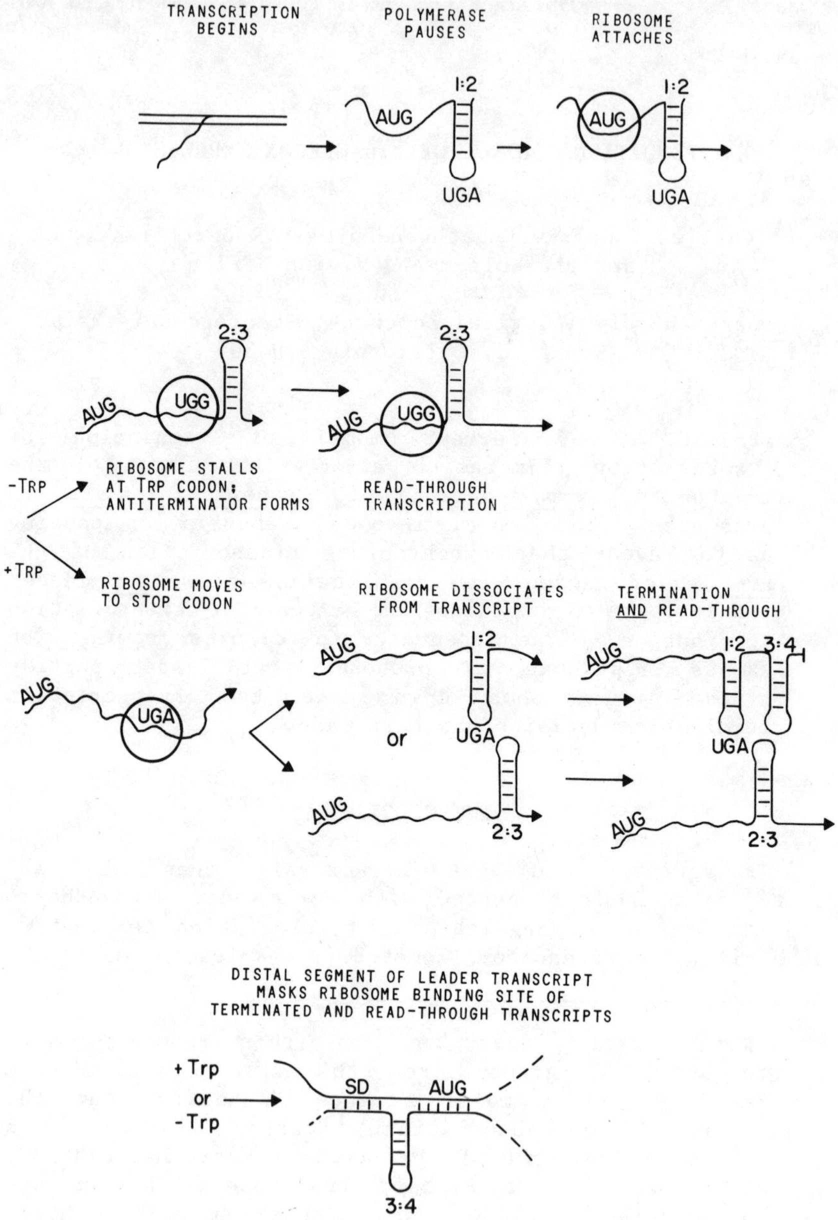

Figure 1 - The presumed sequence of events responsible for attenuation control of transcription of the trp operon of E. coli.

preceding the structural genes of the operon (1, 2). Regulation at this site is accomplished in response to ribosome movement and position on a short peptide coding segment in the leader portion of each transcript. This coding region is rich in codons for the regulating amino acid(s). Inability to translate these codons - due to a deficiency of the charged cognate tRNAs - is believed to result in ribosome stalling at the codons. This stalling facilitates formation of a particular secondary structure in the transcript, termed the antiterminator. When this structure forms it precludes formation of a second RNA structure, the terminator, which is believed to be the transcription termination signal recognized by the transcribing polymerase molecule. If the relevant tRNAs are fully charged, the translating ribosome completes synthesis of the leader peptide, and the RNA terminator forms and causes termination.

Current Model of Attentuation in the <u>trp</u> Operons of Enterobacteria

Our current model of attenuation in the tryptophan (<u>trp</u>) operons of Enterobacteria is summarized in Figure 1. We have incorporated several features in the model that have become evident only recently. We believe that these features explain several poorly understood observations and suggest functional roles for additional segments of the leader transcript. These features include: i) a possible role for transcription pausing in synchronizing translation of the leader peptide coding segment with transcription of the leader region; ii) the role of ribosome dissociation at the leader peptide stop codon in establishing the level of transcriptional read-through <u>in vivo</u>; iii) the role of a complementary RNA segment in blocking multiple rounds of synthesis of the <u>trp</u> leader peptide and iv) the absence of a requirement for a release factor for dissociation of the transcript, template, and polymerase at the <u>trp</u> attenuator.

According to our current view of attenuation in the <u>trp</u> operon, RNA polymerase molecules that have escaped repression begin transcription of the leader region of the operon and continue synthesis until they complete segment 2 of the transcript (see Figure 2). RNA segments 1 and 2 pair and form secondary structure 1:2, causing RNA polymerase to pause (3-5). [The transcribing polymerase presumably is complexed with the nusA protein throughout these stages in chain elongation (6).] When the leader ribosome binding site is synthesized, or soon thereafter, a ribosome binds and initiates synthesis of the leader peptide. The polymerase resumes transcription either spontaneously or as a consequence of the approach of the translating ribosome. When charged tRNA^{Trp} is plentiful the translating ribosome reaches the stop codon and dissociates from the transcript (7). The transcript then forms either structure 1:2 or structure 2:3 (8-9). If 1:2 forms, then 3:4, the terminator, forms immediately thereafter, and transcription is terminated. If structure 2:3, the antiterminator, forms, then the terminator does not form and transcription proceeds to the end of the operon.

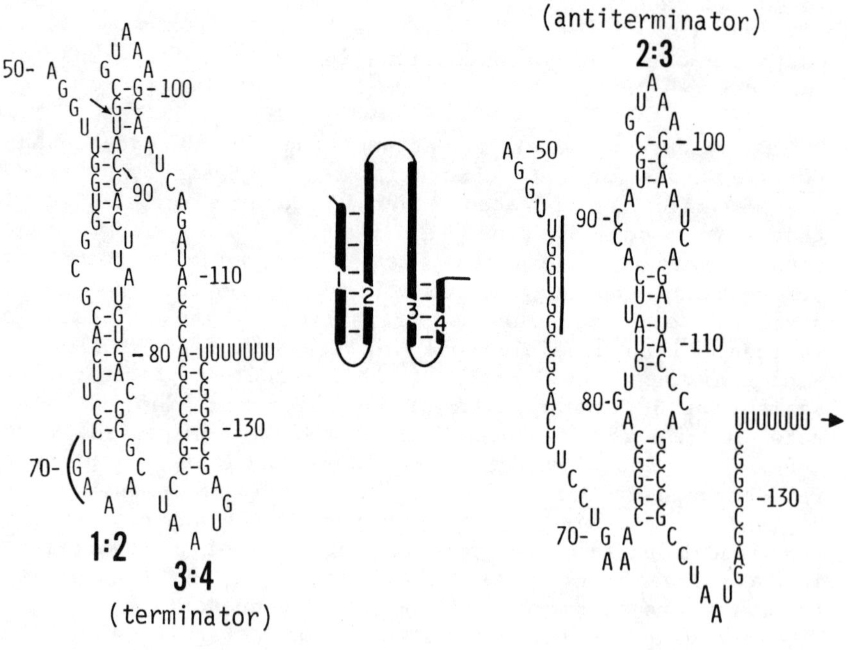

Figure 2. Secondary structures in the trp leader transcript. The alternative secondary structures involving RNA segments designated 1, 2, 3 and 4 are indicated. The translation stop codon UGA and the Trp codons are marked by bars. An arrow at nucleotide 92 marks the 3' terminus of the pause transcript.

When there is a deficiency of charged $tRNA^{Trp}$ the translating ribosome stalls at either of the Trp codons. This results in formation of the antiterminator and transcription continues into the operon. Whether termination at the attenuator occurs or not, the leader ribosome binding site then pairs with a complementary segment of the transcript, and leader peptide synthesis is shut down (10). All of the events depicted in Figure 1 must occur within a few seconds since transcription initiation at the trp promoter occurs about every six seconds in the absence of repression (11, 12). Otherwise, succeeding polymerase molecules could not transcribe the leader region and be regulated properly.

Over the past seven years, many aspects of the above model have been examined and supported by experimental findings. We and/or others have shown that: i) the leader peptide coding region is translated (10); ii) the inability to translate Trp codons, rather than the existence of a tryptophan deficiency, is responsible for relief of termination (13, 14); iii) transcription terminates in the trp leader region at about base pair 140 both in vivo and in vitro (in E. coli) (15, 16); iv) the terminated transcript has extensive secondary structure; both structures 1:2 and 3:4 as drawn (Figure 2) are partially resistant to RNase T1 attack (16, 17). An alternative secondary structure, 2:3, is theoretically possible (Figure 2). This structure has not been demonstrated experimentally, presumably because formation of the other two structures prevents its formation; v) single base pair mutations in the leader region that relieve transcription termination at the attenuator in vivo and in vitro generally occur in the G-C rich region that forms the RNA terminator (18, 19). These changes reduce the stability of the terminator; vi) single base pair mutations that destabilize the antiterminator prevent the relief from termination normally associated with tryptophan starvation (14); vii) mutations altering the leader peptide start codon prevent the tryptophan starvation (termination relief) response, presumably by preventing ribosome movement to the Trp codons of the peptide coding segment present in the transcript (14, 20). These mutations actually increase transcription termination at the attenuator; viii) starvation for arginine as well as tryptophan but not other amino acids in the leader peptide relieves transcription termination in vivo at the E. coli trp attenuator (14). The single

Arg codon in the *trp* leader transcript is the codon following the second Trp codon. If we assume that a ribosome masks ten nucleotides on the 3' side of the codon being read, then stalling over only the Trp or Arg codons could promote formation of the antiterminator secondary structure (1). In *Serratia marcescens*, starvation for His, Trp or Arg relieves termination (8). There is a His codon two codons before the first Trp codon; the Arg codon follows the tandem Trp codons; ix) deletions that remove different segments of the *trp* leader transcript of *Serratia marcescens* affect transcription termination at the attenuator both *in vivo* and *in vitro* in complete accord with the RNA secondary structure model of attenuation (8, 20). The conclusion from these studies is that whenever the RNA terminator forms, it is recognized by the transcribing polymerase as a termination signal. Thus all events that influence attenuation, such as ribosome movement and stalling, simply regulate formation of the terminator; x) the transcription termination complex dissociates spontaneously at the *trp* attenuator *in vitro* (21). Thus no accessory release factor appears to be required in the termination event; xi) mutant RNA polymerases altered in the β subunit behave aberrantly when they terminate transcription at the attenuator (22, 23). The β subunit mutations presumably alter recognition of the RNA terminator. These mutations have no effect on termination when the RNA terminator does not form; xii) a RNA secondary structure, rather than a DNA secondary structure, constitutes the termination signal that RNA polymerase recognizes (16, 24, 25); xiii) when RNA polymerase transcribes the *trp* operon of *E. coli* and other enteric bacteria *in vitro*, polymerase pauses near base pair 90 after synthesizing the first RNA secondary structure, 1:2 (3, 4, 26). It is thought that this pause may serve to synchronize transcription with translation. The pause would ensure that the translating ribosome reaches the Trp codons or the leader peptide stop codon, at a time appropriate to determine which of the alternate RNA secondary structures will form; xiv) *in vitro* addition of an oligomer complementary to RNA segment 1 reduces the pause half-life and relieves transcription termination at the attenuator (5). These effects are presumed to result from the disruption of structure 1:2 by the added oligomer.

In the following sections we will describe more fully aspects of the model of attenuation that have received our recent attention.

Transcription Pausing and its Possible Role in Regulation of Attenuation in the *trp* Operon

When RNA polymerase transcribes the *trp* operon leader region of *E. coli in vitro*, it pauses near base pair 90 after synthesizing the first hairpin secondary structure in the *trp* leader transcript (3, 4, 26). Pausing is enhanced by lowering the concentration of the next nucleoside triphosphate to be added to the growing messenger chain or by adding *E. coli* L-factor (*nusA* protein) (26, 27). Paused complexes formed under either condition do not terminate: i.e., if transcription initiation is inhibited and the transcripts present *in vitro* are examined over a considerable period, every paused complex disappears and pause RNA grows to become a terminated or read-through transcript (26).

We believe that RNA hairpin structure 1:2 is the transcription pause signal. This view is based on *in vitro* transcription experiments in which we added a synthetic DNA oligomer complementary to the proximal segment (1) of the RNA hairpin (5). This complementary DNA oligomer relieved transcription pausing *in vitro* both in the presence and absence of L-factor. The oligomer reduced the stability of the paused complex both during transcription and following formation of paused complexes. Thus the oligomer appears to be capable of disrupting the secondary structure of the pause signal and thereby release the paused complex. When transcription resumes in the presence of the oligomer, high read-through transcription is observed (5). Thus the DNA oligomer not only releases the paused complex but, by virtue of complexing with RNA segment 1, also favors formation of the antiterminator, 2:3 (5, 28). Kinetic analyses of the release of paused complexes by added oligomer suggest that the oligomer interacts with and releases more than 90% of the paused complexes (5). Transcription pausing may be a crucial event in regulation by attenuation. Translation must be coupled to transcription if attenuation is to be effective. If polymerase molecules pause *in vivo* after synthesizing structure 1:2, such pausing could allow sufficient time for a ribosome to bind to the leader peptide start codon and initiate synthesis of the leader peptide. Thus pausing would synchronize transcription and translation by fixing one event, transcription, while allowing the second event, translation, to occur. We have not yet detected transcription pausing in vivo; however, we have performed in vitro transcription-translation experiments that suggest that pausing does in fact occur (29). An interesting but as yet unanswered question is whether the translating ribosome normally releases the paused complex, i.e., whether the act of translation is the event that normally triggers resumption of transcription.

We believe that the pause signal, secondary structure 1:2, and the transcription termination signal, the RNA terminator, may be recognized at the same site on the transcribing RNA polymerase molecule (23). This view is based on analyses with mutants with altered RNA polymerase β subunits. Two types of altered polymerases have been examined; one exhibits enhanced termination whereas the other is defective in termination, both in vivo and in vitro (22, 23). When termination and pausing are assessed independently in vitro with mutant and wild type polymerases and a variety of mutant and wild type templates, we find that the altered polymerases behave comparably in recognizing pause and termination signals; i.e., the altered polymerase that exhibits an extended pause is more proficient in termination at the attenuator and conversely, the altered polymerase that pauses only briefly exhibits decreased transcription termination at the attenuator. These observations support the view that the same site or region of the transcribing polymerase molecule interacts with the pause and termination signals.

Detection of the trp leader peptide

Until recently we were unable to demonstrate synthesis of the trp leader peptides of E. coli and other enteric bacteria. We assumed that our difficulty in detecting these peptides was due to their anticipated lability and the likelihood that leader peptide synthesis was shut down once the transcription termination decision had been made. The latter expectation was based on the finding that in seven enteric species a distal segment of the trp leader transcript was complemetary to the ribosome binding site presumably used in synthesis of the trp leader peptide (Figure 3). Pairing of these segments could theoretically block ribosome binding in all but the initial round of translation. The leader peptide was detected in an *in vitro* coupled transcription-translation system (10). The template used was a self-ligated DNA restriction fragment containing the *trp* promoter and leader region and coding only for the *trp* leader peptide; this fragment lacked the distal segment of the leader region complementary to the ribosome binding site. Synthesis of the *trp* leader peptides of *E. coli* and *Serratia marcescens* was established by demonstrating that only those amino acids predicted to be present in the peptide did in fact label the peptide. As expected, addition of the *E. coli trp* repressor shut off synthesis of the leader peptide. The *E. coli trp* leader peptide was shown to be labile in the S-30 system; it had a half-life of only three to four minutes.

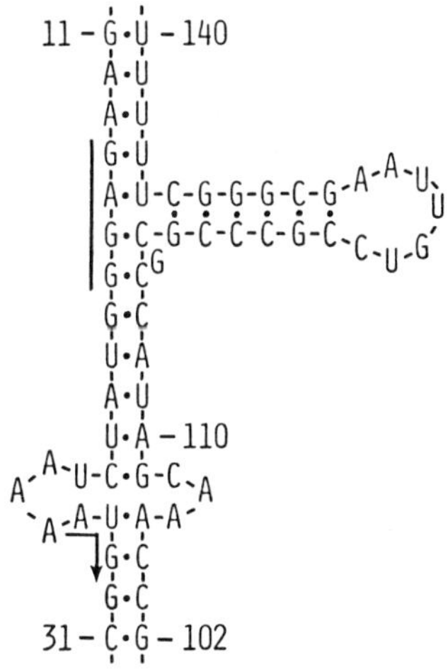

Figure 3. Base pairing around the ribosome binding site for the leader peptide of *Salmonella typhimurium*. The Shine-Dalgarno region and start codon are marked by bars. Nucleotide numbers are from the 5' end of the transcript.

We have used a highly purified DNA-dependent peptide synthesizing system (30, 31) to demonstrate synthesis of the N-terminal di- and tripeptides predicted from the

Salmonella typhimurium trp leader sequence (10). This peptide synthesizing system was also employed to examine the postulated blocking of leader peptide synthesis by a distal RNA segment. Two homologous plasmids were constructed, one containing the downstream complementary region and the second lacking it. When these templates were used to direct *in vitro* dipeptide synthesis, we observed a 10-fold difference in the amount of dipeptide formed (10); the template lacking the downstream complementary region produced the greater amount. We conclude from these studies that the *trp* leader peptide is in fact synthesized, as we suspected from gene fusion analyses (32, 33), and that after the translating ribosome aids in the transcription termination decision, the transcript folds back on itself and shuts off further synthesis of the leader peptide.

How the Steady State Level of Read-through at the *trp* Attenuator May Be Set

Several observations indicate that in bacterial cultures growing in the presence of excess tryptophan there is appreciable read-through at the *trp* attenuator. Point mutations that alter the translation start codon for the leader peptide (14), and deletions that remove the translation initiation region (20), increase transcription termination at the attenuator five- to eight-fold. Similarly, mutations in RNA segment 2 that reduce the stability of structure 2:3 increase transcription termination at the attenuator. These findings suggest that the steady state level of expression of cultures growing in the presence of excess tryptophan is dependent both on translation and secondary structure 2:3 formation.

Recently, additional mutants have been isolated and analyzed (7) which shed light on these observations. These mutants have alterations in RNA segment 1 that reduce the stability of structure 1:2, but do not affect the reading frame of the peptide coding region. When these mutants are grown in the presence of excess tryptophan, transcription termination at the *trp* attenuator is reduced. Thus 1:2 pairing as well as 2:3 pairing appears to be essential in setting the steady state level of read-through in cultures grown in tryptophan. Since both ribosome movement and formation of alternative secondary structures 1:2 and 2:3 are involved in establishing the extent of termination at the attenuator we suggest that during normal transcription and translation the translating ribosome dissociates from the leader peptide stop codon as soon as the leader peptide is synthesized. At this stage in transcription of the leader region, either structure 1:2 or structure 2:3 could form (perhaps an equilibrium mixture is produced with the amount of each species determined by the relative probabilities of forming the two competing structures). If 1:2 forms, then 3:4 would form, and transcription termination would result. If 2:3 forms, then 3:4 would not form, and read-through transcription would occur. Thus ribosome dissociation from the leader peptide stop codon could be responsible for the level of read-through observed in cultures growing with excess tryptophan. Consistent with this interpretation is the behavior of a particular deletion mutant of _Serratia marcescens_; this mutant has an altered leader peptide Shine-Dalgarno region and appears to be slightly defective in translation initiation (20). This mutant has the interesting property of exhibiting somewhat greater termination at the attenuator than wild type when grown in the presence of tryptophan _and_ when it is starved of tryptophan. We believe that because of the defect in this mutant, ribosomes translate only about half of the _trp_ leader transcripts. The half that are untranslated presumably are terminated prematurely.

Spontaneous dissociation of the transcription termination complex at the _trp_ attenuator

In vivo studies have implicated both the _E. coli_ transcription termination protein rho, and the _nusA_ protein, in transcription termination at the _trp_ attenuator (34, 35,). In view of these findings, as well as others (36), we were concerned that a release factor was an essential participant in the transcription termination event at the _trp_ attenuator. Accordingly, we carried out _in vitro_ experiments aimed at determining what cell components, if any, were needed for dissociation of the transcription termination complex at the _trp_ attenuator (21). Earlier experiments had suggested that a stable, non-filterable termination complex was formed _in vitro_ during transcription of the _trp_ leader region. Both labeled template and transcript were retained following transcription of purified DNA restriction fragments containing only the _trp_ promoter-leader region. However, on closer analysis of the transcription termination event, we found that when transcription was carried out in the presence of excess sigma factor, heparin, or transfer RNA, transcription termination complexes were not retained on filters. We also performed reconstitution experiments with _trp_ leader RNA and RNA polymerase and were able to confirm that core polymerase binds non-specifically to RNA or DNA and that this binding can be prevented by sigma factor (37) or heparin. Consistent with our observations is the finding in footprinting analyses with presumed "_trp_ attenuator termination complexes" that, following transcription, polymerase is not at any particular site on the DNA template (38). We therefore conclude that transcript, template and polymerase dissociate spontaneously at the _trp_ attenuator and no accessory factor is needed for this event.

CONCLUSIONS

Our studies on attenuation in the _trp_ operon of Enterobacteria have revealed how the normal cell components that participate in transcription and translation regulate transcription termination. The location of tandem Trp codons in the _trp_ leader transcript relative to the location of transcript segments that can form alternative secondary structures, dictate that ribosome stalling caused by a charged $tRNA^{Trp}$ deficiency will prevent formation of a secondary structure that is recognized as the termination signal. The short, initial segment of the _trp_ transcript therefore contains all the information required for specific termination control. This general pattern is repeated throughout bacterial operons concerned with amino acid biosynthesis (1, 2). Slight modifications thereof are now known to be used in regulation of pyrimidine biosynthesis and ampicillin resistance (39, 40). There is a translation counterpart as well - used in regulating activation of a ribosome binding site (41, 42). In these varied examples of attenuation we see how regulation of gene expression is accomplished in response to signals that the cell might find difficult to recognize if regulation occurred exclusively by transcription initiation control.

REFERENCES

1. Yanofsky C (1981). Attenuation in the control of expression of bacterial operons. Nature 289: 751.
2. Kolter R, Yanofsky C (1982). Attenuation in amino acid biosynthetic operons. Ann Rev Genet 16: 113.
3. Winkler ME, Yanofsky C (1981). Pausing of RNA polymerase during in vitro transcription of the tryptophan operon leader region. Biochem 20: 3738.
4. Farnham PJ, Platt T (1981). Rho-independent termination: dyad symmentry in DNA causes RNA polymerase to pause during transcription in vitro. Nucl Acids Res 9: 563.
5. Fisher R, Yanofsky C (1983). A complementary DNA oligomer releases a transcription pause complex. J Biol Chem (manuscript submitted)
6. Greenblatt J, Li J (1981). Interaction of the sigma factor and the nusA gene protein of E. coli with RNA polymerase in the initiation-termination cycle of transcription. Cell 24: 421.
7. Kolter R, Yanofsky C unpublished.
8. Stroynowski I, Yanofsky C (1982). Transcript secondary structures regulate transcription termination at the attenuator of the trp operon of Serratia marcescens. Nature 298: 34.
9. Stroynowski I, Kuroda M, Yanofsky C (1983). Transcription termination in vitro at the tryptophan operon attenuator is controlled by secondary structures in the leader transcript. Proc Natl Acad Sci USA (manuscript submitted).
10. Das A, Urbanowski J, Weissbach H, Nestor J, Yanofsky C (1983). In vitro synthesis of the tryptophan operon leader peptides of Escherichia coli, Serratia marcescens and Salmonella typhimurium. Proc Natl Acad Sci USA (manuscript submitted).
11. Baker R, Yanofsky C (1972). Transcription initiation frequency and translational yield for the tryptophan operon of Escherichia coli. J. Mol Biol 69: 89.
12. Bertrand K, Yanofsky C (1976). Regulation of transcription termination in the leader region of the tryptophan operon of Escherichia coli involves tryptophan or its metabolic product. J Mol Biol 103: 339.

13. Yanofsky C, Soll L (1977). Mutations affecting $tRNA^{trp}$ and its charging and their effect on regulation of transcription termination at the attenuator of the tryptophan operon. J Mol Biol 113: 663.
14. Zurawski G, Elseviers D, Stauffer G, Yanofsky C (1978). Translational control of transcription termination at the attenuator of the Escherichia coli tryptophan operon. Proc Natl Acad Sci USA 75: 5988.
15. Bertrand K, Korn L, Lee F, Yanofsky C (1977). Heterogenous 3'-OH termini in vivo and deletion mapping of functions. J Mol Biol 117: 227.
16. Lee F, Yanofsky C (1977). Transcription termination at the trp operon attenuators of Escherichia coli and Salmonella typhimurium: RNA secondary structure and regulation of termination. Proc Natl Acad Sci USA 74: 4365.
17. Oxender DL, Zurawski G, Yanofsky C (1979). Attenuation in the Escherichia coli tryptophan operon: role of RNA secondary structure involving the tryptophan coding region. Proc Natl Acad Sci USA 76: 5524.
18. Stauffer GV, Zurawski G, Yanofsky C (1978). Single base-pair alterations in the Escherichia coli trp operon leader region that relieve transcription termination at the trp attenuator. Proc Natl Acad Sci USA 75: 4833.
19. Zurawski G, Yanofsky C (1980). Escherichia coli tryptophan operon leader mutations, which relieve transcription termination, are cis-dominant to trp leader mutations, which increase transcription termination. J Mol Biol 142: 123.
20. Stroynowski I, van Cleemput M, Yanofsky C (1981). Superattenuation in the tryptophan operon of Serratia marcescens. Nature 98: 38.
21. Berlin V, Yanofsky C (1983). Release of transcript and template during transcription termination at the trp operon attenuator. J Biol Chem 258: 1714.
22. Yanofsky C, Horn V (1981). Rifampicin resistance mutations that alter the efficiency of transcription termination at the tryptophan operon attenuator. J Bacteriol 145: 1334.
23. Fisher R, Yanofsky C (1983). Mutations of the β subunit of RNA polymerase alter both transcription pausing and transcription termination in the trp operon leader region in vitro. J Biol Chem (manuscript submitted).

24. Farnham PJ, Platt T (1982). Effects of DNA base analogs on transcription termination at the tryptophan operon attenuator of Escherichia coli. Proc Natl Acad Sci USA 79: 998.
25. Ryan T, Chamberlin M - Personal communication.
26. Fisher R, Yanofsky C. manuscript in preparation.
27. Farnham PJ, Greenblatt J, Platt T (1982). Effects of nusA protein on transcription termination of the tryptophan operon of E. coli. Cell 29: 945.
28. Winkler ME, Mullis K, Barnett J, Stroynowski I, Yanofsky C (1982). Transcription termination at the tryptophan operon attenuator is decreased in vitro by an oligomer complementary to a segment of the leader transcript. Proc Natl Acad Sci USA 79: 2181.
29. Fisher R, Das A, Yanofsky C. unpublished experiments
30. Robakis N, Meza-Basso L, Brot N, Weissbach H (1981). Translational control of ribosomal protein L10 synthesis occurs prior to formation of first dipeptide band. Proc. Natl Acad Sci USA 78: 4261.
31. Cenatiempo Y, Robakis N, Reid B, Weissbach H, Brot N (1982). In vitro expression of Escherichia coli ribosomal protein L10 gene: tripeptide synthesis as a measure of functional mRNA. Arch Biochem Biophys 218: 572.
32. Miozzari G, Yanofsky C (1978). Translation of the leader region of the Escherichia coli tryptophan operon. J Bacteriol 133: 1457.
33. Schmeissner U, Ganem D, Miller J (1977). Genetic studies of the lac repressor II fine structure deletion map of the lac I gene, and its correlation with the physical map. J Mol Biol 109: 303.
34. Korn LJ, Yanofsky C (1976). Polarity suppressors defective in transcription termination at the attenuator of the tryptophan operon of Escherichia coli have altered rho factors. J Mol Biol 106: 231.
35. Ward DF, Gottesman ME (1981). The nus mutations affect transcription termination in Escherichia coli. Nature 292: 212.
36. Fuller RS, Platt T, (1978). The attenuation of the tryptophan operon in E. coli: rho-mediated release of RNA polymerase at a transcription termination complex in vitro. Nucl Acids Res 5: 4613.
37. Hinkle DC, Chamberlin MJ (1972). Studies of the binding of Escherichia coli: RNA polymerase to DNA. J Mol Biol 70: 157.
38. Winkler M, Berlin V, Yanofsky C unpublished.

39. Turnbough CL Jr, Hicks KL, Donahue JP (1983). Attenuation control of pyrBI operon expression in Escherichia coli K-12. Proc Natl Acad Sci USA 80: 368.
40. Jaurin B, Grundstrom T, Edlund T, Normark S (1981). The E. coli β-lactamase attenuator mediates growth rate-dependent regulation. Nature 290: 221.
41. Horinouchi S, Weisblum B (1980). Posttranscriptional modification of mRNA conformation: Mechanism that regulates enrythromycin-induced resistance. Proc Natl Acad Sci USA 77: 7079.
42. Hahn J, Gryczan TJ, Dubnau D (1982). Translational attenuation of ermC: a deletion analysis. Mol Gen Genet 186: 204.

Gene Expression, pages 311-326
© 1983 Alan R. Liss, Inc., 150 Fifth Avenue, New York, NY 10011

CONTROL OF λ INT GENE EXPRESSION BY RNA PROCESSING

Donald Court,[1] Ursula Schmeissner,[2] Susan Bear,[1] Martin Rosenberg,[3] Amos B. Oppenheim,[4] Cecilia Montanez,[5] and Gabriel Guarneros,[5]

[1]Laboratory of Molecular Oncology, National Cancer Institute, National Institutes of Health, Bldg. 41, Room D251, Bethesda, MD 20205. [2]Biogen, S.A., 46 Route des Acacias, 1227 Carouge/Geneva, Switzerland. [3]Department of Molecular Genetics, Smith, Kline, & French Laboratories, Philadelphia, PA 19101. [4]Department of Microbiological Chemistry, The Hebrew University, Jerusalem, Israel. [5]Department of Genetics, Centro de Investigacion y de Estudios Avanzados, A.P. 14-740, Mexico 14, D. F., Mexico.

ABSTRACT The *int* gene of phage λ is transcribed from promoters p_L and p_I with approximately equal efficiency. Nevertheless, nearly all the *int* gene product is produced by translation of the p_I-derived transcripts. The p_L-derived transcripts are a poor source of Int protein even though they are translated efficiently in regions located upstream and downstream from *int*. The critical difference between the two transcripts is their termini: p_I-message terminates 275 nucleotides beyond *int*, while p_L message, whose synthesis depends upon the λ antitermination function, terminates much further downstream. The p_L- but not p_I-derived transcripts contain a site beyond *int* that is susceptible to processing by an endonuclease. Cleavage leaves a 3'OH end that is sensitive to a second ribonuclease that digests the RNA in the direction of *int*. The p_I-transcripts are resistant to the initial endonuclease cleavage and subsequent digestion. Thus a differential mRNA degradation rate explains the different levels of Int protein translated from the two transcripts.

Our evidence for this conclusion is the following: (1) The endoribonuclease has been identified; (2) We have isolated and characterized ribonuclease-resistant mutants of the phage; and (3) We have analysed the cleaved and uncleaved forms of the p_L- and p_I-derived transcripts both in vivo and in vitro.

INTRODUCTION

The bacteriophage λ int gene product, Int, is a type I topoisomerase. Int has the unique property of binding to a specific site, attP, on λ DNA and promoting recombination between this site and a smaller site, attB, in the bacterial chromosome. This reaction completes the integration event during λ lysogeny and establishes λ as a prophage. To excise or recombine out from the chromosome during induction, Int is again used but in combination with another phage function Xis, the product of the xis gene located adjacent to int on the λ genome. Host factors are required for both integration and excision but the direction of the reaction is primarily determined by the presence of Int and/or Xis in the cell (1,2,3).

During infection by λ, int transcription occurs from two distinct promoters, p_L and p_I. The p_L promoter is active immediately after infection and is one of two major λ promoters required for lytic growth. Gene N product controls transcription from p_L by interacting with the transcription complex and preventing transcription termination. This promoter is repressed partially about 8 min. post-infection by the cro gene product (4,5). It is at about this time that the p_I promoter is activated. This activation requires the product of the λ cII gene which facilitates binding of RNA polymerase to p_I (6,7). The cII protein also activates transcription of the p_E promoter from which the transcript for the λ cI repressor is initiated. Control of these promoters by cII ensures that repressor and Int protein, which are the two key proteins required for entrance of λ into the lysogenic mode of growth, are made coordinately (8,9,10).

In addition to the controls exerted at the level of transcription on the int gene, post-transcriptional effects on int expression from both the p_L and p_I transcripts occur. In both cases these effects are exerted from a region that is some 260 bp distal to the int structural gene; this form of control has been termed retroregulation. Interestingly, expression of int from p_L-transcripts is inhibited by retroregulation, while from p_I-transcripts, it is stimulated. This paper reviews what has been published on retroregulation of int from p_L as well as providing new information on retroregulation from the p_I transcript. Other reviews on this subject exist (11,12,13).

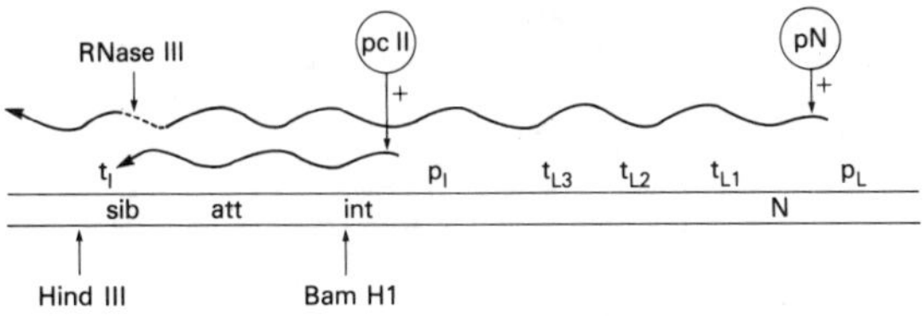

FIGURE 1. Int gene regulation from two transcripts. Transcription (wavy lines) from p_I and p_L is indicated. The dotted line represents processing of p_L-transcripts by RNaseIII at sib. The location of the HindIII site is 250 bp beyond att, and the BamH1 site is 240 bp before att within the int gene.

RESULTS AND DISCUSSION

Transcription and Expression of the int Gene.

Initiation and termination of the p_I transcript. The p_I transcript initiates within the xis gene (6,14) and extends through int, across the attP site, and terminates in the b region (15,16,17). Termination occurs at a rho independent terminator site, t_I (Fig. 1); the structure of this site is shown in Fig. 2. The points of termination, defined both in vivo and in vitro, are located beyond the fifth and sixth uridylate residues in the run of six uridylates (U. Schmeissner, et al., in prep.). Only one gene, int, is contained on this transcript.

Initiation and antitermination of the p_L transcript. The p_L-transcript initiates transcription some 8.2 kb upstream of the int gene. λ gene N product acts at the nutL site near p_L and causes an alteration in the RNA polymerase transcription complex which prevents it from terminating at several terminator sites between p_L and int (Fig. 1)(18,19). This transcript extends through several genes including xis and int. Because of the antitermination property, the p_L-transcript does not stop at t_I but continues on through this site and into the b region (15,U. Schmeissner in prep.). In fact, several λ genes beyond int are expressed from transcripts of p_L (20).

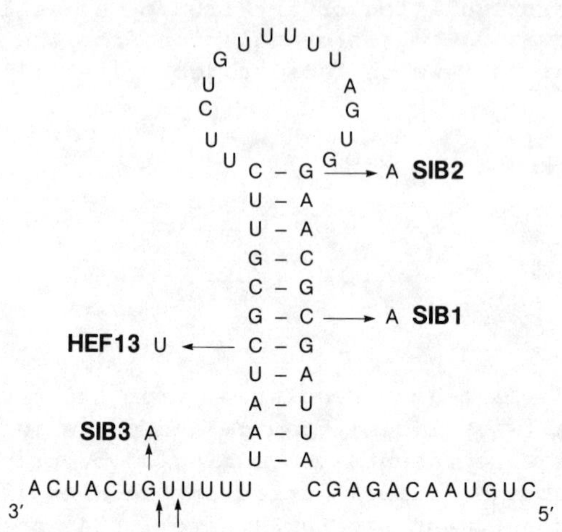

FIGURE 2. A proposed secondary structure for the t_I terminator is shown. Included within the stem are the sib1, sib2, and hef13 mutations; the sib3 mutation is beyond the terminated transcript. Arrows represent the two terminated ends for the p_I-transcript. Note transcripts from p_L pass through this site (U. Schmeissner et al., in prep.).

Thus the p_L-transcript differs from the p_I-transcript in that it has additional RNA coding regions both upstream and downstream of the int gene.

Differential expression of int from p_I and p_L transcripts. After infection, int is expressed primarily from the relatively short p_I-transcript. This was first noticed when it was found that cII mutants were defective in Int synthesis and integrative recombination (21,22,23). Guarneros et al. (24,25) have shown since that there is a several hundred-fold difference in Int synthesis from p_I vs p_L transcripts. Different levels of biological activity are also reflected by measurements of Int antigen using an enzyme linked immunospecific assay (ELISA). Since differences are not caused by the rates of int mRNA synthesis (Table 1) (25,U. Schmeissner et al., in prep.), we are led to the conclusion that post transcriptional controls govern the levels of Int made from these disparate transcripts.

TABLE 1
EXPRESSION OF INT FROM P_L AND P_I

Promoter[a]	Recombination[b]	Antigen[c]	RNA[d]
p_L&p_I	High	High	+
p_I	High	High	+
p_L	Low	Low	+

[a] One or both promoters are active on an infecting phage (see ref. 25 for details). Transcription from p_I alone is achieved by infecting with a nutL mutant; transcription from p_L alone is achieved by infecting with a cII mutant. [b] Integrative recombination is measured (25). The ratio of High to Low levels is about 400. [c] Int antigen was measured by an ELISA using horse radish peroxidase linked to goat anti-rabbit antibody in a competition reaction between purified Int and infected cell extracts for rabbit anti-Int antibody (S. Bear, unpublished). [d] ^{32}P-int (or ^{3}H-int) mRNA made in a short pulse (6,25, U. Schmeissner et al., in prep.).

Retroregulation - Negative control from a site beyond the int gene.

Negative controls on int expression from p_L. Why is the level of Int made from p_L so low as compared to that from p_I? To focus on this problem, Guarneros et al.(25) isolated mutants in λcII$^-$ phage that allowed increased expression from the p_L-transcript. Three mutants called sib were isolated. All mapped genetically to the left of attP indicating that the inhibitor mapped beyond the int gene. To determine if the inhibitor, sib$^+$, was a site or a type of repressor, cis-trans experiments were carried out between sib$^+$ and sib$^-$ co-infecting phage. Surprisingly sib$^+$ exerted its effect on Int expression when it was located on the same chromosome as the int$^+$ phage in the test, i.e., in cis, but the inhibitor had no effect in trans on Int expression from a second sib$^-$ phage (Table 2). An analysis of the changes in sequence caused by the mutations revealed they were located in a small region of 37 bp some 260 nucleotides beyond int which does not encode a protein. Thus from both genetic and sequence data, the inhibitor defined by these mutations encodes a site, like an operator, but located 260 nucleotides distal to its controlled gene (24,25).

TABLE 2

SIB EFFECT ON P_L DEPENDENT INTEGRASE LEVELS

Infecting Phage(s)[a]			Recombination[b]
1.	λ sib$^+$ int$^+$		Low
2.	λ sib$^-$ int$^+$		High
3.	λ sib$^+$/- int$^-$		None
4.	λ sib$^+$ int$^+$ λ sib$^-$ int$^-$	CIS	Low
5.	λ sib$^+$ int$^-$ λ sib$^-$ int$^+$	TRANS	High

[a] All phage carry the cII$^-$ mutation to avoid transcription from p_I. [b] Integrative recombination is measured and the ratio of High to Low values is 400 (25). The result shows no trans inhibition (infection 5). Conclusion: sib$^+$ is an inhibitory site.

RNaseIII processing is correlated with reduced Int levels. How does a site distal to a gene control its expression? First we know from our previous discussion that control is exerted after transcription of the gene. Several general models and predictions for how sib could negatively retroregulate Int synthesis have been discussed in earlier papers (24,25,26,27). Again mutants have been helpful in elucidating what the likely mechanism of control is. In addition to phage mutants, a host mutant, rnc$^-$, has been shown to increase Int expression from p_L (28). This mutant is defective in the endoribonuclease, RNaseIII. Reexamination of the sequence in the vicinity of the sib mutants reveals that the RNA can adopt a secondary structure containing two self-complementary regions (Fig. 3). This structure is similar to known RNA substrates for RNaseIII (29). The sib and hef13 mutations cause mismatches in the stems and decrease the overall stability of the structure. Interestingly the hef13 mutation generates a less disruptive mismatch in the stem than the sib mutations and has an intermediate phenotype between sib$^+$ and sib$^-$ (25). Deletions into the sib region have been generated with Bal31 nuclease

(Fig. 3). Those that enter the proposed secondary structure are sib⁻, while deletion 119, which removes DNA to within one base of this structure, remains sib+ (30). Thus we believe that sib defines an RNaseIII site, and it is the secondary structure of the RNA that is important for processing.

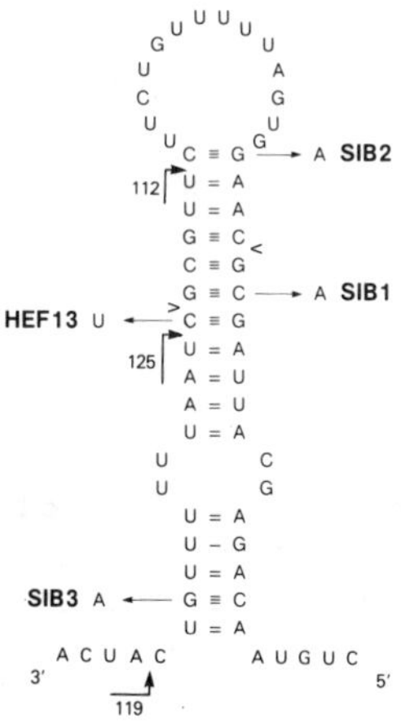

FIGURE 3. The proposed secondary structure of the RNaseIII site at sib is shown. Hyphens between base pairs indicate strength of hydrogen bond formation. The sib mutations (25) and hef13 (31) are shown with their base changes. Bent arrows represent end-points of deletions where DNA to the 3' side of the arrow in each case is deleted (30). Carots indicate the points on the RNA where RNaseIII cuts. Two cut sites were found two base pairs apart in the upper stem (U. Schmeissner et al., in prep.).

U. Schmeissner (in prep.) has shown that processing of the p_L-transcript occurs in the secondary structure of sib both in vivo and in vitro. In vivo processing requires a sib+ infection into a host with normal RNaseIII levels. A correlation exists between the processing event and Int levels in the cell. In phage sib or host rnc mutants, there is no processing and a high level of Int is made. In wild type conditions, processing takes place and very little Int is made. When transcripts of sib+ RNA are treated in vitro with RNaseIII, two processing sites are detected, one on each side of the RNA stem within sib (Fig. 3).

The processed p_L-transcript may be degraded by a 3' to 5' exoribonuclease. Next we must explain how processing at a site 260 nucleotides distal to int can inactivate its expression. A comparison of int mRNA stability between p_L-transcripts from wild type and a sib mutant infection shows that the int mRNA from wild type is more labile (25). Thus we infer that RNaseIII processing at sib destabilizes RNA 5' to the site, possibly by creating a new 3' end that is susceptible to a 3' to 5' exoribonucleolytic activity. As yet no phage or host mutant has been identified for this activity on int, however candidates for such an activity include RNaseII and polynucleotide phosphorylase (PNPase) (29).

Other genes in the p_L operon are not subject to control from sib. The genes located in the b region distal to sib are expressed (32), but a 3' to 5' nuclease would not attack RNA distal to the RNaseIII site, and we would predict that this distal RNA may remain stable even after processing. Unexpectedly retroregulation has little effect upon Xis synthesis (26); the xis gene is located 5' to the sib site (Fig. 5). Two sets of experiments give us insight into this problem. The first set emphasizes moving other genes to a position analogous to int with respect to sib. When this is done, the two genes tested, galK of E. coli (13) and ea22 of λ, (26) come under the control of sib. In both of these cases, the int gene was deleted completely, and thus there is nothing specific or essential about the int sequence in sib regulation.

In a reciprocal set of experiments, one to two kilobases of bacterial DNA was inserted between sib and int; here retroregulation of int was eliminated (Guarneros et al., unpublished). From these different experiments, we suggest that the distance from the processed site to a given gene is critical for retroregulation. Does this mean simply that expression from a distant gene occurs before the 3' to 5' degradation or that the proposed nuclease has encountered a barrier or has fallen from the RNA. Experiments to determine relative half-lives of RNA at different distances from sib should help to answer these questions.

Other reports suggest that the stability of different regions of the p_L-transcripts can be affected by processing events (33) and by phage functions (34). The relevance of these other effects on int mRNA synthesis or stability have not been tested.

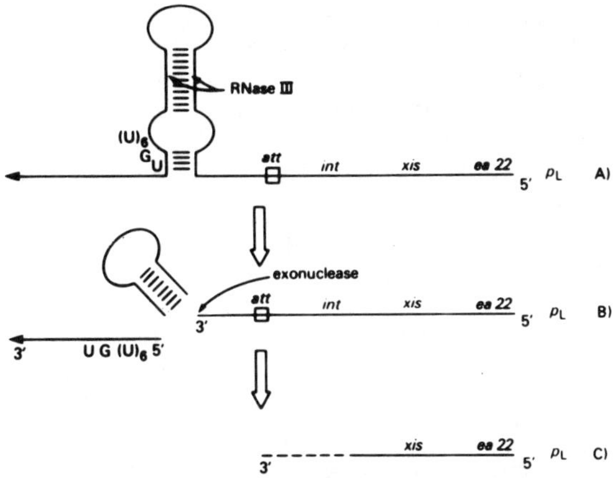

FIGURE 4. Processing and decay of the p_L-transcript negatively retroregulates int synthesis. A. Endonuclease processing by RNaseIII to remove secondary structure. B. Proposed exonuclease attack from 3'OH end of processed transcript. C. Inactivation of the int mRNA region. Note xis and ea22 are not normally affected by retroregulation but when the int xis region is deleted and ea22 is moved adjacent to att, ea22 is retroregulated (26).

Retroregulation - Positive Control from a Site beyond the int Gene.

Transcription termination of the p_I transcript at t_I prevents RNaseIII processing. There is an inverse correlation between endoribonuclease processing of the p_L transcript and Int synthesis. As stated previously high levels of Int are made from the p_I-transcript, therefore we suspect it may not be processed. In fact in vivo analysis of this transcript shows that unlike the p_L-transcript, the terminated p_I-transcript remains unprocessed in normal (sib^+) infections. This is supported also by in vitro experiments which show that the terminated transcript is much more resistant to cutting by RNaseIII than the antiterminated p_L-transcript (U. Schmeissner et al., in prep.). Comparison of the terminated and read-through transcripts (Fig. 2 vs Fig. 3) reveals their difference: the lower stem structure. We note that the p_I transcript lacks the distal sequence to form the lower stem and therefore may not offer a recognizable RNaseIII site. This structure is important for RNaseIII processsing of the p_L read-through transcript: 1. The lower structure is conserved in many RNaseIII sites (29). 2. The sib3 mutation is in this segment (25). This mutation changes a guanadylate residue to an adenylate residue and thereby eliminates effective pairing within the lower stem. RNaseIII processing of the sib region does not occur after infection with this mutant (U. Schmeissner and C. Montanez, unpublished).

Synthesis of Int is reduced by int mRNA degradation after RNaseIII processing of the p_L-transcripts. The fact that unprocessed transcripts of p_L express Int may be due to the long distance between the int gene and the normal 3' end of the p_L-transcript. Expression of int from the p_I-transcript cannot be explained by a distance effect, its 3' end is only 33 nucleotides more distant from the int gene than the processed end of the p_L-transcript. It appears that the exonuclease(s), proposed to degrade the processed RNA, does not affect the terminated RNA. We can speculate that the stable base paired stem of the p_I-transcript or the uridylate residues at its end protect it. Does this mean that any terminated transcript will be resistant to nuclease attack? We do not believe that this is going to be the case and present evidence below which supports our contention.

Certain sib mutants enhance expression of Int from p_L, but reduce expression of Int from p_I (25). An examination of more mutants reveals some new information (G. Guarneros, unpublished). The mutations sib1 and sib2 reduce Int levels from p_I whereas sib3 does not (Table 3). What is the critical difference in the effects of these mutations? The sib1 and sib2 mutations alter the t_I terminator whereas sib3 alters DNA sequence beyond the termination site. When the level of transcription termination was examined in these three mutants, it was found that as expected, sib3 terminates transcription as well as wild type. Unexpectedly sib1 and sib2 terminate transcription well too (G. Guarneros, unpublished). In each, termination occurs as much as 90% of the time when measured in vivo. Thus the sib1 and sib2 mutants terminate, but the terminated transcript fails to express int as well as sib^+ or sib3. The major difference between these two sets is that the mutations in sib1 and sib2 change the RNA sequence and reduce stem stability. Thus the end of the p_I-transcript in the sib1 and sib2 mutants may not be as resistant to the 3' to 5' exonuclease attack as normal p_I-transcript

(G. Guarneros., unpublished). We suspect that terminator structures may play roles in determining the stability of transcripts but that this may depend upon the relative strength of their base paired stems.

TABLE 3
EXPRESSION OF INT IN THE SIB MUTANTS[a]

Sib Allele	Recombination[b] p_L	p_I
+	Low	High
1	High	Low
2	High	Low
3	High	High

[a] Unpublished data of G. Guarneros. [b] Recombination was measured for p_L and p_I independently using mutants to turn-off one or the other promoter (25). High levels from p_L and p_I are comparable, however the low levels found from p_I for sib2 and sib3 are not as low as the sib^+ level from p_L.

Thus from this study, we can suggest three ways in which a gene may be controlled by retroregulation. Transcription terminators may enhance the stabilty of specific upstream mRNAs by blocking the movement of 3' to 5' nucleases; their efficacy may depend upon the ΔG of base paired stem. Processing by endonucleases (25,35,36) can sensitize messenger RNA to exonuclease attack. Finally extension of transcipts to move the 3' end far beyond a gene may allow increased time for expression before 3' to 5' degradation.

Integrative Recombination Regulates int Gene Expression.

Excision of a prophage requires viral Int and Xis products and occurs soon after induction (37). Thus there is an early requirement for Int synthesis to excise the λ genome and allow lytic development to ensue. Whereas synthesis of Int from the p_L promoter is not possible after infection; it is after induction (15). The reason for this is a simple topological consequence of the integration event that formed the prophage. The att site where integration occurs is located between sib and int on the phage DNA (Fig. 1). After integration, sib is recombined to a position far from int in the prophage and therefore cannot exert negative control on the p_L-transcript (Fig. 5). In summary, the int gene is transcribed early both after infection and induction. After infection, sib prevents Int expression from p_L, however, in the prophage after induction, Int is made directly from p_L.

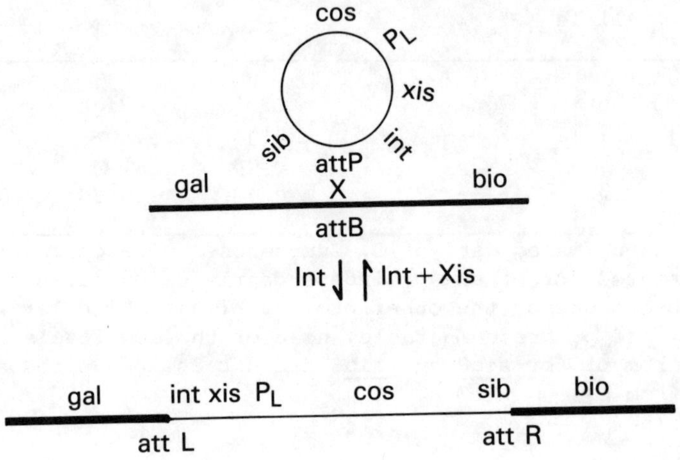

FIGURE 5. The sib site is split from int in the prophage. The circular phage genome recombines via attP into the bacterial attB site between the gal and bio operons on the coli chromosome. This generates the prophage (thin line) with hybrid att sites (L and R) and sib on the opposite side of the prophage from int and p_L.

Temporal Control of Int Synthesis.

Immediately after infection by λ, the lytic promoters including p_L are active but as we have shown int is not expressed. Apparently this negative control on int is not essential for lytic development. All sib mutants express Int early after infection but produce phage bursts equal to wild type λ. After induction of the prophage int is also expressed early to promote excision. Thus synthesis of int early is not harmful to lytic development. Some studies suggest paradoxically that premature synthesis of Int may be harmful for lysogenic development (D. Court, unpublished). Lysogeny by λ requires two events: integration of phage DNA into the coli chromosome and repression of lytic promoters by the λ repressor. The 'decision' to enter the lysogenic mode occurs only after several lytic processes have begun. Early Int synthesis from p_L would allow premature integration of λ DNA before lytic development is shut down. If integration occurs before repression, lytic expression would persist. This would include replication from the phage origin while in the integrated state; such a situation is known to be lethal to the bacterium (38). Thus potential lysogens would be killed before they could be established. λ avoids this problem by preventing Int synthesis from p_L after infection, and by providing int with a second promoter p_I which is activated by cII product. The cII protein also activates repressor synthesis from its p_E promoter, thereby ensuring coordinate synthesis of the two proteins, Int and cI-repressor, required for lysogeny.

ACKNOWLEDGMENTS

Thanks are due to the Consejo Nacional de Ciencia y Technologia for its partial support of C.M. and G.G. Also A.B.O. would like to thank the US-Israel Binational Science Foundation for partial support. R. Weisberg made several helpful comments on the manuscript during its preparation.

REFERENCES

1. Gottesman ME, Weisberg RA (1971). Prophage insertion and excision. in Hershey AD (ed): "The Bacteriophage Lambda," New York: Cold Spring Harbor Press, p 113.
2. Nash H (1981). Integration and excision of bacteriophage λ: The mechanism of conservative site-specific recombination. Ann Rev Genet 15:143.
3. Weisberg RA, Landy A (1983). Site-specific recombination in phage λ. in Hendrix R, Weisberg RA, Stahl F, Roberts J (eds) "Lambda II," New York: Cold Spring Harbor Press, in press.
4. Signer ER (1970). On the control of lysogeny in phage λ. Virology 40:624.

5. Adhya S, Gottesman M, deCrombrugghe B, Court D (1976). Transcription termination regulates gene expression. in Losick R, Chamberlin M (eds) "RNA Polymerase," New York: Cold Spring Harbor Press, p 719.
6. Schmeissner U, Court D, McKenney K, Rosenberg M (1981). Positively activated transcription of λ integrase gene initiates with UTP in vivo. Nature 292:173.
7. Shimatake H, Rosenberg M (1981). Purified λ regulatory protein cII positively activates promoters for lysogenic development. Nature 292:128.
8. Reichardt L, Kaiser A (1971). Control of λ repressor synthesis. Proc Nat Acad Sci USA 68:2185.
9. Echols H, Green L (1971). Establishment and maintenance of repression by bacteriophage lambda: the role of the cI, cII and cIII proteins. Proc Nat Acad Sci USA 68: 2190.
10. Schmeissner U, Court D, Shimatake H, Rosenberg M (1980). Promoter for the establishment of repressor synthesis in bacteriophage λ. Proc Nat Acad Sci USA 77:3191.
11. Gottesman M, Oppenheim A, Court D (1982). Retroregulation: Control of gene expression from sites distal to the gene. Cell 29:727.
12. Court D, Schmeissner U, Rosenberg M, Oppenheim A, Guarneros G, Montanez C (1983). Processing of λ int RNA: A mechanism for gene control. in Schlessinger D (ed) "Microbiology-1983" Washington,DC: American Society of Microbiology, p 78.
13. Rosenberg M, Schmeissner U (1983). Regulation of gene expression by transcription termination and RNA processing. in Safer B, Grunberg-Manago M (eds) "Interaction of Transcriptional and Transclational Controls in the Regulation of Gene Expression" New York: Elsevier/North Holland p 1.
14. Abraham J, Mascarenhas D, Fischer R, Benedik M, Campbell A, Echols H (1980). DNA sequence of regulatory region for integration gene of bacteriophage λ. Proc Nat Acad Sci USA 77:2477.
15. Oppenheim AB, Gottesman S, Gottesman M (1982). Regulation of bacteriophage λ int gene expression. J Mol Biol 158:327.
16. Luk K-C, Dobrzanski P, Szybalski W (1982). Cloning and characterization of the termination site t_I for the gene int transcript in coliphage lambda. Gene 17:259.

17. Mascarenhas D, Trueheart J, Benedik M, Campbell A (1983). Retroregulation: Control of integrase expression by the b2 region of bacteriophages λ and 434. Virology 124:100.
18. Salstrom J, Szybalski W (1978). Coliphage λ nutL: A unique class of mutants defective in the site of gene N product utilization for antitermination of leftward transcription. J Mol Biol 124:195.
19. Lozeron HA, Anevski PJ, Apirion D (1977). Antitermination and absence of processing of the leftward transcript of coliphage lambda in RNaseIII deficient host. J Mol Biol 109:359.
20. Hendrix R (1971). Identification of proteins coded in phage lambda. in Hershey AD (ed) "The Bacteriophage Lambda" New York: Cold Spring Harbor Press p 335.
21. Court D, Adhya S, Nash H, Enquist L (1977). The phage λ integration protein (Int) is subject to control by the cII and cIII gene products. in Bukhari AI, Shapiro JA, Adhya SL (eds) "DNA Insertion Elements, Plasmids and Episomes," New York: Cold Spring Harbor Press p 389.
22. Chung S, Echols H (1977). Positive regulation of integrative recombination by the cII and cIII genes of bacteriophage λ. Virology 79:312.
23. Katzir N, Oppenheim A, Belfort M, Oppenheim AB (1976). Activation of the lambda int gene by the cII and cIII gene products. Virology 74:324.
24. Guarneros G, Galindo JM (1979). The regulation of integrative recombination by the b2 region and the cII gene of bacteriophage λ. Virology 95:119.
25. Guarneros G, Montanez C, Hernandez T, Court D (1982). Posttranscriptional control of bacteriophage λ int gene expression from a site distal to the gene. Proc Nat Acad Sci USA 79:238.
26. Schindler D, Echols H (1981). Retroregulation of the int gene of bacteriophage λ: Control of translation completion. Proc Nat Acad Sci USA 78:4475.
27. Epp C, Pearson M, Enquist L (1981). Downstream regulation of int gene expression by the b2 region in phage lambda. Gene 13:327.
28. Belfort M (1980). The cII-independent expression of the phage λ int gene in RNaseIII defective E. coli. Gene 11:149.
29. Gegenheimer P, Apirion D (1981). Processing of prokaryotic ribonucleic acid. Microbiological Review 45:502.

30. Court D, Huang TF, Oppenheim AB (1983). Deletion analysis of the retroregulatory site for the λ int gene. J Mol Biol 165:in press.
31. Roehrdanz RL, Dove WF (1977). A factor in the b2 region affecting site-specific recombinations in lambda. Virology 79:40.
32. Oppenheim A, Oppenheim AB (1978). Regulation of the int gene and bacteriophage λ: Activation by the cII and cIII gene products and the role of p_I and p_L promoters. Mol Gen Genet 165:39.
33. Wilder DA, Lozeron HA (1979). Differential modes of RNA processing and decay for the major N-dependent RNA transcript of coliphage lambda. Virology 99:241.
34. Court D, de Crombrugghe B, Adhya S, Gottesman M (1980). Bacteriophage lambda hin function: II. Enahanced stability of lambda messenger RNA. J Mol Biol 138:731.
35. Schneider E, Blundell M, Kennell D (1978). Translation and mRNA decay. Mol Gen Genet 160:121.
36. Schlessinger D, Graham MY, Shen V, Tal M (1983). Coupling of lac mRNA formation and decay to translation. in Schlessinger D (ed) "Microbiology-1983" Washington, DC: American Society of Microbiology, in press.
37. Gottesman S, Gottesman ME (1975). Elements involved in site-specific recombination in bacteriophage lambda. J Mol Biol 91:489.
38. Eisen H, Pereira da Silva L, Jacob F (1968). The regulation and mechanism of DNA synthesis in bacteriophage λ. Cold Spring Harbor Symp Quant Biol 33:755.

Gene Expression, pages 327-342
© 1983 Alan R. Liss, Inc., 150 Fifth Avenue, New York, NY 10011

SELF-SPLICING OF THE RIBOSOMAL RNA PRECURSOR OF *TETRAHYMENA*[1]

Paula J. Grabowski, Susan L. Brehm,[2] Arthur J. Zaug, Kelly Kruger,[3] and Thomas R. Cech

Department of Chemistry, University of Colorado
Boulder, Colorado 80309

ABSTRACT The *Tetrahymena thermophila* ribosomal RNA gene (rDNA) contains an intervening sequence (IVS), which is transcribed as a part of the precursor RNA and later removed by splicing. The excised IVS RNA is subsequently converted to a covalent circular form. We review the evidence that led to the conclusion that these reactions occur *in vitro* in the absence of protein. *Tetrahymena* nuclei contain a low steady-state concentration of the unspliced rRNA precursor. We show that this pre-rRNA, synthesized *in vivo*, undergoes autocatalytic excision of its IVS in a guanosine-dependent reaction *in vitro*.

INTRODUCTION

The 26S rRNA-coding region of the rDNA of *Tetrahymena* is interrupted by a 0.4 kb intervening sequence (IVS, or intron). The IVS is transcribed as an integral part of the primary transcript and subsequently removed in the nucleus by splicing (1-3). These two general features are shared with the splicing of mRNA and tRNA precursors (4,5).

[1]This work was supported by grants from the National Institutes of Health (No. GM28039) and the American Cancer Society (No. NP374). T.R.C. is the recipient of a Research Career Development Award from the National Cancer Institute (No. CA00700).

[2]Present address: Department of Microbiology and Immunology, University of California, Berkeley CA 94720.

[3]Present address: Department of Biochemistry, College of Physicians and Surgeons, Columbia University, New York, New York 10027.

Several species of *Tetrahymena*, including *T. thermophila* and *T. pigmentosa* strain 6UM, are known to have rDNA with an interrupted coding region (6). The presence of IVSs in the rDNA of other organisms is not common but is certainly widespread. Nuclear rDNA in *Physarum polycephalum* (7,8), mitochondrial rDNA in yeast (9) and other fungi, and chloroplast rDNA in *Chlamydomonas* (10) are interrupted by intervening sequences. These IVSs all interrupt the rDNA in the 3' half of the large rRNA coding region, but the exact site of the interruption varies.

Both the nuclear and mitochondrial rRNA introns vary widely in size and in nucleotide sequence. They do not contain the consensus sequence found in nuclear mRNA introns at the splice junctions (4,11,12). Instead, the rRNA introns have their own consensus sequences. All IVS-containing pre-rRNAs have a U in the exon immediately preceding the IVS and a G at the 3' end of the IVS (13). More striking are four consensus sequences of length 9-16 nt that are scattered within the body of the intron (14-18). These internal sequences also occur in messenger RNA introns of fungal mitochondrial genes. Two of them, called *box*9 and *box*2 in intron 4 of the yeast *cob* gene, have been extensively studied because they are the sites of cis-dominant splicing-defective mutations (19). On the basis of these genetic studies and the tremendous evolutionary conservation of these sequences, it seems likely that some aspects of the splicing mechanism are conserved between nuclear rRNA and fungal mitochondrial rRNA and mRNA systems.

In this paper, we will first review the evidence that precise splicing of the *Tetrahymena* pre-rRNA can occur in the absence of protein. The observation that *E. coli* plasmid DNA transcripts undergo splicing in a well defined system *in vitro* was the single most compelling evidence. Additional evidence included the resistance of the activity to proteases, ionic detergents and boiling. Some of these experiments are presented here for the first time. It was based on the weight of these several lines of inquiry that we became comfortable with the conclusion that the RNA was self-splicing. In the final section of this paper, we present some new data which, at least to a small extent, bridge the gap between the totally defined *in vitro* splicing system and splicing as it occurs in the living organism.

RESULTS AND DISCUSSION

Structure of the Pre-rRNA That Undergoes Self-Splicing.

We have used an *in vitro* transcription system, described previously (20), to synthesize IVS-containing pre-rRNA from *T. thermophila*. Isolated *Tetrahymena* nuclei were incubated with α-^{32}P-nucleoside triphosphates at low monovalent cation concentration. These conditions allow maximal *in vitro* transcription, but inhibit splicing of newly made pre-rRNA molecules and thereby lead to the accumulation of IVS-containing pre-rRNA. Incubation of this purified RNA *in vitro* with a monovalent cation, a divalent cation and a guanosine compound (Table 1) caused accurate excision of the IVS from the precursor (20). When the unspliced pre-rRNA was subjected to boiling in the presence of SDS and 2-mercaptoethanol, the IVS was not released (P.J.G., unpublished data). This result suggested that the IVS was tightly associated with the pre-rRNA.

TABLE 1
COFACTOR REQUIREMENTS FOR SPLICING[a]

monovalent cation	$(NH_4)_2SO_4$	$\geq$ 75mM
divalent cation	$MgCl_2$	5-10 mM
guanosine cofactor	GTP	$\geq$ 1μM

[a]Other monovalent and divalent cations and guanosine compounds can be substituted for those indicated; similar concentrations are required.

To determine whether the unspliced pre-rRNA was in fact colinear through both splice junctions, we looked for RNA molecules that were protected from nuclease digestion by hybridization to a DNA fragment containing the IVS and portions of adjacent coding sequences (Figure 1). The method used was similar to the S_1 nuclease mapping technique described by Berk and Sharp (21), except that the RNA was labeled, whereas the DNA to which it was hybridized was unlabeled. In this way, we could probe the structure of the unspliced *in vitro* transcription product in the presence of excess unlabeled RNA that had been spliced *in vivo*.

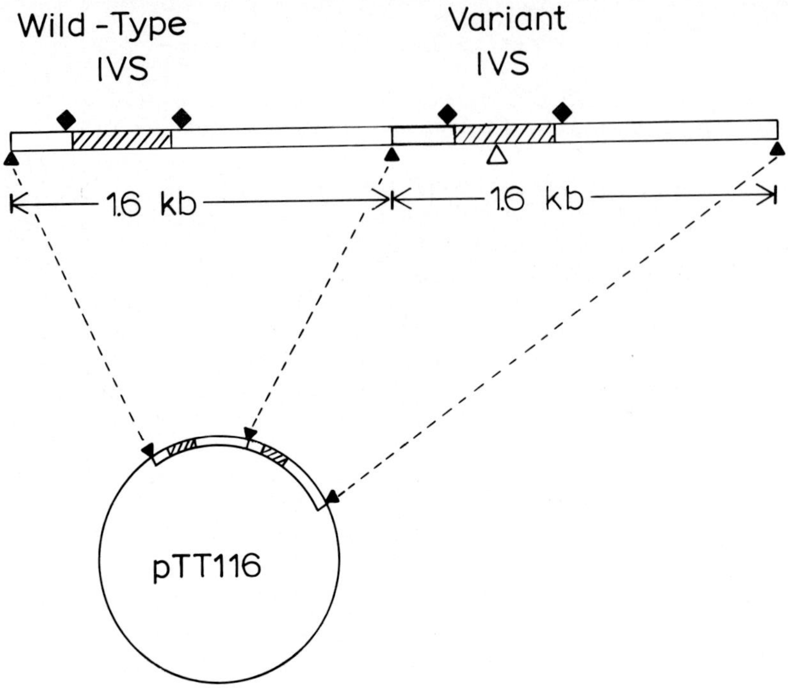

FIGURE 1. Diagram of the two 1.6 kb *Hind* III fragments of *Tetrahymena* rDNA cloned in tandem in the plasmid pTT116 (22). The wild-type 1.6 kb insert is the same as that found in *T. thermophila*, whereas the variant is identical except for a heterogenous region (Δ) near the center of the IVS where it is 7 bp longer (22). Restriction endonuclease sites shown are *Hind* III (▲) and *Hha* I (◆). The 0.5 kb *Hha* I fragment of pTT116 (Wild-type: 484 bp; variant: 491 bp) was used for the nuclease mapping experiment of Figure 2. The *Hha* I fragment contains the IVS (Wild-type: 413 bp; variant 420 bp) and portions of adjacent coding sequences (5' exon: 32 bp; 3' exon: 37 bp).

The unspliced pre-rRNA was hybridized to the 0.5 kb *Hha* I fragment of *T. thermophila* rDNA. This fragment was isolated from the plasmid pTT116, which contains two 1.6 kb *Hind* III fragments of *T. thermophila* rDNA cloned in tandem in pBR313 (Figure 1). (The 0.5 kb *Hha* I fragment is contained in the 1.6 kb *Hind* III fragment). One of the 1.6 kb inserts is identical to the wild-type sequence found in *T. thermophila* rDNA. The other insert (a variant) is also identical except for a region near the middle of the IVS, where it is seven base pairs longer (22). The seven extra base pairs was expected to be a local region of mismatch in the wild-type RNA-variant DNA hybrid near the center of the hybrid (196-208 nt from the 5' end of the RNA portion of the hybrid). In addition, the mismatch should be present in half of the hybrids (see Figure 1).

The RNA-DNA hybrids were digested with RNAase T_1 or S_1 nuclease to remove any single stranded regions of RNA. The nuclease resistant hybrids were then analyzed by electrophoresis on a denaturing polyacrylamide gel with single stranded DNA markers. The results are shown in Figure 2. The major RNA species resistant to digestion by RNAase T_1 and S_1 nuclease, respectively, had sizes of 490 $\pm$ 20 and 480 $\pm$ 19 nt. These are the sizes expected for protection of the pre-rRNA (the IVS and adjacent rRNA segments) by the entire DNA fragment (484 nt; Figure 1). This result is consistent with a structure of the pre-rRNA in which both splice junctions are intact and in which no protein is bound to the RNA at or near the splice junctions. Two RNA species (289 $\pm$ 12 and 216 $\pm$ 9 nt) present in the RNAase T_1 experiment were the sizes expected for cleavage at a G residue within the mismatch of the wild-type RNA-variant DNA hybrid ($\simeq$ 281 and $\simeq$203 nt). Similarly, two RNA species (278 $\pm$ 11, and 197 $\pm$ 8 nt) present in the S_1 nuclease experiment were the sizes expected for hybrids in which the mismatch had been cleaved by the nuclease. Additional minor cleavages, perhaps within A+T-rich regions of the hybrids, produced smaller bands including the one at 148$\pm$6.

IVS Excision and Cyclization Activities Are Resistant to Protease and SDS.

The unspliced pre-rRNA retained the activities necessary to accurately excise and cyclize the IVS after purification by SDS-phenol extraction, heat denaturation and sedimentation in a denaturing sucrose gradient (20). These purification conditions normally insure complete removal of protein. This result demonstrated that the IVS excision and cyclization activities are either tightly bound to the RNA or part of the RNA molecule itself.

Unspliced pre-rRNA was incubated in the presence of proteinase K or pronase as described in the legend to

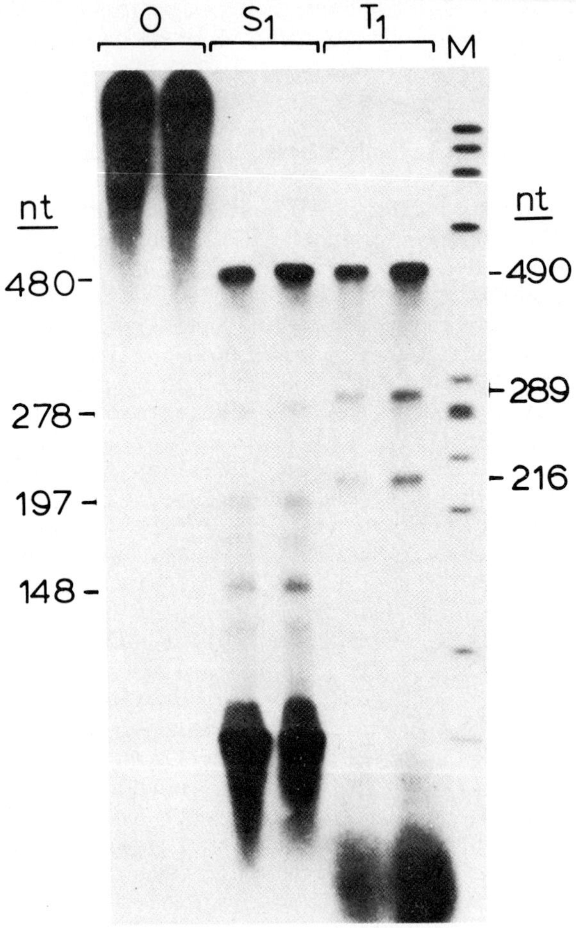

FIGURE 2. The IVS is colinear through both splice junctions in the unspliced pre-rRNA of *Tetrahymena*. Uniformly labeled RNA was synthesized by transcription in isolated *Tetrahymena* nuclei under conditions inhibitory for splicing (20). RNA was hybridized with the 0.5 kb *Hha* I DNA fragment and the resulting hybrids either untreated (0), or digested with RNAase T_1 (T_1) or S_1 nuclease (S_1) under conditions that gave complete degradation of a sample of *Tetrahymena* pre-rRNA. The autoradiograph shows the nuclease-resistant RNAs after electrophoresis on a 4% polyacrylamide, 8 M urea gel run in an oven at 65°C. Denatured ϕX174 *Hae* III DNA fragments (lane M) were used as markers.

Figure 3A. Under these conditions, proteinase K digested 93% of a 1 mg sample of a ^{14}C-labeled mixture, and pronase digested 82% of the protein sample, as determined by trichloroacetic acid solubility. After protease treatment, the solutions were adjusted to splicing conditions (Table 1) and incubated at 30°C. (The protease was not removed prior to the splicing reaction, since the splicing activity was found to be destroyed by phenol/chloroform extraction at this point.) The products of the reaction were analyzed by electrophoresis on a denaturing gel. The results shown in Figure 3A showed that IVS excision and cyclization occurred whether or not the pre-rRNA had been treated with protease. Quantitative densitometry of the autoradiogram revealed that the amount of IVS excised was 89.9% or 99.5% of that which occurred in the control for the proteinase K or pronase sample, respectively. In addition, the above protease treatments were performed after the pre-rRNA was boiled for 5 min in protease buffer. The reactions were then adjusted to splicing conditions and treated as described in the legend to Figure 3A. No decrease in the amount of IVS excision was observed by this treatment (P.J.G., unpublished data).

Unspliced pre-rRNA was also incubated under splicing conditions in the presence of varying concentrations of SDS (Figure 3B). IVS excision was not diminished by concentrations of SDS up to 2%. Quantitative densitometry showed that the presence of SDS in the splicing reaction enhanced the amount of IVS excised by 1.5 fold in the sample containing 2% SDS compared to the control sample that did not contain SDS.

Autoexcision and Autocyclization.

The results summarized in Table 2 strongly suggested that a protein was not involved in IVS-excision or cyclization. We therefore hypothesized that these reactions were mediated by the specific secondary and tertiary structure of the RNA molecule. To test this hypothesis, we constructed a recombinant plasmid (pIVS11) which contained the 1.6 kb *Hind* III fragment of *T. thermophila* rDNA inserted downstream from the *lac* UV5 promoter (23). The 1.6 kb fragment contains the IVS and portions of adjacent coding sequences -- 261 bp of 5' exon and 943 bp of 3' exon. Purified plasmid DNA was transcribed *in vitro* in low salt and in the presence of polyamines with purified *E. coli* RNA polymerase. The *in vitro* made pre-rRNA was purified

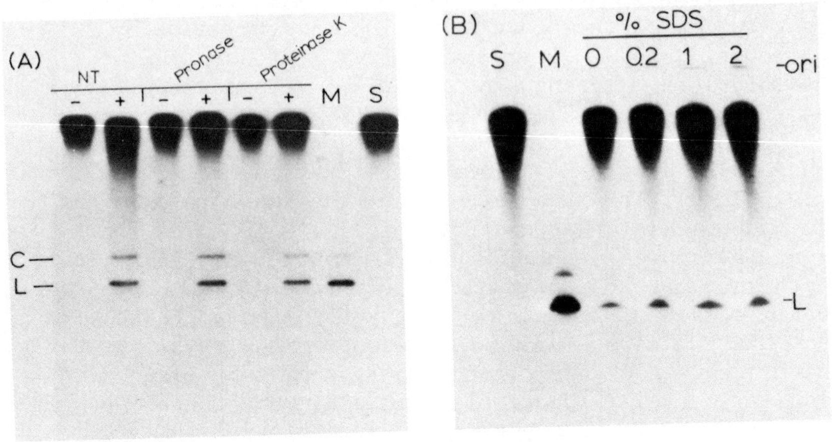

FIGURE 3. IVS excision and cyclization activities are resistant to proteinase K, pronase and SDS. Uniformly labeled pre-rRNA was synthesized by transcription at 30°C in isolated *Tetrahymena* nuclei under conditions inhibitory for splicing (20). (A) RNA purified by SDS-phenol extraction and denaturing sucrose gradient sedimentation was incubated at 37°C for 60 min in the presence of 10 mg/ml pronase (lanes pronase), 0.4 mg/ml proteinase K (lanes proteinase K) or without enzyme (lanes NT) in a buffer containing 50 mM Tris-HCl (pH 7.5), 1 mM EDTA and 0.2% SDS. Immediately after the protease treatments the samples were adjusted to splicing conditions on ice by the addition of a concentrated stock solution of splicing buffer [1X buffer: 10 mM $MgCl_2$, 100 mM $(NH_4)_2SO_4$, 30 mM Tris-HCl (pH 7.5) and 0.1 mM GTP] and incubated at 30°C for 30 min (lanes +) or on ice for 30 min in the presence of EDTA at a concentration of 20 mM (22). Isolated linear and circular IVS RNAs (lane M) and *Tetrahymena* pre-rRNA, untreated following its purification by denaturing sucrose gradient sedimentation (lane S) were used as markers. (B) RNA purified by SDS-phenol extraction and denaturing sucrose gradient sedimentation was incubated in the presence of 0, 0.2, 1 and 2% SDS (lanes 0, 0.2, 1 and 2, respectively) and under splicing conditions at 30° for 30 min in 1X buffer as in (A) except that the Mg^{2+} concentration was 5 mM. Markers (lanes S and M) are as specified in (A).

by SDS-phenol extraction to remove the polymerase. The RNA was then incubated at 30°C under splicing conditions (Table 1) and the products analyzed by denaturing polyacrylamide gel electrophoresis. The results showed a set of discrete high molecular weight transcripts produced during *in vitro* transcription. In those samples that were incubated under splicing conditions, excised linear and circular IVS were produced (23). The plasmid-derived linear and circular IVS RNAs were identified by RNA fingerprinting, by their electrophoretic mobilities on a completely denaturing gel and by the observation that the linear IVS RNA was converted to the circular form by incubation at 39°C in a buffer containing Mg^{+2}(23,24).

TABLE 2
SPLICING/CYCLIZATION ACTIVITY NOT DIMINISHED AFTER:

SDS-phenol extraction
boiling in SDS + 2-mercaptoethanol
electrophoresis in 8M urea
extensive treatment with proteinase K or pronase

The linear IVS RNA was also identified by the addition of a guanosine nucleotide to its 5' end during the IVS excision reaction. RNA was made from pIVS11 in low salt and in the presence of polyamines and tritiated nucleotides. The RNA was deproteinized and residual nucleotides were removed by gel filtration chromatography. The RNA was then incubated under splicing conditions with either α-^{32}P-GTP or γ-^{32}P-GTP, Mg^{+2} and salt. Analysis of the RNA by gel electrophoresis revealed one labeled RNA species, which comigrated with the linear IVS RNA isolated from *Tetrahymena* nuclei (23). Labeling of the linear IVS with γ-^{32}P-GTP is consistent with the finding that the required guanosine cofactor does not provide energy for IVS excision by phosphate bond hydrolysis.

Direct RNA sequencing of the GTP-labeled IVS RNA demonstrated that, except for the additional G at the 5' end, the sequence of the first 39 nt correspond exactly to the known sequence of the IVS DNA (22). Furthermore, the GTP is joined to the RNA by a normal phosphodiester bond, based on the susceptibility of the linkage to cleavage by RNAase T_1 and alkali. Thus, autoexcision of the IVS includes a precise cleavage at the 5' splice junction accompanied by covalent linkage of GTP to the 5' end of the IVS. These results indicate that excision of the IVS from the plasmid transcripts occurs by the same mechanism described for the pre-rRNA isolated from *Tetrahymena* (20).

Exon Ligation.

The fate of the exon sequences could not be directly observed during the *in vitro* splicing reaction due to the complexity of the pattern of plasmid transcription products. Two independent methods were therefore used to investigate whether or not exon ligation occurred during the *in vitro* splicing of pre-rRNA synthesized from pIVS11. First, the ^{32}P-labeled pre-rRNA was hybridized with a DNA fragment that spans the ligation junction and the RNA-DNA hybrids treated with RNAase T_1 or S_1 nuclease as described above. A protected RNA molecule the size of the ligated exons was observed only with the pre-rRNA that had been through the splicing reaction (23).

Secondly, exon ligation was investigated at the nucleotide sequence level by indirect RNA sequencing. A synthetic DNA oligonucleotide primer was hybridized to the 3' exon of the pre-rRNA, either unspliced RNA or RNA that had been spliced *in vitro*. The primer was elongated across the ligation junction using reverse transcriptase in the presence of dideoxynucleotides. The RNA spliced *in vitro* had the same sequence as the mature 26S rRNA of *T. thermophila*, which was sequenced in a parallel experiment (25,26). This result demonstrated at the nucleotide sequence level that the 5' and 3' exons are contiguous at the ligation junction. In addition, the RNAase T_1 oligonucleotide containing the ligation junction was tentatively identified by RNA fingerprinting (25).

Pre-rRNA Made *In Vivo* Can Be Spliced *In Vitro*.

The ability of the plasmid DNA transcripts to undergo autocatalytic splicing *in vitro* provides reasonably strong evidence that there is no obligatory requirement for a protein. This *in vitro* system is so far removed from *Tetrahymena*, however, that there is reason to question whether the reaction occurs in the same manner in the cell.

One way to evaluate the similarity of splicing *in vitro* and *in vivo* is to compare the reaction products. We have used a gel blot hybridization technique (27) to identify the

IVS-containing species of *Tetrahymena* RNA. In addition to the unspliced pre-rRNA, species corresponding in size to the linear and circular IVS excision products were detected in nuclear RNA; none of these was detected in cytoplasmic RNA (28).

A similar experiment is shown in Figure 4AB. In this case, total cellular RNA was separated by sucrose gradient centrifugation, and a portion of each fraction was analyzed by gel blot hybridization. Unspliced pre-rRNA sedimented in the high molecular weight region of the sucrose gradient (fractions 1-3). The linear and circular IVS RNAs were found to cosediment in fractions 8 and 9. Because these two RNAs contain approximately the same number of nucleotides, it is not surprising that they have similar sedimentation rates under non-denaturing conditions. Because of their different topology, however, they have quite different electrophoresis mobilities under denaturing conditions. With the low ionic strength electrophoresis buffer used in the gel shown here, the 399 nt circular IVS had an apparent molecular weight of 750 nt. When the same RNA was subjected to electrophoresis in our standard running buffer, which has a 2.25-fold higher concentration of Tris-borate-EDTA, the circular IVS had an apparent molecular weight of ≃570 nt (data not shown). The sensitivity of the electrophoretic mobility of this RNA to the electrophoresis conditions is a hallmark of its circularity. Under both electrophoresis conditions, both the linear and circular IVS RNAs comigrated with ^{32}P-labeled IVS RNA markers produced by transcription/splicing *in vitro*.

During rRNA splicing *in vitro*, a guanosine nucleotide cofactor acts as an acceptor for the first phosphoester transfer reaction, becoming covalently linked to the 5' end of the excised IVS RNA. In the completely defined *in vitro* system, many guanosine compounds such as GTP, guanosine itself, and trinucleotides ending in guanosine are active as cofactors. All the compounds we have tested that terminate in G-OH are active. We would therefore be interested in knowing what form of the cofactor is used *in vivo*. Judging by the size of the linear IVS RNA in the gel blot hybridization experiments, some possibilities can be eliminated. The acceptor is not some polynucleotide ending in guanosine or some guanylylated polypeptide, but must be a small molecule. Very likely it is guanosine 5'-monophosphate, as is found on the 5' end of IVS RNA produced by transcription/splicing in isolated nuclei (29).

In a further attempt to bridge the gap between

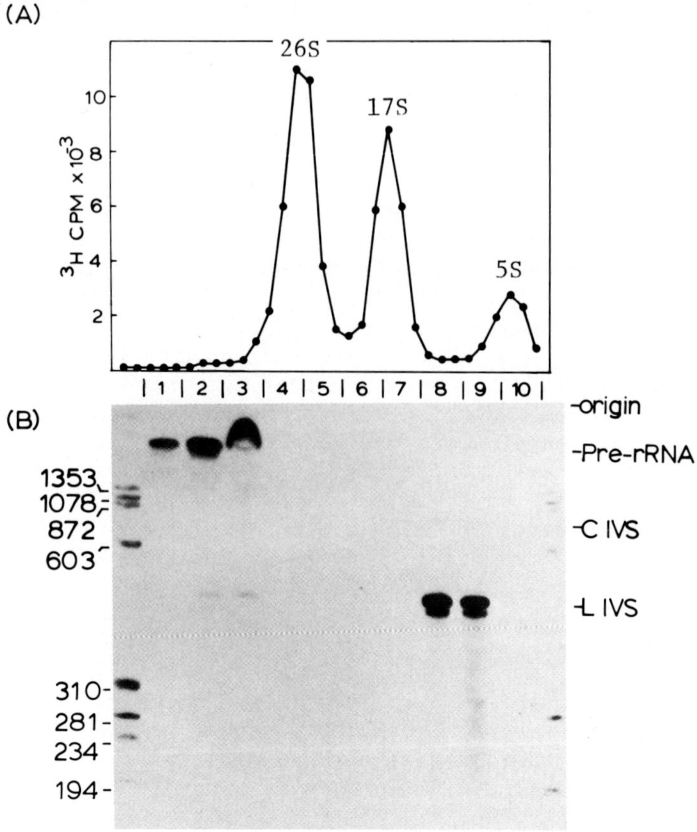

FIGURE 4. Pre-rRNA synthesized *in vivo* undergoes splicing *in vitro*. (A) Rapidly growing *T. thermophila* were labeled for 1 hr with ^{3}H-uridine prior to harvest. Whole cell RNA was isolated by a method designed to prevent autocatalytic RNA splicing or cyclization during the preparation (28). RNA was fractionated on a 15-30% sucrose gradient (10 mM Tris-HCl, 300 mM $NaC_2H_3O_2$, 1 mM EDTA, pH 7.5; gradient centrifuged at 23,000 rpm for 18 hr at 15°C in a Beckman SW27 rotor). (B) RNA from each fraction of the sucrose gradient was separated by electrophoresis on a 4% polyacrylamide, 8 M urea gel run at 30 mA in a 68°C oven. (Large separation between L and C IVS RNAs results from using a running buffer containing 40 mM Tris-HCl, 40 mM boric acid, 1 mM EDTA, pH 8.3.) RNA was then electro-

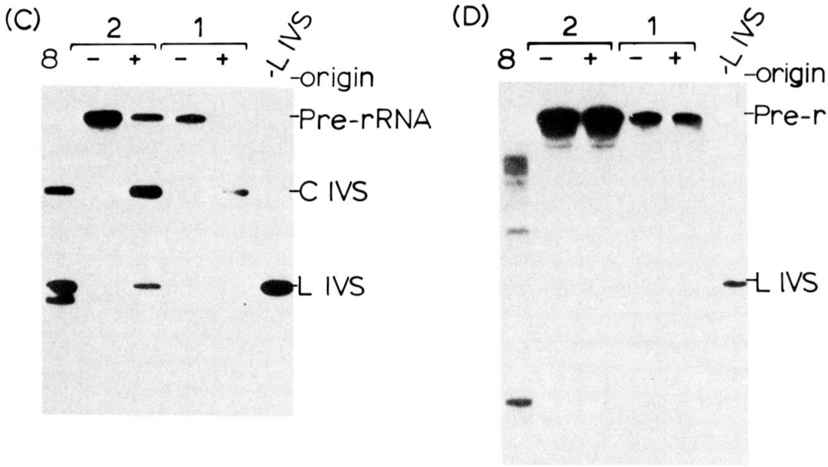

phoretically blotted onto diazotized paper and hybridized with a ^{32}P-labeled DNA probe that is specific for the IVS (the 145 bp Alu I fragment isolated from pTpAA1, obtained from M. Wild; see reference 28 for further description). (C) Gel blot hybridization with IVS probe, as in (B). (Lane 8) RNA from fraction 8 of the sucrose gradient. (Lane 2) RNA from fraction 2 incubated under splicing conditions except that GTP was omitted (-) or in the presence of 100 μM GTP (+). (Lanc 1) RNA from fraction 1, incubations the same as for fraction 2. (Lane L IVS) Purified, ^{32}P-labeled L IVS RNA run as a marker. (D) Same blot shown in (C) after the hybridized DNA was removed and the paper rehybridized with a ^{32}P-labeled DNA probe that is specific for the external transcribed spacer (the 200 bp Alu I-*Hind* III fragment isolated from pTtr1 DNA; this plasmid, a gift of P. Challoner and E. Blackburn, contains the 4.3 *Hind* III fragment from the center of the *T. thermophila* rDNA).

splicing *in vivo* and splicing of pre-rRNA synthesized in a totally artificial system *in vitro*, we examined the ability of pre-rRNA made *in vivo* to be spliced under completely defined conditions *in vitro*. RNA from fractions 1 and 2 of the sucrose gradient (Figure 4A) was incubated for 30 min at 42°C in the presence or absence of GTP. The RNA was again analyzed by gel blot hybridization, using the IVS-specific probe. As shown in Figure 4C, this procedure resulted in the excision of the IVS from the pre-rRNA. Because of the high temperature, most of the excised IVS RNA had been cyclized by 30 min. Based on densitometry of the autoradiograms, we estimate that 81% of the IVS in the pre-rRNA underwent autoexcision.

The gel blot from Figure 4C was heated to remove the hybridized IVS probe and then rehybridized with a DNA probe specific for the external transcribed spacer. As expected, this probe hybridized to the same extent to pre-rRNA independent of whether it was incubated under *in vitro* splicing conditions (Figure 4D). Thus, incubation of the RNA with GTP, $MgCl_2$ and salt did not produce any general breakdown of the molecule, but rather affected only the IVS portion of unspliced pre-rRNA.

From the results of Figures 4C and D, we conclude that pre-rRNA transcribed *in vivo* can be spliced *in vitro*. This RNA is much larger than that transcribed from the *E. coli* plasmid. In particular, the exon that precedes the IVS is ≈5000 nt long in the RNA made *in vivo*, 261 nt long in the plasmid transcripts. In addition, the pre-rRNA made *in vivo* has probably been methylated, while the plasmid transcripts are unmodified. These differences clearly do not prevent the autoexcision reaction from occurring. It would be of interest to see whether the rate of autoexcision of the RNA made *in vivo* is any faster than that of the plasmid transcripts or the RNA synthesized in isolated nuclei.

FURTHER DISCUSSION

The reactions associated with rRNA splicing are cleavage-ligation reactions. The IVS-containing pre-rRNA -- both RNA synthesized *in vitro* from the plasmid and RNA isolated from *Tetrahymena* nuclei -- was found to contain intact splice junctions prior to the *in vitro* splicing reaction. This was determined by nuclease mapping of RNA-DNA hybrids and, in the case of the plasmid transcripts, by indirect RNA sequencing (25,26). In the first cleavage-ligation reaction of rRNA splicing, cleavage at the 5' splice junction is accompanied by the covalent linkage of the guanosine cofactor to the 5' end of the IVS. Second, the 3' splice junction is cleaved and the exons are ligated. In a third cleavage-ligation reaction, the IVS is cyclized with the release of an oligonucleotide (24). Because these reactions occur without an external energy source, such as ATP or GTP hydrolysis, we have proposed a phosphoester transfer mechanism for rRNA splicing in which each cleavage is coupled to a ligation (20,23,24).

The activity or activities responsible for the proposed phosphoester transfers are intrinsic to the RNA molecule. The reactions occur with high specificity and at reasonably high rates *in vitro* in the absence of proteins. It remains possible that, *in vivo*, the reaction rate is enhanced ≃60-fold by proteins acting either stoichiometrically or catalytically. In the related case of mitochondrial mRNA splicing (see Introduction), there is excellent genetic evidence that splicing requires at least one type of protein, the so-called maturase (30). We have speculated (18) that the maturases are specific RNA binding proteins that are necessary for the very large mitochondrial introns to fold into a structure that is self-splicing. Establishing the relationship between these systems will require both the application of genetics to nuclear rRNA splicing and the application of biochemistry to mitochondrial RNA splicing.

ACKNOWLEDGMENTS

We thank Ted Palen for his help with one of the RNA blot hybridization experiments.

REFERENCES

1. Wild, MA and Gall, JG (1979) Cell 16:565-573.
2. Din, N, Engberg, J, Kaffenberger, W and Eckert, W (1979) Cell 18:525-532.
3. Cech, TR and Rio, DC (1979) Proc Natl Acad Sci USA 76:5051-5055.
4. Breathnach, R and Chambon, P (1981) Ann Rev Biochem 50:349-383.
5. Abelson, J (1979) Ann Rev Biochem 48:1035-1069.
6. Din, N and Engberg, J (1979) J Mol Biol 134:555-574.
7. Campbell, GR, Littau, VC, Melera, PW, Allfrey, VG and Johnson, EM (1979) Nucl Acids Res 6:1433-1448.
8. Gubler, U, Wyler, T and Braun, R (1979) FEBS Lett 100:347-350.

9. Bos, JL, Osinga, KA, Van der Horst, G, Hecht, NB, Tabak, HF, Van Ommen, G-JB and Borst, P (1980) Cell 20:207-214.
10. Allet, B and Rochaix, J-D (1979) Cell 18:55-60.
11. Sharp, PA (1981) Cell 23:643-646.
12. Mount, S (1982) Nucl Acids Res 10:459-472.
13. Nomiyama, H, Sakaki, Y and Takagi, Y (1981) Proc Natl Acad Sci USA 78:1376-1380.
14. Burke, J and RajBhandary, UL (1982) Cell 31:509-520.
15. Davies, RW, Waring, RB, Ray, JA, Brown, TA and Scazzocchio, C (1982) Nature 300: 719-724.
16. Michel, F and Dujon, B (1983) EMBO Journal 2:33-38.
17. Waring, RB, Scazzocchio, C, Brown, TA, and Davies, RW (1983) J Mol Biol, in press.
18. Cech, TR, Tanner, NK, Tinoco, I, Weir, BR, Zuker, M and Perlman, PS (1983) Proc Natl Acad Sci USA, in press.
19. DeLaSalle, H, Jacq, C and Slonimski, PP (1982) Cell 28:721-732.
20. Cech, TR, Zaug, AJ and Grabowski, PJ (1981) Cell 27: 487-496.
21. Berk, AJ and Sharp, PA (1977) Cell 12:721-732.
22. Kan, NC and Gall, JG (1982) Nucl Acids Res 10:2809-2822.
23. Kruger, K, Grabowski, PJ, Zaug, AJ, Sands, J, Gottschling, DE and Cech, TR (1982) Cell 31:147-157.
24. Zaug, AJ, Grabowski, PJ and Cech, TR (1983) Nature 301:578-583.
25. Grabowski, PJ (1983) PhD Dissertation, U of Colorado, Boulder CO 80309.
26. Grabowski, PJ and Cech, TR, manuscript in preparation.
27. Alwine, JC, Kemp, DJ and Stark, GR (1977) Proc Natl Acad Sci USA 74:5350-5354.
28. Brehm, SL and Cech, TR (1983) Biochemistry, in press.
29. Zaug, AJ and Cech, TR (1982) Nucl Acids Res 10:2823-2838.
30. Lazowska, J, Jacq, C and Slonimski, PP (1980) Cell 22:333-348.

Gene Expression, pages 343-358
© 1983 Alan R. Liss, Inc., 150 Fifth Avenue, New York, NY 10011

IN VITRO SPLICING OF PURIFIED ADENOVIRAL EARLY mRNA[1]

Carlos J. Goldenberg, Peter DiMaria,
and Scott D. Hauser

Department of Pathology
Washington University School of Medicine
St. Louis, Missouri 63110

ABSTRACT Adenovirus 2 nuclear RNA precursors encoded by early region 2 are spliced *in vitro* by nuclear extracts prepared from MOPC-315 mouse myeloma cells. The nucleotide sequence across the splice junctions in the E_2 RNAs processed *in vitro* was investigated by performing primer extensions in the presence of dideoxynucleotides and direct sequencing on polyacrylamide gels. We conclude that the *in vitro* splicing reaction is accurate and has the same precision as that of *in vivo* E_2 cytoplasmic mRNA prepared from Ad2 infected cells. Approximately 80% of E_2 RNA precursor, on a molar basis, are spliced *in vitro* to a mature RNA. These findings provide evidence that a nuclear extract prepared from MOPC-315 mouse myeloma cells is capable of accurate and efficient splicing of E_2 RNA precursors. The effect of antibodies to U1 snRNP (isolated by Steve Mount and Joan Steitz) was investigated in this *in vitro* splicing system. The splicing reaction was completely inhibited when the extracts were preincubated with antibodies that immunoprecipitate small nuclear ribonucleoprotein particles containing U1 RNA.

[1]This work was supported by research grants from the National Institute of Health (NIH IRO1 AI 19370-01), the American Cancer Society (Wash. Univ. ACS Institutional 43192G), and the Washington University Biomedical Research Support Grant. This study was also supported by Brown Williamson Tobacco Corporation, Philip Morris, Inc., R. J. Reynolds Tobacco Company, and the United States Tobacco Company.

INTRODUCTION

A unique feature of eukaryotic genes is the fact that functional RNA molecules contain covalently linked transcripts of noncontiguous DNA segments. A large number of viral and eukaryotic protein coding genes and also those coding for tRNAs and rRNAs have been shown to possess this mosaic structure (for review see 1-6). Structural studies of several in vivo RNA precursor molecules have provided evidence that a splicing mechanism must function in vivo to excise intervening sequences from the newly formed RNA (for review see 4,5). Although several models for splicing have been proposed (1-6), the biochemical steps and the regulation of this process remain largely unsubstantiated. The study of enzymes and components involved in the splicing reaction requires in vitro cell-free systems which are able to splice mRNA precursors accurately and efficiently. In vitro transcription and splicing of viral (7) and human β globin RNA (8) in the whole cell extracts from HeLa cells was reported. However, the efficiency of the splicing reaction in both systems are extremely low.

Recently (9), we have presented evidence that splicing of purified adenovirus precursor mRNA occurs very efficiently in a whole cell extract prepared from MOPC-315 mouse myeloma cells. The mRNA precursor that we have used for the in vitro splicing reaction is encoded by one of the four regions expressed at early times after Ad2 infection (11,12), early region 2 (E_2). The polypeptide product of E_2 is a 72,000 dalton DNA-binding protein (13-15). E_2 mRNA is copied from the strand transcribed in the leftward direction (16,17). This mRNA consists of three exons and two intervening sequences (18-20). Structural studies of nuclear RNA revealed three major species (Figure 1) (21,22). The largest (28S) precursor contains all the intervening sequences. A 23S species appears to be a processing intermediate lacking the large 5' intervening sequence. The nuclear 20S RNA has a structure identical to the cytoplasmic mRNA and presumably is the direct precursor of the functional mRNA. Studies measuring steady state levels of E_2 nuclear RNA also detected an RNA molecule (26S) in which the smaller 3' intervening sequence has been excised (23).

In this study (also reference 10), we describe a nuclear extract that splices with a very high efficiency purified deproteinized and polyadenylated E_2 mRNA precursors that have been obtained from cultures early in adenovirus infection. We have investigated the precision of the *in vitro* splicing using nucleotide sequencing techniques. The nucleotide sequence across the splice junctions of the *in vitro* processed E_2 RNAs is identical to that of the *in vivo* E_2 cytoplasmic mRNA. Therefore, the nuclear extract prepared from MOPC-315 mouse myeloma cells is capable of efficient and accurate *in vitro* splicing of E_2 mRNA precursors.

Several reports suggest that small nuclear RNAs (snRNAs) might have a role in splicing mRNA precursors (24,25). These studies are based on comparisons of structural homologies between the 5' terminal sequence of snRNAs and splice junctions. The U1 RNA has short sequences complementary to the consensus splice junction sequences. By base pairing, it has been proposed that the U1 RNA aligns the donor and acceptor splice sites in the primary transcripts. Here, we show that antibodies directed against U1 snRNPs, purified in the laboratory of Dr. Joan Steitz, were able to inhibit *in vitro* splicing of E_2 mRNA precursors.

MATERIALS AND METHODS

Cell culture, virus infection, and RNA preparation. Maintenance of KB (human) and MOPC-315 mouse myeloma cell suspension cultures, procedures for Ad2 infections, and isolation of nuclear and cytoplasmic RNA were performed as described (22). Infections were performed in the presence of 25 μg/ml cycloheximide and harvested at 6 hr after infection. High molecular weight poly(A)$^+$ RNA precursors to be used as substrates for *in vitro* incubations were highly purified by fractionation on three successive 15-30% sucrose gradient sedimentations as described previously (9,26).

Preparation of nuclear extracts with splicing activity and in vitro splicing reactions. Nuclear extracts containing splicing activity were prepared as described elsewhere (10). In vitro splicing reactions were performed as described (9,10).

Sequencing by primer extensions. Hybridizations of the *in vitro* E_2 processed RNAs or *in vivo* Ad2 cytoplasmic RNAs to the ^{32}P 5'-end labeled primer DNA fragment, extending from the *XhoI* to the *BalI* site, were performed as described (10). Nucleotide sequencing by chain extensions with dideoxynucleotides was as described (10).

RESULTS

Nuclear Extracts Convert E_2 Nuclear RNA Precursors to the Size of the Mature mRNA.

To assay _in vitro_ splicing activity, we have utilized the procedures outlined in Figure 1. To analyze the RNAs processed _in vitro_, we characterized the RNAs by Northern blots and hybridization to a ^{32}P-labeled 59.5-70.7 DNA fragment cloned in PBR322 which encodes E_2 exons and intervening sequences (Figures 1 and 2). Since the E_2 polyadenylated nuclear RNA contains three large size classes (28S, 26S, and 23S) in addition to molecules the size of cytoplasmic 20S mRNA (Figure 2), this assay for processing activity required a preparation of nuclear RNA precursor that was essentially free of 20S nuclear RNA. To obtain the necessary substrate, polyadenylated nuclear

PROCEDURE FOR DETECTING SPLICING OF REGION 2 RNAs IN VITRO

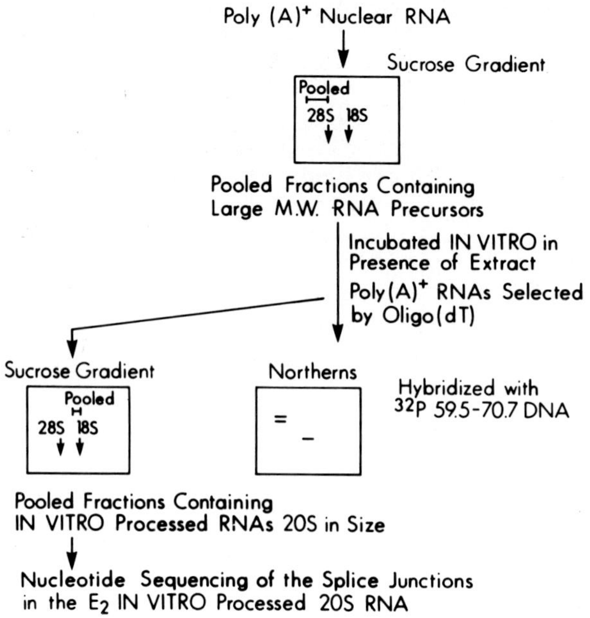

Figure 1. Method for detecting *in vitro* splicing of E_2 RNAs. High molecular weight poly(A)$^+$ RNA precursors are purified by fractionation on 15-30% sucrose gradient. Ribosomal RNA markers (^{14}C) are added to the sample prior to centrifugation. Fractions containing RNA 28S in size are pooled and incubated *in vitro* with the processing extract. The processed RNAs are assayed by Northern blots and then fractionated on sucrose gradients. Fractions containing RNA 20S in size are pooled and the nucleotide sequence of the *in vitro* processed 20S RNAs is then determined.

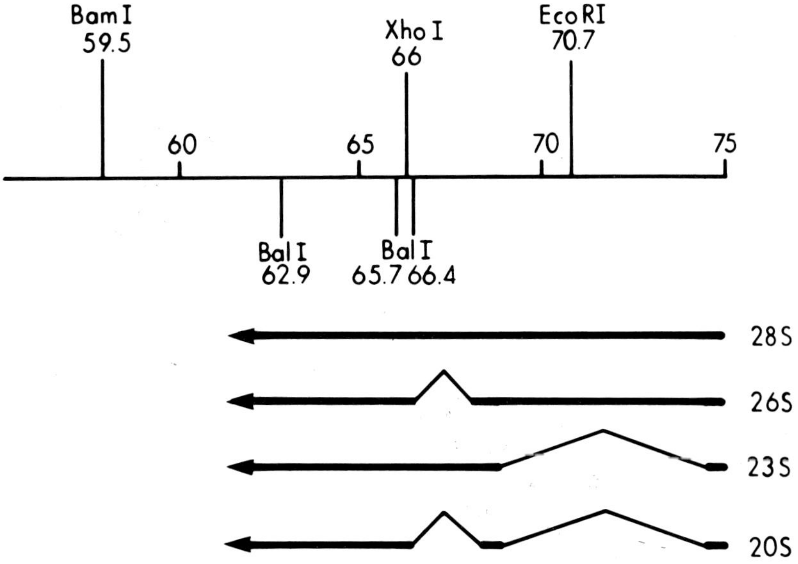

Figure 2. Nuclear poly(A)$^+$ RNA transcribed from E_2 at early times in infection. Solid bars represent the structure of nuclear E_2 RNAs (18,19, 21,22). The arrows indicate the direction of transcription. Caret-shaped symbols indicate sequences covalently joined by splicing. The cytoplasmic mRNA has the structure of the 20S species (22). Relevant cleavage sites of restriction endonucleases EcoRI, BamI, XhoI, and BalI have been reported previously and confirmed at the nucleotide sequence level (35,36). The map positions indicate the relative distance of the cleavage site from the left end of the genome.

RNA was fractionated on a 15–30% sucrose gradient. RNAs 28S and larger in size were pooled, precipitated, and run again twice on 15–30% sucrose gradients. The pooled high molecular weight RNA contained 20S E_2 RNA molecules after the first fractionation on sucrose gradients (Figure 3A, lane 5). However, after the second and third fractionation on sucrose gradients, the high molecular weight nuclear RNA precursors showed no evidence of 20S E_2 RNA (Figure 3A, lanes 2 and 3). Aliquots of this highly purified nuclear RNA precursor were incubated for 50 min at 30°C in incubation buffer without extract. Figure 3A (lane 4) shows that the 28S and 26S RNA species remained intact as judged by their size distribution. When the E_2 precursor RNA was incubated in the presence of nuclear extracts prepared from MOPC-315 mouse myeloma cells and assayed for RNA processing, two new bands of 23S and 20S RNA were generated (Figure 3B, lanes 2 and 3). As reported with the whole cell MOPC-315 extract (9), approximately 80% of the E_2 precursor RNA, on a molar basis, was converted to a 20S species (Figure 3, densitometry scan not shown).

The preparation of the extracts isolated from MOPC-315 mouse myeloma used in the present study differs from that previously reported (9). This extract was prepared from the nuclear fraction of the MOPC-315 cell and was isolated in the presence of Nonidet P-40 detergent. In order to demonstrate that all the RNA processing components were present in the nuclear fraction, the 10,000 g supernatant of the extract utilized for the experiment shown in Figure 3B was assayed for processing activity (see Materials and Methods). After incubation with the post-nuclear fraction, the E_2 RNA precursors remained the same (data not shown), indicating that no major leakage of the processing activities occurred during fractionation.

Nucleotide Sequence of the Splice Junctions in the In Vitro Spliced E_2 RNAs.

In order to investigate the accuracy of the *in vitro* splicing reaction at the nucleotide level, E_2 nuclear RNA was isolated as described in the legend of Figure 3. Since the remainder of the E_2 precursor RNAs after *in vitro* splicing (Figure 3B) may originate bands in a sequencing gel corresponding to the E_2 intervening sequences, the *in vitro* RNA products after incubation with the nuclear extract were fractionated on a 15–30% sucrose gradient. RNAs 20S in size were pooled and the nucleotide sequence of the splice junctions was determined by performing RNA-DNA primer extensions with reverse transcriptase in the presence of 2'-3' dideoxynucleoside triphosphates as specific chain terminators. For this study we have chosen a primer DNA fragment XhoI-BalI of 147 nucleotides, which hybridizes 13 nucleotides downstream from the splice junction between the 5'-end main E_2 mRNA body and the second RNA leader (Figures 2 and 5). The nucleotide sequence of the splice junctions

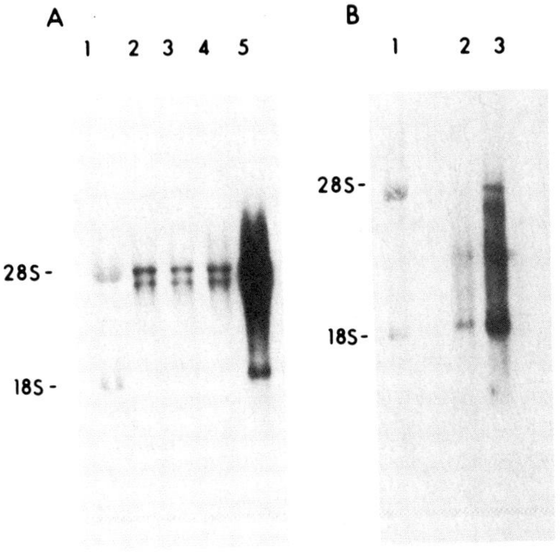

Figure 3. Northern blots for in vitro processed E_2. High molecular weight nuclear poly(A)$^+$ RNA precursors to be used as substrates for in vitro incubations were fractionated on three successive 15–30% sucrose gradient sedimentations (10). Fractions 28S in size and larger were pooled, subjected to electrophoresis on glyoxal/1.1% agarose gels, and transferred to nitrocellulose paper. The nick translated 59.5–70.7 Ad2 DNA fragment cloned in PBR322 (3 x 10^8 cpm/ug) was annealed to the immobilized RNA (37). After hybridization, the strips were exposed to X-ray film with an intensifying screen for 48 hrs at -70°C.

Panel A: Blot hybridization of E_2 poly(A)$^+$ nuclear RNAs after successive fractionations on 15–30% sucrose gradient sedimentation. Lane 1: ribosomal ^{14}C markers (1 x 10^4 cpm). Lane 5: high molecular weight poly(A)$^+$ nuclear RNA pooled from the first sucrose gradient (RNA purified from 6 x 10^7 infected cells). Lanes 2 and 3: nuclear RNA pooled from the second and third sucrose gradients (RNA purified from 6 x 10^7 infected cells). Lane 4: high molecular weight nuclear RNA from the third gradient incubated in vitro in incubation buffer at 30°C for 50 min in the absence of extract as described in the experimental procedure (RNA purified from 6 x 10^7 cells).

Panel B: Blot hybridization of E_2 poly(A)$^+$ nuclear RNA precursors after incubation with a nuclear extract prepared from MOPC-315 mouse myeloma cells. Lane 1: ribosomal ^{14}C markers (1 x 10^4 cpm). Lanes 2 and 3: high molecular weight poly(A)$^+$ nuclear RNAs purified by three successive sucrose gradient sedimentations after in vitro incubation with the nuclear extract as described in the experimental procedure (lane 2: RNA purified from 6 x 10^7 infected cells, and lane 3: RNA purified from 6 x 10^8 infected cells).

of the in vitro processed RNA was directly compared to the E_2 cytoplasmic mRNA isolated from KB cells early during Ad2 infection in the presence of cycloheximide. Figure 4 shows the nucleotide sequence of the splice junctions derived from the cytoplasmic E_2 mRNA (Figure 4A) and the in vitro processed E_2 RNAs (Figure 4B). Both nucleotide sequences were identical demonstrating faithful and accurate in vitro splicing for the splice junction between the 5' acceptor E_2 mRNA body and the 3' donor of the second RNA leader sequence (see Figures 1 and 5). The accuracy of the in vitro splicing for the junction between the second and first leader sequences was analyzed using the primer extension technique in the absence of dideoxynucleotides. Figure 4A and B (lane B) shows the expected cDNA product of 305 nucleotide long, generated by both the in vitro spliced E_2 RNAs and the cytoplasmic E_2 mRNA. The nucleotide sequence of the splice between the first and second leaders was not determined directly, but the appropriate size for the major cDNA product indicates accurate in vitro splicing between the first and second E_2 RNA leaders. The nucleotide sequence of the 5'-end main body and the second and first E_2 mRNA leaders is interpreted in detail in Figure 5.

Anti-RNP Antibodies Inhibit In Vitro Splicing of E_2 mRNA Precursors.

The role of U1 snRNP's (small nuclear ribonucleoprotein particles) in in vitro splicing was investigated. E_2 mRNA precursors were purified as described in the legend of Figure 3 and incubated in the presence of MOPC-315 nuclear extracts which were preincubated for 10 min at 0°C in the presence or absence of antibodies against U1 snRNP's. Antibodies from systemic lupus erythematosus patients that immunoprecipitate U1 snRNP's were purified in the laboratory of Joan Steitz. Figure 6 (lane 5) shows a complete inhibition of in vitro splicing

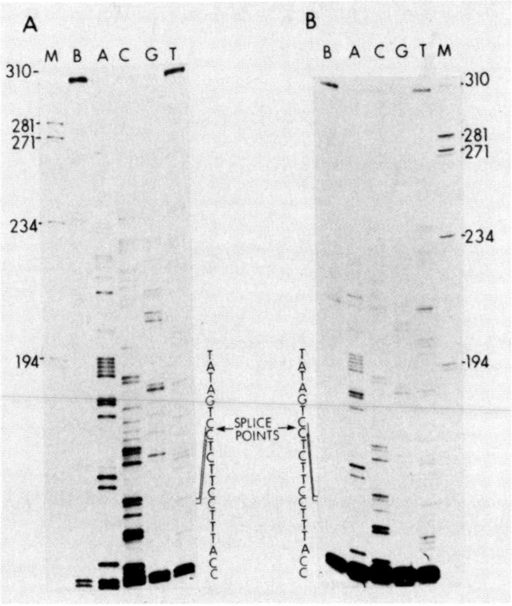

Figure 4. Sequencing of the splice junctions of E_2 RNA precursors processed in vitro. High molecular weight nuclear RNAs were purified from 6 x 10^9 KB cells as described in the legend of Figure 3 and incubated in the presence of a nuclear extract prepared from MOPC-315 cells. Aliquots were tested for in vitro processing (Figure 2B). The in vitro processed RNAs were then fractionated on a 15-30% sucrose gradient sedimentation, and fractions containing RNA 20S in size were pooled and precipitated. The nucleotide sequence of the splice junctions in the in vitro processed RNAs was determined

by performing chain extensions with reverse transcriptase in the presence of 2'-3'-dideoxinucleoside triphosphates. The 147 nucleodite XhoI-BalI DNA primer (Figures 1 and 4) labeled at the 5'-end XhoI site was hybridized and extended as described in the Materials and Methods.

Panel A: Nucleotide sequence of the splice junctions in the in vivo cytoplasmic E_2 mRNA. Exposure to X-ray film with an intensifying screen was for 15 hrs at -70°C.

Panel B: Nucleotide sequence of the splice junctions in the in vitro processed 20S E_2 RNA. Exposure to X-ray film with an intensifying screen was for 7 days at -70°C. A, C, G, and T denote the specific ddNTP inhibitor used in the reaction. B denotes a reaction in the absence of ddNTP's inhibitors, M markers ØX174 DNA 5'-end labeled.

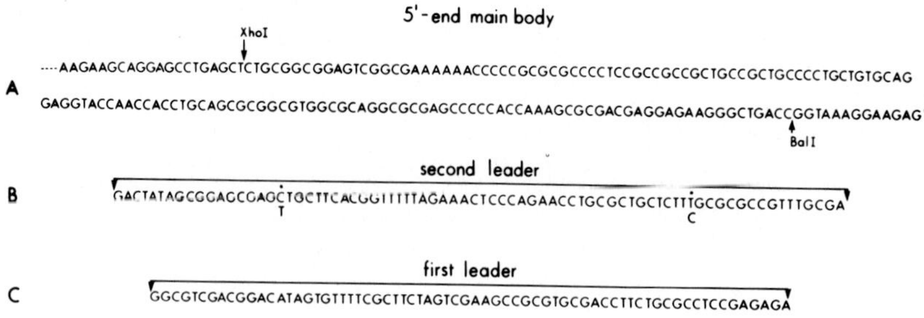

Figure 5. Nucleotide sequence of the splice junctions of the E_2 mRNA. From experiments like the one shown in Figure 3, the nucleotide sequence of the E_2 mRNA second leader and 13 nucleotides from the BalI site to the splice point in the 5'-end main body of the E_2 mRNA were derived. The nucleotide sequence of the 5'-end main body, the second leader and the first leader are indicated in the corresponding RNA sequences (A, B, and C). The positions of the splice points are indicated by vertical arrows. The sites for restriction endonucleases XhoI and BalI are indicated in Panel A. The nucleotide differences present in the Ad5 second leader are indicated underneath the Ad2 sequence. The nucleotide sequence of the Ad2 5'-end main body (36) as well as the first leader and the second leader of Ad5 of the E_2 mRNA have been reported previously (35).

in the presence of anti-U1 snRNP antibodies. No inhibition was observed when the extracts were preincubated in the absence or presence of control antibodies (a non-immune serum) (Figure 6, lanes 3 and 4). We therefore conclude that anti-U1 snRNP antibodies inhibit in vitro splicing of E_2 mRNA precursors.

DISCUSSION

We have shown that adenovirus E_2 precursor RNA was accurately and efficiently spliced in vitro by nuclear extracts prepared from MOPC-315 mouse myeloma cells. The nucleotide sequence analysis of the in vitro E_2 RNA products demonstrated unambiguous evidence that cleavage and ligation have the same precision as occurs in vivo.

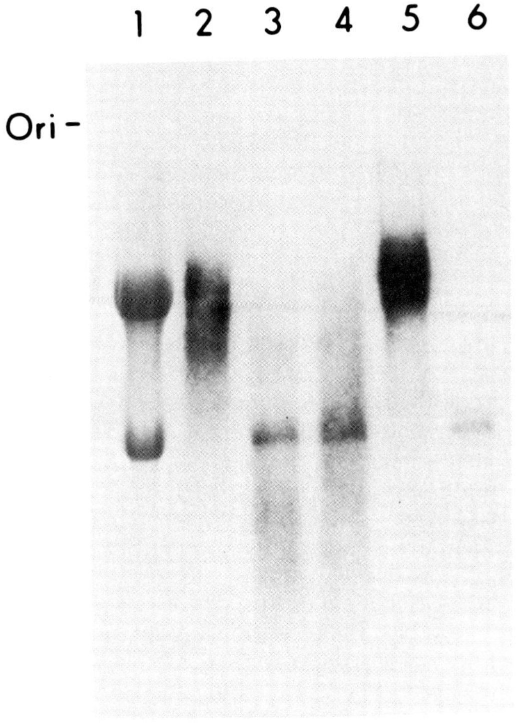

Figure 6. Effect of antibodies to U1 snRNP's on in vitro splicing of E_2 RNAs. High molecular nuclear poly(A)$^+$ RNA precursors were purified as described in

the legend of Figure 3. The nuclear extract was preincubated on ice for 10 min in the presence of antibodies from normal serum, antibodies to U1 snRNP's, or in the absence of antibodies. The amount of antibody was predetermined by immunoprecipitating snRNP's from the nuclear extracts. The processed RNAs were assayed by Northern blots as described in the legend of Figure 3. Each lane RNA purified from 6 x 10^7 infected cells. Lane 1: Ribosomal RNA markers (^{14}C). Lane 2: E_2 nuclear RNAs incubated in absence of nuclear extract. Lane 3: RNAs incubated in the presence of nuclear extract. Lane 4: RNAs incubated in the presence of a nuclear extract preincubated with antibodies from a normal serum. Lane 5: RNAs incubated in the presence of a nuclear extract preincubated with antibodies to U1 snRNP's. Lane 6: Poly(A)$^+$ RNA purified from cytoplasmic early infected cells.

Therefore, we conclude that all the components of the splicing apparatus necessary to recognize and splice E_2 RNAs are present in the nuclear extracts prepared from MOPC-315 cells. For future studies and purification, the nuclear extract may be a valuable source of splicing activity.

The high efficiency of splicing obtained with the nuclear extracts could be due to the fact that we have used polyadenylated mRNA precursor isolated from cultures early in Ad2 infection. This RNA appears to be the precursor to the 72K DNA binding protein mRNA (20-22). RNAs transcribed _in vitro_ by concentrated whole HeLa cell extracts were reported to be spliced _in situ_ with only a very low efficiency, between 2 and 5% (7,8). Other investigators were not able to detect any splicing in very similar _in vitro_ systems (27). The discrepancies in splicing efficiencies between the _in vitro_ system reported here and the coupled _in vitro_ transcription-splicing systems could be due to several factors. The _in vitro_ transcribed RNAs could lack sequences or nucleoside modifications for attaining similar structures to those found _in vivo_ and therefore may not be ideal substrates for the splicing apparatus. For example, specific transcription termination and polyadenylation do not occur in the available eukaryotic _in vitro_ transcription systems (27-29). Alternatively, inactivation of an essential component for splicing or of proteins that organize the precursor RNA into an RNP structure in vivo may occur during preparation of the whole HeLa cell extracts. We are currently testing the ability of the nuclear MOPC-315 extracts to splice run-off RNAs which have been synthesized in in vitro transcription systems. Preliminary results suggest a low efficiency of in vitro splicing (unpublished data). Further experiments on this question are underway. Any manipulations to analyze the in vitro splicing reaction requires relatively rapid and reasonable assays. We have previously utilized assays which required the purification of large amounts of nuclear RNA from infected cells. By using Northern blots, we were able to increase the sensitivity for detection of the in vitro spliced RNAs. This assay will facilitate future characterization of the splicing activities and/or co-factors such as ribonucleoprotein particles containing small nuclear RNAs (snRNP's) that have been implicated in splicing (24,25). Our preliminary results that antibodies to U1 snRNP inhibit splicing in vitro suggest a role for these particles in splicing. It should now be possible to investigate the mechanisms of action of these RNP particles in splicing. Development of in vitro splicing systems are crucial for the understanding of the biochemical steps in RNA processing. For example, major features of the reaction mechanism have been described for the splicing of yeast tRNA precursors (30-32) and the ribosomal RNA precursors of tetrahymena pigmentosa (33,34). Hopefully, further experiments utilizing the in vitro splicing system described here will help to unravel the biochemical steps in mRNA splicing.

ACKNOWLEDGEMENTS

We thank Drs. Steve Mount and Joan Steitz for providing us with purified antibodies to U1 snRNP's and Miss Betsy Klein for typing assistance.

REFERENCES

1. Gilbert W (1978) Why genes in pieces? Nature 271:501.
2. Darnell JE (1978) Implications of RNA-RNA splicing in evolution of eukaryotic cells. Science 202:1257-1260.

3. Crick F (1979) Split genes and RNA splicing. Science 204:264–271.
4. Abelson J (1979) RNA processing and the intervening sequence problem. Ann Rev Biochem 48:1035–1069.
5. Breathnach R, Chambon P (1981) Organization and expression of eukaryotic split genes coding for proteins. Ann Rev Biochem 50: 349–383.
6. Sharp PA (1981) Speculations on RNA splicing. Cell 23:643–646.
7. Weingartner B, Keller W (1981) Transcription and processing of adenoviral RNA by extracts from HeLa cells. Proc Natl Acad Sci USA 78:4092–4096.
8. Kole R, Weissman SM (1982) Accurate in vitro splicing of human β-globin RNA. Nuc Acids Res 10:5429–5445.
9. Goldenberg CJ, Raskas HJ (1981) In vitro splicing of purified precursor RNAs specified by early region 2 of the adenovirus 2 genome. Proc Natl Acad Sci USA 78:5430–5434.
10. Goldenberg, CJ, Hauser, SD (1983) Accurate and efficient in vitro splicing of purified precursor RNAs specified by early region 2 of the adenovirus 2 genome. Nuc Acids Res 11:1337–1348.
11. Flint J (1977) The topography and transcription of the adenovirus genome. Cell 10:153–166.
12. Ziff E (1980) Transcriptional RNA processing by the DNA tumor viruses. Nature 287:491–499.
13. Ginsberg HS, Ensinger MJ, Kauffman RS, Mayer AJ, Landholm V (1974) Cell transformation: A study of regulation with types 5 and 12 adenovirus temperature-sensitive mutants. Cold Spring Harbor Symp Quant Biol 39:419–426.
14. Grodzicker T, Williams J, Sharp PA, Sambrook J (1974) Physical mapping of temperature-sensitive mutations of adenoviruses. Cold Spring Harbor Symp Quant Biol 39:439–446.
15. Lewis JB, Atkins JF, Baum PR, Salem R, Gesteland RF, Anderson CW (1976) Location and identification of the genes for adenovirus type 2 early polypeptides. Cell 7:141–151.
16. Sharp PA, Gallimore PH, Flint SJ (1975) Mapping of adenovirus 2 RNA sequences in lytically infected cells and transformed cell lines. Cold Spring Harbor Symp Quant Biol 39:457–474.
17. Pettersson V, Tibbets C, Philipson L (1976) Hybridization maps of early and late messenger RNA sequences

on the adenovirus type 2 genome. J Mol Biol 101:479-501.
18. Kitchingman GR, Lai S-P, Westphal H (1977) Loop structures in hybrids of early RNA and the separated strands of adenovirus DNA. Proc Natl Acad Sci USA 74:4392-4395.
19. Chow LT, Broker TR, Lewis JB (1979) Complex splicing patterns of RNAs from the early regions of adenovirus 2. J Mol Biol 134:265-304.
20. Berk AJ, Sharp PA (1978) Structure of the adenovirus 2 early mRNAs. Cell 14:695-711.
21. Craig EA, Raskas HJ (1976) Nuclear transcripts larger than the cytoplasmic mRNAs are specific by segments of the adenovirus genome coding for early functions. Cell 8:205-213.
22. Goldenberg CJ, Raskas HJ (1979) Splicing patterns of nuclear precursors to the mRNA for adenovirus 2 DNA binding protein. Cell 16:131-138.
23. Kitchingman GR, Westphal H (1980) The structure of adenovirus 2 early nuclear and cytoplasmic RNAs. J Mol Biol 137:23-48.
24. Lerner MR, Boyle JA, Mount SM, Wolin SL, Steitz JA (1980) Are snRNPs involved in splicing? Nature 283:220-224.
25. Yang VW, Lerner MR, Steitz JA, Flint SJ (1981) A small nuclear ribonucleoprotein is required for splicing of adenoviral early RNA sequences. Proc Natl Acad Sci USA 78:1371-1375.
26. Weber J, Jelinek W, Darnell JE (1977) The definition of a large viral transcription unit late in Ad2 infection of HeLa cells: Mapping of nascent RNA molecules labeled in isolated nuclei. Cell 10:611-616.
27. Handa H, Kaufman RJ, Manley JL, Gefter ML, Sharp PA (1981) Transcription of simian virus 40 DNA in a HeLa whole cell extract. J Biol Chem 256:478-482.
28. Manley JL, Fire A, Cano A, Sharp PA, Gefter ML (1980) DNA-dependent transcription of adenovirus genes in a soluble whole cell extract. Proc Natl Acad Sci USA 77:3855-3859.
29. Weil PA, Luse DS, Segall J, Roeder RG (1979) Selective and accurate initiation of transcription at the Ad2 major late promoter in a soluble system dependent on purified RNA polymerase II and DNA. Cell 18:469-484.

30. Knapp G, Beckman JS, Johnson PF, Fuhrman SA, Abelson J (1978) Transcription and processing of intervening sequences in yeast and tRNA genes. Cell 14:221-236.
31. Knapp G, Ogden RC, Peebles CL, Abelson J (1979) Splicing of yeast tRNA precursors: Structure of the reaction intermediates. Cell 18:37-45.
32. Ogden RC, Beckman JS, Abelson J, Kang HS, Soll D, Schmidt O (1979) In vitro transcription and processing of a yeast tRNA gene containing an intervening sequence. Cell 17:399-406.
33. Cech TR, Zaug AJ, Grabowski PJ (1981) In vitro splicing of the ribosomal RNA precursor of tetrahymena: Involvement of a guanosine nucleotide in the excision of the intervening sequence. Cell 27:487-496.
34. Kruger K, Grabowski PJ, Zaug AJ, Sands J, Gottschling DE, Cech TR (1982) Self-splicing RNA: Autoexcision and autocyclization of the ribosomal RNA intervening sequence of tetrahymena. Cell 31:147-157.
35. Kruijer W, van Schaik FMA, Sussenbach JS (1981) Structure and organization of the gene coding for the DNA binding protein of adenovirus type 5. Nuc Acids Res 9:4439-4457.
36. Kruijer W, van Schaik FMA, Sussenbach JS (1982) Nucleotide sequence of the gene encoding adenovirus type 2 DNA binding protein. Nucl Acids Res 10:4493-4500.
37. Thomas PS (1980) Hybridization of denatured RNA and small DNA fragments transferred to nitrocellulose. Proc Natl Acad Sci USA 77:5201-5205.

Gene Expression, pages 359-370
© 1983 Alan R. Liss, Inc., 150 Fifth Avenue, New York, NY 10011

GENERATION OF HISTONE mRNA-SPECIFIC 3' ENDS BY CLONED MOUSE H4 GENE SEGMENTS INTRODUCED INTO TISSUE CULTURE CELLS

Daniel Schümperli, Erika Lötscher and Claudia Stauber

Institut für Molekularbiologie II der Universität Zürich
Hönggerberg, CH-8093 Zürich

ABSTRACT To study the molecular mechanism responsible for the generation of 3' ends of mammalian histone mRNAs, we have fused the 3' terminal half of the mouse H4 gene, including 230 bp of spacer sequences to the early promoter of SV40. RNA molecules with authentic 3' termini were detected two days after transfection of this plasmid recombinant into hamster or human tissue culture cells. However, a small but significant number of transcripts did not acquire histone mRNA-like 3' ends, as judged by assays for the function of the E.coli galactokinase gene located downstream on the same transcription unit. This notion was substantiated by preliminary results of S1 mapping experiments which allowed us to quantitate the relative numbers of correctly terminated and readthrough transcripts. In addition, we have succeeded in isolating stable cell lines that carry the hybrid SV40:mouse H4:galK transcription unit in either episomal or chromosomal form.

INTRODUCTION

A unique feature of histone genes is that their mRNAs usually lack the typical poly(A) tails found in other eukaryotic mRNAs. Instead, these transcripts have their 3' ends within a region characterized by two highly conserved sequence motifs (1,2, see Fig.1). The first sequence motif includes a 16 base pair (bp) hyphenated inverted repeat that can form a "stem and loop" structure at the 3' end of the mRNA. Nevertheless, it is the sequence *per se* and not just the potential secondary structure which has been conserved between organisms as distant from each other in evolution as sea urchin and man. The second motif lies 6 to 8 bp downstream from the first and is somewhat less well conserved.

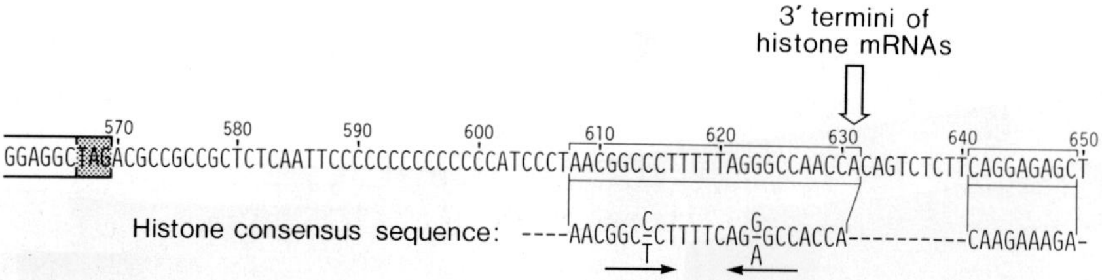

FIGURE 1. Conserved sequence elements at the 3' end of histone genes. Upper line: Nucleotide sequence of the mouse H4 histone gene (4). The end of the H4 coding body and the conserved sequence elements are boxed. Lower line: Consensus sequence of the conserved sequence elements (1,2).

It is still unclear whether the 3' ends of histone mRNAs are sites of transcription termination or if they arise by RNA processing, although extensive searches for longer precursor RNAs in sea urchin blastula (3) or in mouse tissue culture cells (4) have been negative. Since in both of these systems (4,5), the 3' ends of histone mRNAs are located just downstream of the first conserved sequence motif which has certain similarities with bacterial terminators (6,7), it is tempting to assume that those ends arise by a termination mechanism. In a recent study, the conserved 3' sequence elements of the sea urchin H2A gene were subjected to a detailed mutational analysis (8). These studies demonstrated that the dyad symmetry element is essential for the production of authentic 3' ends of H2A gene transcripts in micro-injected frog oocytes, but that spacer sequences downstream of these 3' ends are also required. These results can therefore not rule out that histone gene primary transcripts may extend into spacer DNA for at least a short distance, but that such precursor molecules escaped detection in previous studies.

To address these questions and to help define the sequence requirements for generating histone mRNA 3' ends, we have developed an expression assay based on transfection of tissue culture cells with fragments of a mammalian histone gene cloned into plasmid vectors. Transient gene expression can be monitored in the total cell population at the levels of both RNA and protein synthesis. In addition, the plasmid vectors carry an independent selective gene that allows the isolation of stably transformed cells. These cells may carry the transfected DNA either in episomal or chromosomal form depending on which of two types of vectors is being used. With the help of this system, we have begun to study the putative regulatory sequences involved in the generation of correct 3' ends of a mouse H4 histone mRNA.

TRANSCRIPTIONAL POLARITY OF A 3' TERMINAL MOUSE H4 GENE FRAGMENT

We have previously described the expression of the E.coli galactokinase gene (galK) in mammalian cells (9). On the plasmid recombinant used there, galK is under control of the SV40 early promoter; additional SV40 regulatory sequences for RNA splicing and polyadenylation are positioned downstream of galK to ensure correct RNA maturation. To study the details of the generation of histone mRNA 3' ends with this vector system, we constructed the plasmid pEL1gpt (Fig.2). The galK transcription unit of this vector has been engineered such that the major SV40 early transcription start sites are located some 120 nucleotides upstream of the galK initiation codon. This leader segment contains unique restriction sites for SmaI/AvaI and ClaI, located 50 and 23 nucleotides upstream of the galK initiation codon, respectively. Segments of DNA introduced into this vector at the ClaI site can be tested for transcriptional polarity in two ways: (i) by analysing their effect on galactokinase enzyme production and (ii) by direct mapping of RNA 3' termini by the S1 nuclease technique, using DNA fragments 3' end-labelled at the SmaI/AvaI site as hybridization probes. In addition, the plasmid contains another transcription unit with the dominant selection marker E.coli guanine phosphoribosyltransferase (gpt/Ref.10).

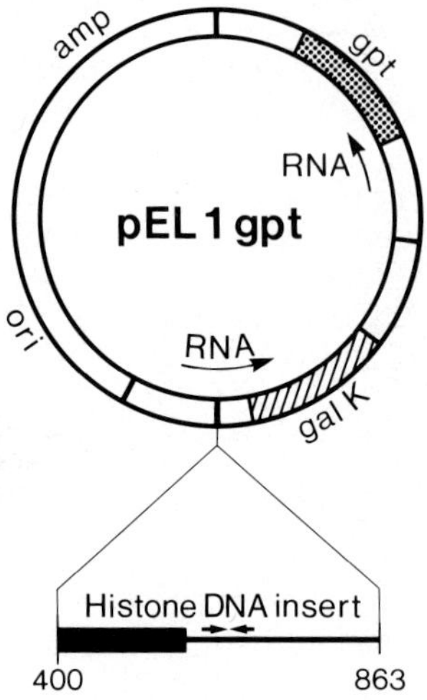

FIGURE 2. Structure of plasmid pEL1gpt. Abbreviations are: galK: coding body of E.coli galactokinase gene, gpt: coding body of E.coli guanine phosphoribosyltransferase gene, amp: ampicillin resistance gene of plasmid pBR322, ori: replication origin of plasmid pBR322. The orientations of the galK and gpt transcription units are indicated by arrows.

For initial studies, we inserted a 463 bp fragment from a cloned mouse H4 histone gene (4) into this vector. This fragment contains the last 166 bp of H4 coding sequence followed by 297 bp of 3' non-coding sequence. The two conserved sequence motifs are located between 208 and 231, and between 240 and 249 bp from the 5' end of this fragment, respectively. This pEL1gpt/463 bp H4 recombinant was introduced into R1610 Chinese hamster cells (11) by the calcium phosphate coprecipitation technique (12). Total cellular RNA was isolated two days later (13) and analysed for histone-specific 3' ends (14,15). A gel autoradiograph from this experiment (Fig.3a) reveals a prominent protected band of 260 nucleotides. This band extends exactly to the 3' boundary of the first conserved motif i.e. the precise location of mouse H4 mRNA 3' ends (4). Interestingly, a minor protected band of 305 nucleotides is also reproducibly observed and probably represents a real RNA species. However, it remains to be seen whether this RNA is relevant to mouse histone mRNA biogenesis. Essentially identical results were also obtained after transfection of the pEL1gpt/463 bp H4 recombinant into HeLa cells (data not shown).

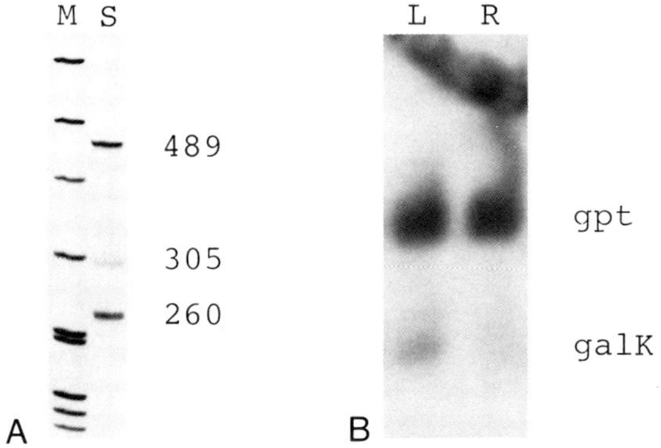

FIGURE 3. Polarity of 463 bp mouse H4 gene fragment. (A) Mapping of RNA 3' termini by S1 nuclease protection. R1610 cells were transfected with the pEL1gpt/463 bp H4 recombinant and total RNA was isolated two days later (13). aproximately 75 μg of RNA were hybridized to 0.03 pmoles of a 3' end labelled 489 bp DNA fragment that spans the termination region. S1 resistant material was analysed on a 8% polyacrylamide/7M urea sequencing gel. M: Hpa II digest of pBR322. S: S1 analysis. Numbers on the right represent the signs of protected fragments in bp. (B) Starch gel analysis of galK and gpt enzyme activities. Protein extracts from 2 day transfected R1610 cells were analyzed on a non-denaturing 14% starch gel. The gel was incubated with a combined reaction mix for galK and gpt enzyme acitivities. The ^{14}C-labelled reaction products were visualized by fluorography. L: Transfection with pEL1gpt, R: Transfection with pEL1gpt/463 bp H4 recombinant. N.B. The R1610 cell line has no endogenous galK or gpt activity.

To measure the effect of the mouse histone DNA fragment on galK expression, we prepared cell extracts from 2 day transfected R1610 cell cultures, subjected them to starch gel electrophoresis and assayed the gel for galK and gpt enzyme activities (Fig.3b). Since R1610 cells lack any endogenous galK or gpt activity (11), this assay reveals only the two plasmid-derived activities, galK being the band which migrates faster. The gpt activity serves as internal standard to reveal any differences in transfection efficiency. As can be seen in Fig.3b, galK expression is drastically reduced but still detectable for the histone DNA recombinant. We find it unlikely that this residual galK expression is due to the introduction of a pseudo-promoter sequence, because sequential deletions of the histone DNA insert lead to an increase rather than to a drop in galK activity (data not shown). Rather, it appears as though enough transcripts traversed the histone DNA insert to ensure galK enzyme production.

These experiments demonstrate that a small segment of histone DNA can be inserted into a different transcription unit and carries sufficient information to generate histone-like 3' ends on transcripts which initiate at a non-histone promoter. However, the polarity conveyed by this fragment is apparently incomplete because expression of the downstream galK gene function is still detectable.

MEASUREMENT OF RATIO BETWEEN TERMINATED AND READTHROUGH TRANSCRIPTS

To define what structures within the H4 DNA insert are responsible for generating histone mRNA-like 3' ends, one must be able to measure precisely the relative numbers of terminated and readthrough RNAs. Given this possibility, one can then alter the histone segment by in vitro mutagenesis and make valid comparisons between the mutated and "wild-type" DNA fragments. To achieve this, we have modified our S1 protection assays such that readthrough RNAs and correctly terminating transcripts are visualized on the same gel and can be related to each other.

When the hybridization probe consists of a DNA fragment from the same vector as the one used to transfect the cells (as in Fig.3a), S1-resistant hybrids between DNA and readthrough RNA are indistinguishable by gel electrophoresis from DNA:DNA hybrids. To circumvent this problem, we have constructed a DNA probe that carries a small deletion in the NH_2-terminal region of the galK coding sequence. Hybrids between this probe and readthrough

RNA will be trimmed by S1 nuclease to the proximal deletion endprint whereas DNA:DNA hybrids will still produce the full-length band. Thus, the bands caused by 3' ends within the histone DNA insert and by readthrough RNA can be quantitated by scanning of gel autoradiographs and compared to each other to obtain a measure of termination efficiency. Preliminary results (E.L. and D.S., unpublished results) show this to be a valid approach and indicate, in full agreement with the galK and gpt enzyme assays, that the termination activity of the 463 bp H4 DNA fragment is only partial. We find approximately 10 to 20% of the S1-protected radioactivity in the readthrough band. Thus it seems that as much as one fifth of the transcripts may escape the machinery that generates histone mRNA-like 3' ends.

Using this approach we have begun to analyse the relative efficiencies of the 463 bp H4 DNA fragment and of smaller fragments obtained by Bal31 deletion (16) in order to identify and delimit the functional elements responsible for 3' end generation.

STABLE CELL LINES CARRYING THE HYBRID SV40:MOUSE H4:galK TRANSCRIPTION UNIT

Transient expression assays after DNA transfection suffer from a number of technical problems or even limitations, perhaps the most important of which is the relatively poor reproducibility of transfection efficiencies. Furthermore, a number of experiments are difficult or even impossible to perform; these include studies on chromatin architecture, DNA replication, gene regulation, kinetics of gene expression and subcellular distribution of gene products. To overcome these problems, we are using the gpt dominant selection marker (10) which is present on pEL1gpt and on all its derivatives, to select cell lines that are stably transformed with the particular plasmid DNA. For instance, R1610 cells transfected with the pEL1gpt/463 bp H4 recombinant produced colonies at a frequency of about 2×10^{-4} when selected in a medium containing xanthine (200μg/ml) and mycophenolic acid (25μg/ml). This demonstrates that the gpt gene can be used to introduce stably a hybrid SV40:mouse H4:galK transcription unit into mammalian tissue culture cells.

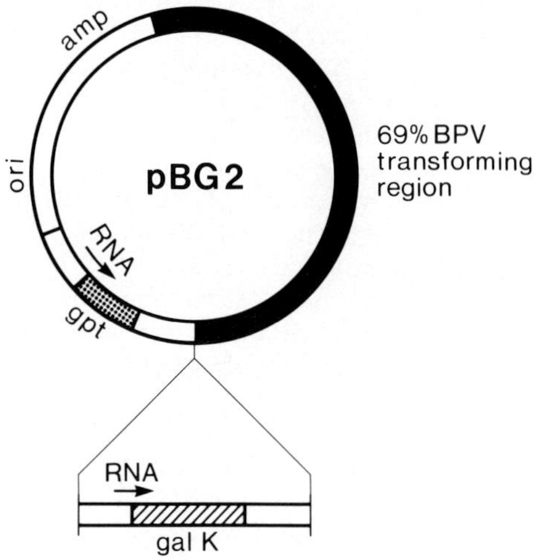

FIGURE 4. Structure of plasmid pBG2. Abbreviations are the same as in Figure 2.

A common problem of such transformation systems is that variable gene copy numbers and variable gene expression, probably reflecting the particular chromosomal environment, render comparisons between different gene variants extremely problematic. A novel vector system which circumvents this problem is based on the fact that Bovine Papilloma virus (BPV) DNA persists in rodent cells as a multicopy extrachromosomal circular DNA molecule (17). A subgenomic fragment of BPV DNA linked to bacterial plasmid DNA has been shown to propagate as a plasmid in both rodent and bacterial cells (18,19). A potential limitation of this system, namely its dependence on the BPV-induced transformed phenotype as the selective marker, can be overcome by the introduction into BPV vectors of the dominant selective marker gpt (20). Such BPV/gpt vector DNA will transform mammalian cells under gpt selection and be maintained in these cells in episomal form. However, this system appears to be susceptible to DNA rearrangements (20).

To test the suitability of BPV/gpt vectors for our purposes, we constructed the plasmid pBG2 (Fig.4). This

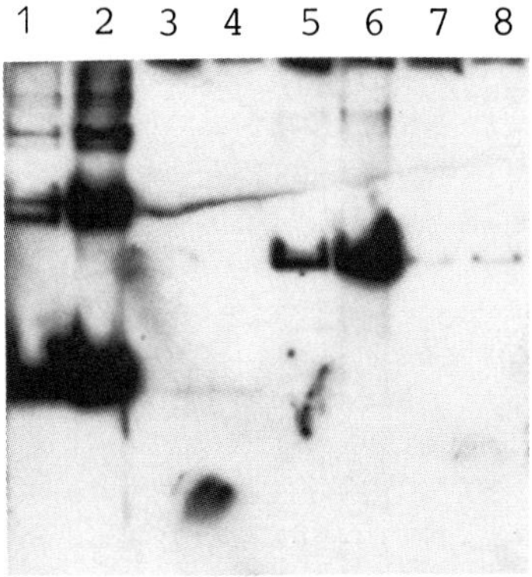

FIGURE 5. Analysis of low molecular weight DNA extracted from pBG2-transformed R1610 hamster cells. The cells were transfected with pBG2 DNA and replated in Dulbecco's modified minimal medium, containing 10% fetal bovine serum, 200 µg/ml xanthine and 25µg/ml mycophenolic acid, 2 days thereafter. Low molecular weight DNA was extracted (23) from two pools of approximately 10 resistant colonies each and analysed on a 0.8% agarose gel. The DNA was transferred to nitrocellulose filters and hybridized with ^{32}P-labelled plasmid DNA (22). Lanes 1 to 4: Uncut DNA; lanes 5 to 8:BamH1-digested DNA; lanes 1 and 5: 0.5ng of pBG2 DNA; lanes 2 and 6: 5 ng of pBG2 DNA; lanes 3 and 7: DNA extracted from pool 1; lanes 4 and 8: DNA extracted from pool 2.

vector contains the following three genetic elements: (i) the origin of replication and the β-lactamase gene of pBR322 (positions 2065 to 4360, ref.21), (ii) the 69% transforming region of BPV DNA (17), and (iii) the gpt transcription unit (10). We transfected R1610 hamster cells with 10µg of pBG2 DNA and 2 days later passaged them into medium containing xanthine and mycophenolic acid as described above. Resistant colonies were obtained at a frequency of 1.6 x 10^{-3}, i.e. about one order of magnitude more frequently than with the pEL1gpt derivative which is lacking the BPV sequences. A Southern blot hybridization (22) of low molecular weight DNA (23) extracted from such cells demonstrates the presence of supercoiled plasmid DNA without apparent sequence rearrangements at 6 or more copies per cell (Fig.5). Transfection of approximately 0.1 ng of this DNA into *E.coli* strain HB101 gave rise to three ampicillin-resistant colonies. Plasmid DNA extracted from these revenant colonies was indistinguishable from the original pBG2 DNA as judged by restriction analyses. These results demonstrate that pBG2 functions as a shuttle vector between bacterial and animal cells.

Passenger DNA can be introduced into pBG2 DNA at a unique BamHI restriction site presumably without disturbing the function of the adjacent BPV and *gpt* elements. We have introduced into this BamH1 site the hybrid SV40:galK transcription units from pEL1gpt and from pEL1gpt/463 bp H4 recombinant, respectively. Preliminary experiments obtained with these multicomponent plasmids indicate that they transform R1610 cells to a gpt^{+} phenotype at a similar frequency as does the original pBG2 plasmid. Using these transformed cells we have begun to study the biochemistry and nucleotide sequence requirements of histone mRNA-specific 3' end generation.

ACKNOWLEDGEMENTS

We thank Stefan Kaiser for technical assistance, Silvia Oberholzer and Fritz Ochsenbein for preparing the manuscript, and Julian Banerji and Jean de Villiers for critical comments. We are especially grateful to Carmen Birchmeier fo her continuous interest and to Professor Max Birnstiel fo his invaluable support and criticism.

REFERENCES

1. Busslinger M, Portmann R, Birnstiel ML (1979). A regulatory sequence near the 3' end of sea urchin histone genes. Nucl Acids Res. 6: 2997.
2. Hentschel CC, Birnstiel ML (1981). The organization and expression of histone gene families. Cell 25: 301.
3. Mauron A, Levy S, Childs G, Kedes L (1981). Monocistronic transcription is the physiological mechanism of sea urchin embryonic histone gene expression. Mol Cell Biol 1: 661.
4. Seiler-Tuyns A, Birnstiel ML (1981). Structure and expression in L-cells of cloned H4 histone gene of the mouse. J Mol Biol 151: 607.

5. Hentschel CC, Irminger JC, Bucher P, Birnstiel ML (1980). Sea urchin histone mRNA termini are located in gene regions downstream of putative regulatory sequences. Nature 285: 147.
6. Pribnow D (1979). The terminator and its function. In Goldberger RF (ed): "Biological Regulation and Development 1" New York: Plenum Press, p.250.
7. Rosenberg M, Court D (1979). Regulatory sequences involved in the promotion and termination of RNA transcription. Ann Rev Genet 13: 319.
8. Birchmeier C, Grosschedl R, Birnstiel ML (1982). Generation of authentic 3' termini of an H2A mRNA in vivo is dependent on a short inverted DNA repeat and on spacer sequences. Cell 28: 739.
9. Schümperli D, Howard BH, Rosenberg M (1982). Efficient expression of E.coli galactakinase gene in mammalian cells. Proc Natl Acad Sci USA 79: 257.
10. Mulligan RC, Berg P (1981). Selection for animal cells that express the Escherichia coli gene coding for xanthine-guanine phosphoribosyltransferase. Proc Natl Acad Sci USA 78: 2072.
11. Thirion JP, Banville D, Noel H (1976). Galactokinase mutants of Chinese hamster somatic cells resistant to 2-deoxygalactose. Genetics 83: 137.
12. Wigler M, Pellicer A, Silverstein S, Axel R, Urlaub G, Chasin L (1979). DNA-mediated transfer of the adenine phosphoribosyltransferase locus into mammalian cells. Proc Natl Acad Sci USA 76: 1373.
13. Scherrer K (1969). Isolation and sucrose gradient analysis of RNA. In Habel K, Salzman NP (eds): "Fundamental Techniques in Virology", New York: Academic Press, p.413.
14. Berk AJ, Sharp P (1978). Spliced early mRNAs of simian virus 40. Proc Natl Acad Sci USA 75: 1274.
15. Weaver RF, Weissmann C (1979). Mapping of RNA by a modification of the Berk-Sharp procedure. Nucl Acids Res 7: 1175.
16. Legerski RJ, Hodnett JL, Gray HB (1978). Extracellular nucleases of pseudomonas BAL 31. III. Use of the double-strand deoxyribonuclease activity as the basis of a convenient method for the mapping of fragments of DNA produced by cleavage with restriction enzymes. Nucl Acids Res 5: 1445.
17. Law MF, Lowy DR, Dvoretzky I, Howley PM (1981). Mouse cells transformed by bovine papillomavirus contain only extrachromosomal viral DNA sequences. Proc Natl Acad Sci USA 78: 2727.

18. DiMayo D, Treisman R, Maniatis T (1982). Bovine papillomavirus vector that propagates as a plasmid in both mouse and bacterial cells. Proc Natl Acad Sci USA 79: 4030.
19. Binetruy B, Meneguzzi G, Breathnach R, Cuzin F (1982). Recombinant DNA molecules comprising bovine papilloma virus type 1 DNA linked to plasmid DNA are maintained in a plasmidial state both in rodent fibroblasts and in bacterial cells. EMBO Journal 1: 621.
20. Law MF, Howard B, Sarver N, Howley PM (1982). Expression of selective traits in mouse cells transformed with a BPV DNA-derived hybrid molecule containing Escherichia coli gpt. In Gluzman Y (ed): "Eukaryotic Viral Vectors," Cold Spring Harbor Laboratory, p. 79.
21. Sutcliffe JG (1979). Complete nucleotide sequence of the Escherichia coli plasmid pBR322. Cold Spring Harbor Symp Quant Biol 43: 77.
22. Southern EM (1975). Detection of specific sequences among DNA fragments separated by gel electrophoresis. J Mol Biol 98: 503.
23. Hirt B (1967). Selective extraction of polyoma DNA from infected mouse cell cultures. J Mol Biol 26: 365.

Gene Expression, pages 371-383
© 1983 Alan R. Liss, Inc., 150 Fifth Avenue, New York, NY 10011

GENETICS OF THE SECRETORY APPARATUS OF E. COLI[1]

Donald Oliver,[2] Carol Kumamoto, Edith Brickman, Susan Ferro-Novick, Jeffrey Garwin,[3] and Jon Beckwith

Department of Microbiology and Molecular Genetics
Harvard Medical School
Boston, MA 02115

ABSTRACT A genetic selection is described for mutants which internalize normally exported proteins in Escherichia coli. In this way, two genes, secA and secB, have been identified, which may code for components of the cell's secretory apparatus. The SecA protein is regulated in response to the secretion needs of the cell. Selection for mutants which suppress a temperature-sensitive secA mutation has resulted in the identification of new genes involved in secretion. Mutations in a previously identified gene, prlA, suppress the temperature-sensitivity of the secA^{ts} mutation. In addition, cold-sensitive mutations in a gene, which we have named secC, have the same effect. SecC cold-sensitive mutants fail to synthesize exported proteins at 23°. The properties of the different mutants are interpreted in terms of a mechanism for protein secretion.

INTRODUCTION

The secretion of proteins in eukaryotic cells and the export of proteins to the cell envelope in gram-negative bacteria share many features in common (1).

[1]This work was supported by a grant from the National Science Foundation, # 79-22624.
[2]Present Address: Department of Microbiology, State University of New York, Stony Brook, NY 11794.
[3]Present Address, Biogen Inc., Cambridge, MA

First, nearly all secreted eukaryotic proteins are synthesized as precursors with signal sequences which are eventually cleaved. The same holds true for all proteins studied so far which are exported to the periplasm or outer membrane of gram-negative bacteria such as Escherichia coli. Secondly, when the gene for rat preproinsulin is cloned into E. coli such that it is transcribed and translated, the bacteria are able to export the protein across the cytoplasmic membrane and correctly process the eukaryotic precursor (2). Conversely, when messenger RNA for the prokaryotic β-lactamase (a periplasmic enzyme) is introduced into an in vitro eukaryotic secretion system, it is incorporated into dog pancreas microsomal vesicles and correctly processed. In this case, the in vitro secretion of the prokaryotic protein is dependent on a specific complex of eukaryotic proteins (SRP) required for secretion (3).

In vitro studies on secretion in the eukaryotic system have revealed a complex machinery required for the process (4,5,6,7). A cytoplasmic particle containing six proteins and a 7S RNA molecule (SRP) appears to recognize signal sequences and transport the polysomes synthesizing them to the membrane of the rough endoplasmic reticulum (RER). There, the SRP interacts with a "docking protein" which is necessary for a productive interaction with the RER. There are also proteins in the membrane, termed ribophorins, which interact with ribosomes.

At the same time, genetic studies in E. coli have suggested the existence of a secretory apparatus in bacteria. Mutants have been detected in a gene termed prlA, which may code for a protein component of this apparatus (8). PrlA mutants restore the secretion of proteins with defective signal sequences. The prlA gene lies at the end of an operon comprised of genes coding for ribosomal proteins. However, the PrlA protein itself does not appear to be a ribosomal component (9). Ito and coworkers using localized mutagenesis, have isolated a conditional-lethal mutant in or close to the prlA gene, which at high termperature causes the accumulation of precursors of normally exported proteins (10). The simplest interpretation of these properties of prlA mutants is that the PrlA protein is part of the secretory apparatus.

In this paper, we will review recent studies in our laboratory on new genes which may code for components of E. coli's secretion machinery. The principle of our approach is to devise a selection for mutants with pleiotropic defects in secretion.

RESULTS AND DISCUSSION

A Genetic Selection for Mutants Defective in the Secretory Machinery

In theory, a selection might be developed for mutants in which a normally exported protein had been internalized. Conceivably, demands could be put on the cell such that it required an enzymatic activity in the cytoplasm which is normally located in the periplasm. However, so far, we have been unable to develop selections of this sort. A general problem in seeking such pleiotropically defective mutants is that they may be lethal to the cell (see below). If this is the case, then any search for such mutants must either involve a screening of conditional lethal mutants or a selection for mutants with slight enough defects in secretion that the cell does not die. Such a selection has fallen into our laps as a result of our studies on gene fusions.

We have constructed gene fusion strains in which the amino-terminal portion of β-galactosidase (β-Gz) is replaced by a large portion of the amino-terminal sequence of the maltose binding protein (MBP) (11). MBP is a periplasmic protein which is synthesized initially with a 26 amino acid signal sequence. The synthesis of MBP is inducible by the addition of maltose to growth media. As a result of the attachment of this portion of MBP to β-Gz, the hybrid protein is associated with the cytoplasmic membrane. In this paper, we will discuss one such fusion strain, MM18, which contains the gene fusion 72-47.

Strain MM18 exhibits two properties which have been important in the genetic studies on secretion. First, when large amounts of the hybrid MBP-β-Gz protein are produced, the cytoplasmic membrane is jammed up to the point where other proteins cannot be exported. In every case in which a specific periplasmic or outer membrane protein has been examined under these conditions, the protein is found in precursor form, its export to its normal location having been blocked (12).

Secondly, the incorporation of the hybrid β-Gz into the cytoplasmic membrane results in aberrant enzymatic properties (13). Normally, the induction ratio for MBP in wild-type cells is 10-20 fold. However, when we examine the induction by maltose of β-Gz in MM18, we obtain a ratio of 600. This high induction ratio is explained by a low uninduced level. It is almost certainly the membrane location of the hybrid protein, not its intrinsic structure, which is responsible for this low activity. When a signal sequence mutation is incorporated into the gene fusion resulting in a cytoplasmic location for the hybrid protein, the uninduced level of β-Gz rises to the expected level. That is, the induction ratio with maltose is about normal.

The result of this low level of β-Gz activity is that MM18 exhibits a Lac⁻ phenotype. However, there is enough β-Gz activity so that bacterial colonies can form on lactose minimal agar after 5 or 6 days. One explanation for the aberrant enzymatic activity of the membrane bound β-Gz is that the enzyme cannot effectively assemble in the membrane. Since β-Gz is a tetramer, composed of identical monomers, it must oligomerize. This may be a particular problem when monomers are present in low concentration in the membrane as in the uninduced MM18 (see Figure 1).

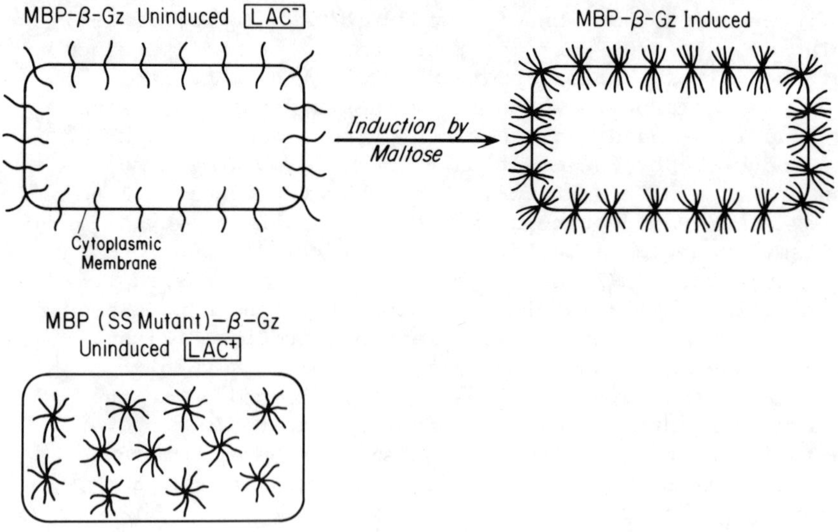

FIGURE 1. Explanation for the Lac⁻ phenotype of MM18. SS mutant refers to a signal sequence mutant.

This Lac$^-$ phenotype of MM18 provides a very sensitive and useful selection for mutants which affect localization of the hybrid protein. Many, if not all, of these will affect export of MBP to the periplasm. That is, selection for Lac$^+$ derivatives of MM18 yields mutants with only slight defects in secretion. As described below, mutants with as little as a 5-10% defect in secretion internalize enough hybrid protein to the cytoplasm to yield β-Gz activity sufficient to increase the growth rate on lactose. We point out that similar phenotypes have been observed for certain gene fusions of β-Gz to the outer membrane protein LamB (14) and the periplasmic protein, alkaline phosphatase (*phoA*, 15). Selections with such a *lamB-lacZ* fusion have not yielded any pleiotropic mutants of the type described below (T. Silhavy, personal communication). Similar studies are in progress with the *phoA-lacZ* fusions.

We proceeded to isolate and characterize Lac$^+$ derivatives of strain MM18. These derivatives fall into two general classes. One class is comprised of mutants in the structural gene for MBP, *malE*, which interfere with the initiation of the export process. Many of these are presumably signal sequence mutants, similar to those characterized earlier (16). We have shown by DNA sequencing that one of these mutations, 4104, does not alter the signal sequence, but rather affects some portion of the mature protein, as yet undetermined.

The second class of mutants is comprised of those due to mutations unlinked to the original gene fusion. It is these mutations which we believe are altering genes coding for components of the cell's secretion machinery. These mutations fall into two genes, which we have termed *secA* and *secB*. Eighty eight percent of these mutations are in (or close to) *secA*, and of these 18% are conditional lethals. Since we did the Lac$^+$ selection at 30°, we were able to detect mutants which had slight defects at 30° and very strong secretion defects at 42°. In constrast, none of the *secB* mutants are conditional lethals.

Properties of SecA and SecB Mutants

We examined the defect in secretion in secA and secB mutants by determining first their effects on secretion of MBP. The secA conditional lethal mutations at 30° and the secB mutations both result in an approximately 5-10% defect in secretion as shown by the accumulation of that percentage of MBP precursor in the cytoplasm (13,17). The $secA^{ts}$ mutant shows almost 100% precursor in a pulse-label after exposure of the strain to higher temperature-either 37° or 42°. Mutations in both loci are pleiotropic. SecA mutations exhibit defects in the export of a number of periplasmic and outer membrane proteins. Certain of the secB mutants accumulate precursor of the outer membrane protein OmpF as well as MBP.

While no conditional lethal mutants were obtained for secB, mutants with a more drastic phenotype have been obtained in the following way. Diploids for the secB region were constructed in which strain MM18 carried an F' factor with a secB mutation. The SecB phenotype is recessive to the $secB^{+}$ wild-type allele present on the chromosome. The Lac^{+} selection was repeated in this strain, yielding mutants in which the chromosomal $secB^{+}$ activity was eliminated. We have found that it is possible to transduce the mutations into an haploid strain on glycerol minimal media at 30°. The haploid secB mutations obtained in this way have the following phenotype. They grow on no other media at 30° or 37° and do not grow on glycerol at 37°. When examined for defects in secretion, these possible secB knock-out mutants accumulate a much greater proportion of OmpF precursor than the original secB mutants. We are in the process of determining whether, in fact, these are secB knock-out mutations.

The secA and secB mutations have been mapped on the E. coli chromosome. The map positions of these genes as well as that of prlA are shown in Figure 2. Each of them corresponds to a newly identified gene in the bacterium.

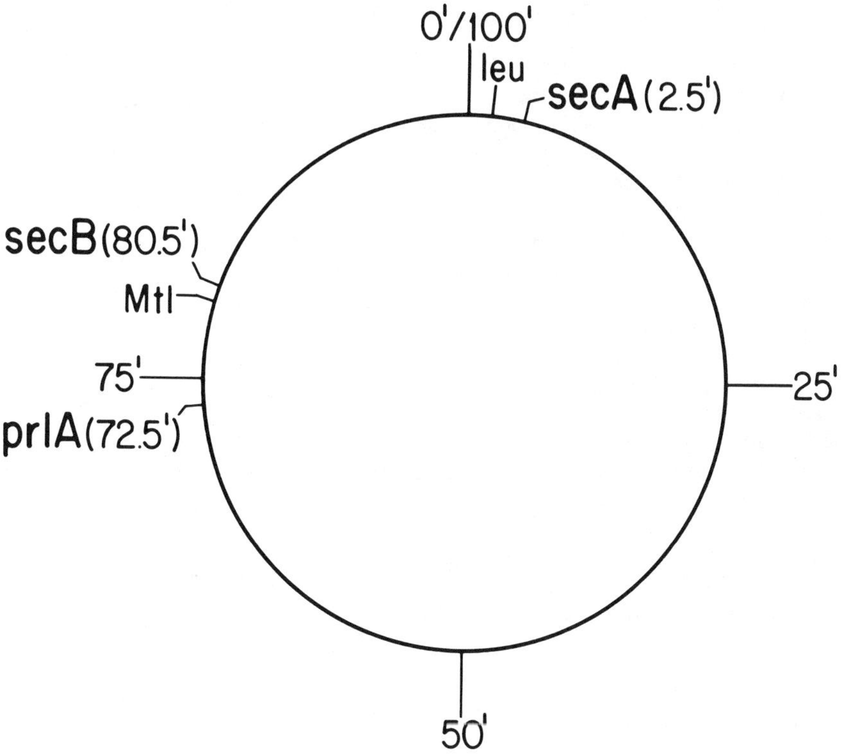

FIGURE 2. Map position of genes involved in secretion. mtl and leu referred to the mannitol catabolism and leucine biosynthesis loci.

We have used the technique first developed in this laboratory (18) to detect the SecA protein. Gene fusions were constructed between secA and lacZ which produced a hybrid protein. The hybrid protein was purified and used to elicit antibody to the SecA antigens present (19). This antibody was used, in turn, to determine the location and other properties of the SecA protein. These results show that secA codes for a 92KD protein which is associated with the cytoplasmic membrane, but can be parially dissociated from the membrane in low salt. Furthermore, we found that the SecA protein is regulated. Under conditions where secretion in E. coli is blocked, there is a derepression in synthesis of SecA protein. It appears that E. coli can regulate its secretion machinery in response to the needs of the cell.

We have used the "blue ghost" technique (20) to isolate an amber mutation in the secA gene. This amber mutation is lethal in the absence of an amber suppressor. By introducing a temperature-sensitive amber suppressor, we can determine the effects on the cell of knocking out the SecA protein. Surprisingly, while the $secA^{ts}$ mutation causes accumulation of precursors of exported proteins, the secA amber at the high temperature results in the absence of synthesis of such proteins. Possible explanations for this finding will be considered in the Discussion.

The antibody to SecA protein has also been used to further establish the role of the gene product. Horiuchi, Caufield, Rhodes, Tai and Davis (in preparation) have defined a set of proteins associated specifically with membrane fractions of Bacillus subtilis which contain polysomes. Evidence suggests that these polysomes are involved in the synthesis of secreted proteins. It was suspected, then, that the proteins associated with this fraction of the membrane were part of the secretory apparatus of this gram-positive bacterium. They have found, in collaboration with us, that antibody to SecA protein of E. coli brings down a protein of similar molecular weight from B. subtilis which is part of this membrane complex. That two such different approaches should have identified proteins with common antigenic determinants reinforced our impression that the SecA protein is part of the cell's secretory apparatus.

Isolation of New Sec Mutations as Suppressors of $SecA^{ts}$

Characterization of a large number of mutants in the Lac^{+} selection with MM18 has yielded mutations in only two genes, secA and secB. From these results, we suspected that other selections would be necessary to detect other genes coding for components of the secretion machinery. An approach originally developed by Jarvik and Botstein (21) offers such a selection. The idea behind this approach is that if two proteins interact in a complex, it may be possible to isolate mutants which cause amino acid alterations in one of these proteins which will suppress the effects of an amino acid change in another. In our case, we considered the possibility that unlinked suppressors of the _secA_^{ts} mutation could be found in genes which code for other proteins involved in secretion.

We have selected revertants of the _secA_^{ts} 51 mutation which will grow at 37°. Many of these are due to mutations in genes unlinked to _secA_. For ease of genetic analysis, we have chosen for further study suppressor mutations which, when separated from the _secA_^{ts} mutation, gave an altered phenotype. Those mutants studied exhibited either slow-growth or a cold-sensitive phenotype.

Transduction with P1 was used to determine whether these suppressor mutations were linked to the previously identified genes, _secB_ and _prlA_. Two of the slow-growing mutants contained lesions near the _prlA_ gene. More careful mapping was done with one of these, which is 99.7% linked to _prlA3_, suggesting that it is an allele of the _prlA_ gene. This mutation unlike previously isolated _prlA_ mutations does not suppress signal sequence mutations.

Two of the cold-sensitive mutations appear to map in a new gene, which we tentatively name _secC_. We have isolated a Tn10 insertion which is linked by P1 transduction to the _secC_ mutations. We are in the process of defining the chromosomal location of these mutations. The properties of _secC_ mutations imply a role for the _secC_ gene product in secretion. A _secC_ mutant was grown at 37°, then switched to 23°, and pulse-labelling of protein was done after one hour. Extracts were then precipitated with antibodies to several secreted proteins and to the cytoplasmic protein, EF-G. In each case, the _synthesis_ of the secreted protein was inhibited at the low temperature. At the same time, the bulk of the _E. coli_ proteins, including EF-G, were still made in normal amounts. These results are similar to those obtained with the _secA_ amber mutant under the nonpermissive conditions.

DISCUSSION

We have identified several genes, mutations in which have pleiotropic effects on secretion in _E. coli_. At least some of these are candidates for genes whose products are part of the cell's secretory machinery.

There are now several reasons for believing that the SecA protein plays a direct role in secretion. The similarity to a _B. subtilis_ protein detected by an entirely different approach and the regulation of SecA protein in response to the secretion needs of the cell add support to this supposition. Further, the finding that mutations in prlA suppress a $secA^{ts}$ mutation indicates that both of these proteins may be part of the same complex. However, care should be taken in over-interpreting the results of such suppressor studies, since suppressors may act by providing an alternate pathway.

The secB and secC genes are less well-defined. Verification of mutations which eliminate secB activity (e.g. nonsense or deletion mutants) will be necessary to interpret our preliminary results. The finding that a cold-sensitive secC mutation and a suppressed secA amber, both, under the non-permissive conditions, fail to synthesize exported proteins suggests another tantalizing link between the eukaryotic and prokaryotic secretion systems. In the eukaryotic system, failure of the SRP complex to interact with the docking protein in the membrane results in a pause early on in the synthesis of secreted proteins. If a similar complex exists in E. coli, the properties of secA and secC mutants would be explained; the products of these two genes may be involved in the SRP-membrane interaction. For instance, they could be proteins comparable to the docking protein and a protein in SRP which makes contact with docking protein. Elimination of either of these two activities would result in the inability of SRP to bind the membrane which would in turn cause an inhibition of synthesis of exported proteins. The $secA^{ts}$ mutation may then be explained as creating an altered SecA protein which permits the SRP-membrane interaction, but does not allow passage of exported proteins through the membrane. Hall and Schwartz (22) have presented evidence that a translation pausing mechanism for secreted proteins does exist in E. coli.

If our interpretation of the secA amber and the secC mutations is correct, these mutants could be used to identify the components of secreted proteins which are involved in the pausing mechanism. Various signal sequence mutants and other structural gene mutations (e.g. 4104, see above) could be tested to see whether they are sensitive to the inhibition of translation caused by the sec mutants. It may also be possible to isolate mutants of a protein such as MBP which are insensitive to the inhibitory effect.

It is worth noting that there was a limitation in our selection for sec mutants with MM18. Since the selection requires that the defect in secretion result in an internalized active β-galactosidase, any mutations which also resulted in a cessation of synthesis of exported proteins would not have been detected. In fact, it would appear from the properties of the secA amber mutation that the $secA^{ts}$ mutations we detected in this selection were a special class. Clearly, mutations which knocked out secC would not have been obtained in the selection with MM18.

The genetic approach to studying the secretion apparatus is essential for a number of reasons. Such studies provide the necessary link between in vitro studies and what is happening inside the cell. By the series of steps described here, mutations can be used to identify genes, which, in turn, permits the identification of gene products. The genetic approach is particularly important in E. coli at this time since no reproducible in vitro system for secretion in bacteria has yet been established.

ACKNOWLEDGMENTS

We thank Theresia Luna for technical assistance and Rosemary Bacco for assistance in the preparation of the manuscript.

REFERENCES

1. Michaelis S, Beckwith J (1982). Mechanism of incorporation of cell envelope proteins in Escherichia coli. Annu Rev Microbiol 36:435.
2. Talmadge K, Kaufman J, Gilbert W (1980). Bacteria mature preproinsulin to proinsulin. Proc Natl Acad Sci USA 77:3988.
3. Miller M, Ibrahimi I, Chang CN, Walter P, Blobel G (1982). A bacterial secretory protein requires signal recognition particle for translocation across mammalian endoplasmic reticulum. J Biol Chem 257: 11860.
4. Kreibich G, Ulrich BL, Sabatini DD (1978). Proteins of rough microsomal membranes related to ribosome binding-I. Identification of ribophorins I and II, membrane proteins characteristic of rough microsomes. J Cell Biol 87:503.

5. Walter P, Blobel G (1981). Translocation of proteins across the endoplasmic reticulum III. Signal recognition protein (SRP) causes signal sequence-dependent and site-specific arrest of chain elongation that is released by microsomal membranes. J Cell Biol 91:557.
6. Walter P, Blobel G (1982). Signal recognition particle contains a 7S RNA essential for protein translocation across the endoplasmic reticulum. Nature 299:691.
7. Meyer DI, Krause E, Dobberstein B (1982). Secretory protein translocation across membranes - The role of the docking protein. Nature 297:647.
8. Emr SD, Hanley-Way S, Silhavy TJ (1981). Suppressor mutations that restore export of a protein with a defective signal sequence. Cell 23:79.
9. Schultz J, Silhavy TJ, Berman ML, Fiil N, Emr SD (1982). A previously unidentified gene in the spc operon of Escherichia coli K12 specifies a component of the protein export machinery. Cell 31:227.
10. Ito K, Wittekind M, Nomura M, Shiba K, Yura T, Miura A, Nashimoto H (1983). A temperature-sensitive mutant of E. coli exhibiting slow processing of exported proteins. Cell 32:789.
11. Bassford PJ, Silhavy TJ, Beckwith JR (1979). Use of gene fusion to study secretion of maltose-binding protein into Escherichia coli periplasm. J Bacteriol 139:19.
12. Ito K, Bassford PJ, Beckwith J (1981). Protein localization in E. coli: is there a common step in the secretion of periplasmic and outer-membrane proteins? Cell 24:707.
13. Oliver DB, Beckwith J (1981). E. coli mutant pleiotropically defective in the export of secreted proteins. Cell 25:765.
14. Hall MN, Schwartz M, Silhavy TJ (1982). Sequence information within the lamB gene is required for proper routing of the bacteriophage λ receptor protein to the outer membrane of Escherichia coli. J Mol Biol 155:93.
15. Michaelis S, Guarente L, Beckwith J (1983). In vitro construction and characterization of phoA-lacZ gene fusions in Escherichia coli. J. Bacteriol 154:356.

16. Bedouelle H, Bassford PJ, Fowler AV, Zabin I, Beckwith J, Hofnung M (1980). Mutations which alter the

function of the signal sequence of the maltose binding protein of Escherichia coli. Nature 285:78.
17. Kumamoto CA, Beckwith J (1983). Mutations in a new gene, secB, cause defective protein localization in Escherichia coli. J Bacteriol 154:253.
18. Shuman HA, Silhavy TJ, Beckwith JR (1980). Labeling of proteins with β-galactosidase by gene fusion. J Biol Chem 255:168.
19. Oliver DB, Beckwith J (1982). Regulation of a membrane component required for protein secretion in Escherichia coli. Cell 30:311.
20. Brown S, Brickman E, Beckwith J (1981). Blue ghosts: a new method of isolating amber mutants defective in essential genes in Escherichia coli. J Bacteriol 146:422.
21. Jarvik T, Botstein D (1975). Conditional-lethal mutations that suppress genetic defects in morphogenesis by altering structural proteins. Proc Natl Acad Sci USA 72:2738.
22. Hall MN, Schwartz M (1982). Reconsidering the early steps of protein secretion. Ann Microbiol (Paris) 133:123.

VI. GENE STRUCTURE AND FUNCTION

Gene Expression, pages 387-401
© 1983 Alan R. Liss, Inc., 150 Fifth Avenue, New York, NY 10011

STRUCTURE AND REGULATORY FEATURES OF A COMPLEX OPERON ENCODING E. COLI RIBOSOMAL PROTEIN S21, DNA PRIMASE, AND THE RNA POLYMERASE SIGMA SUBUNIT[1]

Richard R. Burgess, Zachary F. Burton, Carol A. Gross*, Wayne E. Taylor, and Michael Gribskov

McArdle Laboratory for Cancer Research and the Department of Bacteriology*, University of Wisconsin, Madison, WI 53706

ABSTRACT We summarize the structure and regulatory features of the E. coli sigma operon. Based on DNA sequencing, S1 mapping and cloning of regulatory elements, we have concluded that the genes for ribosomal protein S21 (rpsU), DNA primase (dnaG) and RNA polymerase sigma subunit (rpoD) are cotranscribed and thus in the same operon. Regulatory features include tandem operon promoters (P1 and P2), terminators (t1 and t2), an RNA processing site between dnaG and rpoD, and a promoter (Px) located very near P1 but oriented in the opposite direction. Transcripts initiated at Px contain an open reading frame (orfX) that encodes an unidentified protein. In addition three minor promoters (Pa, Pb, and Phs) have been identified within dnaG. Phs is activated upon temperature upshift to allow the heat shock synthesis of sigma. The sequences of the six promoters in the operon are compared. Codon usage preference and ribosome binding sites of the genes are compared and related to the differential expression of dnaG and rpoD. We discuss how regulatory features can explain the observed differential expression and discoordinate regulation of operon genes.

[1]This work was supported by NIH grants CA-07175, CA-23076, and GM 28575.

INTRODUCTION

The sigma operon of E. coli is the only operon so far reported that encodes proteins essential for translation, replication, and transcription. Since the protein products of the three operon genes, S21, DNA primase, and sigma, are all involved in the initiation of their respective biosynthetic process, regulation of the sigma operon may be important for balanced cell growth (1,2,3). The structure of the sigma operon has been determined using a number of techniques. The entire operon has been sequenced (1,2,4,5). S1 nuclease mapping has shown that rpsU, dnaG, and rpoD are cotranscribed (1). Nuclease mapping studies have also been used to identify regulatory features for the operon and to show that these features operate in vivo (1). Promoter and terminator cloning studies have been used to establish the biological activity of these operon features (1,2,6).

In this paper we summarize the structure and regulatory features of the sigma operon and discuss how these features contribute to the various levels of operon complexity observed. We also present sequence comparisons of the ribosome binding sites and promoters found in the region of the operon, analyses of the AT content, and codon usage of the sequenced region. Finally, we summarize our recent discovery of a minor promoter within dnaG that is transiently activated by temperature upshift, resulting in the heat shock synthesis of sigma.

RESULTS AND DISCUSSION

Operon Complexity.

We find it useful to consider three levels of operon complexity, illustrated in Figure 1. At the simplest level the genes in an operon are coordinately regulated and equally expressed. The promoter region before the first gene may be complex, with sites for positive or negative control, perhaps containing multiple promoters or an attenuator. However, if all the genes in the operon are induced or repressed together as the cell goes from growth condition A to condition B, they are coordinately regulated. At the next level of complexity, operon genes are differentially expressed, but since they are still coordinately regulated, their relative levels remain unchanged with changing conditions. Finally an operon may be capable of discoordinate regulation, resulting in a change in the relative expression of the genes as growth conditions change. Discoordinate regulation can occur if regulable sites of transcription termination or initiation are located within an operon, or by regulable changes in the stability or translational efficiency of various operon mRNAs.

As we will discuss, in the sigma operon three functionally diverse proteins are grouped together, presumably to allow their coordinate regulation under some conditions. However the operon contains a variety of regulatory features that help explain its differential expression and at which discoordinate regulation can be effected.

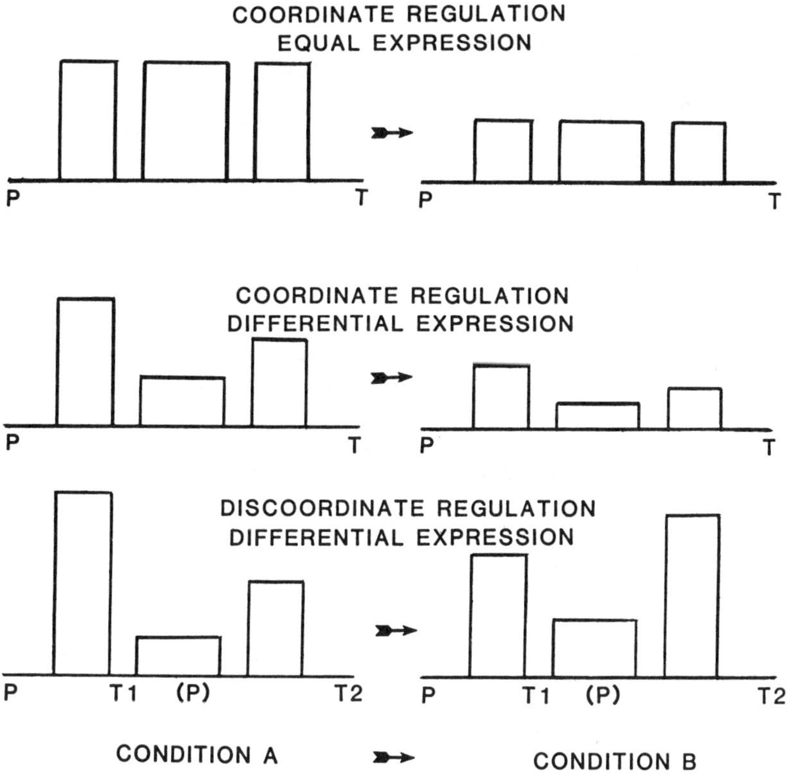

FIGURE 1. Schematic to Illustrate Levels of Operon Complexity. Each operon contains three genes, whose level of expression is represented by the height of the open bar, a promoter, P, one or more terminators, T, and in some cases a minor promoter, (P).

Sigma Operon Structure and Regulatory Features

We have recently determined the structure of the sigma operon using a combination of DNA sequencing of over 6000 base pairs, S1 nuclease mapping of the start and stop points of *in vivo* RNA synthesized from this region, and promoter and termination subcloning (1,6). Our sequence analysis shows that the genes for the ribosomal protein S21 (*rps*U), DNA primase (*dna*G) and the RNA polymerase sigma subunit (*rpo*D) are contiguous and transcribed in the same direction, clockwise with respect to the *E. coli* genetic map. Low resolution S1 mapping of *in vivo* transcripts identified a 4.3 kb transcript that encodes S21, primase, and sigma. Thus these three genes form an operon. In addition we observed a 0.43 kb transcript coding for S21 and a 2.1 kb transcript coding for sigma. High resolution S1 mapping allowed us to locate precisely 5' and 3' ends of RNA from this region and to identify some of the regulatory features of the operon. These features include tandem promoters at the beginning of the operon (P1 and P2), terminators (t1 and t2), an RNA processing site between *dna*G and *rpo*D, and a promoter (Px) located very near P1 but oriented in the opposite direction. In addition three minor promoters, Pa, Pb, and Phs, are located within *dna*G. One of these, Phs, is transiently activated when cells are shifted to higher temperature. This information is summarized schematically in Figure 2.

A long open reading frame (ORF) has been located next to the sigma operon (Z. Burton, unpublished results). Since the identity of this putative gene is unknown, we will refer to it as *orf*X. *Orf*X codes for a 337 amino acid protein of molecular weight 36,008. The RNA that contains *orf*X is synthesized in vivo (1). We speculate that *orf*X may be involved in the regulation of the sigma operon. This idea is based on the following considerations. First Px overlaps P1, suggesting that these two transcription units do not function independently. Second, a number of operons are regulated by a protein transcribed from a nearby divergent transcription unit (e.g., maltose, biotin, arabinose). Such a regulatory role for the *orf*X product in vivo has yet to be determined.

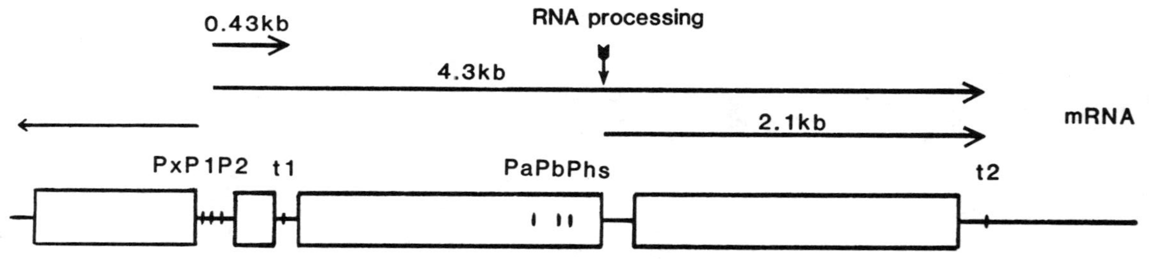

orfX	rpsU	dnaG	rpoD	GENE
?	S21	DNA primase	Sigma	PROTEIN
(36008)	(8369)	(65563)	(70263)	(MW)
?	translation	replication	transcription	FUNCTION
?	50000	50	3000	PER CELL

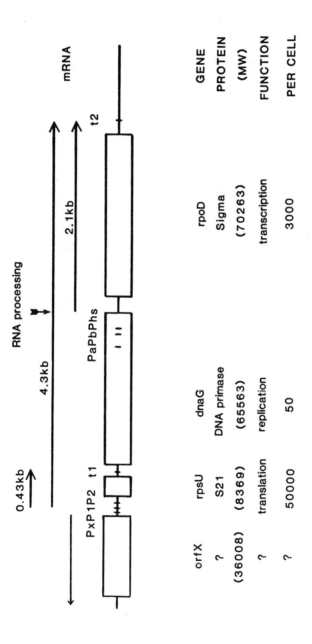

FIGURE 2. Summary of Sigma Operon. Indicated are sizes of transcripts, gene organization, regulatory features as described in text, and information about the genes.

Promoters. The locations of six promoters found in the sigma operon are indicated in Figure 2. P1 and P2 are the major operon promoters. Px is the divergent promoter, overlapping P1 and responsible for the transcription of orfX. Only one other region of the operon contains sequences that can provide even weak promoter function when cloned in front of a promoterless galactokinase gene in the promoter cloning vector pKO1 (7). This 250 bp region within dnaG contains three minor promoters, which we call Pa, Pb, and Phs (W. Taylor, unpublished results). Phs is responsible for the heat shock synthesis of sigma described below.

TABLE 1

SIGMA OPERON PROMOTER SEQUENCES

Promoter	Sequence	-35 region	spacer	-10 region
P1	GTAAAACTTTGTTCGCCCCTGGAGAAAGCCTCGTGTATACTCCTCACCCT	x x x	17	x
P2	GGCCGCGGTGCTTTACAAAGCAGCAGCAATTGCAGTAAAATTCCGCACCA	x	18	x
Px	GGCGAACAAAGTTTTACATCAACCCGCATTGGTCCTACACTGCGCGGTAA	x	17	x x
"CONSENSUS"		TTGACA	17	TATAAT
(Pa)	CGCCCTGTTCCGCAGCTAAAACGCACGACCATGCGTATACTTATAGGGTT	xx xx	17	x
(Pb)	AGCCAGGTCTGACCACCGGGCAACTTTTAGAGCACTATCGTGGTACAAAT	xxx x	18	xx
(Phs)	ATGCTGCCACCCTTGAAAAACTGTCGATGTGGGACGATATAGCAGATAAG	x	17	x xx

The DNA sequence of these six promoters are compared in Table 1. None of these promoters shows a perfect match with the consensus promoter sequence (8,9) either in the -35 region or in the -10 region although all have spacings between these two regions of 17 or 18 base pairs. Pa and Pb have particularly poor -35 regions while Phs differs from the consensus promoter at the -10 region in two of the most highly conserved bases. The two clusters of promoters fall in the two most AT rich regions of the sequenced DNA (Figure 3), consistent with previous observations of promoters being located in AT rich regions (10).

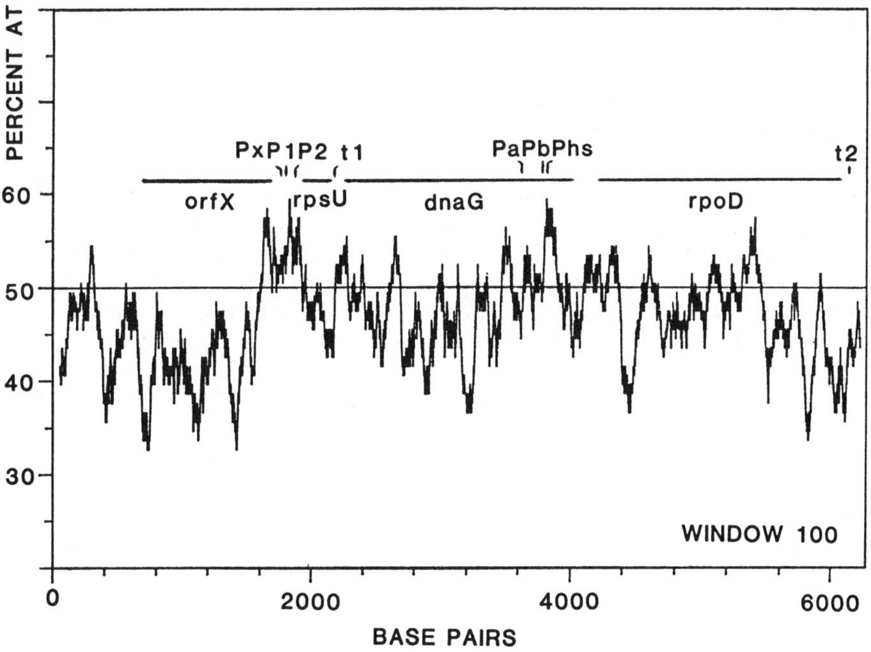

FIGURE 3. AT-Rich Regions of the Sigma Operon. The percent AT of the DNA averaged over 100 base pairs is plotted versus position in the sequenced region. The positions of genes and regulatory features are indicated. The AT content for the whole region is 46.3%. OrfX, rpsU, dnaG, and rpoD coding regions have AT contents of 42.3, 48.1, 47.7, and 47.1 percent, respectively. The Px, P1, P2 region is 53.6% AT while the Pa, Pb, Phs region is 51.1% AT.

Terminators. Terminators t1, located between rpsU and dnaG, and t2, located 60 to 80 base pairs after rpoD, both show typical rho-independent terminator characteristics (8). They both allow formation of an RNA secondary structure with a stable stem and loop immediately followed by a U-rich run in which in vivo termination occurs.

Other Regulatory Features. We have identified an RNA processing site just downstream from dnaG (1). We have proposed that endonucleolytic cleavage at this position is associated with the rapid degradation of dnaG mRNA by 3' exonucleases in a process analogous to retroregulation of the int gene of bacteriophage lambda (11). We believe that degradation of processed dnaG mRNA is one mechanism for limiting primase expression.

Smiley et al. (5) and Lupski et al. (2) have also sequenced much of the sigma operon and have identified by sequence comparison two additional possible regulatory sites. The first is a site similar to the phage lambda nut site involved in antitermination. This site consists of a 5 base pair stem with a GAAAA in the loop and is located within rpsU. The second is a sequence very much like the lexA protein binding site which would partially overlap the P2 promoter. LexA is a regulatory protein that binds near to the promoters of a family of genes and represses their synthesis (12). In addition, it has been pointed out to us (David Mount, personal communication) that three more sequences similar to lexA binding site sequences (CTG N_{10} CAG) are present in the sigma operon. One is 350 base pairs upstream of Pa, one overlaps Pa, and one is 100 base pairs downstream of Phs. It is not known if lexA is actually able to bind to these sites or to play any regulatory role in the operon.

Sigma Operon Regulation.

Coordinate Regulation. Since the three genes rpsU, dnaG, and rpoD are cotranscribed, they are by definition in an operon and are subject to coordinate regulation. Each of the three proteins plays a role in the initiation of its respective process (1,2,3): translation, replication, and transcription. This suggests a rationale for why these three diverse proteins would be grouped in one operon. The regulation of this operon may be of central importance in regulating the growth of the cell. Such coordinate regulation could be exerted by activating or repressing one or both of the tandem operon promoters, P1 and P2, and possibly by activating or repressing the divergent promoter, Px.

Differential Expression. The three proteins are present in dramatically different amounts in the cell, in the ratio of S21:primase:sigma of 1000:1:60. This is an extreme example of differential expression. A likely explanation for the higher expression of rpsU relative to rpoD is the presence of the terminator, tl. If only one RNA in 16 was able to transcribe through tl this differential expression could be explained. But why is the expression of dnaG so much lower than rpoD which follows it in the operon? There seem to be several factors that may effect low expression of dnaG. The first is mRNA stability. If most transcripts are processed at the major RNA processing site just after dnaG and then the dnaG mRNA rapidly degraded, the level of dnaG mRNA would be much lower than that of rpoD mRNA. The second factor may be translational efficiency. Table 2 shows the ribosome binding site sequences of rpsU, dnaG, rpoD and orfX. While rpsU and orfX appear to show considerable similarity to the "consensus" ribosome binding site (13), and rpoD appears to show moderate similarity, dnaG appears to show a poor match with the consensus in either of the two alignments shown. In the upper alignment the match with the consensus sequence is better but the AUG is too close. In the lower alignment, the distance to the AUG is normal but the sequence match is very poor.

TABLE 2
COMPARISON OF RIBOSOME BINDING SITE SEQUENCES

"CONSENSUS"	UAAGGAGGUGAUCNNNAUG
rpsU	UUAAUCAAAGGUGAGAGGCACAUGCCGGUA
dnaG	GCUAAAAAUCGGGGCCUAUGGCUGGA
	or GCUAAAAAUCGGGGCCUAUGGCUGGA
rpoD	UAAGUGUGUGGAUUACCGUCUUAUGGAGCAA
orfX	GUAAUAAAGCGAGGUAAAACAAGUCAUGCGUGUA

Finally it is possible that primase levels are low because of slow translation due to the abundance of rarely used codons in its coding sequence (1,14). E. coli uses synonomous codons (those specifying the same amino acid) in a distinctly non-random fashion. In general, strongly expressed genes use primarily those codons corresponding to the most abundant tRNA, whereas moderately or weakly expressed genes use synonomous codons more randomly, in rough proportion to the corresponding tRNA abundance (15). In Figure 4 we plot the codon preference of four different reading frames across the sequenced region. The codon preference is a measure of how much more frequently a particular codon is used in the RNA of a highly expressed E. coli protein than it would be used in a random RNA sequence of the same base composition. When codon preference is averaged over 50 codons and plotted, noncoding regions give lines fluctuating around zero (dotted line) and reading frames encoding proteins give lines above the zero line. The distance above the zero line indicates the degree of similarity to a strongly expressed gene's codon usage: weakly expressed genes (e.g. dnaG) are only slightly above background and strongly expressed genes (rpsU and rpoD) are much higher. OrfX appears to be intermediate. We routinely use codon preference plots such as shown in Figure 4 to locate likely genes in sequenced DNA, to estimate levels of expression, and to correct DNA sequencing errors (a shift into a parallel open reading frame due to a single base insertion or deletion is easily detected).

Discoordinate Regulation. A perplexing feature of the sigma operon is that, although the three genes are co-transcribed, there are several conditions under which they are discoordinately regulated. These conditions include stringent response induced by amino acid starvation, phage lambda infection, treatment of the cells with the antibiotic rifampicin, and temperature upshift (heat shock). Below we speculate on how the regulatory features identified in the sigma operon might function to allow this discoordinate regulation.

During the stringent response the level of expression of ribosomal proteins decreases several-fold while the level of sigma does not. Primase levels have not been measured under these conditions. Since stringent control of rRNA synthesis has recently been shown to be exerted at the level of initiation at one of the two tandem promoters (16), it

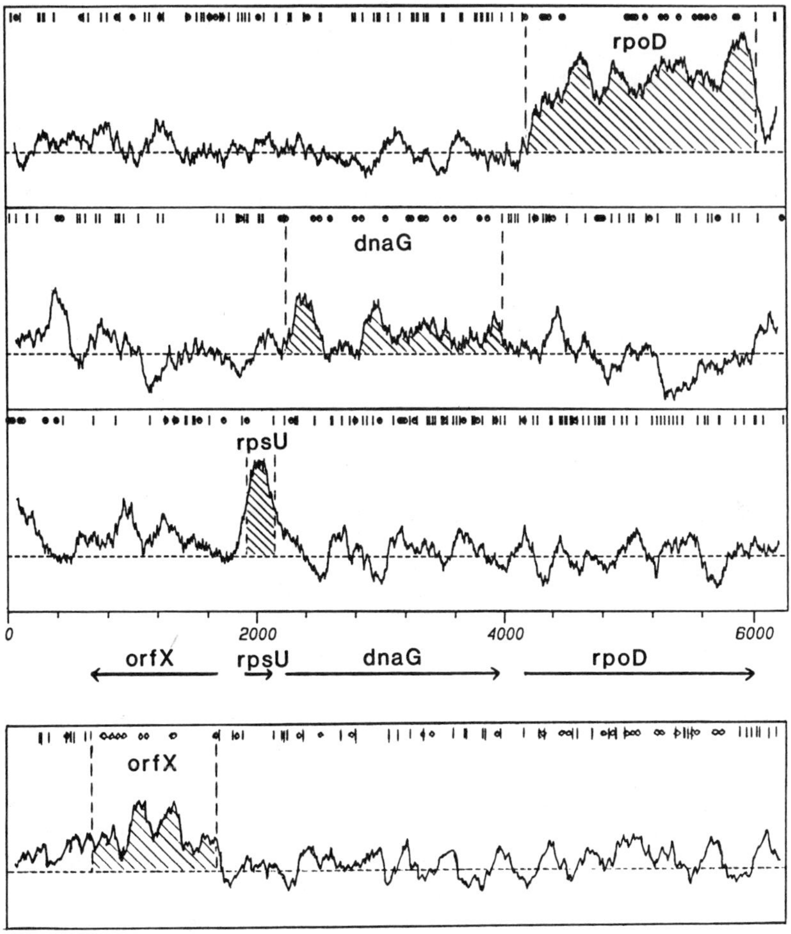

FIGURE 4. Codon Preference Plot of the Sigma Operon. The top three panels show the codon preference index of the three rightward reading frames plotted versus position within the sequenced DNA. The bottom panel is the leftward reading frame containing orfX. The gene name is given above the shaded region indicating the coding region for that gene. At the top of each panel initiation codons (O) and termination codons (|) are indicated for that reading frame. More details of this plot will be presented elsewhere (M. Gribskov and R. Burgess, unpublished results).

seems possible that during the stringent response, initiation of transcription at one of the major operon promoters might be similarly inhibited. One mechanism capable of maintaining sigma at normal levels would be to have a compensating increase in read-through efficiency at terminator tl so that rpoD mRNA levels would remain the same.

It has been reported that sigma is overproduced several-fold after phage lambda infection, apparently due to the action of the antitermination factor N (17). S21 and DNA primase were not measured. Lupski et al. (2) have proposed that the site they identified in the rpsU gene, similar to the nut site required for N-action in antitermination of lambda transcription, plays a role similar to lambda nut in causing the sigma overproduction after lambda infection.

The synthesis of all four of the RNA polymerase subunits has been shown to be transiently stimulated soon after addition of the antibiotic rifampicin to growing E. coli cells (18). Recently, it has been observed that rifampicin can stimulate transcription in vitro past a terminator just preceding the genes for the β and β' subunits of RNA polymerase (19). Rifampicin could similarly stimulate sigma synthesis by increasing transcription past tl. Alternatively, rifampicin could activate one of the three minor promoters just preceding rpoD. This mechanism would be analogous to the activation of Phs after temperature upshift discussed below.

Not much is known about the regulation of primase since there is so little in the cell. If its level is low primarily because its mRNA is unstable after RNA processing between dnaG and rpoD, one can imagine that the level of primase could be discoordinately regulated by decreasing the fraction of transcripts that are processed.

Activation of the Minor Promoter Phs Causes Heat Shock Synthesis of Sigma. While the possible modes of discoordinate regulation discussed above are speculative and remain to be tested, we have recently obtained evidence that the heat shock synthesis of sigma is due to the transient activation of promoter Phs after shifting cells to a higher temperature (Wayne Taylor and David Straus, unpublished results) Sigma is one of several proteins in E. coli which respond to a temperature upshift from 30° to 42° by a rapid 5 to 20 fold increase in rate of synthesis, peaking around 5 minutes after the upshift, and then rapidly returning to a new steady state rate of synthesis (often several fold higher than at 30°) by 10-15 minutes (6,20).

Although the details of these studies will be presented elsewhere, the conclusions can be summarized as follows. When an 89 bp long DNA fragment containing the promoter Phs is placed in front of a promoterless galactokinase gene in the promoter cloning vector pKO1 (7), it provides weak promoter function. When cells containing this plasmid are shifted from 30° to 42°, galactokinase synthesis is stimulated as in a typical heat shock response. S1 mapping experiments reveal that a major new RNA 5' end, appropriate for a start at the Phs promoter, appears soon after temperature upshift and with the same kinetics expected if the heat shock synthesis of protein were regulated at the level of transcription from this promoter. The transient appearance of this start can be seen both in cells with a plasmid containing Phs, and in cells lacking the plasmid (but at a reduced level consistent with the single copy of the sigma Phs in the genomic DNA). Bal 31 deletion analysis of the Phs promoter region indicates that sequence information just upstream from the -35 region of Phs is necessary for the heat shock response.

The mechanisms by which Phs is rapidly turned on and off following heat shock are not yet known. The fact that the Phs sequence is such a poor match to the "consensus" bacterial promoter sequence and the implication of upstream sequence information in its regulation suggests that additional factors may be needed both for the activation of Phs and for its subsequent repression.

These results have important implications for gene regulation. They illustrate that minor promoters, even promoters located within gene coding regions, may be activated to effect discoordinate regulation of genes that are normally coregulated.

The sigma operon demonstrates a balance between the apparent need for coordinately regulating key components of the cellular synthetic machinery and the occasional need for discoordinate regulation of these genes. In doing so it has developed a richness of regulatory possibilities and has provided us with an excellent system for studying regulatory mechanisms.

ACKNOWLEDGMENTS

We appreciate the excellent technical assistance of Kathy Watanabe in the DNA sequencing and S1 mapping and thank Kristi Traeder for her skillful typing of this manuscript. We thank Jim Lupski, Bob Smiley, Nigel Godson, and Roger McMacken for communicating information to us before it was published. We also acknowledge the help of John Devereux and Paul Haeberli of the University of Wisconsin Genetics Computer Group for assembling a collection of software for manipulating DNA, RNA, and protein information. This facility has been a major aid in the sigma operon sequencing project and its analysis, especially in generating the data presented in figures 3 and 4. This software operates on a VAX computer and is available on request from the University of Wisconsin Genetics Computer Group, c/o John Devereux, Laboratory of Genetics, University of Wisconsin, Madison, WI 53706.

REFERENCES

1. Burton ZF, Gross CA, Watanabe KK, Burgess RR (1983). The operon that encodes the sigma subunit of RNA polymerase also encodes ribosomal protein S21 and DNA primase in E. coli K12. Cell 32:335.
2. Lupski JR, Smiley BL, Godson GN (1983). Regulation of the rpsU-dnaG-rpoD macromolecular synthesis operon and the initiation of DNA replication in E. coli K-12. Mol Gen Genet 189:48.
3. Wold MS, McMacken R (1982). Regulation of expression of the E. coli dnaG gene and amplification of the dnaG primase. Proc Nat Acad Sci USA 79:4907.
4. Burton Z, Burgess RR, Lin J, Moore D, Holder S, Gross CA (1981). The nucleotide sequence of the cloned rpoD gene for the RNA polymerase sigma subunit from E. coli K12. Nucl Acids Res 9:2889.
5. Smiley BL, Lupski JR, Svec PS, McMacken R, Godson GN (1982). Sequences of the E. coli dnaG primase gene and regulation of its expression. Proc Natl Acad Sci USA 79:4550.
6. Gross CA, Burton Z, Gribskov M, Grossman A, Liebke H., Taylor W, Walter W, Burgess RR (1982). Genetic, functional, and structural analysis of sigma subunit: a heat shock protein. In Rodriguez R, Chamberlin M (eds): "Promoters: Structure and Function," New York: Praeger, p 252.

7. McKenney K, Shimatake H, Court D, Schmeissner U, Rosenberg M (1981). A system to study promoter and terminator signals recognized by E. coli RNA polymerase. In Chirikjian J, Papas T (eds): "Gene Amplification and Analysis 2," Amsterdam: Elsevier/North Holland Press, p 383.
8. Rosenberg M, Court D (1979). Regulatory sequences involved in the promotion and termination of RNA transcription. Ann Rev Biochem 13:319.
9. Hawley DK, McClure WR (1983). Compilation and analysis of E. coli promoter DNA sequences. Nucl Acids Res (in press).
10. Vollenweider HJ, Fiandt M, Szybalski W (1979). A relationship between DNA helix stability and recognition sites for RNA polymerase. Science 205:508.
11. Gottesman M, Oppenheim A, Court D (1982). Retroregulation: control of gene expression from sites distal to the gene. Cell 29:727.
12. Little JW, Mount DW (1982). The SOS regulatory system of E. coli. Cell 29:11.
13. Stormo GD, Schneider TD, Gold LM (1982). Characterization of translational initiation sites in E. coli. Nucl Acids Res 10:2971.
14. Konigsberg W, Godson GN (1983). Evidence for use of rare codons in the dnaG gene and other regulatory genes of E. coli. Proc Natl Acad Sci USA 80:687.
15. Ikemura T (1982). Correlation between the abundance of yeast transfer RNAs and the occurrence of the respective codons in protein genes. J Mol Biol 158:573.
16. Glaser G, Sarmientos P, Cashel M (1983). Functional interrelationship between two tandem E. coli ribosomal RNA promoters. Nature 302:74.
17. Nakamura Y, Yura T (1976). Induction of sigma factor synthesis in E. coli by the λ N gene product of bacteriophage lambda. Proc Natl Acad Sci USA 73:4405.
18. Hayward RS, Fyfe SK (1978). Non-coordinate expression of the neighboring genes rplL and rpoB,C of E. coli. Mol Gen Genet 160:77.
19. Fukuda R, Nagasawa-Fujimori H (1983). Mechanism of the rifampicin induction of RNA polymerase β and β' subunit synthesis in E. coli. J Biol Chem 258:2720.
20. Gross CA, Grossman AD, Liebke H, Walter W, Burgess RR (1983). Effects of the mutant sigma allele rpoD800 on the synthesis of specific macromolecular components of the E. coli K12 cell. J Mol Biol (in press).

Gene Expression, pages 403-412
© 1983 Alan R. Liss, Inc., 150 Fifth Avenue, New York, NY 10011

ORGANIZATION AND EXPRESSION OF BACTERIOPHAGE T7 DNA

F. William Studier and John J. Dunn

Biology Department, Brookhaven National Laboratory
Upton, New York 11973

Bacteriophage T7 is a virulent phage that infects *E. coli*. The seventh of the set of T phages focussed on by the Cold Spring Harbor group of phage workers in the 1940's, T7 has been the object of considerable genetic and biochemical investigation. The T7 virion contains a single molecule of linear, double-stranded DNA, and the nucleotide sequence of the entire 39,936 base pairs has now been determined (1). All of the known T7 genes and genetic signals have been located within the nucleotide sequence, and several previously unsuspected genes have been revealed. Our current understanding of the genetic organization of T7 DNA and the overall pattern of gene expression during T7 infection has been discussed in detail in relation to the nucleotide sequence (1). The present article is a condensation of a more general review of the organization and expression of T7 DNA (2), and is meant to give an overview and introduction to the system. More detailed discussion, experimental details, and references to the pertinent literature will be found in references 1 and 2.

ARRANGEMENT OF T7 GENES

T7 is a good example of the rule that viruses carry genetic information very efficiently, presumably because they pack the maximum amount of useful information into a DNA whose size is limited by a virion of fixed size. T7 appears to specify 50 genes whose coding sequences are closely packed but generally not overlapping in the DNA. In eight instances the termination codon for one coding sequence overlaps the initiation codon for the next, and in more than half the cases, coding sequences of adjacent genes are separated by less than 20 base pairs. All but two of the noncoding gaps are smaller than 100 base pairs, and the two that are larger contain origins of replication. All but three of the gaps of 25 base pairs or larger contain one or more known promoters, transcription termination sites, RNase III cleavage sites, or origins of replication.

The positions of the 50 closely packed coding sequences are drawn to scale in Figure 1. On the scale of this figure, only the two largest noncoding gaps between coding sequences are apparent. The longest sequences that do not code for any protein are found at the ends of the DNA. If these are excluded, 95% of the internal sequence codes for proteins.

The efficient packing of coding sequences in T7 DNA argues strongly that all 50 of the closely packed genes are expressed, although genetic or biochemical evidence for the expression of only 38 of them is available so far. A search of the nucleotide sequence has also identified five potential genes whose coding sequences would overlap one or more of the 50 closely packed genes in a different reading frame. Mutations that appear to lie in one of these overlapping genes have recently been isolated, so it seems possible that T7 may express as many as 55 different genes.

The T7 genes fall naturally into three groups, according to their functions, and each group is expressed coordinately during infection. The early, or class I genes are transcribed by the host RNA polymerase, and include functions to overcome host restriction and to convert the metabolism of the host cell to the production of T7 proteins. The class II genes are the next to be expressed, and include functions involved in DNA metabolism. The class III genes are the last to be expressed, and include genes for proteins of the phage particle and functions involved in maturation and packaging of the DNA. In Figure 1, the early genes are located to the left of TE, the class II genes between TE and ø6.5, and the class III genes to the right of ø6.5.

TRANSCRIPTION OF T7 DNA

During infection, T7 DNA is transcribed entirely from left to right, first by the host RNA polymerase and then by newly made T7 RNA polymerase. All of the major promoters and trancription termination sites for both the host and phage RNA polymerases have been identified in the nucleotide sequence, and the overall pattern of transcription is summarized in Figure 1.

The early region is transcribed by E. coli RNA polymerase from a set of three strong promoters, A1, A2, and A3, located about 125 base pairs apart in the noncoding region just ahead of the first T7 gene. Transcription from these early promoters terminates at a site, designated TE, near position 19 in T7 DNA. (Position in T7 DNA is given in units of 1% the length of the entire molecule, beginning at the genetic left end. A T7 unit corresponds to 399.36 base pairs.)

One of the early genes specifies the T7 RNA polymerase, a single-chain enzyme containing 883 amino acids (Mr = 98,000). This enzyme is very specific for its own promoters, which have a highly conserved sequence of 23 continuous base pairs corresponding to nucleotides -17 to +6 relative to the start of the RNA chain. Seventeen different promoters for T7 RNA polymerase are found in T7 DNA, and they have been classified into three groups. Class III promoters direct the transcription of class III genes; they are the strongest promoters, and have identical sequences over the entire 23 base-pair promoter sequence. Class II promoters direct the transcription of class II genes, and this transcription also continues across part of the class III region; these promoters differ from the 23 base-pair class III promoter sequence in from 2 to 7 positions, and are somewhat weaker promoters. The øOL and øOR promoters, near the left and right ends of the DNA, are thought to be associated with origins of replication.

A single termination signal for T7 RNA polymerase, designated Tø, is found near position 60 in the DNA, just behind the gene for the major capsid protein of the virion. Termination at Tø is not completely efficient, and in fact, completely efficient termination at Tø would be lethal to T7: two genes that specify proteins of the tail structure of the virion are transcribed entirely by readthrough of

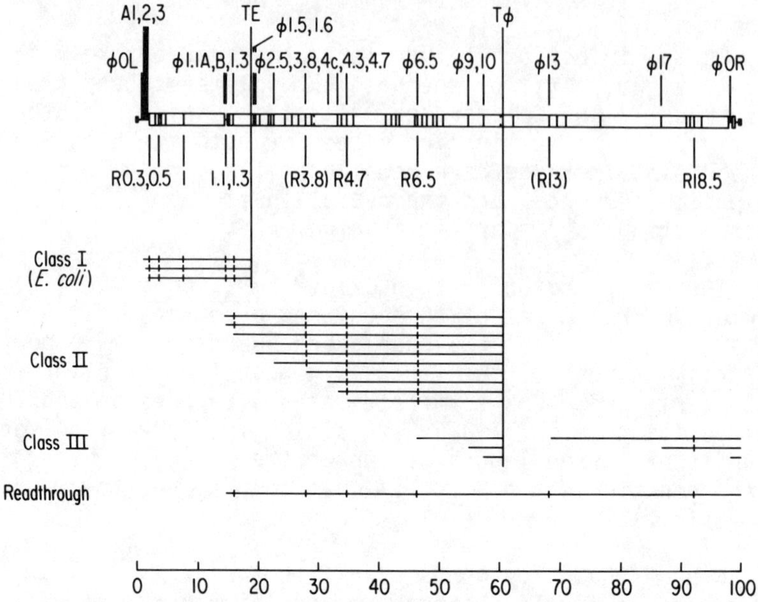

FIGURE 1. Synthesis and processing of T7 mRNAs. The T7 genes are represented by open boxes, and the terminal repetitions by the smaller, closed boxes at the ends of the molecule. Positions of transcription signals are indicated above the genes, and positions of RNase III cleavage sites are indicated below the genes. A1, A2 and A3 are the strong early promoters and TE is the transcription termination signal for E. coli RNA polymerase. The 17 promoters for T7 RNA polymerase are designated by ø, followed in most cases by the number of the gene first transcribed from the promoter; øOL and øOR refer to promoters thought to be associated with origins of replication near the left and right ends of the DNA. Tø is the transcription termination signal for T7 RNA polymerase. RNase III cleavage sites are designated R, followed by the number of the first gene to the right of the cleavage site. Cleavage at R3.8 and R13 is not completely efficient, which is indicated by the parentheses. The primary transcript from each promoter is indicated by a horizontal line, and the sites of RNase III cleavage by the short vertical lines. The expected class I, II and III transcripts are shown, as is the longest readthrough transcript.

Tø. The location of Tø is apparently part of a trancriptional strategy that ensures production of large amounts of the major capsid protein of T7. Transcription from 10 class II promoters and three class III promoters crosses the gene for this protein.

Contrary to what might have been expected, the weaker class II promoters are utilized before the stronger class III promoters during infection. This pattern of promoter utilization may be the consequence of the mode of entry of T7 DNA into the cell. Several lines of evidence suggest that T7 DNA enters the cell from left to right, and that the process of entry may take as long as 40% of the latent period. Thus, at the time T7 RNA polymerase is made, class II promoters may be the only ones accessible to the enzyme. In fact, it seems probable that T7 DNA is drawn into the cell by the act of transcription itself, first by the host RNA polymerase and then by T7 RNA polymerase. This mode of entry appears to be an important part of the mechanism for controlling gene expression during T7 infection.

T7 RNA polymerase transcribes its own DNA very efficiently but is virtually inactive on unrelated DNAs. This is presumably because T7 RNA polymerase interacts very specifically with its promoter sequence, and this promoter sequence is large enough to be unlikely to occur by chance in an unrelated DNA. This specificity permits T7 to switch all transcription from host DNA to its own DNA simply by making T7 RNA polymerase and inactivating the host polymerase. In fact, T7 specifies two proteins capable of inactivating E. coli RNA polymerase, a protein kinase that is specified by an early gene, and an inhibitor that inactivates the host polymerase by binding to it, specified by a class II gene.

RNase III CLEAVAGE SITES

Both early and late T7 RNAs are cut at specific sites by a host enzyme, RNase III, to produce the mRNAs observed during infection. Five RNase III cleavage sites have been identified in the early RNAs and five in the late RNAs (Fig. 1). Eight of these sites seem to be cut very efficiently, probably as soon as the site is exposed during transcription, but two of the sites are cut less efficiently. The combined effects of multiple promoters, termination sites and RNase III cleavage sites produces a rather large set of T7 mRNAs (Fig. 1). The relative amounts of individual mRNAs differ widely, because of differences in promoter strengths and in the number of promoters that direct transcription across different regions of the DNA, and because some chains escape termination or RNase III cleavage.

Although most T7 mRNAs are the products of RNase III cleavage, the purpose of such cleavages is not well understood. Most T7 proteins seem to be made with about the same efficiency whether or not the RNAs have been cut by RNase III, although there are a few exceptions. One possibility is that the RNase III cleavage sites have something to do with the extraordinary stability of T7 mRNAs in the cell. The RNA around each cleavage site can be arranged in a characteristic pattern of base pairing. The site of cleavage is at a point that leaves a potentially stable base-paired structure at the 3' end of each RNA. Perhaps these base-paired structures interfere with exonucleolytic degradation from the 3' end and are an important reason for the stability of T7 mRNAs. Accumulation of stable RNAs may be part of the strategy by which T7 directs ribosomes to its own mRNAs, and perhaps this is the reason RNase III cleavage sites are such a prominent feature of T7 transcription.

SYNTHESIS OF T7 PROTEINS

Coding sequences for all of the T7 proteins except T7 RNA polymerase are found as part of one or more polycistronic mRNAs. Nevertheless, there are only one or two examples where T7 mutations have a polar effect, and almost all of the T7 proteins would appear to be initiated independently. In fact, the coding sequences for all of the known and predicted T7 proteins are preceded by recognizably good ribosome-binding and initiation sites. Because of the close packing of the T7 genes, all or part of the ribosome-binding and initiation sequence frequently lies within the coding sequence of the previous gene.

Different T7 proteins are produced at vastly different rates, and these differences in rate do not seem to be accounted for solely by differences in mRNA concentrations. It seems clear that some messages are translated much more efficiently than others. Furthermore, coordinate synthesis of the three classes of T7 proteins during infection appears to involve some type of control at the translational level, and it seems likely that later classes of mRNA outcompete earlier ones for translation by ribosomes. However, the nucleotide sequence has so far provided little indication of what factors might be responsible for relative efficiency of protein synthesis.

Certain T7 genes specify pairs of overlapping proteins that are produced by translating the same mRNA but using different initiation or termination sites. The T7 primase gene produces two protein chains (566 and 503 amino acids) in about equal amounts. These two proteins are made from separate ribosome-binding and initiation sites, located 63 codons apart in the same reading frame, which direct translation to a single termination codon. Both protein chains are found in purified preparations of the primase, but whether both are needed for proper function is not yet known.

The gene for the major capsid protein, on the other hand, produces a large amount of a protein that is 344 amino acids long and a small amount of a protein that is 397 amino acids long. These two proteins begin at the same initiation codon, but the minor protein is produced by a translational frameshift just before the termination codon, which allows an additional 53 amino acids to be added before a termination codon is reached in the new reading frame. Both the major and minor proteins are incorporated into phage capsids, but again, it is not yet known whether both are needed. At least one other pair of T7 proteins also appears to be produced by the translational frameshifting mechanism.

T7 clearly uses translational mechanisms to generate pairs of proteins that differ only at their amino or carboxyl ends. It will be interesting to see what the significance of this is for the functioning of the proteins, and whether such mechanisms are widespread in other systems.

ORIGINS OF DNA REPLICATION

The first round of replication of parental T7 DNA begins in a noncoding gap just after the T7 RNA polymerase gene, near position 15 in the early region of T7 DNA. This region contains two promoters for T7 RNA polymerase, the ø1.1A and B promoters. A secondary origin, used when the above origin is deleted, appears to be associated with the øOL promoter for T7 RNA polymerase near the left end of the DNA. Other origins of replication have been identified by testing the ability of cloned fragments of T7 DNA to serve as origins of replication in a plasmid during T7 infection. This assay has identified relatively strong origin activity associated with the øOR promoter near the right end of the DNA and with the ø13 promoter near position 68. How these origins are used during infection, and whether they have different functions in the process of replication and maturation of T7 DNA, is not known.

The origins of T7 DNA replication all contain a promoter for T7 RNA polymerase, and this enzyme is needed for proper initiation of replication in vitro. Perhaps T7 uses the specificity of T7 RNA polymerase for its own promoters as a way to direct replication to its own DNA, in much the same way as it directs transcription to its own DNA. T7 specifies virtually all of the proteins needed for its replication apparatus. Thus, if T7 were to inactivate the host replication apparatus, all replication in the cell would be directed to T7 DNA. Host DNA replication is known to stop during T7 infection, but the mechanism of shutting it off has not been elucidated.

NONCODING SEQUENCES AT THE ENDS OF T7 DNA

The longest stretches of T7 DNA that do not code for any proteins are at the ends of the molecule, about 2.3% of the molecule at the left end and 1.0% at the right end. The first and last 0.4% of the molecule contain a perfect direct repeat of 160 base pairs. T7 DNA is replicated as concatemers, that is, long molecules containing tandemly repeated T7 genomes. At the junction between adjacent genomes in a concatemer, the noncoding regions that will ultimately be at the ends of the mature DNA flank a single copy of the terminal repetition. The terminal repetition of mature DNA is generated during maturation and packaging.

A substantial portion of the noncoding region at the left end of the mature DNA molecule is taken up by the three strong promoters for E. coli RNA polymerase and the RNase III cleavage site just ahead of the coding sequence for the first gene. Ahead of these promoters is the secondary origin of replication that is associated with the øOL promoter for T7 RNA polymerase.

A striking feature is found in the sequences adjacent to the the terminal repetition, at both ends of the DNA: a regular array of 12 short repeated sequences, arranged in two sets of six, occupies about 160 base pairs of sequence. In each set of six, the sequence CCTAAAG, or variants of it, is repeated at 11-13, or in one case 15 base-pair intervals, and the two sets of six are separated by 28 or 29 base pairs. The array of repeats at the left end of the DNA is much more perfect and regular than that at the right end, and the array at the right end probably would not have been identified without the array at the left end as a model. In concatemers, these arrays of repeated sequences flank the single copy of the terminal repetition. However, even though they are located symmetrically about the terminal repetition, the repeated sequences have the same polarity relative to the genome, and are therefore not truly symmetrical about the terminal repetition. The location of these arrays suggests that they have a role in some reaction involving the ends of T7 DNA, possibly concatemer formation, maturation, packaging, or the initial stages of entry of T7 DNA into the cell.

CONCLUSIONS

T7 genes are packed very efficiently into the T7 DNA molecule, and are arranged in the order of their expression during infection. All transcription is from left to right, and the entry of T7 DNA into the cell may be coupled to transcription of the DNA. This mode of entry appears to be important in the orderly expression of T7 genes. A central feature of T7 infection is the specificity of T7 RNA polymerase for its own promoters, which permits all transcription and replication in the cell to be directed to T7 DNA. Most T7 mRNAs are the products of specific cleavages by E. coli RNase III, and the base-paired structures that remain at the 3' ends of the RNAs after cleavage or transcription termination may be responsible for the extraordinary stability of the T7 mRNAs. Most T7 proteins seem to be synthesized independently, each from its own ribosome-binding and initiation site. Certain genes specify pairs of overlapping proteins, by initiating synthesis at two separate intitiation sites in the same reading frame, or by shifting the reading frame during translation. Noncoding sequences at the ends of the DNA contain a terminal direct repeat of 160 base pairs and flanking arrays of short repeated sequences, which presumably have some role in reactions involving the ends of the DNA.

ACKNOWLEDGEMENT

This research was carried out at Brookhaven National Laboratory under the auspices of the United States Department of Energy.

REFERENCES

1. Dunn JJ, Studier FW (1983). The complete nucleotide sequence of bacteriophage T7 DNA and the locations of T7 genetic elements. J Mol Biol 166:
2. Studier FW, Dunn JJ (1983). Organization and expression of bacteriophage T7 DNA. Cold Spring Harbor Symp. Quant. Biol. 47: 999.

Gene Expression, pages 413-435
© 1983 Alan R. Liss, Inc., 150 Fifth Avenue, New York, NY 10011

THE ROLE OF MINI-CHROMOSOMES AND GENE TRANSLOCATION IN THE EXPRESSION AND EVOLUTION OF VSG GENES

Piet Borst*, André Bernards*, Lex H.T.Van der Ploeg*, Paul A.M.Michels*, Alvin Y.C.Liu*, Titia De Lange*, Paul Sloof, David C.Schwartz** and Charles R.Cantor**

Section for Medical Enzymology, Laboratory of Biochemistry, University of Amsterdam, P.O.Box 60.000, 1005 GA Amsterdam, The Netherlands and **College of Physicians and Surgeons, Columbia University, Department of Human Genetics and Development, 701 West 168th Street, New York, N.Y. 10032, U.S.A.

ABSTRACT

The surface coat of African trypanosomes, like Trypanosoma brucei, consists of a single protein, the variant surface glycoprotein (VSG). By switching from the synthesis of one VSG to the next, trypanosomes change the antigenic nature of their surface and escape immune destruction. We have shown that one route for VSG gene activation involves the duplication and transposition of a silent basic copy (BC) gene into an expression site, yielding an expression-linked copy that is transcribed. The expression site is located at the end of a chromosome. The 5' 35 nucleotides of the mature messenger RNA are not encoded in the transposed segment and are derived from a mini-exon in the expression site.

Another group of VSG BC genes is activated by a

---

* Present address: Antoni van Leeuwenhoekhuis, The Netherlands Cancer Institute, Plesmanlaan 121, 1066 CX, Amsterdam, The Netherlands.

Abbreviations: BC, basic copy; bp, base pair(s); C.F., culture form; DA, duplication-activated; ELC, expression-linked copy; mRNA, messenger RNA; NDA, non-duplication-activated; VAT, variant antigen type; VSG, variant surface glycoprotein.

second route that does not involve gene duplication. Like the VSG gene expression site, such genes are at the end of a chromosome. We interpret our recent results to mean that these genes are transposed into the expression site by chromosome end exchange or by a telomeric gene conversion. A single expression site could, therefore, control expression of the 1000-odd VSG genes in a mutually exclusive fashion.

To analyse the number of chromosome ends present per trypanosome we have used a new electrophoretic method, pulsed field gradient agarose gel electrophoresis, to fractionate large DNA from gently lysed trypanosomes. This method allows sizing of DNA in the 100 to 2000-kb range and blot hybridization analysis of the fractionated DNA. We conclude from our preliminary analysis that T. brucei contains at least 40 mini-chromosomes in the 100-kb range; at least seven chromosomes in the 200 to 700-kb range; at least one chromosome with an apparent size of 1800 kb; and DNA that hardly leaves the slot. The mini-chromosomes strongly hybridize with a telomere-specific DNA probe and with a probe that recognizes a sequence common to many VSG genes. The role of mini-chromosomes, the arguments for a single expression site, the hazardous life at chromosome ends and the possible advantages of having two different routes for VSG gene activation - one for all genes, one for telomeric genes - are discussed.

INTRODUCTION

The surface coat which covers trypanosomes during their stay in the bloodstream and tissues of their mammalian hosts [1] consists of a single protein, the Variant Surface Glycoprotein or VSG [2]. Each trypanosome can synthesize at least 100 different VSGs [3], but at least 1000 separate VSG genes are present in the genome [4]. Repeated switching of the VSG gene expressed gives rise to the variation of surface antigens that allows the trypanosome to maintain a chronic infection in the face of effective host antibody formation (see refs 5-7).

To make efficient use of this large repertoire of VSG genes, the trypanosome has evolved rather unusual methods to control VSG gene expression. One method involves the duplication of a silent Basic Copy gene (BC gene) and the transfer of the duplicate to an expression site to yield an Expression-Linked extra Copy (ELC) of the gene [8-10]. BC and ELC can be distinguished in blots of nuclear DNA by large differences in their surrounding DNA sequences. The

BC gene is silent because it is incomplete: it lacks a mini-exon sequence of 35 base pairs (bp) present at the 5'-end of all VSG messenger RNAs (mRNAs) examined [11-14]. This mini-exon (and the promoter, presumably adjacent to it) must be provided by the expression site and is apparently added on to the body of the mRNA by splicing. This mode of VSG gene activation is illustrated in Fig. 1. As the trypanosome switches coats, the incoming VSG gene duplicate apparently displaces the resident ELC from the expression site [14,15], ensuring that only one VSG gene is transcribed at a time. Our data on the nature of the expression site and the process of duplication-transposition have recently been summarized [16,17].

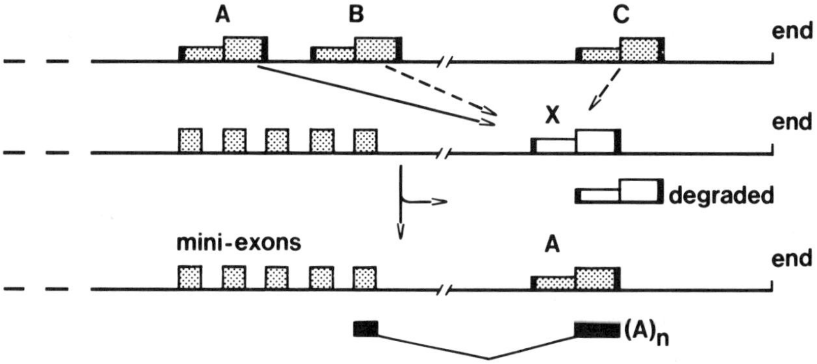

Fig. 1. Some VSG genes are activated by a duplicative transposition to an expression site that lies at the end of a chromosome. The incoming gene A displaces gene X from the expression site, possibly by a gene conversion process that utilizes short blocks of sequence homology at the edges of the transposed segment (black bars) to align the gene. The displaced gene X is degraded. The mini-exons in the expression site provide the 5' 35 nucleotides of the mature mRNA. Presumably, this occurs by splicing of long precursor RNAs as indicated, but this remains to be verified. The mini-exons are present in long tandem arrays of a 1.35-kb repeat, as schematically indicated, but whether all these mini-exons are adjacent to functional transcription starts is not certain (see ref. 17).

The first analyses of VSG genes in 1979-1980 already showed that gene duplication is not a prerequisite for the activation of all VSG genes [18-21]. Some VSG genes were activated without detectable duplication and this sub-set of genes was invariably characterized by local gene rearrangements [18]. These rearrangements were limited to an area downstream of the gene and were unrelated to gene activity. This showed that another pathway for VSG gene activation must be present, one that does not involve gene duplication. We refer to these genes as non-duplication activated (NDA) genes.

Three observations have thrown light on this second pathway:

1. The expression site for duplication-activated (DA) genes was found to reside next to a discontinuity in the DNA [10,15,22], recently shown to be a chromosome end [16,17,23]. A remarkable property of this end (and of other chromosome ends in trypanosomes) is that it grows by about 10 bp per cell division. Growth presumably occurs at the very end and may result from the mechanism employed to replicate the ends. The increase in length is balanced by occasional deletions in the telomeric segment [16,17,23].

Following the characterization of the expression site, it was soon realized that the NDA genes characterized by us (the 221 gene) and by Williams and co-workers, are also located next to a discontinuity in the DNA [24,25]. In fact, the discontinuity adjacent to the 221 gene moves away from the gene at a rate of 7 bp per cell division, showing that this gene lies next to a chromosome end [23]. Chromosome growth and deletions in the telomeric segment readily explain the 'gene rearrangements' [18] observed downstream of NDA genes.

2. The VSG 221 gene, the NDA gene studied by our group, was found to give rise to a mRNA with the same 35-nucleotide sequence at its 5'-end as the DA genes that procure this sequence by being transferred into the expression site [12]. This 35-nucleotide mini-exon is not present within 8 kb of the main exon of the 221 gene [26]. This suggested that the 221 gene also moves into the expression site to acquire an upstream mini-exon [16].

3. An analysis of the expression site and the various ELCs contained in it has shown that the distance between ELC and the nearest upstream 35-bp mini-exon sequence can be more than 30 kb [13,16,17]. This implies that VSG mRNA synthesis may involve very long precursors and that the acquisition of an upstream mini-exon by a telomeric gene

might involve a crossover more than 30 kb upstream of the gene and, hence, not detectable in our blots of nuclear DNA.

On the basis of these observations we have proposed that NDA genes employ the same expression site as the DA genes, but enter this site by means of a reciprocal translocation between NDA gene telomere and expression-site telomere, rather than via a duplicative transposition of the gene [16,17]. This proposal is illustrated in Fig. 2 and is supported by several indirect lines of evidence, summarized by Borst et al. [17]. Direct evidence for the postulated chromosome translocation is still lacking, however.

In this paper we present our recent attempts to verify chromosome translocation by a new method that allows separation and analysis of DNA molecules in the 1000-kb range. We also consider the question why trypanosomes use two different routes for the activation of VSG genes.

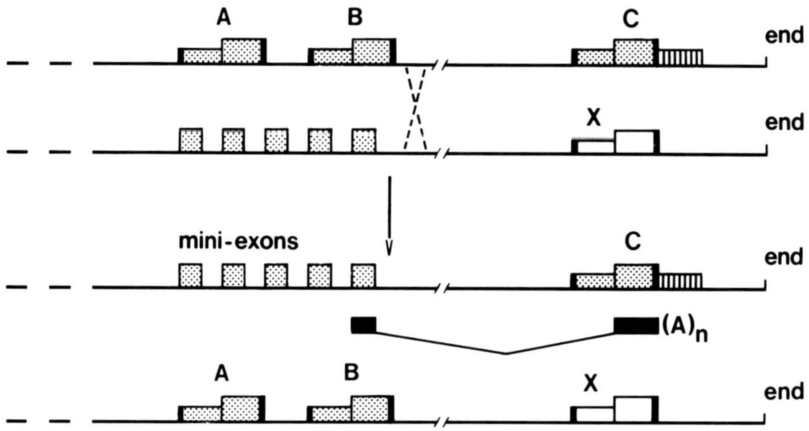

Fig. 2. Speculative scheme showing how telomeric VSG genes could be activated by a reciprocal translocation. See Fig. 1 and text for explanation.

A MINI-CHROMOSOMAL DNA FRACTION, CONTAINING SATELLITE DNAs AND VSG GENES

If several VSG genes are located next to chromosome ends, it becomes of interest to know the number of trypanosome chromosomes. This question can not be studied by standard cytology, because trypanosomes do not condense their chromosomes in any phase of their cell cycle. In initial experiments we have, therefore, tried to separate chromosomes either on sucrose gradients or in dilute agarose gels at very low voltage gradients. This led to the separation of the DNA from gently lysed trypanosomes in three fractions: large (>350S), middle-sized (60-250S) and small (<60S) DNA. The large DNA fraction contains the bulk of the DNA, most VSG genes and the ribosomal genes. The middle-sized DNA was found to contain all of the 177-bp satellite DNA [27] and part of two repetitive sequences which are associated with VSG genes: the 70-bp repeats found at the 5'-edge of the transposed segment of many VSG genes [14]; and the 3'-end of VSG gene 221, which contains sequence elements common to all VSG gene 3'-ends [7,21,28-30].

Our first report of these results [16] coincided with a paper by Williams et al. [25] who reported that one of their telomeric VSG genes was located on a mini-chromosome of 80 kb. Further analysis of our middle-sized DNA fraction revealed the presence of one VSG gene in our strain as well [17]. Both the telomeric 221 gene and the telomeric expression site were found in the large DNA fraction, however [16]. These results will be presented in detail elsewhere [31].

Our inability to resolve the large DNA fraction either on sucrose gradients or in gels, made it necessary to look for other methods to sub-fractionate large DNA. We therefore turned to a new method recently developed by two of us (D.C.S. and C.R.C.) to separate DNA molecules in the 100- to 2000-kb range [34].

FRACTIONATION OF LARGE TRYPANOSOME DNA BY PULSED FIELD GRADIENT GEL ELECTROPHORESIS

Size-fractionation of DNA in standard agarose gels is minimal when the size of the DNA exceeds 100 kb [32,33,39]. This has been attributed to long DNA moving through the gel like a snake through grass [32,33], but this is probably an over-simplification (cf. ref. 34). To overcome the size limitation of standard agarose gels, Schwartz et al. [34]

have developed a system in which DNA molecules are stretched by a gradient field and forced to turn corners by subjecting them alternately to pulses of current in two different directions, perpendicular to each other. The rate at which the molecule re-orients when the direction of the field is changed is dependent on the length of the molecule at least up to 2000 kb and a size separation has been obtained in the 100 to 2000-kb range. The principle of the method is depicted in the left-hand panel of Fig. 3; its application to trypanosome DNA in the right-hand panel of Fig. 3. The figure shows the ethidium bromide-stained gel, after running DNA from gently-lysed and deproteinized trypanosomes next to a mixture of phage T7 DNA, phage T2 DNA and phage G DNA. Three trypanosome lysates are shown, the bloodstream variants 118 and 221 and coat-less cultured trypanosomes. At this pulse time of 43 sec the trypanosome DNA separates into four zones: a big blob of DNA which has hardly moved from the slot; a large chromosome band which may contain several chromosomes and which migrates with an apparent size of 1800 kb; a zone with approximately seven chromosomes with a length ranging from 700-200 kb and a mini-chromosomal fraction containing probably over 40 molecules migrating in between the marker DNAs from phage T7 and phage T2. The mini-chromosomal area can be separated into a large number of individual bands at shorter pulse times (not shown).

Fig. 4 shows DNA fractionated at longer pulse times (90 sec). Under these conditions the mini-chromosome area and most of the middle-sized chromosomes are compressed into one broad band. Blots of this gel were hybridized with various DNA probes; two of the resulting autoradiograms are shown in Fig. 4. From those and similar experiments we conclude that the complex patterns in Figs 3 and 4 are no simple artefacts of the electrophoretic procedure and probably reflect the size distribution of chromosomal DNA in the intact trypanosome. The absence of random degradation follows from the observation that the probes for the 118 (Fig. 4) and 221 (not shown) BC genes only hybridize with the large DNA. The interpretation that the multitude of bands in the 100-kb range indeed represents a large array of mini-chromosomes, is supported by the predominant hybridization of a telomere probe with the smaller material (not shown). This probe contains repetitive DNA and in nuclear DNA blots it detects many bands that are preferentially shortened by treatment of large DNA with BAL-31 nuclease prior to restriction enzyme digestion (unpublished).

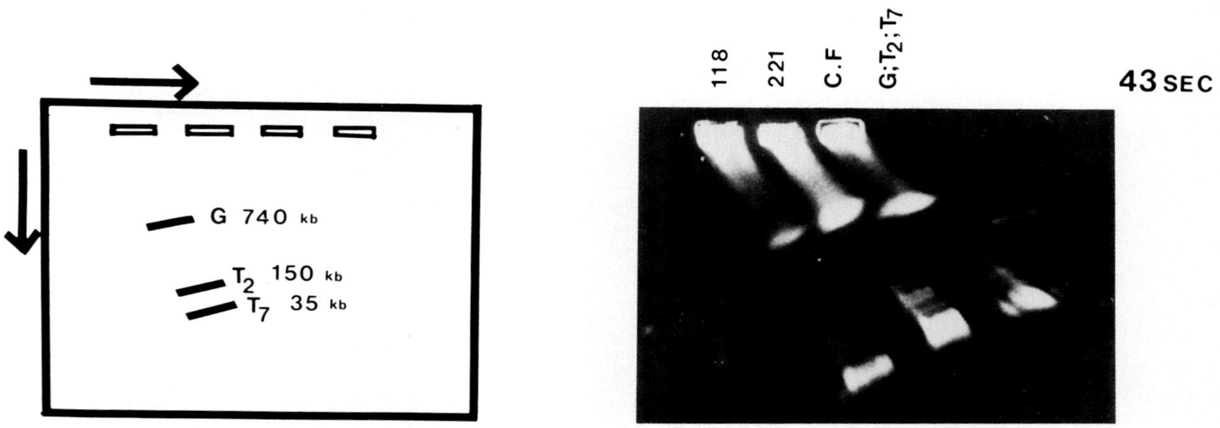

Fig. 3. Pulsed field gradient gel electrophoresis of trypanosome DNAs and marker DNAs. The left-hand panel schematically indicates the direction of the pulses and the separation obtained with a mixture of DNA from bacteriophages T7, T2 and G. The right-hand panel shows the ethidium bromide-stained agarose gel after electrophoresis of the marker DNAs and three trypanosome samples: Variant Antigen Type (VAT) 118, VAT 117 and culture form (C.F.) DNA. Trypanosomes were lysed and deproteinized in agarose blocks to avoid shear degradation. The pulse time was 43 sec. See ref. 34 for details of the method.

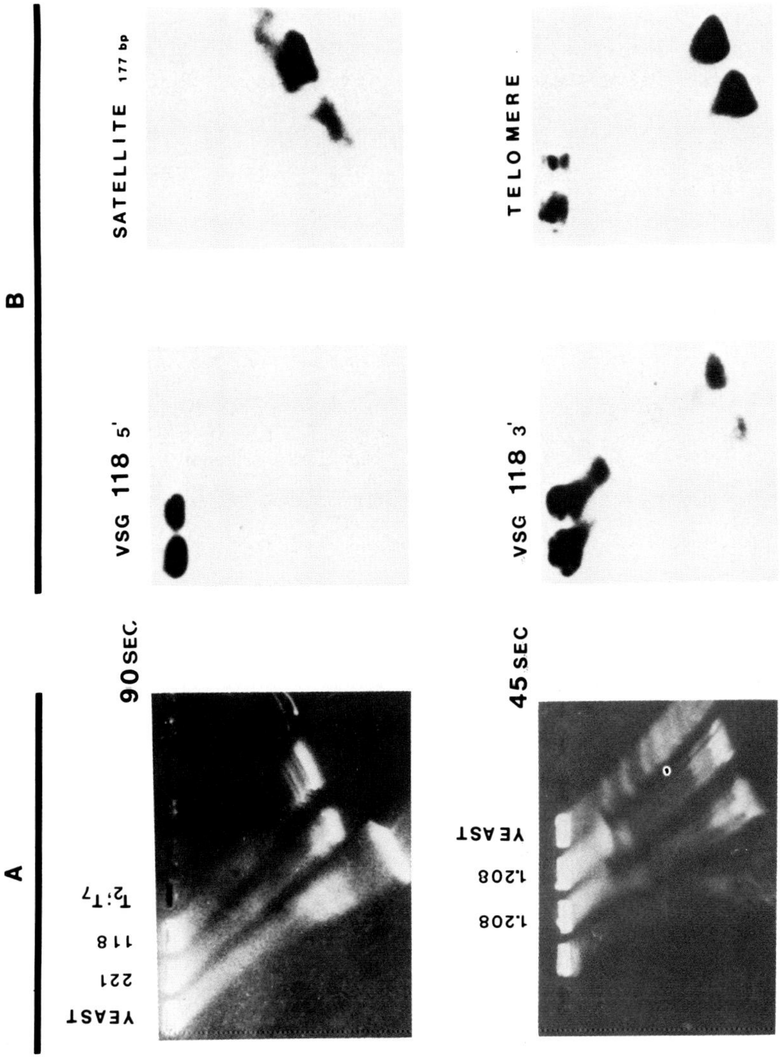

Fig. 4. (See legend, page 422.)

The multitude of bands in the 100-kb region probably does not result from asynchrony of chromosome end growth in different trypanosomes in the population. We have not seen asynchrony in recently cloned trypanosomes, like the ones used for the experiments in Fig. 3. This remains to be verified, however, by showing the presence of unique genes in each of these mini-chromosomal bands.

Although a large fraction of the DNA remains in or close to the slot, this is not merely present in trypanosomes that have not lysed properly, because satellite DNA hybridization with this fraction is very weak and may be entirely due to non-specific binding of label to the trypanosome debris in the slot (Fig. 4). Non-specific ethidium bromide binding might also be responsible for part of the prominent staining in the slot and this remains to be tested by determining the amount of DNA in each of the gel fractions.

On the basis of these considerations we tentatively conclude that T. brucei contains a large array of mini-chromosomes, at least 40, but possibly as many as 100. These mini-chromosomes contain at least three types of sequences: telomeric DNA, the 177-bp satellite DNA and sequences that hybridize to the 3'-end of VSG genes under non-stringent conditions (not shown). This is compatible with the idea that these mini-chromosomes serve as a vehicle to expand the number of telomeric VSG genes. The satellite DNA may merely be used to pad the mini-chromosomes to give them a size that allows efficient segregation during mitosis.

Some of the chromosomes of T. brucei appear to be rather large. The band at about 1800 kb migrates slower than the largest yeast chromosome and the specific

==

Fig. 4. Pulsed field gradient gel electrophoresis of trypanosome DNAs and marker DNAs. Panel A shows the ethidium bromide-stained gels; panel B autoradiograms of blots of the gels in A hybridized with various probes. The DNA samples were from Saccharomyces (Yeast), VATs 221, 118 and 1.208 and bacteriophages T2 and T7. The pulse times used are indicated. The hybridization probes were half-gene cDNA probes of VSG 118 mRNA (see ref. 10), a cloned repeat of the 177-bp satellite DNA of T. brucei [27] and a cloned nuclear DNA segment from T. brucei that hybridizes with chromosome telomeres (telomere probe). See Fig. 3 for further details.

retention of the genes for the 118 and 221 BC genes close to the slot suggests that some of the DNA is even larger, although unusual topological constraints (branching, circularity) cannot be ruled out. Since we have not found yet a probe that hybridizes exclusively to the 1800-kb band, we cannot rule out that the chromosomes in this band are in part retained near the slot.

Experiments to test our model for chromosome end exchange are in progress. Thusfar we have found no significant hybridization of the mini-chromosome area with a mini-exon probe under conditions where such hybridization would have been detected if every mini-chromosome carried at least one mini-exon.

LIFE AT ENDS IS HAZARDOUS FOR GENES IN TRYPANOSOMES

In an early survey, carried out before we realized that the 221 gene lies at a chromosome end, we observed that this gene was absent in all the 12 other African trypanosome strains checked, whereas the 117 VSG gene was present in nearly all of them [35]. Such marked evolutionary instability was recently also observed for a second telomeric VSG gene (designated 1.1005 in our trypanosome stock). The hazards of life at ends are underlined by two additional observations:

1. We have analysed trypanosome populations from the second wave of parasitaemia in two rats infected with variant 221a. Both populations had switched expression from VSG gene 221 to other VSG genes and both had lost the 221 gene [17].

2. We have isolated a trypanosome clone that expresses VSG gene 118 (called variant 118c) and this clone contains an extra sequence (called TIS3) behind the 118 ELC in the expression site [16,17]. TIS 3 is unstable behind the 118 ELC because we have isolated two variants from the same parasitaemic peak that have lost part of TIS 3.

Variant 118c contains two other copies of TIS3 and both are close to ends, as judged from BAL-31 nuclease experiments. In other variants of our strain that do not have TIS3 in the expression site, the copy number of TIS3 varies between 1 and 3.

We have recently found that the parental copy of TIS3 is actually located downstream of a VSG gene (designated 1.1006), between the gene and the barren region adjacent to the chromosome end. Our unpublished sequence analysis has shown that TIS3 has moved into the expression site together with the 3'-end of the adjacent VSG gene. The crossover must have occurred in an area of 16 bp just upstream of the 14-mer sequence common to the ends of VSG mRNAs (cf. ref. 7).

The marked instability of genes at ends could be explained in four ways:

a) Chromosome ends are intrinsically unstable and tend to undergo frequent internal duplications and deletions, possibly arising from the presence of sub-telomeric repetitive DNA stretches.

b) Most telomeric VSG genes are on mini-chromosomes and these do not segregate properly during mitosis, leading to change in copy number of such genes. Like a), this process would not generate new telomeric genes; moreover, the mini-chromosomal VSG gene studied by Williams et al. [25] does not seem to be mitotically unstable.

c) Many chromosome ends contain VSG genes that are able to act as acceptor for chromosome-internal VSG genes transferred to the telomere by duplicative transposition. The loss of VSG genes in such acceptor-competent telomeres is, therefore, not a net loss but an exchange of one VSG gene for another. As one will usually not have a probe for the incoming gene, the exchange is only detected by the loss of the displaced gene.

We expect that acceptor-competent telomeres can enter the expression site by a reciprocal translocation, as depicted in Fig. 2. If the upstream sequences of the different acceptor-competent telomeres differ and if the translocation can take place far upstream from the VSG gene, there may seem to be several functional expression sites in one trypanosome. This has in fact been observed by Longacre et al. [36] (although they interpret their data differently).

d) There is an active process of gene conversion between non-homologous telomeres. If this is random, the copy number of a telomeric gene will fluctuate around 1. If there is bias, certain telomeric genes may increase their representation at the cost of others.

There are not enough experimental data yet to judge the relative contribution of the four processes discussed to the instability of telomeric genes. The loss of the 221 gene, referred to above, is not simply the result of process (c), because we have shown that in this case the gene is lost together with its adjacent upstream sequences (unpublished). Since the 221 gene is on large DNA [31], its loss is not due to the loss of a mini-chromosome. Process (d) seems to provide the most plausible explanation in this case. Even if this process would take place at the same rate as VSG gene switching (i.e. about 10^{-5} per generation), it would only become visible when one selects for a trypanosome that has lost the expression of VSG 221. This selection is only possible in trypanosomes that initially expressed this gene. Hence, the loss of the 221 gene in trypanosomes that have switched from 221 expression to expression of another VSG gene, is fully compatible with the stability of the gene when it is not expressed. Whether the telomere that contains the 221 gene is acceptor-competent, is not known. We have not seen a displacement of the 221 gene by another VSG gene without the simultaneous loss of upstream sequences, but only two relapse populations have been checked.

Telomeric gene conversion also provides the obvious explanation for the observations on TIS3. Both its entry into the expression-site telomere and its changing copy number is not easily accounted for by any other process. Finally, our recent (unpublished) observations on variant 221b can also be accounted for by telomeric gene conversion. Variant 221b expresses the 221 gene but (in contrast to variant 221a) contains an extra copy of this gene that seems to reside in the only expression-site telomere that we have seen in our trypanosome strain. The 221b ELC is unusual, however, in that the duplicated segment contains the 221 gene plus the adjacent 3' sequence. This contrasts with the DA genes studied thusfar, in which the transposed segment stops at the 3'-end of the gene. The 221b variant may, thereforc, have arisen by gene conversion of the expression site by the 221-containing telomere. Conversion may have started just upstream of the 221 gene and spread down to the chromosome end as illustrated in Fig. 5.

Process (d) can only explain the introduction of new genes at ends if the expression-site telomere participates in the gene conversion. On the basis of the results of Longacre et al. [36] we think, however, that process (c) must also contribute.

In a recent paper Laurent et al. [44] have reported results on the activation of the AnTat 1.3 gene that are also readily explained by telomeric gene conversion. They observed that the restriction map of the BC of the AnTat 1.3 gene looks like the ELCs analysed in their stock. This BC gene is telomeric; it is flanked on both sides by a barren region devoid of restriction enzyme recognition sites; and, most importantly, the restriction sites upstream of the 5' barren region are identical to those found upstream of their ELCs. This is readily explained by a telomeric gene

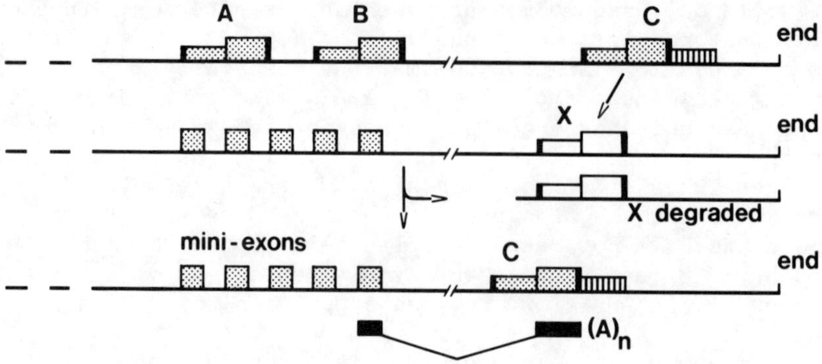

Fig. 5. Some telomeric genes can be activated by a telomeric gene conversion. See Fig. 1 and text for explanation.

conversion in which the expression site telomere acts as donor and the gene conversion has spread more than 20 kb upstream of the ELC itself (our interpretation).

INTERMEZZO: THE ORDER OF VSG GENE EXPRESSION IN A CHRONIC INFECTION

In a chronic infection in mammals, different variant antigen types (VATs) appear in an imprecisely predictable order. Some VATs always appear early, others semi-late or late in infection, but within - say - the early group the order varies in different infections (see refs 7 and 37) for references to the original literature). When trypanosomes in the course of a chronic infection are transferred by syringe to a fresh animal, the programme is 're-set', i.e. early VATs reappear. Since the trypanosome has no way of knowing that it is transferred to a fresh rabbit, this result indicates that in a chronic infection 'early' variants continuously reappear as a result of switching, but that these fail to be detected later in the infection because they are immediately eliminated by antibodies.

The ordered programming of VSG gene expression is also manifest in the metacyclic VAT repertoire that develops in the salivary gland of the tsetse fly. This repertoire consists of a limited number of VATs that differ from the very early ones found in a chronic infection, started with trypanosomes from another mammal. The metacyclic repertoire is strain-specific and its composition is only influenced to a limited degree by the nature of the variant antigen population injected in the fly.

Two factors appear to contribute to the programmed appearance of VATs: differences in growth rates of different variants and programmed switching. We are especially interested in programmed switching and we have considered elsewhere different ways in which this programming could operate [37]. Here we discuss two puzzling observations made by other groups that seemed initially difficult to explain by the single expression site model proposed by us. Recent experiments have shed some light on these observations and these will be briefly considered here.

The first observation was made by Miller and Turner [41] who found that, in a different trypanosome strain, some VATs tend to appear after another with high probability. This observation is not easily explained by the duplication-transposition of genes into a single expression site. Either one would have to assume that two genes can be transposed in tandem and that rearrangements in the expression site will then allow the expression of one gene after the other, or one would have to postulate that the presence of one gene in the expression site would preferentially lead to a transposition of the other gene. Since there are no obvious preferred homologies among genes that are expressed in the same period of a chronic infection, it seems rather unlikely that the order in which genes are transferred to the expression site could be based on homology between the genes.

Recently, we have found a VSG gene couple in our <u>T. brucei</u> strain 427 that is also expressed in sequence with high probability, exactly as the couples described by Miller and Turner [41]. These coupled genes were detected in the course of experiments designed to obtain evidence for the telomeric translocation model of VSG gene expression. Variant 118b switched to expression of the MITat 1.8 gene. The MITat 1.8 variant contained a lingering 118b ELC, as predicted for the reciprocal translocation model, but we have not verified yet whether the MITat 1.8 gene is a telomeric gene. We then found that the MITat 1.8 variant obtained in this way switches back with very high frequency to expression of the 118 gene, presumably by reversing the reciprocal translocation that inactivated the 118 ELC. In a mixture of the MITat 1.8 and the 118b variants, the 118b rapidly outgrows MITat 1.8. Hence, the 118b gene appears with high frequency when the MITat 1.8 is switched off, at least when a MITat 1.8 variant that contains a lingering 118 ELC is used. These results suggest the following explanation for the Miller-Turner experiments: The frequency at which telomeres undergo reciprocal translocation is dependent on the degree of homology in the region upstream of the telomeric VSG gene. Telomeres that have a long stretch of homology in this region will undergo frequent reciprocal translocation. If one of the telomeres is located in the expression site it will switch with high frequency to the gene that lies on the other telomere and vice versa.

The second puzzling observation was made by Hajduk and Vickerman [42] and concerns the metacyclic repertoire. They infected a tsetse fly with a homogeneous population of variant A, allowed this fly (after a suitable period) to infect a rabbit and analysed the bloodstream VATs that came up directly after the metacyclic VATs had disappeared. They found that these early bloodstream VATs always contain VAT A, even if this is a late VAT. VAT A was not represented in the metacyclic repertoire. Obviously, the genetic programme has a memory function. Hajduk (personal communication) has suggested that this experiment is best explained by the presence of two expression sites in the trypanosomes, one for the bloodstream repertoire and one for the metacyclic repertoire. Although this is a reasonable explanation for the memory function, we can now also offer a simpler alternative: In recent experiments Parsons et al. [43] have studied the fate of the ELC when trypanosomes switch off VSG synthesis upon entering the tsetse fly mid-gut. This process can be simulated by bringing the trypanosomes in culture at 25°C. Although VSG synthesis is absent in these cultured trypanosomes, they had apparently retained the ELC of the original VAT population in the same expression site. If this expression site is also used for the metacyclic repertoire, one would expect the ingested VAT to be displaced by other metacyclic antigenic types in this expression site. However, if one of the metacyclic genes were a telomeric one that enters the expression site by reciprocal translocation, this fraction of the trypanosomes would retain a lingering ELC of the gene expressed in the ingested VAT. If this lingering ELC is reactivated when the metacyclic repertoire is switched off, the ingested VAT will appear as an early gene in the bloodstream repertoire.

What could be the origin of the metacyclic repertoire if there is not a separate expression site? The simplest explanation is that a modified enzyme system is used in the tsetse fly to activate VSG genes. This modified enzyme system would have a higher affinity for metacyclic genes than for the bloodstream repertoire. This model predicts that there is only a single expression site that is used both for the metacyclic and the bloodstream repertoires and that a gene expressed both in the metacyclic repertoire and in the bloodstream repertoire could be one and the same gene. Experiments are in progress to test this prediction.

WHY TWO DIFFERENT ROUTES FOR VSG GENE ACTIVATION?

Although many details remain to be filled in, it is clear that there are at least two different routes for the activation of VSG genes. The first involves the duplicative transposition of a gene into an expression site at the end of a chromosome. The second is still undefined, but appears to be restricted to telomeric genes. For this discussion we shall assume that such genes enter the expression site by reciprocal translocation or gene conversion; that there is only a single functional expression site; that there are many (i.e. some 100) telomeric VSG genes, most of which are located on mini-chromosomes; and that VSG genes can move in and out of telomeres by one of the processes discussed in the previous sections.

Even with these simplifying assumptions, the process of VSG gene switching remains pretty complex. Why is that? Why should there be a second route for VSG gene activation when the first route looks adequate? The following explanations can be considered:

1. One of the routes is an evolutionary remnant. We have previously argued that the VSG gene switching system may have evolved from a process akin to the mating-type switch [37]. Mating-type is wide spread among protists, it involves surface characteristics and in the two cases studied in detail - Saccharomyces and Schizosaccharomyces - switching requires gene rearrangements (see refs 38 and 40). In Saccharomyces the expression site (MAT) for mating-type is in the middle of chromosome III, the basic copy genes for mating-type information (HML and HMR) are near the ends of this chromosome. It is easy to envisage a reversal of roles, in which the expression site function moves to a chromosome end, with silent copies in the middle and at the other end. By gene duplication (internally) and emigration of genes to other telomeres, the present gene distribution in trypanosomes could have arisen. The only extra step required would be the introduction of telomeric gene exchange. The implication is that the duplicative transposition of chromosome-internal genes would be an evolutionary remnant. This does not seem a realistic

proposal. From an analysis of cosmid clones Van der Ploeg et al. [4] have estimated that there are some 1000 VSG genes. Since telomeric genes are selected against in the cloning procedure, most of these genes must be chromosome-internal. An evolutionary remnant of 1000 genes, occupying 10% of the genome, seems far-fetched.

The alternative - that telomeric genes are evolutionary relics - does not seem acceptable either, because of the indication that many of the mini-chromosomes may just be present to generate ends for VSG genes.

2. The second pathway is an accidental consequence of the tendency of chromosome telomeres to undergo gene exchange processes. Even if this were the case (there is no evidence for or against this), the question remains why the expression site is located at a chromosome end at all. If this were deleterious, there is no reason why the expression site would not have moved to a chromosome-internal position, the position where the expression site for mating-type information in yeast is actually located now.

3. There are two routes for VSG gene activation, because this is an advantageous arrangement. Two versions of this explanation can be considered:

i) The telomeric route of VSG gene activation is required to create a reservoire of VSG genes that evolve more rapidly than the bulk of the genes. Under field conditions the selective pressure to evolve new VSG genes must be very high, because all suitable mamalian hosts are chronically infected and only new variant antigens will allow super-infection. If replication of telomeres were error-prone, the telomeric position could be used to allow selective mutagenesis [35]. If there is a natural tendency of telomeres to exchange genes, a second route for VSG gene activation would be created and the maintenance of this route would be ensured by the continuing advantage of the new VSG genes arising from telomeric mutagenesis. Since there is no simple route back from telomere to a chromosome-internal location, it will be important to have many telomeres available for VSG gene location. This may be the selective force maintaining a large number of mini-chromosomes. This explanation hinges on the assumption that telomeric genes evolve more rapidly than other ones. The only evidence to support this is our finding of multiple restriction site polymorphisms in and 5' of the 221 gene [26]. It remains to be seen, however, whether these polymorphisms are due to modification or mutation of nucleotides.

ii) The telomeric route for VSG gene activation is

required to allow changes in the order in which VSG genes are expressed. For a discussion of this idea we refer to Borst et al. [17].

We prefer explanation (ii), but none of the explanations offered here can be ruled out nor is compelling enough to convince us that we know why trypanosomes handle VSG gene activation in such a complex fashion. Continuing efforts to relate the biology of trypanosome infection in the field to the molecular biology of VSG gene activation in the lab should eventually allow the unraveling of this complex story.

ACKNOWLEDGEMENTS

We thank Mrs M.M.W.Van der Bijl, Mr A.H.Schinkel, Mr C.R.Lincke, Mr J.M.Kooter, Mr P.A.J.Leegwater and Mr F.Eijgenraam for contributing to the unpublished experiments mentioned. This work was supported in part by grants from the UNDP/World Bank/WHO Special Programme for Research and Training in Tropical Diseases ((TYR)T16/181/T7/34) and from the Foundation for Fundamental Biological Research (BION), which is subsidized by The Netherlands Organization for the Advancement of Pure Research (ZWO). The collaborative experiments in New York were made possible by a Short-term Travel Grant (ASTF 3958) from the European Molecular Biology Organization (EMBO) to Mr L.H.T.Van der Ploeg.

REFERENCES

1. Vickerman K (1969). On the surface coat and flagellar adhesion in trypanosomes. J.Cell Sci. 5, 163.
2. Cross, GAM (1975). Identification, purification and properties of clone-specific glycoprotein antigens constituting the surface coat of Trypanosoma brucei. Parasitol. 71, 393.
3. Capbern A, Giroud C, Baltz T, Mattern P (1977). Trypanosoma equiperdum: Etudes des variations antigéniques au cours de la trypanosome expérimentale du lapin. Exptl.Parasitol. 42, 6.
4. Van der Ploeg LHT, Valerio D, De Lange T, Bernards A, Borst P, Grosveld FG (1982). An analysis of cosmid clones of nuclear DNA from Trypanosoma brucei shows that the genes for variant surface glycoproteins are clustered in the genome. Nucl.Acids Res. 10, 5905.
5. Vickerman K (1978). Antigenic variation in trypanosomes. Nature 273, 613.

6. Englund PT, Hajduk SL, Marini JC (1982). The molecular biology of trypanosomes. Ann.Rev.Biochem. 51, 695.
7. Borst P, Cross GAM (1982). Molecular basis for trypanosome antigenic variation.Cell 29, 291.
8. Hoeijmakers JHJ, Frasch ACC, Bernards A, Borst P, Cross GAM (1980). Novel expression-linked copies of the genes for variant surface antigens in trypanosomes. Nature 284, 78.
9. Pays E, Van Meirvenne N, Le Ray D, Steinert M (1981). Gene duplication and transposition linked to antigenic variation in Trypanosoma brucei. Proc.Natl.Acad.Sci. U.S. 78, 2673.
10. Van der Ploeg LHT, Bernards A, Rijsewijk FAM, Borst P (1982). Characterization of the DNA duplication-transposition that controls the expression of two genes for variant surface glycoproteins in Trypanosoma brucei. Nucl.Acids Res. 10, 593.
11. Van der Ploeg LHT, Liu AYC, Michels PAM, De Lange T, Borst P, Majumder HK, Weber H, Veeneman GH, Van Boom J (1982). RNA splicing is required to make the messenger RNA for a variant surface antigen in trypanosomes. Nucl.Acids Res. 10, 3591.
12. Boothroyd JC, Cross GAM (1982). Transcripts coding for different variant surface glycoproteins of Trypanosoma brucei have a short, identical exon at their 5'-end. Gene 20, 279.
13. Michels PAM, Liu AYC, Bernards A, Sloof P, Van der Bijl MMW, Schinkel AH, Menke HH, Borst P, Veeneman GH, Tromp MC, Van Boom JH (1983). Activation of the genes for variant surface glycoproteins 117 and 118 in Trypanosoma brucei. J.Mol.Biol., in press.
14. Liu AYC, Van der Ploeg LHT, Rijsewijk FAM, Borst P (1983). The transposition unit of VSG gene 118 of Trypanosoma brucei: Presence of repeated elements at its border and absence of promoter-associated sequences. J.Mol.Biol., in press.
15. Bernards A, Van der Ploeg LHT, Frasch ACC, Borst P, Boothroyd JC, Coleman S, Cross GAM (1981). Activation of trypanosome surface glycoprotein genes involves a gene duplication-transposition leading to an altered 3'-end. Cell 27, 497.
16. Borst P, Bernards A, Van der Ploeg LHT, Michels PAM, Liu AYC, De Lange T, Sloof P, Veeneman GH, Tromp MC, Van Boom JH (1983). DNA rearrangements controlling the expression of genes for variant surface antigens in trypanosomes. In: 5th John Innes Symp. on Biological

Consequences of DNA Structure and Genome Arrangement, Norwich, Croom Helm, London, in press.

17. Borst P, Bernards A, Van der Ploeg LHT, Michels PAM, Liu AYC, De Lange T, Sloof P (1983). Gene rearrangements controlling the expression of surface antigen genes in trypanosomes. In: Proc. Cetus-UCLA Meeting on Molecular Biology of Host-Parasite Interactions (Agabian N, Eisen H, Eds), Park City, Utah, in press.

18. Williams RO, Young JR, Majiwa PAO (1979). Genomic rearrangements correlated with antigenic variation in Trypanosoma brucei. Nature 282, 847.

19. Williams RO, Young JR, Majiwa PAO, Doyle JJ, Shapiro SZ (1981). Contextural genomic rearrangements of variable-antigen genes in Trypanosoma brucei. Cold Spring Harbor Symp.Quant.Biol. 45, 945.

20. Majiwa PAO, Young JR, Englund PT, Shapiro SZ, Williams RO (1982). Two distinct forms of surface antigen gene rearrangements in Trypanosoma brucei. Nature 297, 514.

21. Borst P, Frasch ACC, Bernards A, Van der Ploeg LHT, Hoeijmakers JHJ, Arnberg AC, Cross, GAM (1981). DNA rearrangements involving the genes for variant antigens in Trypanosoma brucei. Cold Spring Harbor Symp.Quant.Biol. 45, 935.

22. De Lange T, Borst P (1982). Genomic environment of the expression-linked extra copies of genes for surface antigens of Trypanosoma brucei resembles the end of a chromosome. Nature 299, 451.

23. Bernards A, Michels PAM, Lincke CR, Borst P (1983). Growth of chromosome ends in multiplying trypanosomes. Nature, in press.

24. Bernards A (1982). Transposable genes for surface glycoproteins in trypanosomes. TIBS 7, 253.

25. Williams RO, Young JR, Majiwa PAO (1982). Genomic environment of T. brucei VSG genes: Presence of a mini-chromosome. Nature 299, 417.

26. Bernards A, De Lange T, Michels PAM, Huisman MJ, Borst P (1983). Surface antigen gene 221 of Trypanosoma brucei is activated without gene duplication and lies near a chromosome end. Cell, submitted.

27. Sloof P, Bos JL, Konings AFJM, Menke HH, Borst P, Gutteridge WE, Leon W (1983). Characterization of satellite DNA in Trypanosoma brucei and Trypanosoma cruzi. J.Mol.Biol., in press.

28. Boothroyd JC, Cross, GAM, Hoeijmakers JHJ, Borst P (1980). A variant surface glycoprotein of Trypanosoma

brucei is synthesized with a hydrophobic carboxy-terminal extension absent from purified glycoprotein. Nature 288, 624.
29. Rice-Ficht AC, Chen KK, Donelson JE (1981). Sequence homologies near the C-termini of the variable surface glycoproteins of Trypanosoma brucei. Nature 294, 53.
30. Majumder HK, Boothroyd, JC, Weber H (1981). Homologous 3'-terminal regions of mRNAs for surface antigens of different antigenic variants of Trypanosoma brucei. Nucl.Acids Res. 9, 4745.
31. Sloof P, Menke HH, Caspers MPM, Borst P (1983). Size fractionation of Trypanosoma brucei DNA: Localization of the 177-bp repeat satellite DNA and a variant surface glycoprotein gene in a mini-chromosomal DNA fraction. Nucl.Acids Res., submitted.
32. Fisher MP, Dingman CW (1971). Role of molecular conformation in determining the electrophoretic properties of polynucleotides in agarose-acrylamide composite gels. Biochemistry 10, 1895.
33. Aaij C, Borst P (1972). The gel electrophoresis of DNA. Biochim.Biophys.Acta 269, 192.
34. Schwartz DC, Saffran W, Welsh J, Haas R, Goldenberg M, Cantor CR (1983). New techniques for purifying large DNA's and studying their properties and packaging. Cold Spring Harbor Symp.Quant.Biol. 47, in press.
35. Frasch ACC, Borst P, Van den Burg J (1982). Rapid evolution of genes coding for variant surface glycoproteins in trypanosomes. Gene 17, 197.
36. Longacre S, Hibner U, Raibaud A, Eisen H, Baltz T, Giroud C, Baltz D (1983). DNA rearrangements and antigenic variation in Trypanosoma equiperdum: Multiple expression linked sites in independent isolates of trypanosomes expressing the same antigen. Mol.Biochem.Parasitol., in press.
37. Borst P (1983). Antigenic variation in trypanosomes. in: Mobile Genetic Elements (Shapiro J, Ed.), Academic Press, New York, pp. 621.
38. Haber JE (1983). Mating type genes of Saccharomyces cerevisiae. in: Mobile Genetic Elements (Shapiro J, Ed.), Academic Press, New York, pp. 560.
39. Fangman WL (1978). Separation of very large DNA molecules by gel electrophoresis. Nucl.Acids Res. 5, 653.
40. Beach D, Nurse P, Egel R (1982). Molecular rearrangement of mating-type genes in fission yeast. Nature 296, 682.

41. Miller EN, Turner MJ (1981). Analysis of antigenic types appearing in first relapse populations of clones of *Trypanosoma brucei*. Parasitology 82, 63.

42. Hajduk SL, Vickerman K (1981). Antigenic variation in cyclically transmitted *Trypanosoma brucei*. Variable antigen type composition of the first parasitaemia in mice bitten by trypanosome-infected *Glossina morsitans*. Parasitology 83, 609.

43. Parsons M, Nelson R, Agabian N (1983). Genomic organization of variant surface glycoprotein genes in T. brucei procyclic culture forms. J.Cell.Biochem. Suppl. 7A, 22.

44. Laurent M, Pays E, Magnus E, Van Meirvenne N, Matthyssens G, Williams RC, Steinert M (1983). DNA rearrangements linked to expression of a predominant surface antigen gene of trypanosomes. Nature 302, 263.

Gene Expression, pages 437-444
© 1983 Alan R. Liss, Inc., 150 Fifth Avenue, New York, NY 10011

MOLECULAR CLONING OF HUMAN LYMPHOKINES

H. Cheroutre, R. Devos, G. Plaetinck, S. Scahill, W. Degrave, J. Tavernier, Y. Taya, and W. Fiers

Laboratory of Molecular Biology, State University of Ghent
9000 Ghent, BELGIUM

ABSTRACT Poly A^+ RNA isolated from mitogen-induced human splenocytes was purified by oligo-d(T)-cellulose chromatography and further fractionated by sucrosegradient centrifugation. The translational capacity of each fraction was evaluated in a wheat-germ extract. By assaying the incubation medium of *Xenopus laevis* oocytes microinjected with mRNA, fractions containing mRNA coding for IFN-γ (immune interferon) and IL-2 (interleukin 2) were identified. Double stranded cDNA was then synthesized and cloned into the eukaryotic expression vector pSV529. To identify the recombinant plasmids containing either the IFN-γ cDNA or IL-2 cDNA, mRNA derived from mitogen-induced splenocytes was hybridized to plasmid-inserted cDNA bound onto nitrocellulose filters. The hybrid-selected mRNA was translated in oocytes and assayed for antiviral activity or T-cell growth activity. These procedures allowed us to identify plasmids containing the entire γ-IFN cDNA sequence and plasmids containing the human IL-2 specific cDNA sequence. The IFN-γ containing recombinant plasmid was expressed by transfected AP8 monkey cells. Cotransformation of this recombinant plasmid together with a plasmid containing a mouse dihydrofolate reductase (DHFR) gene pAdD26SV(A)-3 has been used to establish Chinese hamster ovary (CHO) cell lines which secrete high levels of Hu IFN-γ .

INTRODUCTION

Proteins secreted by lymphocytes, differing from

immunoglobulins and acting as effector molecules on the immune system are defined as "lymphokines". Although some lymphokines and their producer cells have been characterized, the knowledge of their field of action and their various biological activities is still very incomplete. Here we report on the molecular cloning of two human lymphokines, immune interferon (IFN-γ) and interleukin-2 (IL-2), both released by T-lymphocytes in response to several mitogens or antigens.

Interferon-γ (Type II) shares antiviral and anti-cell proliferating activity with the viral induced interferon-β and interfereon-α (Type I). However, IFN-γ apparently acts more as an immunoregulatory molecule (1-3). It is found that in some immune deficiencies, the production of IFN-γ is reduced (4), while in some autoimmune diseases it seems to increase (5).

IL-2, also known as T-cell growth factor (TCGF), allows for the long-term proliferation of activated immunocompetent T-cells (6) and NK cells (7). This finding has stimulated several areas of immunological research. The use of IL-2 makes it possible to select, in vitro, single clones of T-cell subpopulations with known antigen specificity and to use them as immunological tools in cell typing and perhaps in immunotherapy. (8-9).

METHODS

Spleen Cultures

Human spleens, preferably from traumatic cases, were used for large scale preparation of lymphocytes (10). Splenocytes were incuced with SEA (11) and cultured for 72 h at 37°C for optimal production of IFN-γ. Induction of IL-2 was optimal in the presence of PHA and TPA (12) during 24 h at 37°C.

Isolation and Translation of RNA

Total RNA was isolated from mitogen-induced splenocytes and the poly A^+RNA was purified on an oligo-d(T)-cellulose column (10). Poly A^+ RNA was further fractionated on a 5-20% sucrose gradient. In order to evaluate the translational capacity of the mRNA, aliquots of each fraction were translated in a wheat-germ extract in the presence of ^{35}S-methionine followed by an analysis of the product on a 12.5% SDS-polyacrylamide gel. To detect the fraction which contained IFN-γ mRNA or IL-2 mRNA, we injected each fraction into <u>Xenopus laevis</u> oocytes (13). After incubation for 72 h at 20° C, we assayed the incubation medium for its antiviral activity or T-cell growth activity. The antiviral assay for IFN-γ was performed on T21 cells (14) using EMC virus as the challenge virus (15). For assaying IL-2, mononuclear cells isolated from human peripheral blood were used as target cells. These cells were made sensitive to IL-2 by stimulation with PHA. Growth was determined by incorporation of tritiated thymidine (^{3}H-TdR) (16).

Construction of cDNA Plasmids and Screening of the Clones

cDNA was synthesized from sucrose gradient purified mRNA fractions under optimal conditions using AMV reverse transcriptase followed by second strand synthesis, S-I treatment and sizing on a 4% polyacrylamide gel (17). Ds DNA of an appropriate length was isolated, purified and tailed with oligo dG. The cloning vector pSV529 (18) was tailed with oligo dC at its unique Bam H1 site and annealed with the ds DNA. This recombinant vector was used to transform either <u>E. coli</u> strain DH1 or strain HB101. CsCl-gradient purified plasmid DNA, derived from groups of 50 clones, was digested with Bam H1 to excise the inserted DNA. After purification on a sucrose gradient, this insert DNA was bound on nitrocellulose filters and hybridized with SEA-induced or PHA-TPA induced splenocyte poly A^{+}RNA (19). The eluted mRNA was injected into oocytes of <u>Xenopus laevis</u> and the oocyte incubation medium was assayed for antiviral or T-cell growth activity.

Restriction Enzyme Analysis and DNA Sequence Determination

Plasmid DNA and sucrose gradient purified insert DNA was analyzed by cleavage with different restriction endonucleases. The nucleotide sequence was determined by the chemical degradation method of Mascam and Gilbert (20).

RESULTS AND DISCUSSION

Incubation of human splenocytes during 72 h in the presence of SEA was found to be optimal for the production of IFN-γ. For production of IL-2, however, TPA and PHA were used as a mitogen and cultured for 24 h. RNA was isolated from the splenocytes as described. About 30 mg total RNA was obtained per spleen, this in turn yielded approximately 1 mg poly A^+RNA (of which more than 60% is still rRNA) after one single oligo-(dT)-cellulose chromatographic step. We subsequently analyzed the activity of the sucrose gradient purified mRNA fractions by microinjecting them into *Xenopus laevis* oocytes and assaying the secreted translation products. Antiviral activity was found in the fractions of mRNA which sediment around 13S; the IL-2 activity was found among the fractions which contain mRNA which sediment around 10S (Fig. 1).

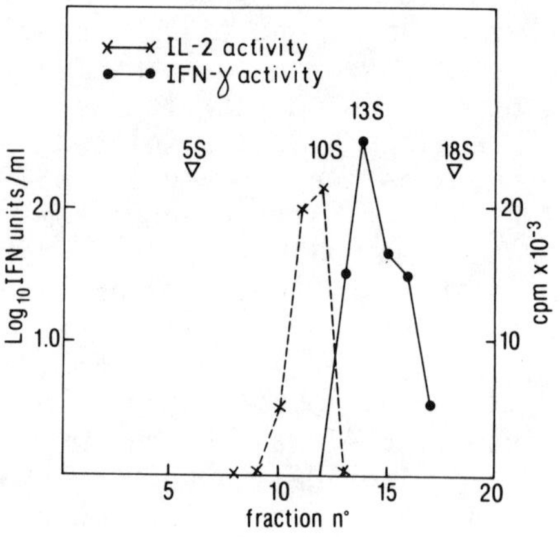

FIGURE 1. Antiviral activity and H-thymidine (^{3}H-TdR) incorporation obtained after microinjection of *Xenopus laevis* oocytes with splenocyte derived poly A^+ RNA, fractionated on a non-denaturing sucrose gradient. IFN-γ activity is expressed in laboraory units. IL-2 activity is expressed in counts per min.

The length of the mRNA used for reverse transcription was determined by electrophoresis of the corresponding cDNA in an alkaline agarose gel (21). A length between 900 and

1400 nucleotides was found for the IFN-γ mRNA and we found a length between 600 and 1200 bp for IL-2 mRNA. cDNA was synthesized on 40 μg poly A^+ RNA using oligo d(T) as primer, followed by second strand synthesis and SI treatment. dsDNA was sized on a 4% polyacrylamide gel. We eluted three groups of dsDNA ranging in length from between 1400-1250 bp (G2), 1250 and 1000 bp (G3) and 1000-800 bp (G4) for IFN-γ. In the case of IL-2, we eluted four groups between 1300-1100 bp (M1), 1100-750 bp (M2), 750-600 bp (M3) and 600-500 bp (M4). This dsDNA was inserted into the unique Bam H1 site of pSV529 DNA and used for transformation of *E. coli* strain DH1 or strain HB101. Plasmid DNA was prepared from mixtures of 50 individual clones. Next, a positive screeing of the clones by cDNA-mRNA hybridization followed by translation in oocytes was performed. To enhance the sensitivity of this method, only the sucrose gradient purified Bam H1 excised insert DNA was bound to nitrocellulose filters. A schematic summary of the screening procedure is given in Table 1.

TABLE 1
METHOD OF SCREENING FOR AN INDIVIDUAL cDNA CLONE

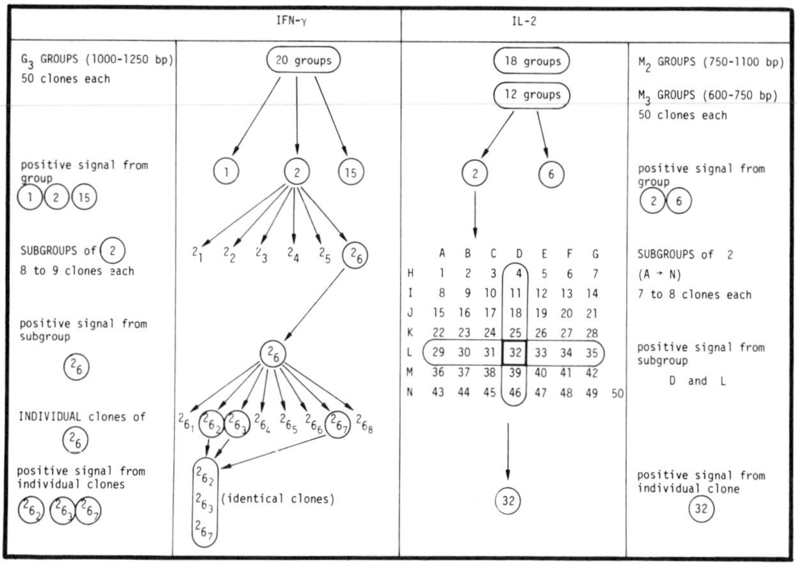

In the case of IL-2, we used a "grid-method" which allowed us to isolate directly an individual clone. A complete IFN-γ cDNA sequence was identified by hybridization of the remaining clones using a ^{32}P-labelled internal fragment derived from a positive clone. Insert DNA of the positive clones was analyzed by digestion with restriction enzymes and the nucleotide sequence of the cDNA insert was determined (22) by the chemical degradation method of Mascam and Gilbert. In the case of IL-2 cDNA, we could not find an initiation start codon which indicated that the insert DNA was not full size. We are now screening a human genomic library in order to pick up the missing sequences.

We transfected AP8 monkey cells with the IFN-γ clone and characterized the resulting product in order to prove that the isolated clone did actually code for human IFN-γ.

We also cotransformed Chinese hamster ovary (CHO) cells, deficient in dihydrofolate reductase ($DHFR^-$) (24), with the plasmid pAdD26SV(A)-3 (25) which encodes a selectable DHFR marker, and the plasmid pSV2-IFN-γ which encodes Hu IFN-γ. We then selected the transformants by means of the marker gene and examined them for their ability to express the IFN-γ gene by the assay system described above. Several high producers were selected and their production of Hu IFN-γ significantly exceeded that of stimulated lymphocytes.

We also isolated the human IFN-γ chromosomal gene from a human DNA library cloned in λ Charon 4A phage (23). The data obtained revealed the presence of three introns in the the IFN-γ chromosomal gene. A schematic representation of the organization of the IFN-γ gene is given in Figure 2.

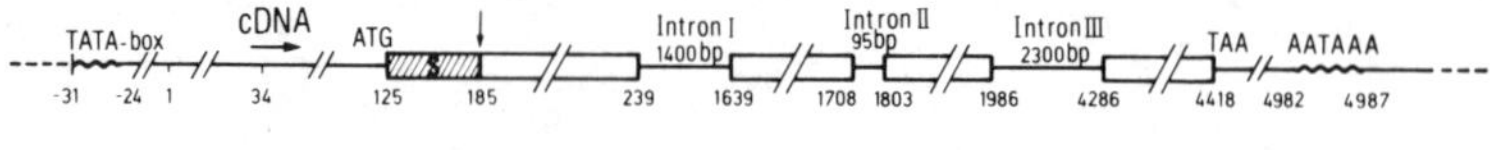

Figure 2. Schematic representation of the organization of the IFN-γ gene. (The boxed parts are translated regions.)

ACKNOWLEDGEMENTS

We thank the surgeons of UCL-St. Luc Hospital, Brussels; the Academisch Ziekenhuis, Ghent; the Erasmus Hospital, Brussels; the St. Pierre Hospital, Brussels and the St. Jan Algemeen Ziekenhuis, Brugge for procuring human spleens for us. We are also grateful to W. Burm, C. Seurinck and J. Vanderheyden for carrying out the interferon assay, the plasmid preparation and the tranfection experiments. F. Shapiro is acknowledged for assistance with the manuscript and W. Drijvers for the illustrations.

REFERENCES

1. Sonnenfeld G, Mandel AD, Merigan TC (1977). Cell Immun. 34:193-206.
2. Vizelizier JL, Chan EL, Allison AC (1977). Clin. Exp. Immun. 30:299-304.
3. Wallach D. Human Lymphokines, Biological Immune Response Modifiers. Khan A, Hill NO (eds), New York: Academic Press pp. 435-449.
4. Epstein LB, Ammann AJ (1974). J. Immunol. 112:617-626.
5. Hooks JJ, Moutsapoulos HM, Geis SA, Stahl NI, Decker JL Notkins AL (1979). New Engl. J. Med. 301:5-8.
6. Morgan DA, Ruscetti FW, Gallo RC (1976). Science 193:1007.
7. Henney CS, Kuribayashi K, Kern D, Gillis S (1981). Nature 291:335.
8. Bonnard GD, Yasaka K, Jacobson D, Fine RL, Morgan DA Herberman RB (1979). Cell Biology and Immunology of Leukocyte Function. Proceedings of the 12th Leukocyte Culture Conference. New York: Academic Press, p. 569.
9. Ruscetti FW, Gallo RC (1981). Blood 57:379.
10. Devos R, Cheroutre H, Taya Y, Fiers W (1982). J. of Interferon Research 2(3):409-420.
11. Wallace DM, Hitchcock MJM, Reber SB, Berger SL (1981). Biochem. Biophys. Res. Commun. 100:865-871.
12. Moretta A, Colombatti M, Chapius B (1981). Clin. Exp. Immunol. 44:262-269.
13. Cavalieri RL, Havell EA, Vilček J, Pestka S (1977). Proc. Natl. Acad. Sci. USA 74:3287-3291.
14. Tan YH, Schneider EL, Tischfield Y, Epstein CJ, Ruddle FH (1974). Science 186:61-63.

15. Berger SL, Hitchcock MJM, Zoon KC, Birkenmeier CS, Friedman RM, Chang EH (1980). J. Biol. Chem. 255:2955-2961.
16. Gillis S, Fern MM, Ou W, Smith KA (1978). J. Immunol. 120:2027-2032.
17. Devos R, Van Emmelo J, Contreras R, Fiers W (1979). J. Mol. Biol. 128:595-619.
18. Gheysen D, Fiers W (1982). J. of Mol. and Appl. Genetics 1(no. S):385-394.
19. Parnes JR, Velan B, Felsenfeld A, Ramanathan L, Ferrini U, Appella E, Seidman JG (1981). Proc. Natl. Acad. Sci. USA 78:2253-2257.
20. Maxam AM, Gilbert W (1977). Proc. Natl. Acad. Sci. USA 74:560-564.
21. McDonell MW, Simon MN, Studier FW (1977). J. Mol. Biol. 110:119-146.
22. Devos R, Cheroutre H, Taya Y, Degrave W, Van Heuverswyn H, Fiers W (1982). Nucl. Acids Research 10(8):2487-2501.
23. Taya Y, Devos R, Tavernier J, Cheroutre H, Engler G, Fiers W (1982). The EMBO Journal 1(8):953-958.
24. Urlaub G, Chasin LA (1980). Proc. Natl. Acad. Sci. USA 77:4216-4220.
25. Kaufman RJ, Sharp PA (in press). Mol. Cell Biol.

Gene Expression, pages 445-455
© 1983 Alan R. Liss, Inc., 150 Fifth Avenue, New York, NY 10011

THE CHICKEN H2b GENE FAMILY

David K. Grandy[†], James D. Engel[§],
and Jerry B. Dodgson[†]

[†]Departments of Microbiology and
Public Health and Biochemistry
Michigan State University
East Lansing, Michigan, 48824

[§]Department of Biochemistry,
Molecular and Cell Biology
Northwestern University
Evanston, Illinois, 60201

ABSTRACT

This paper describes recent progress in our analysis of the chicken H2b histone gene family. Using a previously identified and sequenced H2b gene we have shown that chickens contain 7-8 H2b histone genes per haploid genome. The nucleotide sequence of a second H2b histone gene is also described. A comparison between this DNA sequence and that of the previously sequenced chicken H2b gene, H2b-1a, revealed that both genes are continuous and code for identical proteins 125 amino acids long. In contrast the non-coding flanking sequences 5' to the two genes share little sequence homology. A small set of sequences has been conserved, however, whose elements are found associated with certain other eucaryotic genes. One of these conserved sequences, ATTTGCATA, is positioned 26 bp 3' to the CCAAT box and 6 bp 5' to the TATA box in both genes. An analysis of DNA sequence data from other species suggests that this nine nucleotide sequence is unique to members of the H2b gene family and may serve a gene-specific regulatory function.

INTRODUCTION

The histones form the primary protein constituent of chromatin in eucaryotic organisms. Therefore, the regulation of the structure and expression of the histone proteins represent possible controlling elements at which major alterations in total cellular gene expression could occur. The synthesis of histone proteins is generally under cell cycle control (1,2) with the exception of rapidly dividing sea urchin embryos (3). The expression of certain histones is also developmentally regulated. The two best documented examples of this are metazoan spermatogenesis (4,5) and the maturation of nucleated erythrocytes in birds, fish, reptiles and amphibians (6-9).

The possible importance of histone-DNA interactions in achieving cellular differentiation together with the existence of developmental programs regulating the expression of certain histone genes has led us to examine the fine structure of these genes in the chicken. The examination of a wide variety of histone gene-containing chicken DNA λCharon 4A recombinants showed that members of the five primary histone gene families (H1, H2a, H2b, H3 and H4) were often but not always linked in chicken chromosomal DNA (10-13), but not in any ordered or repeating manner such as had previously been observed in invertebrate histone gene families (1, 14,15). Moreover, restriction mapping of these recombinants demonstrated that each histone gene was located in a unique region of chicken chromosomal DNA that bore no obvious (large scale) resemblance to the DNA surrounding other members of the same family.

We have therefore chosen to make a more thorough investigation of all members of a specific histone gene family, the H2b histone genes, to see if these genes are expressed in an equivalent or differential manner and to examine their nucleotide sequence for elements which might regulate this expression. We initially reported the first complete sequence of a vertebrate H2b gene (16). This paper describes the sequence of a separate unlinked H2b histone gene and the comparison of the two H2b gene structures. Similar (probably identical or allelic) chicken H2b gene sequences have been recently described by Harvey et al. (17).

RESULTS AND DISCUSSION

Analysis of the Complete Chicken H2b Gene Family.

Our initial molecular cloning results (10) were in approximate agreement with the initial report (18) that chickens contain about 10 members of each primary histone gene family per haploid genome. We have analyzed the repetition frequency of the H2b gene family in greater detail by Southern blotting analysis of total chicken DNA using our previously isolated and sequenced chicken H2b gene (H2b-1a) (16)as hybridization probe. Total chicken DNA was cut with EcoRI, electrophoresed on 0.7% agarose, blotted, and hybridized to a labeled H2b histone gene BstEII fragment internal to the coding regions of H2b-1a (Figure 1). Six primary bands are observed, one or two of which seem to contain two gene equivalents, linked on a single DNA fragment or on fragments of identical size. One faint band of considerably less than one gene equivalent is also observed. Thus it appears that there are about 7-8 H2b histone genes per haploid chicken genome. (The chicken DNA used was sperm DNA from two or three roosters of an inbred line provided by the USDA Regional Poultry Research Labs, East Lansing, Michigan, so it is unlikely that any of the bands represent heterozygous polymorphisms.)

Construction of a Chicken Histone H2b Gene-Containing Plasmid.

The isolation of the λ Charon 4A histone gene recombinants λCH1a and λCH2e have been described (10), as has the subclonong and sequence analysis of the H2b gene on λCH1a (H2b-1a) (16). The restriction map of λCH2e is shown in Figure 2A. The 3.5 kb* EcoRI fragment containing H2b sequences (as identified previously, 10) was subcloned by standard procedures (10). A partial restriction map of this plasmid, designated pRR2e-3.5, is shown in Figure 2B.

DNA sequence analysis of pRR2e-3.5.

The nucleotide sequence of the H2b gene and its 5' flanking sequence on pRR2e-3.5 was determined by the method

*The abbreviations used are: kb, kilobase pairs; bp, base pairs.

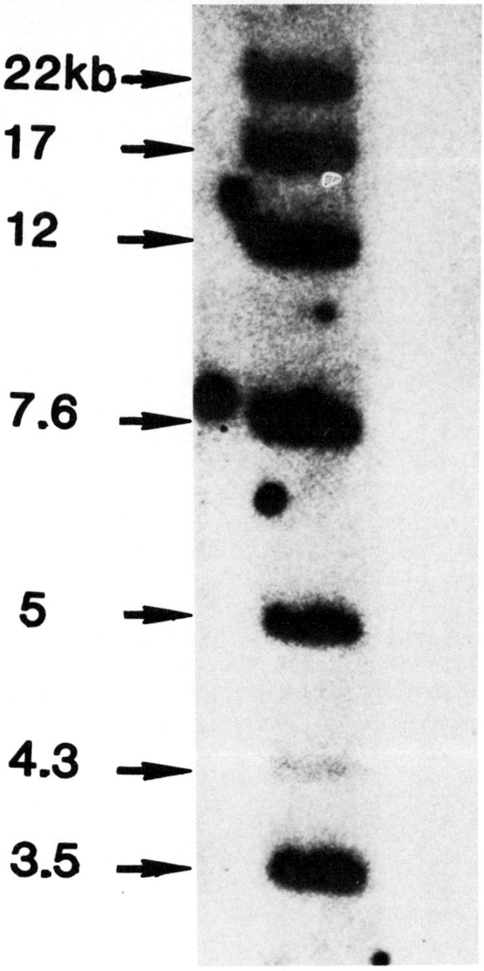

Figure 1. Identification of the chicken H2b histone gene family. Chicken sperm DNA was cut with EcoRI, electrophoresed, transferred to nitrocellulose and hybridized to a nick-translated 0.3 kb BstEII DNA fragment internal to the coding region of H2b-1a.

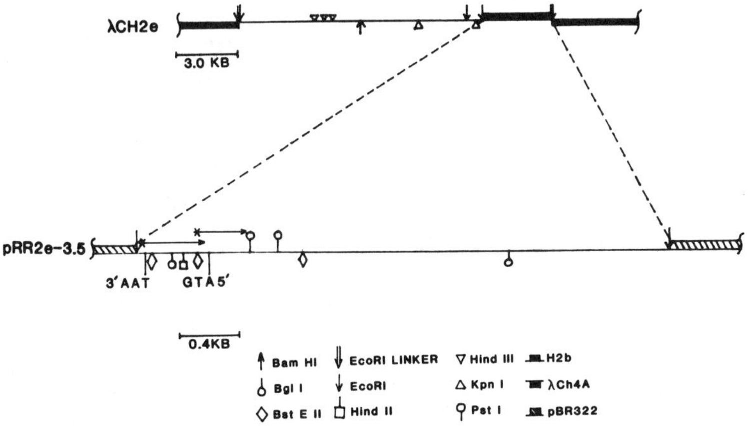

Figure 2. Restriction maps of λCH2e and pRR2e-3.5. A, a portion of the clone λCH2e containing the chicken DNA insert is shown. B, a partial restriction map of the recombinant plasmid pRR2e-3.5. The arrows indicate the direction and extent to which the DNA sequence was read from individual experiments. The H2b coding region is positioned on this map within the initiation codon (ATG) and the termination codon (TAA) as designated.

of Maxam and Gilbert (19) as modified by Smith and Calvo (20) and is presented in Figure 3. This H2b gene was identified by the 125 amino acid sequence it predicts, which is identical to that of the chicken H2b gene previously isolated (on plasmid pKR1a-1.3) (16). A comparison of the two H2b genes reveals that their coding sequences differ in eight nucleotides (Figure 3). These point mutations are clustered at the 5' end of the H2b-2e gene in codons 1, 4, 7, 10 and 14 and at the 3' end in codons 96, 112 and 114. In each case the nucleotide polymorphism occurs in the degenerate third position of the codon and is silent with respect to the amino acid sequence. The effect of these

```
                                          1a 5' AATTGTTATCCAGTAGCAGCGAGGGATTTGTGATGGACCA
                                          2e -200 CGAATCGAATTTGAACCAATGAAAAGCAGTTATAAGGGAG

-160
ATTACAGAGCCGTTCGTATCGAGCGACACGTCACGGACTCTGTTATCCAATCAGAGAGCAGATACAGAAGGCACTCGATT
GAAAGGCGTGTTCTCGAGTTCCGACCAATGAAAGAGTGCGAAAGGAATGCTTCTCATTTGCATAGAGGGGCTATAAATAA

-80
TGCATACTGCCCCTATAAATAGGCGAGCAGTGCTCGCAGCCGGCACTCCGCTGCGCCGAAGGGATCGTGGAGAGTTCGAC
ATGCCTACGACCCCTTCGTTTCCATTCAGCGTCTCCTGGTCGTTTTTGTTCGCCTCGCTTCGTGAGCGCGTTGTGCCACT

          T           C           A           C               C
+1   ATG CCC GAG CCG GCT AAG TCC GCG CCC GCC CCG AAG AAG GGC TCT AAG AAG GCG GTC ACC
         Pro Glu Pro Ala Lys Ser Ala Pro Ala Pro Lys Lys Gly Ser Lys Lys Ala Val Thr

+61  AAG ACC CAG AAG AAG GGC GAC AAG AAG CGC AAG AAG AGC CGC AAG GAG AGC TAC TCG ATC
     Lys Thr Gln Lys Lys Gly Asp Lys Lys Arg Lys Lys Ser Arg Lys Glu Ser Tyr Ser Ile

+121 TAC GTG TAC AAG GTG CTG AAG CAG GTG CAC CCC GAC ACG GGC ATC TCG TCC AAG GCC ATG
     Tyr Val Tyr Lys Val Leu Lys Gln Val His Pro Asp Thr Gly Ile Ser Ser Lys Ala Met

+181 GGC ATC ATG AAC TCG TTC GTC AAC GAC ATC TTC GAG CGC ATC GCC GGC GAG GCG TCG CGC
     Gly Ile Met Asn Ser Phe Val Asn Asp Ile Phe Glu Arg Ile Ala Gly Glu Ala Ser Arg

                                                                      A
+241 CTG GCG CAC TAC AAC AAG CGC TCG ACC ATC ACG TCG CGG GAG ATC CAG ACG GCC GTG CGG
     Leu Ala His Tyr Asn Lys Arg Ser Thr Ile Thr Ser Arg Glu Ile Gln Thr Ala Val Arg

                                                      C       T
+301 CTG CTG CTG CCC GGC GAG CTG GCC AAG CAC GCG GTC TCG GAG GGC ACC AAG GCG GTC ACC
     Leu Leu Leu Pro Gly Glu Leu Ala Lys His Ala Val Ser Glu Gly Thr Lys Ala Val Thr

                              G   3'
+361 AAG TAC ACC AGC TCC AAG TAA +381
     Lys Tyr Thr Ser Ser Lys
```

Figure 3. The nucleotide sequence of two chicken H2b histone genes. The DNA sequence corresponding to the mRNA strand is presented for both genes. The predicted amino acid sequence is given below the sequence of H2b-2e. The entire 5' region flanking H2b-1a is shown above the 2e sequence. The underlined sequences are those which are shared by both 1a and 2e and are discussed in the text. Only those bases which are unique to H2b-1a are shown above the protein-coding sequences of H2b-2e.

changes on the primary structure of the H2b-2e gene is to increase the number of rare CpG dinucleotides to 40 compared to 37 in H2b-1a and to reduce the number of A or T residues in the third codon position from 4 in 1a to 2 in 2e. As described previously (13,16), most histone genes contain a very high percentage of G-C base pairs in degenerate sites leading to the high %GC of the coding sequence and the comparatively high frequency of CpG dinucleotides. In addition to the silent coding-sequence polymorphisms the stop codon of H2b-2e differs from that of H2b-1a; the former being UAA (ochre) and the latter being UAG (amber).

Sequence conservation in the 5' gene-flanking regions of H2b-1a and H2b-2e.

In an attempt to identify conserved sequences which may play a role in regulating the expression of chicken H2b histone genes we have examined the noncoding 5' regions flanking the two genes. It is apparent from the sequence data (Figure 3) that H2b-1a and H2b-2e present very divergent 5' gene-flanking sequences. However, contained within these divergent sequences are blocks of homology which have been conserved. The general lack of homology outside of the coding sequences and consensus promoter sequences suggests that individual H2b histone gene-specific probes for transcribed sequences can be constructed to analyze the expression of specific chicken H2b genes. Further work along these lines is in progress. By analogy with other histone genes (14) a putative mRNA initiation locus or cap site, CA$^{C}_{T}$TC, has been identified in the 5' region of both genes. This "cap box" is located 32 bp and 53 bp upstream of the initiation codon and 22 bp and 23 bp downstream of the TATA box (see below) in H2b-1a and H2b-2e respectively. The precise locations of the sites at which the transcripts are initiated have not yet been determined. The AT-rich block which bears a very strong resemblance to the Hogness (21) or TATA box consensus sequence associated with histones and other genes transcribed by RNA polymerase II (14, 21) is identified in H2b-1a and H2b-2e by the sequence, CTATAAATA.

Another distinct nucleotide sequence often found associated with RNA polymerase II-transcribed genes is given by the canonical sequence CCAAT (14, 22). Both H2b genes in fact contain two such elements in their promoter regions. The first such element is found in both H2b genes 42 bp 5' to the TATA box. The second CCAAT box is located 43 bp and

44 bp 5' to the first CCAAT box in H2b-1a and H2b-2e respectively. The significance of a repeated CCAAT box is unknown, however similar duplications have been observed in an embryonic α-globin gene (23) and in certain fetal γ-globin genes (22). A summary of these consensus sequences and their locations in the flanking regions of H2b-1a and H2b-2e is presented in Figure 4.

In addition to the putative sequence elements recognized by the RNA polymerase II transcription apparatus a sequence of nine nucleotides, ATTTGCATA, has been identified 26 bp 3' from the 3'-most CCAAT boxes and 6 bp 5' to the TATA box in both H2b gene-flanking regions. The function, if any, of this AT-rich sequence is unknown, however it may serve some H2b gene-specific regulatory role. This supposition is supported by the fact that the sequence has been well conserved through evolution. An analysis of the DNA sequence data available for a variety of H2b genes reveals that this sequence occurs in the sea urchin species *S. purpuratus* (1, 24) and *P. miliaris* (1, 25, 26) and a nearly identical sequence has been found 6 bp 5' to the TATA box in an *X. laevis* H2b gene (27). A summary of these homologies is presented in Figure 5. A similar observation has been made by Harvey et al. (17) in their analysis of chicken H2b gene sequences.

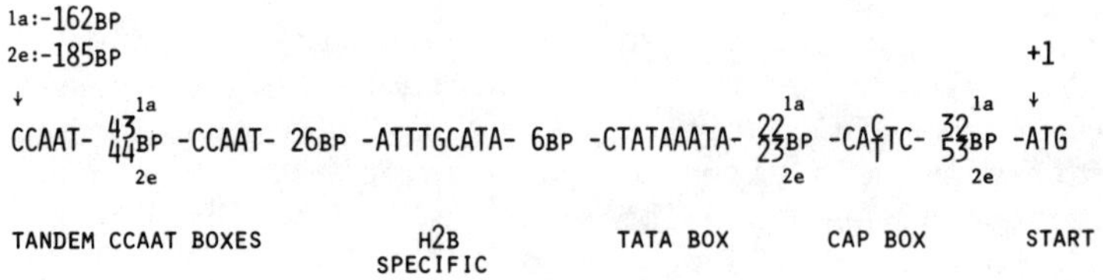

Figure 4. Sequence and spacing homologies found in the 5' flanking regions of two chicken H2b genes. This is a summary of the conserved sequences found in both genes and discussed in the text.

CH H2B-1A	CCAAT···	26BP	···ATTTGCATA···	6BP	···CTATAAATA···	59BP	···ATG
CH H2B-2E	CCAAT···	26BP	···ATTTGCATA···	6BP	···CTATAAATA···	81BP	···ATG
X. LAEVIS	CCAGT···	26BP	···ATTTGCATG···	6BP	···CTATAAATA···	59BP	···ATG
S. P.-2	CCAAT···	22BP	···ATTTGCATA···	30BP	··TATAAAAAG···	96BP	···ATG
P. M. H19	CCAAT···	22BP	···ATTTGCATA···	32BP	··TATAAAAAG···	96BP	···ATG
P. M. H22	CCAAT···	31BP	···ATTTGCATA···	26BP	··TATAAAGAG···	97BP	···ATG

H2B SPECIFIC
CONSENSUS SEQUENCE: ATTTGCAT$^{A}_{G}$

Figure 5. A conserved nine nucleotide sequence unique to H2b genes. A comparison of the available DNA sequence data from several species reveals the presence of a well conserved sequence element located between the CCAAT and TATA boxes in the noncoding 5' region flanking H2b genes. Sequences are from chicken (see figure 3), S. purpuratus (1, 24), P. miliaris (1, 25, 26) and X. laevis (27).

The consensus sequence, GATCC, which has been found associated with several histone genes (14) was not found associated with H2b-1a or H2b-2e. This sequence is also missing from the 5' region flanking Drosophila (14) and Xenopus (27) H2b genes and therefore its significance is unclear.

ACKNOWLEDGMENTS

This work was supported by Grant GM 29687 from the National Institutes of Health and by Grant 1-759 from the March of Dimes Birth Defects Foundation. Journal Article No. 10823 from the Michigan Agricultural Experiment Station.

REFERENCES

1. Kedes LH (1979). Histone genes and histone messengers. Ann Rev Biochem 48:837.
2. Hereford LM, Osley MA, Ludwig JR III, McLaughlin CS (1981). Cell cycle regulation of yeast histone mRNA.

Cell 24:367.
3. Woodland HR (1980). Histone synthesis during the development of Xenopus. FEBS Letters 121:1.
4. Geraci G, Lanciere M, Marchi P, Noviello L (1979). The sea urchin (Sphaerechinus granularis) codes different H2b histones to assemble sperm and embryo chromatin. Cell Diff 8:187.
5. Seyedin SM, Kistler WS (1979). H1 histone subfractions of mammalian testes. 2. Organ specificity in mice and rabbits. Biochemistry 18:1376.
6. Molgaard HV, Perrucho M, Ruiz-Corrillo A (1980). Histone H5 messenger RNA is polyadenylated in chicken erythrocytes. Nature (London) 283:502.
7. Lewin B (1980). "Gene Expression 2." New York: Wiley-Interscience, pp 311-312.
8. Destree OHS, Hoenders HJ, Moorman AFM, Charles R (1979). Histone of Xenopus laevis erythrocytes Purification and characterization of the lysine-rich fractions. Biochim Biophys ACTA 577:61.
9. Risley MS, Eckhardt RA (1981). H1 histone variants in Xenopus laevis. Dev Biol 84:79.
10. Engel JD, Dodgson JB (1981). Histone genes are clustered but not tandemly repeated in the chicken genome. Proc Nat Acad Sci USA 78:2856.
11. Harvey RP, Wells JRE (1979). Isolation of a genomal clone containing chicken histone genes. Nucl Acids Res 7:1787.
12. Harvey RP, Krieg PA, Robins AJ, Coles LS, Wells JRE (1981). Non-tandem arrangement and divergent transcription of chicken histone genes. Nature (London) 294:49.
13. D'Andrea R, Harvey RP, Wells JRE (1981). Vertebrate histone genes: nucleotide sequence of a chicken H2a gene and regulatory flanking sequences. Nucl Acids Res 9:3119.
14. Hentschel CC, Birnstiel ML (1981). The organization and expression of histone gene families. Cell 25:301.
15. Lifton RP, Goldberg ML, Karp RW, Hogness DS (1977). The organization of the histone genes in Drosophila melanogaster: functional and evolutionary implications. Cold Spring Harbor Symp Quant Biol 42:1047.
16. Grandy DK, Engel JD, Dodgson JB (1982). Complete nucleotide sequence of a chicken H2b histone gene. J Biol Chem 257:8577.
17. Harvey RP, Robins AJ, Wells JRE (1982). Independently evolving chicken histone H2b genes: identification of a ubiquitous H2b-specific 5' element. Nucl Acids Res 10:7851.

18. Crawford RJ, Krieg P, Harvey RP, Hewish DA, Wells JRE (1979). Histone genes are clustered with a 15-kilobase repeat in the chicken genome. Nature (London) 279:132.
19. Maxam AM, Gilbert W (1980). Sequencing end-labeled DNA with base-specific chemical cleavages. Methods Enzymol 65:499.
20. Smith DR, Calvo JM (1980). Nucleotide sequence of the E. coli gene coding for dihydrofolate reductase. Nucl Acids Res 8:2255.
21. Salser W (1978). Globin mRNA sequences: analysis of base pairing and evolutionary implications. Cold Spring Harbor Symp Quant Biol 42:985.
22. Efstradiatis A, Posakony JW, Maniatis T, Lawn RM, O'Connell C, Spritz RA, DeRiel JK, Forget BG, Weissman SM, Slightom JL, Blechl AE, Smithies O, Baralle FE, Shoulders CC, Proudfoot NJ (1980). The structure and evolution of the human β-globin gene family. Cell 21:653.
23. Engel JD, Rusling DJ, McCune KC, Dodgson JB (1983). Unusual structure of the chicken embryonic α-globin gene, π'. Proc Natl Acad Sci USA 80:1392.
24. Sures I, Lowry J, Kedes LH (1978). The DNA sequence of sea urchin (S. purpuratus) H2a, H2b and H3 histone coding and spacer regions. Cell 15:1033.
25. Busslinger M, Portmann R, Irminger JC, Birnstiel ML (1980). Ubiquitous and gene-specific regulatory 5' sequences in a sea urchin histone DNA clone coding for histone protein variants. Nucl Acids Res 8:957.
26. Schaffner W, Kung G, Daetwyler H, Telford J, Smith HO, Birnstiel ML (1978). Genes and spacers of cloned sea urchin histone DNA analyzed by sequencing. Cell 14:655.
27. Moorman AFM, DeBoer PAJ, DeLaaf RTM, Destree OHJ (1982). Primary structure of the histones H2a and H2b genes and their flanking sequences in a minor histone gene cluster of Xenopus laevis. FEBS Letters 144: 235.

Gene Expression, pages 457-465
© 1983 Alan R. Liss, Inc., 150 Fifth Avenue, New York, NY 10011

FUNCTIONAL ANALYSIS OF HUMAN GLOBIN GENE MUTANTS[1]

Stuart H. Orkin

Division of Hematology-Oncology, Children's Hospital, Dana-Farber Cancer Institute, Department of Pediatrics, Harvard Medical School, Boston, Mass. 02115

ABSTRACT Naturally occurring mutations of the human β-globin gene lead to the reduced synthesis of β-globin that characterizes β-thalassemia. Through combined analysis of DNA polymorphisms and gene cloning numerous different mutant alleles have been identified. Study of several of these in a transient, heterologous expression system has afforded various insights into aspects of RNA processing and gene transcription.

INTRODUCTION

β-thalassemia is an inherited disorder in which the synthesis of β-globin polypeptide is reduced or absent from an affected allele (1). The condition is clinically heterogeneous and distributed worldwide. In view of this, it seemed likely that the molecular basis for the disease was also heterogeneous. This has proved to be the case. In order to define the spectrum of mutations leading to the β-thalassemic phenotype it was necessary to develop a systematic approach to the dissection of the disorder. Through classification of alleles on the basis of various DNA polymorphisms detectable in the β-globin gene cluster by restriction mapping mutant genes within a population that are likely to differ from one another can be identified (2,3). Upon gene cloning in bacteriophage mutant β-globin genes can be isolated for detailed structural analysis by DNA

[1]This work was supported by grants from the National Institutes of Health and the Birth Defects Foundation, as well as a Research Career Development Award of the NIH.

sequencing and for functional study by various expression assays. This general approach to the molecular analysis of mutations causing the disease has led to characterization of defects involving virtually all aspects of gene expression (3). In this brief review I will focus on those newly discovered β-gene mutants that shed particular light on either RNA processing or gene transcription.

RNA PROCESSING MUTANTS

RNA transcripts from globin genes initially contain intervening sequences that are precisely excised during RNA processing. Certain sequences features have been identified at the borders of the coding blocks (or exons) and intervening sequences. These include the 5'-GT and 3'-AG dinucleotides, and general consensus sequences for these regions (4). Mutations that affect the invariant GT at the 5'- or donor splice site prevent normal RNA splicing at the exon-intervening sequence junction (5,6,7). Deletion of the 3'-AG plus its preceding pyrimidine tract completely prevented splicing at the end of the first intervening sequence in a recently discovered β-thalassemic gene (8). These mutant genes reflect those most impaired in splicing functions. Several examples of new splicing sites within intervening sequences have also been observed in thalassemia due to the creation of new internal GTs or AGs in the appropriate sequence context within an intervening sequence (7,9,10,11).

Of great interest are two thalassemia genes that represent a new class of RNA processing mutant. Rather than containing nucleotide substitutions at the splice junctions themselves, two genes revealed changes within the consensus sequence of the IVS-1 donor site at nucleotides 5 and 6 (numbering the GT as 1 and 2, respectively) (3) (see Figure 1). The phenotype produced by these genes is especially informative. On the one hand, the flexibility of consensus sequences of normal genes implied that these nucleotide substitutions might be inconsequential. On the other hand, their discovery in genes isolated from patients with disorders of β-globin production suggested true functional significance. Expression of these genes in the transient, heterologous cell system described by Treisman et al (6) revealed that the mutations reduce the efficiency of splicing at the normal donor site and also activate cryptic donor-like splice sites within exon-1 and within IVS-1 (see

Figure 1) (7). Thus, these nucleotide substitutions

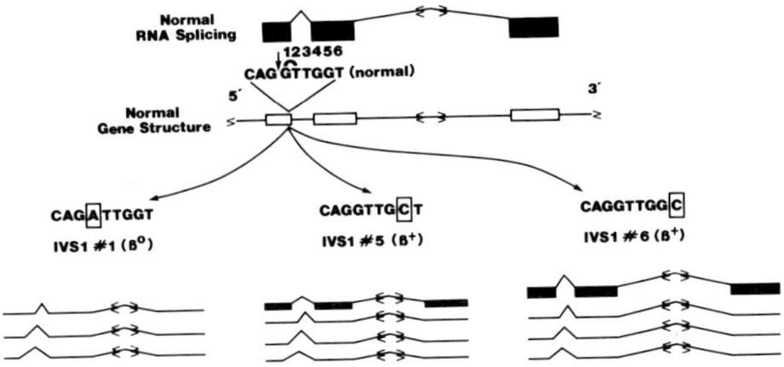

Figure 1. Nucleotide substitutions of the 5'-IVS-1 consensus sequence. The normal gene sequence is shown in the center with the pattern of RNA processing shown at the top. Below are the three point mutations that have been found in this region (3,7) and their associated RNA processing patterns. The heights of the coding blocks depict the relative level of normal β-RNA. See reference 7 for details.

adversely affect the manner by which the splicing apparatus discriminates between the normal and preexisting cryptic sites located elsewhere within the RNA transcript. Based on the relative amounts of normally processed globin RNA produced from these genes it appeared that the position 5 mutation was more functionally significant than the position 6 change. This prediction of the heterologous expression system is borne out by observation of the phenotypes of patients homozygous for these two different alleles. Those with the position 5 gene are typical thalassemia patients with a transfusion requirement due to very reduced β-globin synthesis. However, individuals with the position 6 mutant gene have especially mild disease due to only moderate reduction in β-globin production. These findings provide striking correlation between a heterologous test expression system and the in vivo phenotype produced by these mutations within an erythroblast.

Another novel class of RNA processing mutant is represented by a gene encoding a β-globin variant known as β^E. Mutation of G→A at exon nucleotide 129 (within codon 26) leads to the production of the β^E-globin chain characterized by a glutamic acid to lysine substitution. For some time it has been appreciated that this variant is underproduced in erythroid cells, i.e. it has a thalassemic phenotype (12,13). We sought an explanation for this association by analysis of cloned β^E-globin genes. The entire DNA sequence of a β^E-gene showed no additional mutations (14). Therefore, the phenotype associated with the variant must be attributed solely to the codon 26 mutation. In the heterologous expression system RNA transcripts were processed abnormally in two ways: (i) excision of IVS-1 sequences was delayed and (ii) alternative splicing into exon-1 occurred neighboring the codon 26 mutation itself (14) (Figure 2). Inspection of the DNA sequence of the exon demonstrated that the codon 26 mutation changed a preexisting donor-like sequence to one that more closely resembled an authentic splice site (Figure 3). In addition, this cryptic site coincided precisely with one of the exon sites utilized in processing transcripts from mutant genes with substitutions within the IVS-1 splice junction or consensus sequence. Thus, the β^E-mutation activated a preexisting cryptic splice site and reflected a new variety of gene dysfunction: abnormal RNA processing due to an exon mutation. More recently, a β-thalassemic gene with a mutation in codon 24 has been shown to have similar molecular pathology (15). These mutations illustrate again how subtle nucleotide changes can alter relative splicing at normal and cryptic sites.

TRANSCRIPTION MUTANTS

Site-specific and deletion mutagenesis of various eukaryotic genes have identified DNA sequences located upstream or 5'- that are essential for high level transcription (16). These include a proximal region at about -30 (relative to the cap site), the ATA box homology, and a more distal segment from -70 to -90 in which the CAT box homology is situated (17). Three different β-globin genes from thalassemic individuals have been found with nucleotide

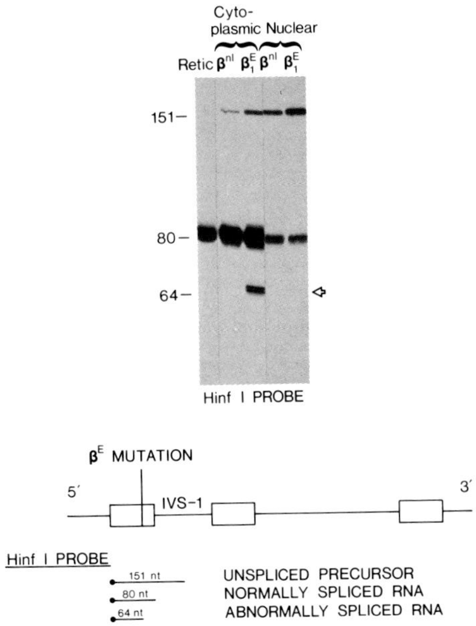

Figure 2. Abnormal processing of β^E-RNA transcripts. Normal and β^E-globin genes were transfected into HeLa cells. Nuclear and cytoplasmic RNA fractions were prepared after 24 hr and subjected to S1 nuclease mapping with the probe shown at the bottom. Note the presence of abnormally spliced RNA arising from the β^E-gene and increase in precursor containing IVS-1 sequences. See reference 14 for details.

substitutions in the upstream sequences. The first, a C→G change at position -87 discovered in a Mediterranean patient, is located within a conserved sequence ACACCC just upstream from the CAT box (3). When this gene is introduced into HeLa cells in the transient expression assay, it directs 10-fold less globin RNA than the normal gene (7). Truncation to position -128 has established that the phenotype is produced by the -87 change rather than a substitution further upstream. Two independent substitutions at position -28 have been observed: A→C in the gene of a Kurdish patient (18); and A→G in a Chinese (19). The

Consensus Sequence: $^{A}_{C}$AG↓GT$^{A}_{G}$AGT

Normal IVS-1 Donor: CAG↓GTTGGT

Exon-1 Cryptic Donors:

GTG↓GTGAGG —— Beta^{E} ——→ GTG↓GTAAGG

AAG↓GTGAAC

IVS-1 Cryptic Donor:

AAG↓GTTACA

Figure 3. Normal and cryptic donor sites in exon-1 and IVS-1. The consensus sequence for a donor site is shown at the top. Below the normal IVS-1 donor site the cryptic donors contributing to the abnormal RNA species illustrated in Figure 1 are displayed. The mutation in the β^{E}-gene alters one of these cryptic sites to a donor-like sequence that can be utilized at a low frequency.

latter gene has been introduced into HeLa cells and directs 3-5-fold less RNA than normal (19). In vivo β-RNA is about 10-fold reduced relative to α-RNA in erythroid cells. The identification of transcription mutants in β-thalassemia strengthens the belief that the DNA sequences previously identified only by in vitro manipulations of eukaryotic genes and expression in heterologous cells are, in fact, physiologically significant in vivo. These findings support the further use of heterologous cell expression systems for the characterization of naturally occurring mutant genes. The existence of sequence-specific transcription initiation factors, as that found in HeLa cells by Davison et al (20), may explain the similar behavior of the -28 A→G mutant gene in the test system and the erythroblast. The mechanism by which the -87 change exerts its down-regulation effect on transcription is unknown at the present time.

CONCLUSIONS

Studies of globin genes found in thalassemic individuals has provided novel mutant genes for functional studies. These substrates for molecular analysis have demonstrated the complexity of RNA processing and the exquisite sensitivity with which the splicing apparatus normally discriminates between appropriate and inappropriate splicing sites. Furthermore, we have been led to discover how coding region mutations may affect RNA processing. The transcription mutants have provided the best evidence yet for the functional significance of controlling sequences identified in heterologous systems. These results demonstrate the relevance of the upstream sequences to expression in the normal cellular environment. Consideration of rather surprising mutants, such as the -87 gene, may provide insights into those sequence features that are recognized during the binding of RNA polymerase or other proteins to DNA in the transcription process. The extent to which additional naturally occurring mutant genes will increase our understanding of gene expression may depend on the diligence with which we search for new candidates and explore the precise pathologic consequences of each new allele.

ACKNOWLEDGMENTS

The studies briefly summarized here were performed in collaboration with Haig H. Kazazian, Jr. and members of his laboratory at the Johns Hopkins University School of Medicine and with Richard Treisman and Tom Maniatis at Harvard University. I am indebted to them for their valuable contributions.

REFERENCES

1. Weatherall DJ, Clegg JB (1982). Thalassemia revisited. Cell 29:7-9.
2. Antonarakis SE, Boehm CD, Giardina PJV, Kazazian HH Jr (1982). Nonrandom association of polymorphic restriction sites in the β-globin gene cluster. Proc Natl Acad Sci USA 79:137-141.
3. Orkin SH, Kazazian HH Jr, Antonarakis SE, Goff SC, Boehm CD, Sexton JP, Waber PG, Giardina PJV (1982). Linkage of β-thalassaemia mutations and β-globin gene polymorphisms with DNA polymorphisms in the human β-globin gene cluster. Nature 296:627-631.

4. Breathnach R, Chambon P (1981). Organization and expression of eucaryotic split genes coding for proteins. Annu Rev Biochem 50:349-383.
5. Treisman R, Proudfoot NJ, Shander M, Maniatis T (1982). A single base change at a splice site in a β^0-thalassemic gene causes abnormal RNA splicing. Cell 29:903-911.
6. Felber B, Orkin SH, Hamer DH (1982). Abnormal RNA splicing causes one form of α-thalassemia. Cell 29: 895-902.
7. Treisman R, Orkin SH, Maniatis T (1983). Specific transcription and RNA splicing defects in five cloned β-thalassaemia genes. Nature, in press.
8. Orkin SH, Sexton JP, Goff SC, Kazazian HH Jr (1983). Inactivation of an RNA acceptor splice site by a small deletion in β-thalassemia. J Biol Chem, submitted.
9. Busslinger M, Moschonas N, Flavell RA (1981). β^+-thalassemia: aberrant splicing results from a single point mutation in an intron. Cell 27:289-298.
10. Fukamaki Y, Ghosh PK, Benz EJ Jr, Reddy VB, Lebowitz P, Forget BG, Weissman SM (1982). Abnormally spliced messenger RNA in erythroid cells from patients with β^+-thalassemia and monkey kidney cells expressing a cloned β^+-thalassemia gene. Cell 28:585-593.
11. Spence SE, Pergolizzi RG, Donovan-Pelluso M, Kosche KA, Dobkin CS, Bank A (1982). Five nucleotide changes in the large intervening sequence of β-globin gene in a β^+-thalassemia patient. Nucleic Acids Res 10:1283-1290.
12. Benz EJ Jr, Berman BW, Tonkonow BL, Coupal E, Coates T, Boxer LA, Altman A, Adams JG (1981). Molecular analysis of the β-thalassemia phenotype associated with inheritance of hemoglobin E. J Clin Invest 68:118-126.
13. Traeger J, Wood WG, Clegg JB, Weatherall DJ (1980). Defective synthesis of HbE is due to reduced levels of β^E-mRNA. Nature 288:497-499.
14. Orkin SH, Kazazian HH Jr, Antonarakis SE, Ostrer H, Goff SC, Sexton JP (1982). Abnormal RNA processing due to the exon mutation of the β^E-globin gene. Nature 300:768-769.
15. Humphries RK, Ley TJ, Goldsmith ME, Cline A, Kantor JA, Nienhuis AW (1982). Silent nucleotide substitution in β^+-thalassemia gene activates a cryptic splice site in globin RNA coding sequence. Blood 50:54a.

16. McKnight SL, Kingsbury R (1982). Transcriptional control signals of a eucaryotic protein-coding gene. Science 217:316-324.
17. Efstratiadis A, Posakany JW, Maniatis T, Lawn RM, O'Connell C, Spritz RA, deRiel JK, Forget BG, Weissman SM, Slightom JL, Blechl AE, Smithies O, Baralle RE, Shoulders CC, Proudfoot NJ (1980). The structure and evolution of the human β-globin gene family. Cell 21: 653-668.
18. Poncz M, Ballantine M, Solowiejczyk D, Barak I, Schwartz E, Surrey S (1982). β-thalassemia in a Kurdish Jew. J Biol Chem 257:5994-5996.
19. Orkin SH, Kazazian HH Jr, Cheng T-C, Sexton JP, Goff SC, in preparation.
20. Davison BL, Egly J-M, Mulvihill ER, Chambon P (1983). Formation of stable preinitiation complexes between eukaryotic class B transcription factors and promoter sequences. Nature 301:680-686.

Gene Expression, pages 467-479
© 1983 Alan R. Liss, Inc., 150 Fifth Avenue, New York, NY 10011

COORDINATE AMPLIFICATION OF METALLOTHIONEIN I AND II GENE SEQUENCES IN CADMIUM-RESISTANT CHO VARIANTS[1]

C. E. Hildebrand, B. D. Crawford,[2] M. D. Enger, B. B. Griffith,[2] J. K. Griffith,[2] J. L. Hanners, P. J. Jackson, J. Longmire, A. C. Munk, J. G. Tesmer, R. A. Walters

Genetics Group, Los Alamos National Laboratory
Los Alamos, NM 87545

and

R. L. Stallings

University of Texas System Cancer Center
Smithville, TX 78957

ABSTRACT Cadmium-resistant (Cd^r) variants of the Chinese hamster cell line, CHO, have been derived by stepwise selection for growth in medium containing $CdCl_2$. These variants show coordinately increased production of both metallothionein (MT) I and II and were stably resistant to Cd^{2+} in the absence of continued selection. Genomic DNAs from these Cd^r sublines were analyzed for both MT gene copy number and MT gene organization, using cDNA sequence probes specific for each of the two Chinese hamster isometallothioneins. These analyses revealed coordinate amplification of MT I and II genes in all Cd^r variants which had increased copies of MT-encoding sequences. *In situ* hybridization of an MT-encoding probe to mitotic chromosomes of a Cd^r variant, which has amplified MT

[1]This work supported by the United States Department of Energy and the Los Alamos National Laboratory.

[2]Present address: Department of Cell Biology
Cancer Research and Treatment Center
University of New Mexico School of Medicine
Albuquerque, NM 87131

genes at least 14-fold, revealed a single chromsomal site of hybridization. These results suggest that the isoMTs constitute a functionally related gene cluster which amplifies coordinately in response to toxic metal stress.

INTRODUCTION

The phenomenon of gene amplification has been observed in both bacteriophage and bacteria [reviewed in (1)] and in eukaryotic cells _in vivo_ and in culture [reviewed in (2)]. Specific gene amplification has been reported for developmentally regulated genes (3-6), genes conferring drug resistance (7,8), cellular homologs of viral oncogenes (9), and genes conferring resistance to heavy metal toxicity (10-12).

In the context of resistance to heavy metal toxicity, we have examined the regulation of metallothionein gene expression in the cadmium sensitive (Cd^s) Chinese hamster line, CHO, and in Cd-resistant (Cd^r) variants derived from these cells. Production of high affinity metal-binding proteins, the metallothioneins (MTs), has been shown to be a major factor contributing to the stable Cd^r phenotype (10-16). It is of interest that synthesis of two major isometallothioneins is induced coordinately in Cd^r variants in response to subtoxic exposures to $CdCl_2$ or $ZnCl_2$ (15). In this paper we examine the role of amplification of the genes encoding the two major isoMTs in regulation of isoMT expression and in conferring the Cd^r phenotype. Results of preliminary studies on the chromosomal organization of these genes, examined in one Cd^r variant, may provide insight concerning the coordinate regulation of the isoMTs and the stability of the Cd-resistance phenotype.

RESULTS

Cd^{2+}-resistant CHO sublines have been derived by exposure of the CHO cell in monolayer culture to stepwise increases in $CdCl_2$ levels (13,14). After continuous growth in medium containing stepwise increases in Cd^{2+} concentration Cd-resistant (Cd^r) variants were cloned and characterized for (a) stability of the Cd^r phenotype in the absence of selective pressure, (b) Cd^{2+} uptake and intracellular partitioning, and (c) metallothionein (MT) expression (13, 14, 16). Phenotypic characteristics of the CHO cell and four Cd^r variants resistant to $CdCl_2$ concentrations from 2 to 200 μM in cell culture medium are summarized in Table I. All of the Cd^r variants are stably resistant to Cd^{2+} during long-term growth (up to 135 population doublings) in the absence of selective pressure.

TABLE I

Phenotypic Characteristics of CHO Cells and Cd^r CHO Variants

Cell Line	Toxic Threshold for $CdCl_2$ Exposure (μ$\underline{M}$ $CdCl_2$)	Basal MT Synthesis Rate[b]	Maximally Induced MT Synthesis Rate[b]
CHO	0.2	--[c]	--[c]
Cd^r2C10	2.0	--[c]	28.3
Cd^r20F4	26	--[c]	60.6
Cd^r30F9	40	--[c]	41.7
Cd^r200T1	145	40	320

[a] Methods for the measurement of toxic threshold for $CdCl_2$ exposure have been published (14).

[b] Relative MT synthesis rate measurement and calculation has been described (15).

[c] MT synthesis was undetectable in these cells under the conditions given.

Increased production of MT in response to Cd^{2+} exposure (both rate of induction and maximal level of expression) is a major factor in development of increased cellular Cd^{2+}-resistance. The Cd^r30F9 variant was an interesting exception to the correlation between increased MT expression and increased Cd^{2+}-resistance (14). Nondenaturing polyacrylamide gel electrophoresis revealed coordinate induction of both major isoMTs in all Cd^r variants (Fig. 1, 15). Given the metal-loading capacity and molecular weight of MT, measurements of Cd^{2+} uptake into MT indicate that in maximally-induced Cd 200T1 cells, MT could represent at least 2% of total cytoplasmic protein.

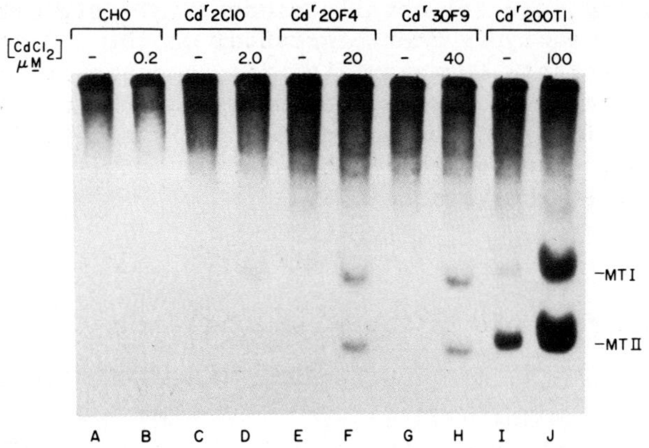

FIGURE 1. Nondenaturing polyacrylamide gel electrophoresis of cytoplasmic fractions from CHO cells and Cd^r variants. Cells were exposed to indicated concentrations of $CdCl_2$ and pulse-labeled with ^{35}S-cysteine for 30 minutes prior to the time of maximal MT synthesis (14-16).

To determine whether the MT overproduction phenotype of of the Cd^r variants was a consequence of genotypic alteration in the genes encoding either or both of isoMTs, genomic DNA from CHO cells and each of the Cd^r variants was purified and analyzed for MT gene organization and copy number. These analyses used specific MT-encoding and 3' non-coding sequence probes derived from recombinant cDNA clones for the two major Chinese hamster isoMTs [pCHMT1 and pCHMT2, (17)]. Figure 2 diagrams the strategy for derivation of DNA sequence probes for MT protein encoding regions and for 3' non-coding regions specific for MTI and MTII genes. In control experiments, filter hybridization analyses of linearized plasmid pCHMT1 and pCHMT2 DNAs using the MT protein coding region probe derived from pCHMT2 provided cross-hybridization with pCHMT1 under conditions of high stringency ($\geq$ 80% homology cut-off) as expected from the 81% nucleotide sequence homology in the protein coding region. In contrast, under the same hybridization stringency conditions, probes derived from the 3'-untranslated regions of pCHMT1 and pCHMT2 showed hybridization to their homologous plasmids but no cross-hybridization to their nonhomologous plasmids. The properties of these pCHMT1- and pCHMT2-derived cDNA probes permitted analysis of both the organization and dosage of MT genes in the CHO cells and in the Cd^{2+}-resistant variants.

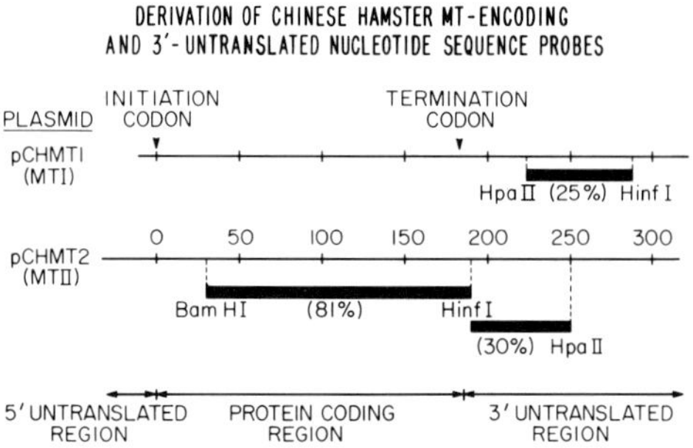

FIGURE 2. Strategy for derivation of Chinese hamster MT-encoding and 3'-untranslated nucleotide sequence probes. Heavy solid lines indicate regions used for probe construction. Numbers in parentheses under heavy lines represent sequence homology between pCHMT1 and pCHMT2 in the designated regions (17).

Sequence representation of MT structural genes in CHO cells and the CHO Cd^r variants were estimated 1) by nucleic acid reassociation kinetic analyses using the MTII-encoding structural gene probe (Fig. 2) as tracer driven by genomic DNAs isolated from the different cell lines, and 2) by Southern blotting and filter hybridization analysis using the same probe and DNAs from the same variants. The reassociation kinetics of Chinese hamster pCHMT2 DNA with total genomic DNA from the CHO cells and each of the cadmium resistant variant cells is shown in Fig. 3. Similar results were been obtained using an independently purified set of genomic DNAs and a 63 nucleotide HinfI-HpaII restriction fragment from pCHMT2 (Fig. 2) as tracer. When genomic DNA was omitted from the reaction mixes, the rate of pCHMT2 DNA self-reassociation was slow relative to that measured in the presence of CHO DNA indicating that tracer self reassociation does not significantly affect the hybridization kinetics observed in Fig. 3. Based upon extrapolation from Cot 1/2 values of the respective hybridization reactions, MT-like gene sequences are amplified approximately 1x, 7x, 3x, and 14x in the genomes of Cd^r2C10, Cd^r20F4, Cd^r30F9, and Cd^r200T1 respectively, relative to their abundance in the genome of the cadmium-sensitive, parental CHO cell. By comparison with the rate of reaction of the slowest kinetic component in 300 nucleotide long total, genomic Chinese hamster DNA (10), we estimate that the complement of MT genes in CHO cells is near single-copy levels.

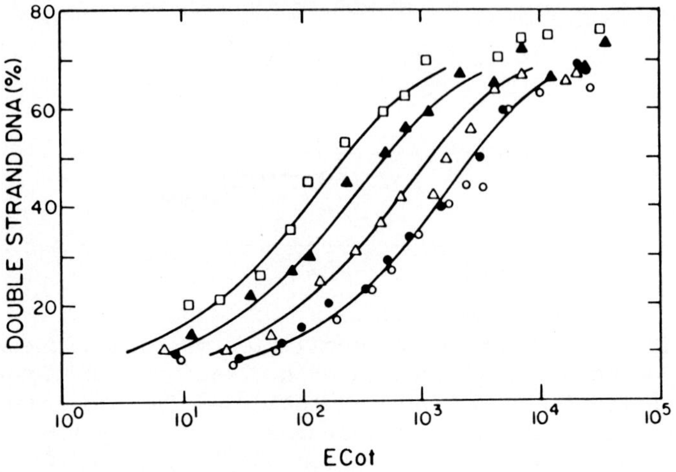

FIGURE 3. Nucleic acid reassociation kinetic analyses of pCHMT2 with genomic DNAs from CHO (○), Cd^r2C10 (●), Cd^r20F4 (▲), Cd^r30F9 (△), and Cd^r200Tl cells (□). Methods for extraction, shearing, annealing, and hydroxylapatite chromatography of Chinese hamster cell DNA have been described (10).

To investigate further how the amplified MT genes are organized in the genomes of the Cd^r variants, electrophoretically-resolved, restriction endonuclease digests of genomic DNA from CHO cells and each of the Cd^r variants were analyzed by filter hybridization with the MT-coding region probe (Fig. 2). Two major bands of hybridization, at ~2.3 kb and ~17-19 kb, were observed in the HindIII digests of Cd^r20F4, Cd^r30F9 and Cd^r200Tl. This result suggested that these two hybridization bands may represent MTI and II genes since MTI and II coding regions share extensive homology (17). Further, the results in Fig. 4 indicate that the MT genes are amplified coordinately.

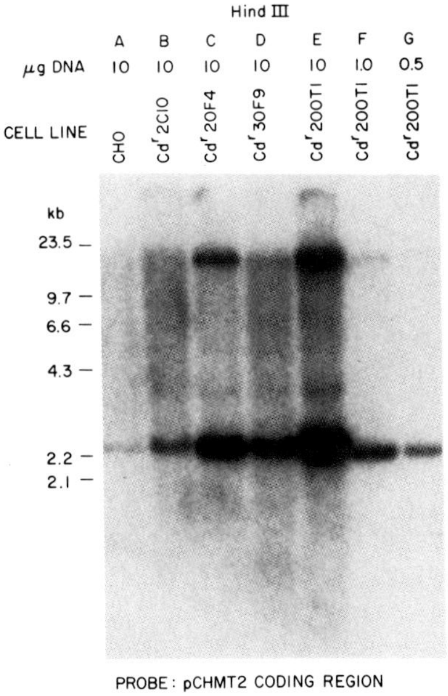

FIGURE 4. Filter hybridization analyses of genomic DNAs from CHO cells and Cd^r variants. DNA was digested with HindIII according to conditions specified by the supplier, electrophoretically resolved in a 1% agarose gel, transferred to Gene Screen filters (New England Nuclear Corp.), and hybridized with the pCHMT2 protein-coding region probe (Fig. 2). Hybridization was for 24-48 hr at 60°C according to procedures specified for Gene Screen by New England Nuclear Corp. Filter washes were also performed according to procedures specified by the supplier. In independent digests the high molecular weight fragment (17-19 kb) is also observed in CHO and Cd^r2C10 cells. The lower intensity observed for the 17-19 kb fragment may reflect comparative specificity of the pCHMTII probe for the 2.3 kb fragment. Alternatively, less efficient transfer of high molecular weight DNA fragments may explain this result.

To test the hypothesis of coordinate amplification, genomic DNA from the Cd^r200T1 was digested with different restriction endonucleases (HindIII, BamHI, and EcoRI) and analyzed by filter hybridization with both MT-encoding structural gene probe and the MTI- and MTII-specific cDNA probes obtained from the 3' untranslated regions of pCHMT1 and pCHMT2 inserts. The results of these filter hybridization analyses (Fig. 5A) reveal two primary bands of hybridization (of approximately equal intensity) with the MTII protein-encoding region probe in each of the digests described above. In contrast, when digested with Hind III or BamHI Cd^r200T1 DNA yielded two fragments which hybridized differentially to the MTI- and MTII-specific 3' untranslated region probes (Fig. 5B) probes. These fragments were of the same apparent size as the HindIII and BamHI fragments which hybridized with MT coding region probes (Fig. 5A). The two fragments generated by Eco RI did not show differential hybridization to the isoMT-specific probes. The results with HindIII and BamHI demonstrate that the genomic DNA sequences encoding MTI and MTII are amplified coordinately in the Cd^r200T1 subline. Further the similarity of fragment sizes hybridizing the MT-protein coding region probe in both HindIII (Fig. 4) and BamHI (data not shown) digests of all Cd^r variants indicates that the MTI and II genes are amplified coordinately in all of these cell lines. In independent filter hybridization measurements, comparison of hybridization of the MTII encoding probe with CHO genomic DNA and dilutions of Cd^r200T1 genomicDNA revealed that the MT gene copy number in Cd^r200T1 may be as high as 40-50 fold greater than the copy number in the CHO cells (Fig. 4).

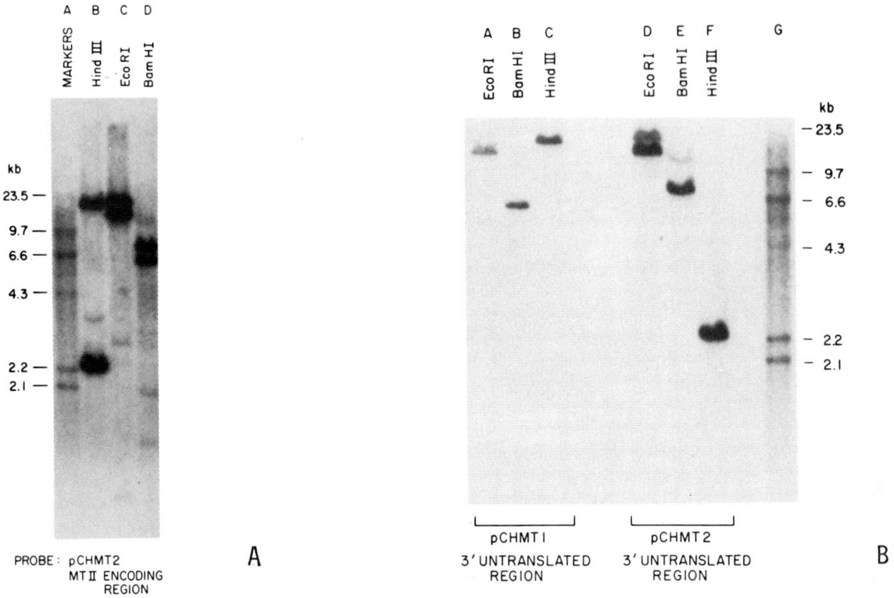

FIGURE 5. Analyses of genomic DNA from Cd^r200T1 digested with the indicated restriction endonuclease, electrophoretically resolved in a 1% agarose gel, transferred to filter and hybridized with MT coding region probe (A, lanes B-D) or 3' untranslated region probe from pCHMT1 (B, lanes A-C) or 3' untranslated region probe from pCHMT2 (B, lanes D-F). Lane G contains ^{32}P-labeled HindIII digested markers. Filter hybridization and wash procedures were described in the legend to Fig. 4.

The coordinate amplification of the MTI and MTII encoding genes in Cd^r200T1, coupled with the observed stability of the Cd^r phenotype suggested that the amplification event(s) may be localized to a specific chromosome(s). In situ hybridization (21,22) of the MTII protein coding region probe with mitotic chromosomes from the Cd^r200T1 cell line revealed a single chromosomal site of hybridization (Fig. 6). This observation supports the proposal that MT genes are closely linked. Interestingly, further cytogenetic analyses have shown that the region of hybridization in these cells corresponds to a translocation breakpoint on a large rearranged chromosome in the Cd^r200T1 karyotype. Further molecular genetic and cytogenetic analyses are directed toward revealing the genomic arrangement(s) and chromosomal location(s) of the isoMT genes in the Chinese

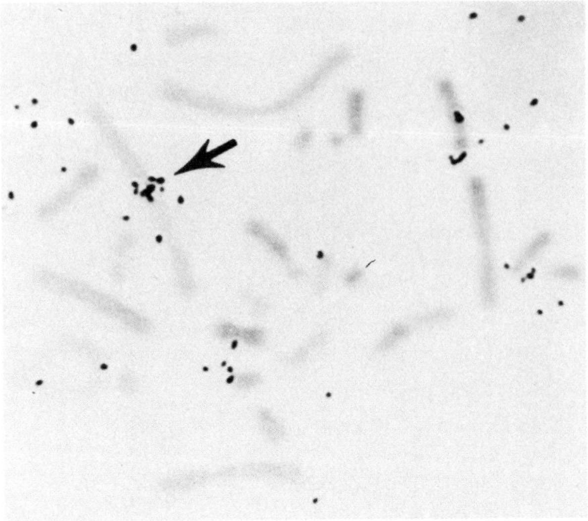

FIGURE 6. *In situ* hybridization of mitotic chromosomes from the Cd^r200T1 cell with an MTII-encoding sequence probe. ^{125}IdCTP-labeled MTII-encoding probe was hybridized with the Cd^r200T1 metaphase plates according to established procedures (21,22). Arrow designates consistently labeled region on a large, rearranged chromosome.

DISCUSSION

Specific gene amplification is one mechanism by which eukaryotic cells respond to metabolic stress [reviewed in (2)]. In the cases studied most extensively, inhibitors of specific target enzymes have been used to select cells containing increased copies of the gene(s) encoding the target enzyme (7,8). In other cases, cellular resistance to metabolic inhibitors arises through amplification of genes encoding products not directly related to the action of the inhibitor (18-19). Cellular resistance to heavy metal toxicity falls into the latter category.

Increased production of metallothionein in Cd-resistant cells has been identified as a major factor in conferring the resistant phenotype (13-16). By virtue of their high affinity Cd-binding capacity, cytoplasmic MTs sequester the toxic Cd^{2+} ion, thereby reducing access of this inhibitory heavy metal to intracellular targets (14, 16). In this study, increased production of both major isoMTs was observed in Cd^r Chinese hamster cells. In the clonal variants resistant to 20-200 μM Cd^{2+} (Cd^r20F4, Cd^r30F9, Cd^r200T1), coordinate amplification of both MTI- and MTII-encoding genes occurred at levels ~7-, 3-, and 14-fold, respectively, above the gene copy number in CHO by reassociation kinetics analyses, and up to 50-fold in Cd^r200T1 by filter hybridization analyses. Paradoxically, both the degree of gene amplification, as well as MT induction capacity, do not correspond quantitatively with increased Cd-resistance suggesting that other mechanisms, possibly metal-inducible (20), operate in conferring Cd^{2+}-resistance (14,16).

The stability of the Cd^r phenotype reported here for Cd^r-CHO variants suggested that the amplified MT genes were stably chromosome-associated (2). In all Cd^r variants which maintain amplified MT genes, the G1 DNA content is indistinguishable from the parental CHO cell (data not shown). This finding is in contrast to that of Gick and McCarty (12) who demonstrated that Cd^r CHO cells with amplified MT genes also displayed an abundance of tetraploid cells, as well as partial instability of the Cd^r phenotype. In the context of the relationship between stability of the Cd^r phenotype and chromosomal or extrachromosomal location of amplified genes, the _in situ_ hybridization analyses shown here indicate a chromosomally-integrated site for the amplified MT genes. This localization of MT-encoding sequences to a single chromosome is consistent with our observation of the coordinate amplification of Chinese hamster MTI and II genes suggesting their close linkage. Further analyses of the organization of the unit(s) of amplification, the linkage of the MT genes, and the chromosomal location of the MT genes in both CHO and euploid Chinese hamster cells are in progress.

ACKNOWLEDGMENTS

The authors thank Myrna Jones and Cleo Naranjo for skillful technical assistance in various phases of this work. The expert secretarial and editorial assistance of Monica Fink is gratefully acknowledged.

REFERENCES

1. Anderson RP, Roth JR (1977). Tandem genetic duplications in phage and bacteria. Ann Rev Microbiol 31:473.
2. Schimke RT (ed) (1982): "Gene Amplification," Cold Spring Harbor Laboratory.
3. Brown DD, Dawid I (1968). Specific gene amplification in oocytes. Science 160:272.
4. Nagel W (1978). "Endopolyploidy and polyteny in differentiation and evolution," Amsterdam: Elsevier/North-Holland Biomedical Press.
5. Spradling AD, Mahowald AP (1980). Amplification of genes for chorion proteins during oogenesis in Drosophila melanogaster. Proc Natl Acad Sci 77:1096.
6. Zimmer WE, Schwartz RJ (1982). Amplification of chicken actin genes during myogenesis. In Schimke RT (ed): "Gene Amplification," Cold Spring Harbor Laboratory, p 137.
7. Schimke RT, Alt FW, Kellems RE, Kaufman R, Bertino JR (1978). Amplification of dihydrofolate reductase genes in methotrexate resistant cultured mouse cells. Cold Spring Harbor Symp. Quant. Biol. XLII:649.
8. Wahl GM, Padgett RA, Stark GR (1979). Gene amplification causes overproduction of the first three enzymes of UMP synthesis in N-(phosphonacetyl)-L-aspartate-resistant hamster cells. J Biol Chem 254:8679.
9. Chattopadhyay SK, Chang EH, Landes MR, Ellis RW, Scolnick EM, Lowy DR (1982). Amplification and rearrangement of onc genes in mammalian species. Nature 296:361.
10. Walters RA, Enger MD, Hildebrand CE, Griffith JK (1981). Genes coding for metal induced synthesis of RNA sequences are differentially amplified and regulated in mammalian cells. In Brown DD, Fox CF (eds): "Developmental Biology Using Purified Genes,".
11. Beach LR, Palmiter RD (1981). Amplification of metallothionein-I gene in cadmium-resistant mouse cells. Proc Natl Acad Sci 78:2110.
12. Gick GG, McCarty KS, Sr (1982). Amplification of the metallothionein-I gene in cadmium- and zinc-resistant Chinese hamster ovary cells. J Biol Chem 257:9049.
13. Hildebrand CE, Tobey RA, Campbell EW, Enger MD (1979). A cadmium-resistant variant of the Chinese hamster (CHO) cell with increased metallothionein induction capacity. Exptl Cell Res 124:237.

14. Enger MD, Ferzoco LT, Tobey RA, Hildebrand CE (1981). Cadmium resistance correlated with cadmium uptake and thionein binding in CHO variants Cd^r20F4 and Cd^r30F9. J Toxicol Environ Health 7:675.
15. Hildebrand CE, Enger MD (1980). Regulation of Cd^{2+}/Zn^{2+}-stimulated metallothionein synthesis during induction, deinduction, and superinduction. Biochem 19:5850.
16. Hildebrand CE, Griffith JK, Tobey RA, Walters RA, Enger MD (1982). Molecular mechanisms of cadmium detoxification in cadmium-resistant cultured cells: role of metallothionein and other inducible factors. In Foulkes EC (ed): "Biological Roles of Metallothionein," New York: Elsevier North-Holland, Inc. p 279.
17. Griffith BB, Walters RA, Enger MD, Hildebrand CE, Griffith JK (1983). cDNA cloning and nucleotide sequence comparison of Chinese hamster metallothionein I and II mRNAs. Nucleic Acids Res 11:901.
18. Biedler JL (1982). Evidence for transient or prolonged extrachromosomal existence of amplified DNA sequences in antifolate-resistant, vincristine-resistant, and human neuroblastoma cells. In Schimke RT (ed): "Gene Amplification." Cold Spring Harbor Laboratory, p 39.
19. Kuo T, Pathak S, Ramagli L, Rodriguez L, Hsu TC (1982). Vincristine-resistant Chinese hamster ovary cells. In "Gene Amplification," Cold Spring Harbor Laboratory, p 53.
20. Griffith JK, Enger MD, Hildebrand CE, Walters RA (1981). The differential induction by cadmium of a low complexity RNA class in cadmium resistant and cadmium sensitive mammalian cells. Biochem 20:4755.
21. Harper ME, Ullrich A, Saunders GF (1981). Localization of the human insulin gene to the distal end of the short arm of chromosome 11. Proc Natl Acad Sci 78:4458.
22. Wahl GM, Vitto L, Padgett RA, Stark GR (1982). Single copy and amplified CAD genes in Syrian hamster chromosomes localized by a highly sensitive method for in situ hybridization. Mol Cell Biol 2:308.

Gene Expression, pages 481-490
© 1983 Alan R. Liss, Inc., 150 Fifth Avenue, New York, NY 10011

HOMOLOGY MATRIX COMPARISON OF HUMAN AND MURINE CLASS II ANTIGENS *

Magali ROUX-DOSSETO, Charles AUFFRAY, James W. LILLIE, Alan J. KORMAN and Jack L. STROMINGER

Department of Biochemistry and Molecular Biology
Harvard University
Cambridge, Massachusetts, 02138

The human HLA-D region codes for the class II antigens involved in allogenic proliferation and antigen presentation by macrophages. These molecules are composed of two non-covalently associated polypeptide chains, the heavy (α) and the light (β) subunits. The HLA-DR molecules are the immunodominant components of these antigens and account for the HLA-D/DR typing. To date, 8 alleles have been defined. The molecular basis of this extreme polymorphism is mainly confined to the β subunit of the DR antigens as well as their mouse counterparts, the I-E antigens (for review, see reference 1). Subsequently, the use of oligospecific sera allowed the identification of several segregant series of supertypic DR groups. Two loci are believed to be coding for these sets of antigens (reviewed in reference 2). Structural analysis of DC-1 series antigens clearly established that these molecules are encoded by a locus distinct from DR and suggested that they are the human equivalent of murine I-A antigens (3). In mouse, the α and β chains of I-A molecules are polymorphic. In man, similar features were reported for DC antigens at the protein (4) and the DNA (5,6,7) levels. Another segregant series of HLA-linked specificities detected by secondary response using primed lymphocyte typing (PLT) reagents has been designated "SB" (8,9). These antigens display limited polymorphism, and 6 alleles have been defined so far (9,10). Despite a func-

*Supported by grants from NIH (AM 30241) and the Kroc Foundation.

tional (9) and structural (11) homology with the DR products, the SB antigens are encoded by a separate locus which has been mapped between DR and GLO (8,9). Thus, from genetic and serological data, the D region appears as a multigenic system. A schematic map of the human MHC is presented in figure 1.

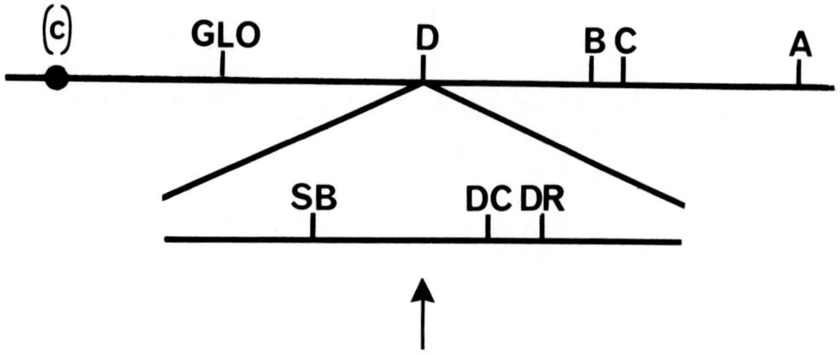

FIGURE 1. Genetic map of the human MHC. The arrow indicates the recombination event leading to the definition of SB locus. The relative order of the DR and DC loci is not known. Distances are not to scale (c: centromere).

Because class II antigens account for a small part of the membrane proteins on B cells (0.01 - 0.05%), mostly amino terminal sequence data are available. Such limited knowledge together with the absence of appropriate recombinants have hampered an accurate mapping of the HLA-D region. Recently, the use of molecular cloning unlocked the situation, allowing the rapid accumulation of structural data. Thus, the complete amino acid sequence for human and murine class II antigen α chains have been deduced from genomic or cDNA clone sequences (12-16). Likewise, cDNA clones encoding the human class II β chains have been isolated (17, 18, Roux-Dosseto et al., in preparation).

We present here preliminary observations concerning the evolutionary relationship of the class II molecules as assessed by statistical comparisons of their amino acid sequences.

RESULTS AND DISCUSSION

Matrix Homology Comparison of Human and Murine Class II Antigen α Chains.

Because of the high level of polymorphism of class II antigens, they provide a suitable model to investigate the possible evolutionary mechanisms by which constrained sequences diverge. Recently, we described the isolation and the characterization of genomic (13) and cDNA (14) clones coding for the α chains of DR and DC molecules. Similarly, the DNA sequences of Eα and Aα chains have been reported (15,16). In order to analyze the homology existing between different class II α chains in human and mouse, we examined them two by two using a two dimensional homology matrix program.

The six maps presented in Fig. 2 display a complex pattern of homology corresponding to two types of segments: those characteristic of the two subsets of class II antigens and those corresponding to possible alloantigenic sites; and two regions of higher homology.

One of the two most conserved regions is the second extracellular domain extending from residue 88 to residue 181 in DCα and from residue 85 to residue 178 for the other α chains compared in this study. This domain exhibits one disulfide bridge and displays statistically significant homology with the constant region of the immunoglobulins (13,19) as well as with β_2-microglobulin and the α_3 domain of the class I heavy chain (20). Thus, multiple sequence constraints may account for the conservation featuring this domain. The second homologous region is confined to the transmembrane portion of the α subunits (which may be involved in highly specific interactions with the corresponding portion of the appropriate light chains as well as other membrane proteins). Careful examination of the letter assignment to these sequences points out higher conservation displayed by the homologous map (i.e.: DCα / I-Aα, DRα / I-Eα, fig. 2b,e) than by the heterologous map (i.e.: DCα / I-Eα, DRα / I-Aα, figure 2 c, f), suggesting that these sequences may reflect the "A"- or "E"-ness of the class II antigen α chains.

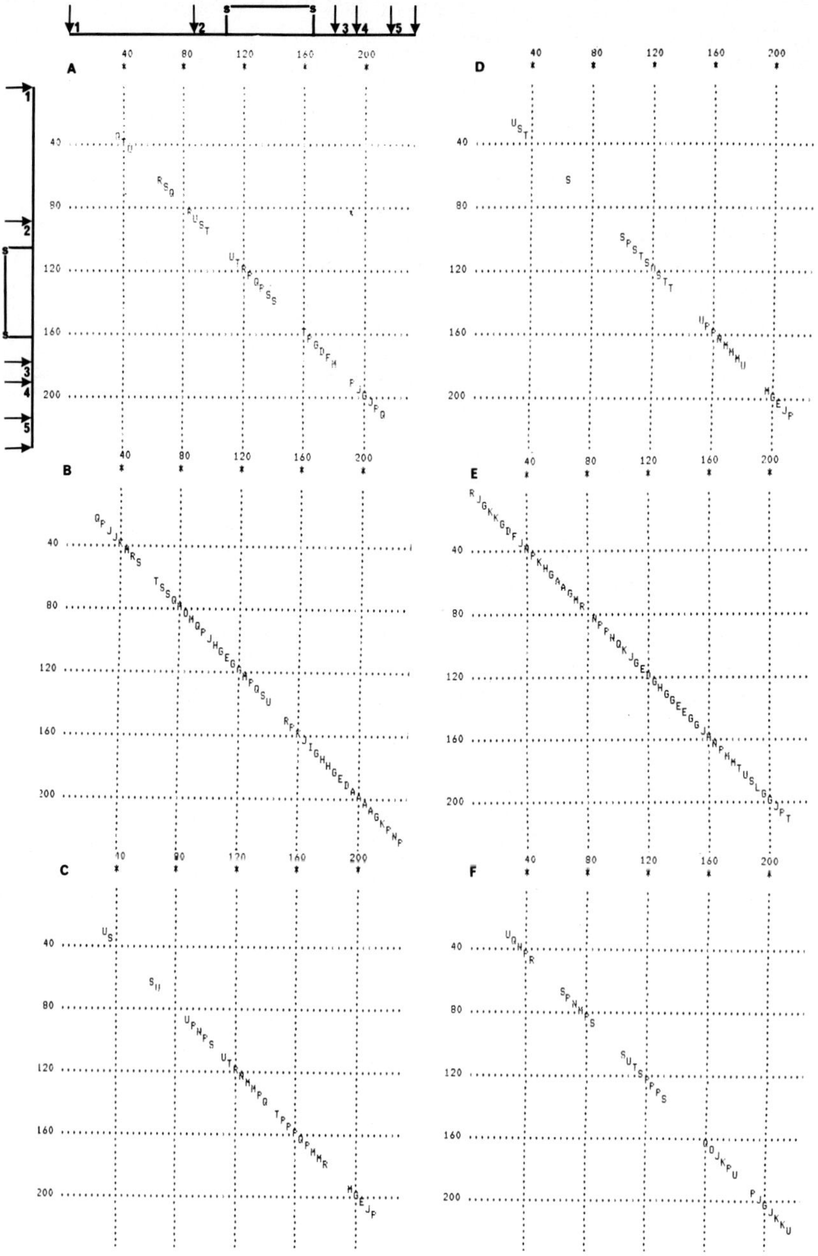
A
B
C
D
E
F
40
80
120
160
200

Three other segments appear to have the same characteristics: one corresponding to the junction between the first and the second extracellular domains, the other two to the connecting peptide and the cytoplasmic tail respectively. Whereas the heterologous maps exhibit no homology within these three segments (figure 2a,c,d,f), the DCα / I-Aα map displays a remarkable similarity (figure 2b). The same situation is encountered by comparing the junctions between the extracellular domains of the DRα and I-Eα sequences (figure 2e). However, the cytoplasmic tail of the DRα chain fails to display any homology when compared to the carboxyl terminal portion of I-Eα chain sequences.

On the other hand, quite divergent fragments have been found whatever the sequences compared, excluding the DR/I-E map; they are clustered within the portions spanning from about residue 135 to residue 160 in the second Ig-like domain. It is noteworthy that they are located at a similar place in each domain where alloantigenic sites have been reported for murine and human class I heavy chain antigens (21, 22). Also, low sequence homology is obvious at the amino terminus of protein sequences (residues 1-25). Since the DCα and I-Aα

FIGURE 2. Matrix homology comparison of human and murine class II antigen α chain sequences. The homology is measured for each position over a string of 17 amino acids (8 on each side) and the value obtained represented by a letter: A = 100% or 99%, B = 98% or 97% ... U = 60% and 59% (minimum value plotted). The matrixes have been compressed 4 times for presentation, with the highest value of each group of 4 amino acids presented. A full description of the homology matrix program can be found in J. Pustall and F.C. Kafatos, Nucl. Acid Res. 10: 4765-4782. A schematic representation of class II antigen α chains points out the organization of domains: 1: first extracellular domain, 2: second extracellular domain, 3: connecting peptide, 4: transmembrane region, 5: cytoplasmic tail; and the position of the disulfide bond. Comparisons are as follows (horizontal axis/vertical axis): A: DCα / DRα, B: DCα / I-Aα, C: DCα / I-Eα, D: I-Aα / I-Eα, E: DRα / I-Eα, F: DRα / I-Aα.

chains are polymorphic whereas DRα and I-Eα chains are extremely conserved, these observations suggest these segments are part of the alloantigenic sites.

Thus our finding that inter-species comparisons of the corresponding class II molecules display higher homology than intra-species comparisons supports the idea that the duplication event that generated the DR/I-E and DC/I-A -α chain loci took place before the species separation. Furthermore, it serves to caution against predicting genetic evolution on the basis of inaccurate comparisons.

Matrix Homology Comparison of Human Class II Antigen β Chains.

Recently the complete extracellular amino acid sequence of a DRβ chain has been reported (23). In addition, the protein sequence deduced from a full length cDNA clone (17) has been found to match with the amino terminal sequence of the DCβ chain (24). By probing a human cDNA library with the I-Aβ gene, we have isolated a cDNA clone encoding a class II antigen β chain, (M. Roux-Dosseto et al., manuscript in preparation). The cDNA insert of this clone, provisionally called DYβ is 780 b.p. long and contains 78% of the DNA coding sequence for the β chain of a class II antigen distinct from DR and DC and the complete 3' end untranslated region of the mRNA, as assessed by DNA sequencing. Because the sequence of the murine β chains have not yet been determined, it was not possible to perform murine/human comparisons. In order to evaluate the relatedness of DYβ with DRβ and DCβ chains, we compared their sequences by using the matrix homology program. The homology maps presented in figure 3 show two highly conserved regions. The first conserved region corresponds to a stretch extending from about position 40 to 70, in the first extracellular domain. The second homologous region fits with the Ig-like second extracellular domain (residues 92-192). Nevertheless, the maps exhibit local discripancy in the position of the sequence where a putative allogenic cluster has been located for class I and class II antigen α chains (see above). Whereas residues 145-166 display the low homology expected for a variability cluster, in the DCβ/DRβ (figure 3c) and DYβ/DRβ (figure 3b) maps, the DYβ/DCβ corresponding portion (fig. 3a) exhibits high value assignments (ECBBCE). Because products of different loci are being considered, it is possible that a segmental exchange has occurred between the DYβ and the DCβ genes.

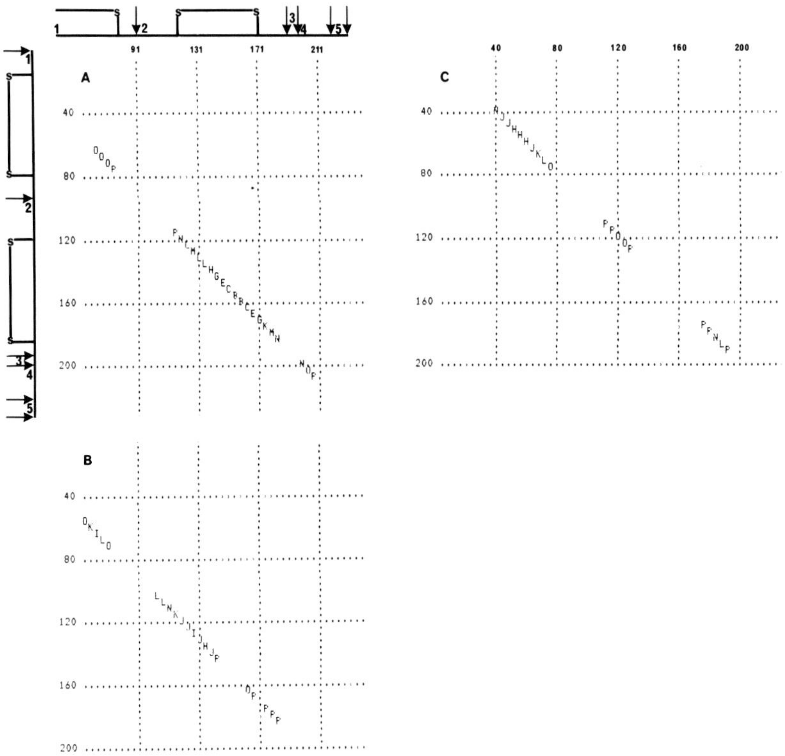

FIGURE 3. Matrix homology comparison of human class II antigen β chain sequences. For legend see figure 2. Homology measured on a string of 24 amino acids, minimal value plotted: P (70%). A: DYβ / DCβ, B: DYβ / DRβ, C: DC β / DRβ.

More divergent sequences are located within five additional segments. Two of them fall within probably allogenic areas, at the beginning (residues 1-20) and the end (residues 75-90) of the first extracellular domain, as reported above for the α chain. This particular location for extensive diversity may be a general feature of MHC antigens because similar observations were made for the first domain of class I antigens (21,22). The two following low homology sequences are located at the junction between domains. Since there is a correlation between domains and exons in the α chain genes, it is possible that these variations are due to splicing variations which do not affect the overall structure of the molecule. The last diversity segment seems clustered to the cytoplasmic tail. As displayed by the homology matrix comparisons performed between the heterologous class II antigen α chains in human and mouse (figure 2a,c,d,f), the DYβ/DCβ map (figure 3a) exhibits no conservative sequence in the cytoplasmic region which appears to be a locus specific segment.

Because no sequence data are available for the carboxy-terminus of DR light chains, we cannot evaluate the relatedness of DYβ to DRβ in this portion. Nevertheless, Southern hybridization with DYβ of genomic DNA from DR and DC-null variants (25) displays identical patterns to those obtained with the parental cell lines that express both DR and DC antigens (Roux-Dosseto et al., manuscript in preparation). In contrary, parallel hybridization experiments with the DCβ probe present noteworthy differences. Together, these findings with the amino sequence comparisons allow us to identify a third class II β chain locus. Evaluation of its relationship to either a second supertypic specificity encoding subregion or to the SB locus is in progress.

ACKNOWLEDGEMENTS

We thank Joanne Kuo for expert technical assistance and Alice Furumoto-Dawson for manuscript editing.

REFERENCES

1. Shackelford, D.A., Kaufman, J.F., Korman, A.J. and Strominger, J.L. (1982). Immunol. Rev. 66: 133-187.

2. Tanigaki, N. and Tosi, R. (1982) Immunol. Rev. 66: 5-37.

3. Bono, M.R. and Strominger, J.L. (1982) Nature 299: 836-838.

4. de Kretser, T.A., Crumpton, M.J., Bodmer, J.G. and Bodmer, W.F. (1982) Eur. J. Immunol. 12: 214-221.

5. Auffray, C. Ben-Nun, A., Roux-Dosseto, M., Germain, R.N., Seidman, J.G. and Strominger, J.L. (1983) EMBO J. 2: 121-124.

6. Wake, C.T., Long, E.O. and Mach, B. (1982) Nature 300: 372-376.

7. Böhme, J., Owerbach, D., Denaro, M., Lernmark, A., Peterson, P.A. and Rask, L. (1983) Nature 301: 82-84.

8. Mawas, C., Charmot, M., Sivy, M., Mercier, P., Tongio, M. and Hauptmann, G. (1978) J. Immunogenet. 5: 383-395.

9. Shaw, S., Pollack, M.S., Payne, S.M. and Johnson, A.H. (1980) Human Immunol. 1: 177-185.

10. Bourgue, F., Charmot, D., Mercier, P., Tongio, M.M., Shaw, S. and Mawas, C. (1982) Tissue Antigens 20: 254-259.

11. Katovich-Hurley, C. Shaw, S., Nadler, L., Schlossman, S. and Capra, J.D. (1982) J. Exp. Med. 156: 1557-1562.

12. Larhammar, D., Gustafsson, K., Claesson, L., Bill, P., Wiman, K., Schenning, L., Sundelin, J., Widmark, E., Peterson, P.A. and Rask, L. (1982) Cell 30: 153-161.

13. Korman, A.J., Auffray, C., Schamboeck, A. and Strominger, J.L. (1982) Proc. Natl. Acad. Sci. USA 79: 6013-6017.

14. Auffray, C., Korman, A.J., Roux-Dosseto, M., Bono, R. and Strominger, J.L. (1982) Proc. Natl. Acad. Sci. USA 79: 6337-6341.

15. McNicholas, J., Steinmetz, M., Hunkapiller, T., Jones, P. and Hood, L. (1982) Science 218: 1229-1232.

16. Benoist, C.O., Mathis, D.J., Kanter, M.R., Williams II, V.E. and McDevitt, H.O. (1983) Proc. Natl. Acad. Sci. USA 80: 534-538.

17. Larhammar, D., Schenning, L., Gustafsson, K., Wiman, K., Claesson, L., Rask, L. and Peterson, P.A. (1982) Proc. Natl. Acad. Sci. USA 79: 3687-3691.

18. Long, E.O., Wake, C.T., Strubin, M., Gross, N., Accola, R.S., Carrel, S. and Mach, B. (1982) Proc. Natl. Acad. Sci. USA 79: 7465-7469.

19. Kaufman, J.K. and Strominger, J.L. (1982) Nature 297: 694-696.

20. Kaufman, J.F. and Strominger, J.L. (1983) J. Immunol. 130: 808-817.

21. Nathenson, S.G., Vehara, H., Ewenstein, B.M., Kindt, T.J. and Coligan, J.E. (1981) Ann. Rev. Biochem. 50: 1023-1052.

22. Lopez de Castro, J.A., Strominger, J.L. Strong, D.M. and Orr, H.T. (1982) Proc. Natl. Acad. Sci. USA 79: 3813-3817.

23. Kratzin, H., Yang, C.-Y., Götz, H., Pauly, E., Köbel, S., Egert, G., Thinnes, F.P., Wernet, P., Altevogt, P. and Hilschmann (1981) Hoppe-Seyler's Z. Physiol. Chem. 362: 1665-1669.

24. Bono, M.R. and Strominger, J.L. (1983) Immunogenetics, in press.

25. Kavathas, P., Bach, F.H. and DeMars, R. (1980) Proc. Natl. Acad. Sci. USA 77: 4251-4255.

Gene Expression, pages 491-510
© 1983 Alan R. Liss, Inc., 150 Fifth Avenue, New York, NY 10011

EXPRESSION OF THE GLUCAGON GENE IN FETAL BOVINE PANCREAS[1]

Linda C. Lopez, Marsha L. Frazier, Chung-Jey Su, Ashok Kumar,[2] and Grady F. Saunders

Department of Biochemistry, The University of Texas System Cancer Center M.D. Anderson Hospital and Tumor Institute, Houston, Texas 77030

ABSTRACT Glucagon is a 29 amino acid polypeptide hormone produced in the α cells of the pancreatic islets. As with other polypeptide hormones, the synthesis of glucagon is thought to involve a larger precursor, which is then enzymatically cleaved to the functional form. To investigate the biosynthetic pathway of glucagon, synthetic oligodeoxynucleotide probes of 14-nucleotides (a 14-mer) and 17-nucleotides (a 17-mer) complementary to codons specifying a unique sequence of mature glucagon were synthesized. The ^{32}P-labeled-14-mer was hybridized with size-fractionated fetal bovine pancreatic poly(A^+)RNA bound to nitrocellulose. RNA fractions of ~14S were found to hybridize specifically, resulting in an ~10-fold enrichment for these sequences. These poly(A^+)RNAs were translated in a cell-free system and the products analyzed by gel electrophoresis. The translation products were found to be enriched for a protein of ~21 kd, the suggested size of mammalian preproglucagon. These enriched RNA fractions were used to construct a complementary DNA (cDNA) library in plasmid pBR322.

[1]This work was supported by grants from the Kroc Foundation, Robert A. Welch Foundation, U.S. Public Health Service (CA 16672), and the Redfern Fund. L.C.L. is a recipient of a Rosalie B. Hite Predoctoral Fellowship.

[2]Department of Anatomy, University of Texas Health Science Center, San Antonio, Texas.

The screening of duplicate colony filters with ^{32}P-labeled-17-mer and ^{32}P-labeled-17-mer primed cDNA indicated that 26 of 3100 colonies screened were possible glucagon clones.

Analysis of the cDNA suggested that the glucagon messenger RNA encoded an initial translation product of 180 amino acids with an M_r = 21 kd. This polypeptide contained a 20 amino acid signal peptide followed by glicentin, a 69 amino acid protein containing the glucagon moiety. Following glicentin are two additional glu-agon-related sequences; all three glucagon-like peptides are flanked by paired basic amino acids (lys-arg or arg-arg) characteristic of prohormone processing. Based upon the amino acid sequence of preproglucagon deduced from the nucleotide sequence of the cloned cDNA, a scheme for the post-translational processing of fetal bovine pancreatic preproglucagon has been proposed.

INTRODUCTION

Glucagon, a 29 amino acid residue single chain polypeptide hormone secreted by the α cells of the Islets of Langerhans in the endocrine pancreas, binds to specific receptors on the plasma membrane, leading to the activation of adenylate cyclase (1). In mammals, the primary physiological target of this activation is the liver, where it affects the metabolism of glucose by inhibiting glycogen synthesis, stimulating glycogenolysis, and enhancing gluconeogenesis (2,3).

It had been demonstrated very early that extracts of gastrointestinal tissue (4) and blood (5) were associated with the glucagon-like ability to activate hepatic adenylate cyclase. The development of the radioimmunoassay for glucagon (6), and the discovery that much glucagon-immunoreactive material extracted from pancreas and the gut was immunochemically distinct from glucagon led to an upsurge in interest in substances related to glucagon.

The significance and biosynthetic origins of the glucagon-like proteins is understood poorly. Molecules of various sizes, all larger than glucagon and displaying glucagon-like immunoreactivity, i.e. putative glucagon precursors, have been isolated from the pancreatic tissues of many species. Additional high molecular weight forms are found in gastric and intestinal mucosa (7,8), and in plasma (9). Some of these forms differ in size and immunoreactivity from those located in the pancreas (10). Glucagon-like immunoreactive polypeptides of at least nine different molecular weights, ranging from 3.5 kd to 200 kd have been found in tissues of various species. The relationship between the different glucagon-like molecules in pancreas and gut has been most clearly documented in rabbit, where glucagon and four related higher molecular weight forms have been characterized (10).

Glucagon is a member of a family of structurally related peptides which includes secretin (11), glicentin (12), gastric inhibitory peptide (GIP) (13), vasoactive intestinal peptide (VIP) (14), prealbumin (15), peptide HI-27 (PHI-27) (16), and a growth hormone releasing factor (hpGRF) (17). Based upon amino acid sequence homologies these peptides are considered to be part of a family, but glicentin is worthy of separate consideration. A possible proglucagon fragment isolated from porcine intestine, glicentin is a 69 amino acid polypeptide containing the glucagon sequence in residues 33-61 (12); amino acid sequence 1-30 of glicentin represents the glicentin-related pancreatic peptide (GRPP) isolated from porcine pancreas (18).

Studies of anglerfish glucagon messenger RNAs (mRNAs) have begun to more specifically elucidate the nature of the glucagon precursors. At least two distinct glucagon precursors of 14.5 kd and 12.5 kd are found in anglerfish islets (19,20). The mRNAs coding for the anglerfish preproglucagons have been found to contain sequences which suggest polypeptides that could give rise to glucagon and a second glucagon-like peptide by simple enzymatic cleavage (21, 22). Experiments on isolated rat islets have identified a possible glucagon precursor which exhibits prohormone-like kinetics of formation and processing. The 18 kd polypeptide observed following a short pulse of radioactive amino acids disappeared during a brief chase period with the concommitant formation of two species of 13 kd and 10 kd. Immunoprecipitation of these newly synthesized islet proteins showed that treatment of the 18 kd protein with trypsin and carboxypeptidase B led to the appearance of a polypeptide the size of mature glucagon (3.5 kd) which was recognized by glucagon carboxy-terminal specific antiserum (23). Allowing for a signal peptide, which would be co-translationally cleaved and therefore not detected in this system, this data suggests that mammalian preproglucagon may be ~21 kd in size.

Isolation and characterization of eukaryotic mRNAs is important for the understanding of eukaryotic gene expression. The application of recombinant DNA technology, where the mRNA of interest is cloned in bacteria in the form of a of a complementary DNA (cDNA) copy, has increased our understanding of the expression, regulation, and evolution of several eukaryotic genes. The first applications of these techniques were to mRNAs obtained from specialized tissues where the mRNAs occur as very abundant species, or can be induced to a relatively high level.

However, many mRNAs cannot be purified and cloned on the basis of their abundance, and the study of these genes requires more sensitive isolation and detection procedures. One alternative enrichment procedure is by immunoprecipitation of polyribosomes where the specific antibodies recognize immunological determinants on the nascent polypeptide chains. This approach was used to purify three low abundance hepatic mRNAs (24). This approach is limited, however, by the availability of antibodies capable of recognizing the initial translation product, a difficulty encountered in the cloning of gastrin, where the available antisera seem to be directed against only the mature hormone (25). A technique has been developed that permits the identification of clones for genes expressed at low levels. If part of the amino acid sequence of a protein is known, a specific oligodeoxynucleotide (oligomer) can be synthesized whose sequence is deduced from this unique protein sequence. This oligomer can be utilized directly as a hybridization probe, or as a primer for the enzymatic synthesis of a specific cDNA probe. It is this approach which we have used to identify and to characterize the cDNA to fetal bovine pancreatic glucagon mRNA.

METHODS

Two oligodeoxynucleotide mixtures were synthesized as described (26,27) to correspond to the region of glucagon coding for amino acids 24-29 (a 17-base oligomer) (fig. 1) or amino acids 24-28 (a 14-base oligomer).

Poly(A^+)RNA extracted from second trimester fetal bovine pancreas as previously described (28-30) was fractionated by centrifugation on 15-30% linear sucrose gradients. RNA from each fraction was dot-blotted onto nitrocellulose (31), and hybridized as described (32) with the glucagon-specific 14-mer which had been phosphorylated with [γ-^{32}P] ATP (33). In addition, RNA from each fraction was translated in a cell-free system (34) and the products analyzed by electrophoresis on 15% $NaDoSO_4$ (SDS)-polyacrylamide gels (35). RNAs suggested to be enriched for glucagon-coding

```
        1       2       3       4       5       6       7
H2N - his - ser - gln - gly - thr - phe - thr -
8       9       10      11      12      13      14      15
ser - asp - tyr - ser - lys - tyr - leu - asp -
16      17      18      19      20      21      22      23
ser - arg - arg - ala - gln - asp - phe - val -
24      25      26      27      28      29
gln - trp - leu - met - asn - thr - COOH
 ↓     ↓     ↓     ↓     ↓     ↓
5'- CAA - UGG - UUA - AUG - AAU - AC_ -3' mRNA
      G           G           C          sequence
                 CUX
 ↓     ↓     ↓     ↓     ↓     ↓
3'- GTT - ACC - AAX - TAC - TTA - TG_ -5'  17-mer
      C           G           G
```

FIGURE 1. The amino acid sequence of mammalian glucagon and the sequence of the oligodeoxynucleotide probe (17-mer). The underscored residues in glucagon indicate the region from which the synthetic oligomer was deduced. Sequence of the mRNA and cDNA corresponding to glucagon amino acid residues 24-29 are shown (where X=A,G,C,T/U). All possible ambiguities were included in the oligomer mixture.

sequences were used in the preparation of a cDNA library as described by Land et al. (36) (fig. 2). cDNAs were inserted into the Pst I site of plasmid vector pBR322 as illustrated and replicated in host Escherichia coli RRI.

The 14-mer used in hybridization experiments with poly (A^+)RNA apparently cross-hybridized with E. coli DNA when used to screen the cDNA library (data not shown). Therefore, a 17-mer was synthesized and used in subsequent experiments. Use of the 17-mer permitted an increase in stringency that eliminated the difficulties encountered with the use of the 14-mer.

The cloned library of cDNAs representative of an enriched population of glucagon-coding mRNAs was screened by the Grunstein and Hogness colony assay (37). Utilizing a double-positive hybridization scheme, one set of filters was hybridized with ^{32}P-labeled-17-mer as described (38), and the duplicate set of filters was hybridized with a ^{32}P-labeled cDNA probe prepared by primer extension of the 17-mer with the enriched glucagon-coding mRNA as template (39). Colonies hybridizing with both probes were selected as possible glucagon-coding clones.

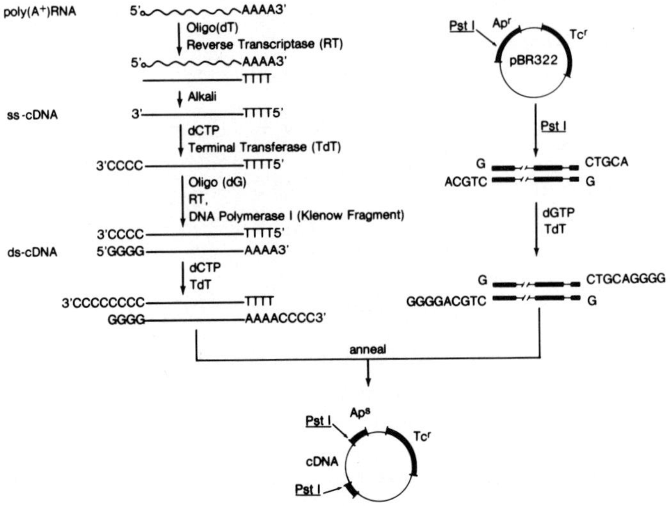

FIGURE 2. General scheme for the synthesis of double strand cDNA complementary to poly(A^+)RNA by the method of Land et al. (36) and it's insertion into plasmid vector pBR322. This procedure selects for the synthesis and cloning of full-length cDNA copies of mRNA.

Primary sequence analysis of the cloned glucagon cDNA was by the dideoxy chain-termination method of Sanger (40) utilizing the M13 bacteriophage cloning vector (41).

RESULTS

Enrichment of Poly(A^+)RNA for Glucagon-Coding Sequences.

Sucrose gradient sedimentation of poly(A^+)RNA from fetal bovine pancreas (fig. 3) resulted in an ~10-fold enrichment for sequences corresponding to an RNA size of ~14S.

One μg of poly(A^+)RNA from each RNA fraction was translated in a cell-free system in the presence of ^{35}S-methionine and the products were analyzed by SDS-PAGE. 5.0×10^4 cpm of the products from each sample were analyzed by electrophoresis on 6-15% discontinuous polyacrylamide gels in the presence of 1% SDS. The products of translation of fractions 10 and 11 (~14S) were found to

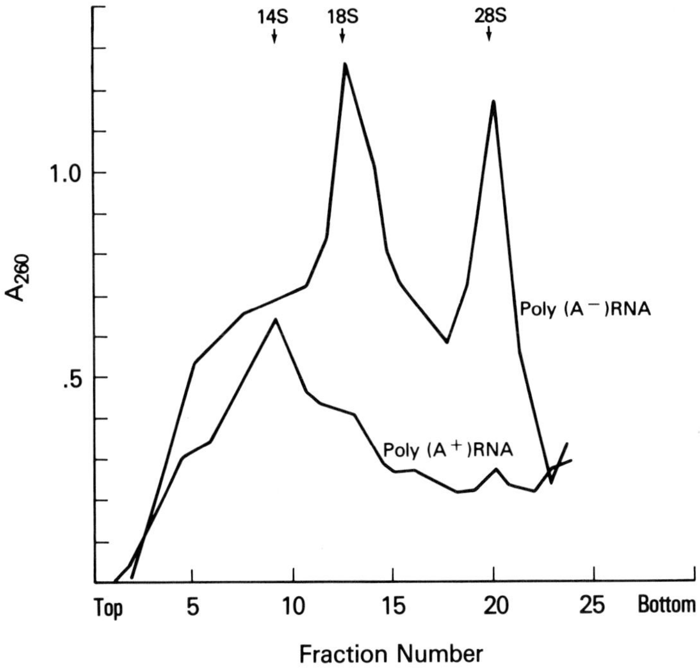

FIGURE 3. Sucrose gradient sedimentation of poly(A^+) RNA obtained from second trimester fetal bovine pancreas was fractionated by centrifugation on 15-30% sucrose gradients. Poly(A^-)RNA was applied to a second gradient in order to estimate sedimentation values.

be enriched for polypeptides in the range of ~21 kd, the suggested size of mammalian preproglucagon (fig. 4).

An RNA dot blot was prepared by spotting 1.25 μg of fractions 7-16 from the sucrose gradient fractionated poly (A^+)RNA onto nitrocellulose. The blot was hybridized with 7.5 pmol of ^{32}P-labeled-14-mer (specific activity 2.0 x 10^6 cpm/pmol) for 20 hours at 28°C. Hybridization to RNA dots was visualized by autoradiography with overnight exposure (fig. 5).

Preparation and Screening of the Fetal Bovine Pancreas cDNA Library.

The size-fractionated poly(A^+)RNA (fractions 10 and 11) that exhibited hybridization with the glucagon-specific

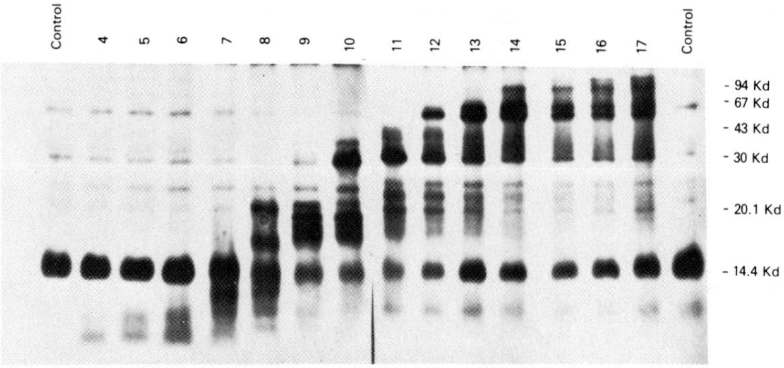

FIGURE 4. Translation in a cell-free system of poly(A^+) RNA fractionated by sucrose gradient sedimentation and analysis of the products by SDS-PAGE. RNA fractions 10 and 11 were found to be enriched for polypeptides in the range of ~21 kd, the suggested size of mammalian preproglucagon. Polypeptide standards of known sizes were run in a parallel lane. Controls were the products of cell-free translation without any added RNA.

oligomer (fig. 5) and whose cell-free translation products were enriched for polypeptides of the expected size of mammalian preproglucagon (fig. 4) were pooled and used in the preparation of a cDNA library as described by Land et al. (36) (fig. 2). A total of 3100 tetracycline resistant (tet^r) transformants were selected; >95% of these were ampicillin-resistant (amp^r) and were probably recombinants.

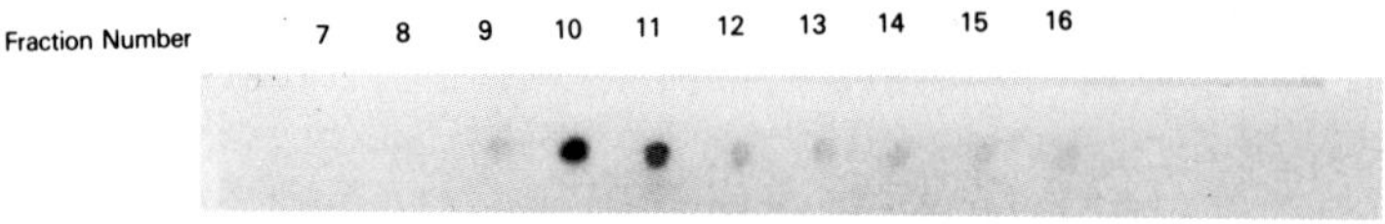

FIGURE 5. Hybridization of ^{32}P-labeled-14-mer to poly(A^+)RNA bound to nitrocellulose. Size-fractionated poly(A^+)RNA (fractions 7 to 16) was bound to a nitrocellulose filter and hybridized with ^{32}P-labeled-14-mer. Hybridization at RNA dots was visualized by autoradiography with overnight exposure.

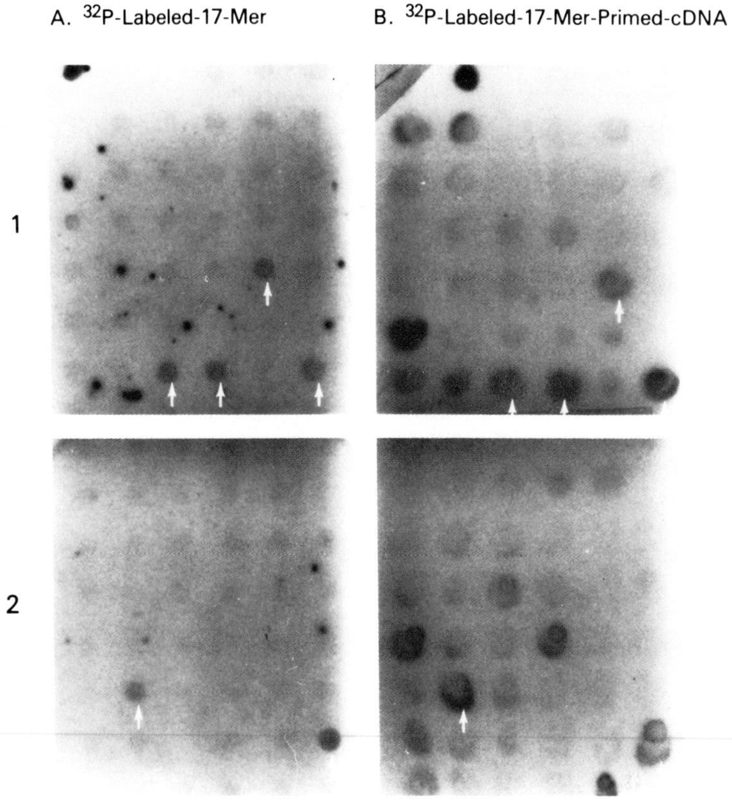

FIGURE 6. Representatives of colony screening of the fetal bovine pancreas cDNA library. Duplicate sets of 62 filters were hybridized overnight with ^{32}P-labeled-17-mer or ^{32}P-labeled (17-mer-primed) cDNA. Hybridization was visualized by autoradiography. One and two are representative duplicate filters showing positive signals (arrows).

The cDNA library was screened as described in Methods and DNA from 26 colonies showed hybridization with both probes. Two sets of representative filters are shown in fig. 6. The arrows indicate colonies hybridizing with both probes that were selected as positives. Each screening approach also indicated hybridizing colonies unique to the individual probes. These false-positives were usually generated with the ^{32}P-labeled-17-mer when cellular debris adhered to the filter. False-positives with the cDNA probe may have been due to hybridization with cDNAs not specifically 17-mer primed. The conditions by which the cDNA probe was synthesized permits priming of cDNA to RNAs by the 17-mer or by small fragments of RNA or DNA in the RNA preparation. In addition, it is possible that RNA self-priming can occur. An experiment was performed which addresses this question. ^{32}P-labeled-cDNA was synthesized by primer-extension of the oligomer and bovine pancreatic poly(A^+)RNA (with [α-^{32}P] dCTP in the reaction mix) and used as a hybridization probe with Northern blots of a similar RNA population. The ^{32}P-labeled-cDNA hybridized with the full range of RNAs (data not shown), with the strongest signal at the most abundant RNAs. This does not preclude using cDNAs generated in this way as a specific hybridization probe, but caution in interpretation of the results is warranted. In many cases, the strongest signals in the screening were at colonies which did not exhibit hybridization with the ^{32}P-labeled-17-mer. For this reason, we selected the double-positive selection scheme previously described.

Plasmid DNA from the 26 positive colonies was prepared by the minilysate procedure (42). From the electrophoretic mobility of the DNA on 1% agarose gels, all but one of the plasmids contained uniformly large inserts. The DNA was transferred to nitrocellulose and hybridized with ^{32}P-labeled-17-mer. Hybridization of the plasmid DNA bands was visualized by autoradiography. All of the recombinant plasmid DNAs hybridized with the 17-mer with the exception of the single plasmid containing a smaller insert (data not shown). The interpretation of this analysis was that of the 26 clones selected in the initial screening procedure, 25 were likely to be true positives. From the number of colonies screened, this reflects an abundance of 0.8% (25/3100 colonies detected). Since the poly(A^+)RNA used in the preparation had been enriched ~10-fold for glucagon mRNA, this reflects an abundance of roughly 0.1% for glucagon mRNA in second trimester fetal bovine pancreas.

Six of the 25 positive colonies were selected at random and the plasmid DNA digested with restriction endonuclease _Pst I_ to release the inserts. The DNA fragments were analyzed by electrophoresis on 1.5% agarose gels with _Alu I_- and _Taq I_-digested pBR322 DNA included as size standards (data not shown). Four of the cDNA inserts were ~1200 bp in length, the other 2 were ~850 bp (data not shown).

Later results revealed that the sequence in the glucagon mRNA complementary to the oligomer is ~800 nucleotides from the poly-A tail. Since truncation of cloned mRNAs prepared by the Land technique is likely to be at the 5'-end, it is probable that ~850 bp is the minimum size of the glucagon cDNA insert expected to be detected by hybridization with the oligomer.

Primary Sequence Analysis of the Cloned Glucagon mRNA.

Primary sequence analysis was by the dideoxy chain-termination method of Sanger utilizing the M13 bacteriophage vector. Reproduced is the primary structure of the cloned pancreatic glucagon mRNA and the deduced amino acid sequence of preproglucagon (fig. 7) (43). Analysis of the primary sequence of the cloned DNA indicates that it codes for an initial translation product of 180 amino acids with a M_r = 21 kd. The coding sequence of 540 nucleotides is preceded by a 5'-untranslated region of 90 nucleotides, and followed by a longer 3'-untranslated region of 471 nucleotides, resulting in a total of 1101 nucleotides. The remainder of the inserted DNA is the dG- and dC-homopolymeric extensions and a poly-A tract at the 3'-terminus. The first initiation codon, ATG (methionine), in the longest open reading frame of the nucleotide sequence is located 90 nucleotides from the 5'-end of the cDNA. The initial methionine is followed by 19 predominantly hydrophobic amino acids, which is characteristic of the signal peptide sequences found in secretory proteins. The first arginine following the putative signal peptide is likely to be the initial amino acid residue of proglucagon. Amino acids 1-69 (designated by arrows) is a sequence which shares 94% homology with the putative proglucagon fragment, glicentin. Glucagon is contained within the sequence at residues 33-61, flanked by paired basic amino acid residues (lys, arg) characteristic of prohormone processing (44). Following the octapeptide extension at the carboxy-terminus of glucagon are two additional glucagon-like peptides (GLP-I and GLP-II) of 37 and 33 amino acids, respectively. Like glucagon, these peptides are flanked by paired basic residues, and are separated from each other by a 13 amino acid peptide sequence. GLP-II is followed immediately by an ochre stop codon, 471 nucleotides of 3'-untranslated sequence, and the poly-A tract characteristic of most eukaryotic mRNAs (45). Twenty nucleotides upstream from the poly-A tract is the sequence AATAAA thought to represent a polyadenyaltion signal (46). Although a second possible poly-

BOVINE PANCREATIC PREPROGLUCAGON

```
5'...AACAGAGCTCAGGACACTGCACACCCAAACGAGGGCTCACTCTCTCTTCACCTGCTCTGTTC     62
                                -20
                                met lys ser leu tyr phe val ala gly
CACCTCCTGGTGTCAGAAGGCAGCAAAA    ATG AAA AGC CTT TAC TTT GTG GCT GGA    117
    -10                                    -1 ↓ Glicentin
leu phe val met leu val gln gly ser trp gln arg ser leu gln asn thr
TTG TTT GTA ATG CTG GTA CAA GGC AGC TGG CAA CGT TCC CTT CAG AAC ACA    168
            10                                      20
glu glu lys ser ser ser phe pro ala pro gln thr asp pro leu gly asp
GAG GAG AAA TCC AGT TCA TTC CCA GCT CCC CAG ACC GAC CCG CTC GGC GAT    219
                        30          Glucagon                    40
pro asp gln ile asn glu asp lys arg his ser gln gly thr phe thr ser
CCA GAT CAG ATC AAT GAA GAT AAG CGC CAC TCG CAG GGC ACA TTC ACC AGT    270
                                    50
asp tyr ser lys tyr leu asp ser arg arg ala gln asp phe val gln trp
GAC TAC AGC AAG TAC CTG GAC TCC AGG CGT GCC CAG GAC TTC GTG CAG TGG    321
        60                                 ↓ 70         Glucagon-
leu met asn thr lys arg asn lys asn asn ile ala lys arg his asp glu
TTG ATG AAT ACC AAG AGA AAC AAG AAT AAC ATT GCC AAA CGT CAT GAT GAA    372
like peptide I      80                                      90
phe glu arg his ala glu gly thr phe thr ser asp val ser ser tyr leu
TTT GAG AGA CAT GCT GAA GGG ACC TTT ACC AGT GAT GTA AGT TCT TAT TTG    423
                                100
glu gly gln ala ala lys glu phe ile ala trp leu val lys gly arg gly
GAA GGC CAA GCT GCC AAG GAA TTC ATT GCT TGG CTG GTG AAA GGC CGA GGA    474
    110                                     120
arg arg asp phe pro glu glu val asn ile val glu glu leu arg arg arg
AGG CGA GAT TTC CCA GAA GAA GTC AAC ATC GTT GAA GAA CTC CGC CGC AGA    525
Glucagon-like peptide II                                140
his ala asp gly ser phe ser asp glu met asn tyr val leu asp ser leu
CAC GCC GAT GGC TCT TTC TCT GAT GAG ATG AAC ACT GTT CTC GAT AGT CTT    576
                            150
ala thr arg asp phe ile asn trp leu leu gln thr lys ile thr asp arg
GCC ACC CGA GAC TTT ATA AAC TGG TTG CTT CAG ACG AAA ATT ACT GAC AGG    627
160
lys OC
AAG TAA   GTGTGTCATTCATTACTCAAGATCATCTTCACAATATCACCTGCCAGCCATGTGGGAT   691

GTTTGAAATTTTAAGTTCTGTAAATTTAACAGCTGTATTCTAAAGCCATATTGCTTGCATGCAAATA   758

AATAAATTTCCTTTTAATATTGTATAACCAAAAGATTATAAATTGAATACACCATTGTCAAAATAGT   825

GCTAAAATATCAGCTTTAAAATATGTTAATTCAGAATTCTCTTTCTTTTCTTCTGCTAACCTGCTTA   892

GCAATGAAATTATTTCTCTGTGATATAATTTGTATATATAAATTACTCCAATCACAACATATTTGCA   959

TTATAATAAGATAAGGGGAAGGACTGGTAGCCACAGTTGTGAAATGGGAAAGAGAATTTTCTTCTTG  1026

AAACTTTTGTCATAAAAATGCTCAACTTTCAGTATATAAAAGATAAACTAAATAAAAATTTCAAGCT  1093

TCTTCATCAAAAAAA...3'                                                 1100
```

FIGURE 7. Primary structure of cloned bovine pancreatic glucagon mRNA and the amino acid sequence of preproglucagon. Glucagon and two glucagon-like peptides (amino acid residues 33-61, 72-108, and 126-158, respectively) are underscored. Basic dipeptides which are potential proteolytic processing sites are enclosed in boxes. The sequence of glicentin (residues 1-69) is indicated by vertical arrows. Underscored AATAAA sequences in the 3'-untranslated region are characteristic sites involved in polyadenylation of eukaryotic mRNAs, but only the site at nucleotides 1076-1081 appears to be utilized (from Lopez et al. (43)).

adenylation site is located ~120 nucleotides downstream from the stop codon, there is no evidence that this site is utilized; hybridization of ^{32}P-labeled nick-translated glucagon cDNA insert with Northern blots of fetal pancreas poly (A^+)RNA indicated a single hybridizing RNA species of ~1250 nucleotides (data not shown).

DISCUSSION

Analysis of the primary sequence of the cloned glucagon mRNA indicated that it codes for an initial translation product of 180 amino acids with a M_r = 21 kd. From the deduced amino acid sequence of preproglucagon it was determined that the glucagon moiety was contained within a sequence encoding glicentin, a polypeptide previously isolated from porcine intestines. Furthermore, two additional glucagon-like peptides, designated GLP-I and and GLP-II, follow the glicentin sequence. Therefore, bovine preproglucagon contains three glucagon-related peptides in tandem. The structure of the glucagon mRNA suggests a tandem triplication of the primary glucagon sequence. GLP-I and GLP-II are similar but not identical to the other members of the glucagon-related peptide family. The function, if any, of GLP-I and GLP-II is not known.

These results are somewhat surprising for several reasons. It had previously been demonstrated by Lund et al. (21) that the 124 amino acid anglerfish preproglucagon I (AFG I) shared little homology with glicentin (other than the internal glucagon moiety), and contained only two glucagon-related peptides. This divergence in prohormone structure differs from that observed for proinsulin and prosomatostatin (47,48), whose structural organizations are conserved between fish and mammals. The 56 amino acid difference between bovine and fish preproglucagon I is accounted for by the presence in the bovine sequence of an additional 7 amino acids between glucagon and GLP-I and the 49 carboxy-terminal amino acids which comprise GLP II and it's flanking region. Anglerfish preproglucagon II (AFG II) (22), with 122 amino acids, is similar in structure to AFG I, but has lost one amino acid residue in the signal peptide and one in the region linking glucagon and the glucagon-like peptide. Fig. 8 is an illustration of the amino acid sequences of bovine preproglucagon and AFG I and AFG II. A comparison of the bovine to the anglerfish I and II preproglucagon sequences, respectively, indicates that they share 20% and 35% homology

BOVINE AND ANGLERFISH I AND II PREPROGLUCAGONS

```
          -20                                     -10
     5'.. ATG AAA AGC CTT TAC TTT GTG GCT GGA TTG TTT GTA ATG CTG GTA CAA GGC AGC
BG        met lys ser leu tyr phe val ala gly leu phe val met leu val gln gly ser
AFGI       *   *  arg ile his ser leu  *   *  ile leu leu val  *  gly leu ile gln
AFGII      *  thr  *   *  his ser leu  *   *   *  leu leu leu met --- ile ile gln
              1                                   10
          TGG CAA CGT TCC CTT CAG AAC ACA GAG GAG AAA TCC AGT TCA TTC CCA GCT CCC
          trp gln arg ser leu gln asn thr glu glu lys ser ser ser phe pro ala pro
          ser ser cys arg val leu met gln  *  ala asp pro  *   *  ser leu glu ala
          ser ser trp gln met pro asp gln asp pro asp arg asn  *  met leu leu asn
                      20                                      30
          CAG ACC GAC CCG CTC GGC GAT CCA GAT CAG ATC AAT GAA GAT AAG CGC CAC TCG
GLUCAGON  gln thr asp pro leu gly asp pro asp gln ile asn glu asp lys arg his ser
          asp ser thr leu lys asp glu  *  arg glu leu ser asn met  *   *   *   *
          glu asn ser met  *  thr glu  *  ile glu pro leu asn met  *   *   *   *
                              40                                      50
          CAG GGC ACA TTC ACC AGT GAC TAC AGC AAG TAC CTG GAC TCC AGG CGT GCC CAG
          gln gly thr phe thr ser asp tyr ser lys tyr leu asp ser arg arg ala gln
          glu  *   *   *  ser asn  *   *   *   *   *   *  glu asp  *  lys  *   *
          glu  *   *   *  ser asn  *   *   *   *   *   *  glu thr  *   *   *   *
                                      60                                      70
          GAC TTC GTG CAG TGG TTG ATG AAT ACC AAG AGA AAC AAG AAT AAC ATT GCC AAA
          asp phe val gln trp leu met asn thr lys arg asn lys asn asn ile ala lys
          glu  *   *  arg  *   *   *   *  asn  *   *  ser gly val ala gly --- lys
           *  phe  *   *   *   *  lys  *  ser  *   *   *  gly leu phe --- --- arg
                                              80
          CGT CAT GAT GAA TTT GAG AGA CAT GCT GAA GGG ACC TTT ACC AGT GAT GTA AGT
GLP-I     arg his asp glu phe glu arg his ala glu gly thr phe thr ser asp val ser
          arg --- --- --- --- --- ---  *   *  asp  *   *   *   *   *   *   *   *
          arg --- --- --- --- --- ---  *   *  asp  *   *  tyr  *   *   *   *   *
              90                                      100
          TCT TAT TTG GAA GGC CAA GCT GCC AAG GAA TTC ATT GCT TGG CTG GTG AAA GGC
          ser tyr leu glu gly gln ala ala lys glu phe ile ala trp leu val lys gly
           *   *   *  lys asp  *   *  ile  *  asp  *  val asp arg  *  lys ala  *
           *   *   *  gln asp  *   *   *   *  asp  *  val ser  *   *  lys ala  *
                      110                                     120
          CGA GGA AGG CGA GAT TTC CCA GAA GAA GTC AAC ATC GTT GAA GAA CTC CGC CGC
          arg gly arg arg asp phe pro glu glu val asn ile val glu glu leu arg arg
          gln val  *   *  glu AM
           *   *   *   *  gly OC
                              130                                     140
          AGA CAC GCC GAT GGC TCT TTC TCT GAT GAG ATG AAC ACT GTT CTC GAT AGT CTT
GLP-II    arg his ala asp gly ser phe ser asp glu met asn thr val leu asp ser leu
                                      150                                     160
          GCC ACC CGA GAC TTT ATA AAC TGG TTG CTT CAG ACG AAA ATT ACT GAC AGG AAG TAA..3'
          ala thr arg asp phe ile asn trp leu leu gln thr lys ile thr asp arg lys OC
```

FIGURE 8. Comparison of bovine and anglerfish I and II preproglucagons. Amino acid sequences are aligned and gaps are introduced to maximize homology. Asterisks (*) indicate residues homologous to the bovine sequence. Bovine glucagon, GLP-I and GLP-II are underscored, and basic dipeptides (arg, lys) which are potential proteolytic processing sites are enclosed in boxes. BG, AFG I, and AFG II are bovine, anglerfish I and anglerfish II preproglucagons, respectively.

in the signal peptide region, 17% and 10% homology in the GRPP region, 69% and 76% homology between the glucagon moieties, and 61% and 71% homology between the second glucagon-like peptides. The sequences joining fish and bovine glucagon (amino acid residues 64-69 in bovine) and the glucagon-like peptides which follow are not conserved between the two species in size or sequence and may serve only as spacers. The conservation of the second glucagon-like peptides between the two species is strongly suggestive of a biological function, but this peptide has not been previously identified and no function is known.

Based upon the amino acid sequence of preproglucagon which was deduced from the amino acid sequence of it's cloned cDNA, a scheme for the post-translational processing of pancreatic preproglucagon is proposed (fig. 9). This processing scheme is consistent with the results of Patzelt et al. (23) who analyzed glucagon processing in cultured rat islets, and Moody et al. (18) (49) who determined that a peptide with

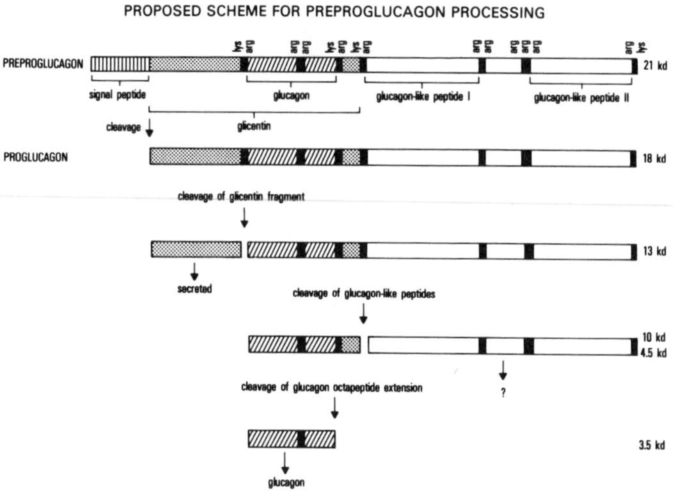

FIGURE 9. Proposed scheme for bovine pancreatic preproglucagon processing. This biosynthetic pathway for the processing of pancreatic preproglucagon is based upon our deduced precursor amino acid sequence and the results of Patzelt et al. (23) and Moody et al. (18,49) (from Lopez et al. (43)). Estimation of the M_r of the precursors are from Patzelt et al. (23)

glicentin-like immunoreactivity and lacking glucagon-like immunoreactivity was secreted synchronously with glucagon. It is suggested that the initial translation product of M_r=21 kd contains a 20 amino acid signal peptide which is co-translationally cleaved to produce an 18 kd proglucagon. This fragment is cleaved at the lys-arg sequence (residues 31, 32) in glicentin, releasing the GRPP peptide (residues 1-30) and the 13 kd precursor. Cleavage at the lys-arg sequence (residues 70,71) which follows the octapeptide extension at the carboxy-terminus of glucagon in the 13 kd precursor, produces a 4.5 kd precursor and a 10 kd fragment containing GLP-I and GLP-II. Proteolytic processing at the lys-arg sequence (residues 62, 63) of the 4.5 kd precursor results in 3.5 kd mature glucagon. The fate of the 10 kd fragment containing GLP-I and GLP-II is unknown. It is possible that the variety of possible processing sites is involved in a tissue-specific expression of the glucagon gene. At least one of the possible processing sites (residues 31,32) seems to be cleaved in a tissue-specific manner, resulting in the production of glicentin in the intestine and GRPP and glucagon in the pancreas (18,20,49).

The numerous processing sites in mammalian preproglucagon have the potential for generating a variety of polypeptides, but the biological effects of the various forms have not been established. The advantage in encoding a precursor with multiple copies of a similar peptide is not clear. Theoretically, differential processing of the 21 kd preproglucagon could selectively mask or free biological effector sites and/or receptor binding sites resulting in a wide variety of biologically active polypeptides, or coordinate the synthesis of functionally related peptides. In any case, information is accumulating that polyproteins can be multifunctional polypeptides that may be important regulatory elements of complex biological responses in eukaryotes.

Following completion of this manuscript, the sequence of hamster preproglucagon was reported (50). The structures of of bovine and hamster preproglucagons are highly conserved.

ACKNOWLEDGMENTS

We wish to thank Dr. M. L. Caruthers for the gift of the synthetic 14-mer oligodeoxynucleotide probe.

REFERENCES

1. Birnbaumer L (1973). Hormone sensitive adenylyl cyclases useful models for studying hormone receptor functions in cell-free systems. Biochim Biophys Acta 300:129.
2. Sokal JE (1973). In Meth in Invest and Diag Endo Berson SA, Yalow RS (eds) New York, Am. Elsevier Pub Co, p 901.
3. Hers HG (1976). The control of glycogen metabolism in the liver. Ann Rev Biochem 45:167.
4. Sutherland EW, de Duve C (1948). Origin and distribution of the hyperglycemic-glycogenolytic factor of the pancreas. J Biol Chem 175:663.
5. Makman MH, Makman RS, Sutherland EW (1958). Presence of a glucagon-like material in blood of man and dog. J Biol Chem 233:894.
6. Unger RH, Eisentraut AM, McCall MS, Keller S, Lang HC, Madison LL (1959). Glucagon antibodies and their use for immunoassay for glucagon. Proc Soc. Exp Biol Med 102:621.
7. Valverde I, Rigopoulou D, Marco J, Faloona GR, Unger RH (1970). Characterization of glucagon-like immunoreactivity (GLI). Diabetes 19:614.
8. Srikant CB, McCorkle K, Unger RH (19/6). Characteristics of tissue IRGs in the dog. Metabolism 25:1403.
9. Valverde I, Villanueva ML, Lozano I, Marco J (1974). Presence of glucagon immunoreactivity in the globulin fraction of human plasma ("big plasma glucagon"). J Clin Endocr Metabol 39:1090.
10. Tager HS, Markese J (1979). Intestinal and pancreatic glucagon-like peptides. Evidence for identity of higher molecular weight forms. J Biol Chem 254:2229.
11. Mutt V, Jorpes JE, Magnusson S (1970). Structure of porcine secretin: The amino acid sequence. Eur J Biochem 15:513.
12. Thim L, Moody, AJ (1981). The primary structure of porcine glicentin (proglucagon). Regulatory Peptides 2:139.
13. Brown JC (1971). A gastric inhibitory peptide I: The amino acid composition and the tryptic peptides. Can J Biochem 49:255.
14. Mutt V, Said SI (1974). Structure of the porcine vasoactive intestinal octacosapeptide: The amino acid sequence. Use of Kallikrein in its determination. Eur J Biochem 42:581.

15. Jornvall H, Carlstrom A, Pettersson L, Jacobsson B, Persson M, Mutt V (1981). Structural homologies between prealbumin, gastrointestinal prohormones and other proteins. Nature 291:261.
16. Tatemoto K, Mutt V (1981). Isolation and characterization of the intestinal peptide porcine PHI (PHI-27), a new member of the glucagon-secretin family. Proc Natl Acad Sci 78:6003.
17. Spiess J, Rivier J, Thorner M, Vale W (1982). Biochem 21:6037.
18. Thim L, Moody AJ (1982). Purification and chemical characterization of a glicentin-related pancreatic peptide (proglucagon fragment) from porcine pancreas. Biochim Biophys Acta 703:134.
19. Lund PK, Goodman RH, Habener JF (1981). Pancreatic preproglucagons are encoded by two separate mRNAs. J Biol Chem 256:6515.
20. Shields D, Warren TG, Roth SE, Brenner MJ (1981). Cell-free synthesis and processing of multiple precursors to glucagon. Nature (London) 289:511.
21. Lund PK, Goodman RH, Dee PC, Habener JF (1982). Pancreatic preproglucagon cDNA contains two glucagon-related coding sequences arranged in tandem. Proc. Natl Acad Sci 79:345.
22. Lund PK, Goodman RH, Montminy, MR, Dee PC, Habener JF (1983). Anglerfish islet preproglucagon II. Nucleotide and corresponding amino acid sequence of the cDNA. J Biol Chem 258:3280.
23. Patzelt C, Tager HS, Carroll RJ, Steiner DF (1979). Identification and processing of proglucagon in pancreatic islets. Nature 282:260.
24. Kraus JP, Rosenberg LE (1982). Purification of low-abundance messenger RNAs from rat liver by polysome immunoadsorption. Proc Natl Acad Sci 79:4015.
25. Noyes BE, Mevarech M, Stein R, Agarwal KL (1979). Detection and partial sequence analysis of gastrin mRNA by using an oligodeoxynucleotide probe. Proc Natl Acad Sci 76:1770.
26. Matteuccim MD, Caruthers MH (1981). Synthesis of deoxyoligonucleotides on a polymer support. J Am Chem Soc 103:3185.
27. Beaucage SL, Caruthers MH (1981). Deoxynucleotide phosphoramidites - a new class of key intermediate for deoxypolynucleotide synthesis. Tetrahedron Lett 22:1859.

28. Lomedico PT, Saunders GF (1976). Preparation of pancreatic mRNA: cell-free translation of an insulin-immunoreactive polypeptide. Nucleic Acids Research 3:381.
29. Frazier ML, Montagna RA, Saunders GF (1981). Insulin gene expression during development of the fetal bovine pancreas. Biochemistry 20:367.
30. Aviv H, Leder P. (1972). Purification of biologically active globin messenger RNA by chromotography on oligo-thymidylic acid-cellulose. Proc Natl Acad Sci 69:1408.
31. Thomas PS (1980). Hybridization of denatured RNA and small DNA fragments transferred to nitrocellulose. Proc Natl Acad Sci 77:5201.
32. Gubler U, Kilpatrick DL, Seeburg PH, Gage LP, Udenfriend S (1981). Detection and partial characterization of proenkephalin mRNA. Proc Natl Acad Sci 78:5484.
33. Maxam AM, Gilbert W (1979). Sequencing end-labeled DNA with base specific chemical cleavages. Methods in Enzymol 65:499.
34. Murphy EC Jr, Arlinghaus RB (1978). Cell-free synthesis of Rauscher murine leukemia virus "gag" and "gag-pol" precursor polyproteins from virion 35 S RNA in a mRNA-dependent translation system derived from mouse tissue culture cells. Virology 86:329.
35. Laemli, UK (1970). Cleavage of structural proteins during the assembly of the head of bacteriophage T_4. Nature 227:680.
36. Land H, Grez M, Hauser J, Lindenmaier W, Schultz G (1981). 5' terminal sequences of eucaryotic mRNA can be cloned with high efficiency. Nucleic Acids Research 9:2251.
37. Grunstein M, and Hogness DS (1975). Colony hybridization: a method for the isolation of cloned DNAs that contain a specific gene. Proc Natl Acad Sci 72:3961.
38. Southern EM (1975). Detection of specific sequences among DNA fragments separated by gel electrophoresis. J Mol Biol 98:503.
39. Agarwal KL, Brunstedt J, Noyes BE (1981). A general method for detection and characterization of an mRNA using an oligonucleotide probe. J Biol Chem 256:1023.
40. Sanger F, Nicklen S, Coulson AR (1977). DNA sequencing with chain terminating inhibitors. Proc Natl Acad Sci 74:5463.

41. Messing J, Crea R, Seeburg P (1981). A system for shotgun DNA sequencing. Nucleic Acids Res 9:309.
42. Birnboim HC, Doly J (1979). A rapid alkaline extraction procedure for screening recombinant plasmid DNA. Nucl. Acids Res 7:1513-1522.
43. Lopez LC, Frazier ML, Su CJ, Kumar A, Saunders GF (1983). Mammalian pancreatic preproglucagon contains three glucagon-related peptides. Proc Natl Acad Sci (in press).
44. Steiner DF, Quinn PS, Chan SJ, Marsh J, Tager HS (1980). Processing mechanisms in the biosynthesis of proteins. Ann NY Acad Sci 343:1.
45. Lewin B (1975). Units of transcription and translation: the relationship between heterogeneous nuclear RNA and messenger RNA. Cell 4:11.
46. Proudfoot NJ, Brownlee GG (1976). 3' noncoding region sequences in eucaryotic mRNA. Nature 263:211.
47. Shen LP, Pictet RL, Rutter WJ (1982). Human somatostatin I: sequence of the cDNA. Proc Natl Acad Sci 79:4575.
48. Chan SJ, Emdin SO, Kwok SCM, Kramer JM, Falkmer S, Steiner DF (1981). Messenger RNA sequence and primary structure of preproinsulin in a primative vertebrate, the Atlantic hagfish. J Biol Chem 256:7595.
49. Moody AJ, Holst JJ, Thim L, Jensen SL (1981). Relationship of glicentin to proglucagon and glucagon in the porcine pancreas. Nature 289:514.
50. Bell GI, Santerre RF, Mullenbach GT (1983). Hamster preproglucagon contains the sequence of glucagon and two related peptides. Nature 302:716.

VII. GENE TRANSFER

Gene Expression, pages 513-522
© 1983 Alan R. Liss, Inc., 150 Fifth Avenue, New York, NY 10011

EXPRESSION OF AN IMMUNOGLOBULIN LIGHT CHAIN GENE IN ESCHERICHIA COLI

Michael A. Boss and Spencer Emtage

Celltech Ltd., 244-250 Bath Road, Slough SL1 4DY, Berks, England.

ABSTRACT A gene for murine lambda light chain immunoglobulin has been inserted into a bacterial expression plasmid containing the *Escherichia coli* tryptophan promoter and a bacterial ribosome binding site. Induction of transcription from the tryptophan promoter results in a significant level of light chain polypeptide synthesis. Fractionation of *E.coli* extracts showed that the recombinant protein was found exclusively as an insoluble product. Extraction with urea rendered the light chain protein soluble and amenable to further purification.

INTRODUCTION

The expression of a number of eucaryotic genes in *E.coli* has been demonstrated (1). Immunoglobulin genes and their products are one of the most extensively studied gene families. Immunoglobulin polypeptides are synthesised *in vivo* with an amino-terminal signal peptide which is cleaved to yield the mature protein. These genes have previously only been expressed at a low level in *E.coli* as modified forms consisting of incomplete amino-terminal fusion proteins (2,3). We describe here the reconstruction of a gene for murine lambda light chain immunoglobulin from chemically synthesised oligodeoxyribonucleotides and cloned DNA copies of pre-light chain mRNA. This gene has been cloned into a bacterial expression plasmid containing the *E.coli* tryptophan promoter and a bacterial ribosome binding site so that the lambda light chain is synthesised as a native protein lacking a eucaryotic leader but possessing an amino-terminal methionine residue. We report significant expression of immunoreactive lambda light chain protein.

METHODS

Plasmid pABλ1-15 (4) was a gift from Drs A. Bothwell and D. Baltimore. Materials were purchased as follows: antisera (Miles), MOPC104E a $\mu\lambda_1$ myeloma protein (Bionetics), restriction enzymes (New England Biolabs), deoxyribonuclease 1 (Sigma), calf intestinal alkaline phosphatase and S_1 nuclease (Boehringer Mannheim). Cloning procedures were as described (5). Oligodeoxyribonucleotides were synthesised by the phosphotriester procedure (6) and had the sequence 5'-GATCAATGCAGGCTGTTGTG-3' (R45) and 5'-ATTCCTGAGTCACAACAGCC -3' (R44). Plasmids were transformed into *E.coli* K12 strain HB101 or DH1 (5) and grown in L broth and 100 μg carbenicillin per ml.

Inductions were carried out by 1:100 dilution of overnight cultures of HB101 into M9 medium (5) supplemented with glucose, vitamin B_1, carbenicillin, leucine and proline. Cultures were shaken at 37° to mid-log phase of growth and pulse labeled with 24 μCi/ml [^{35}S] methionine for 5-20 minutes. Labeled cells were harvested by centrifugation and lysed by boiling in 1% SDS and diluted in a buffer consisting of 50mM Tris pH8, 0.15M NaCl, 0.1mM EDTA and 2% Triton X-100. Samples were immunoprecipitated by addition of 5 μl normal or immune serum to aliquots of labeled *E.coli* extracts, and immune complexes isolated by binding to *Staphylococcus aureus* cells. Sodium dodecyl sulfate-polyacrylamide gel electrophoresis was performed as described (7).

Cell lysis was achieved using lysozyme/sodium deoxycholate and deoxyribonuclease treatment (8). The resulting suspension was centrifuged at 10,000 g for 15 minutes at 4° to yield a pellet and a soluble fraction. Immunoprecipitations were carried out by solubilising the pellet as described above. For further purification of lambda light chain the cell debris were dissolved in 8M urea and dialysed overnight at 4° against 50mM Tris pH8, 0.1M NaCl, 1mM EDTA, 10% glycerol or 1M acetic acid. After dialysis the heavy precipitate of insoluble *E.coli* proteins was removed by centrifugation.

RESULTS AND DISCUSSION

We chose to express the lambda gene in *E.coli* by direct expression of the gene lacking the eucaryotic signal peptide but containing a methionine initiator residue at the amino-terminus (met-lambda). The approach used for bacterial synthesis of met-lambda was to reconstruct the gene *in vitro* from restriction fragments of a cDNA clone and to utilise synthetic DNA fragments for insertion into the bacterial plasmid pCT54 (8). This vector contains the *E.coli* trp promoter, operator and leader ribosome binding site; in addition 14 nucleotides downstream of the ribosome binding site is an initiator ATG followed immediately by EcoR1 and HindIII sites and the terminator for *E.coli* RNA polymerase from bacteriophage T7 (9).

As a source of light chain we used a plasmid pABλ1-15 which contains a full-length λ_1 light chain cDNA cloned into the PstI site of pBR322 (4). In order to create a HindIII site 3' to the end of the lambda gene for insertion into the HindIII site of pCT54, the cDNA was excised from pABλ1-15 using Pst1. The cohesive ends were blunt ended using the Klenow fragment of DNA polymerase and synthetic HindIII linker molecules of sequence 5'-CCAAGCTTGG-3' ligated. The DNA was digested with HindIII and the 850bp lambda gene isolated by gel electrophoresis and cloned into HindIII cut pAT153 to yield plasmid pATλ1-15. The 3' end of the lambda gene was isolated from pATλ1-15 by HindIII plus partial Sac1 digestion as a ~630bp SacI-HindIII fragment (2 in Figure 1). The HindIII cohesive end was dephosphorylated by calf intestinal alkaline phosphatase during isolation of the fragment to prevent unwanted ligations at this end in subsequent reactions. A HinfI restriction site is located between codons 7 and 8 of the lambda sequence. The 5' end of the lambda gene was isolated as a 148bp HinfI to SacI fragment (1 in Figure 1). Two oligodeoxyribonucleotides were designed to restore codons 1-8, and to provide an initiator ATG as well as BclI and HinfI sticky ends. pCT54 was cut with both Bcl1 and HindIII and the resulting linear molecules isolated, mixed together with the two oligodeoxyribonucleotide linkers R44 and R45 and both fragments 1 and 2, and ligated using T4 ligase (Figure 1). The mixture was used to transform *E.coli* DH1 to ampicillin resistance. Recombinant clones in pCT54 were identified by hybridisation of DNA from replica plated colonies on nitrocellulose to a nick-translated probe derived from the pATλ1-15 insert. A clone was identified which hybridised to lambda cDNA and

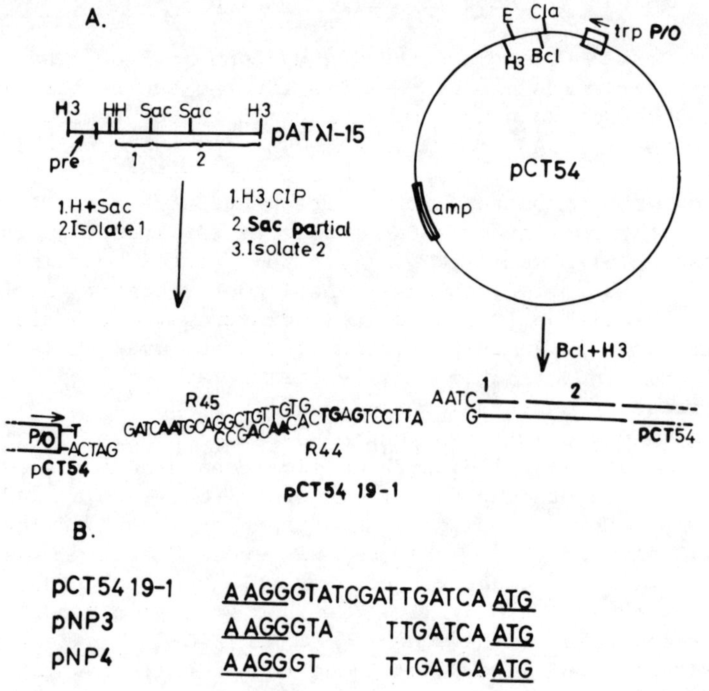

FIGURE 1. Construction of plasmids for the direct synthesis of lambda light chain in *E.coli*. Plasmid pATλ1-15 contains the lambda gene inserted into the HindIII site of pAT153. A. A 5' HinfI - SacI fragment 1 was isolated by polyacrylamide gel electrophoresis. The 3' fragment 2 of the gene was isolated as a SacI - HindIII fragment. pCT54 was cut with Bcl1 + HindIII and the lambda gene fragments together with oligodeoxyribonucleotides R45 and R44 ligated to yield plasmid pCT54 19-1. B. Digestion of pCT54 19-1 with Cla1 and S_1 nuclease to produce plasmids pNP3 and pNP4. E, EcoR1; H, Hinf1; H3, HindIII.

also showed the predicted restriction fragment pattern. This plasmid (designated pCT54 19-1) was sequenced from the ClaI site and shown to have the anticipated sequence except that there was a mutation of the fourth codon from CTG to ATG, changing the amino acid at this point from valine to methionine.

pCT54 had been constructed to include two restriction sites (Bcl1 and Cla1) between the Shine-Dalgarno (SD)

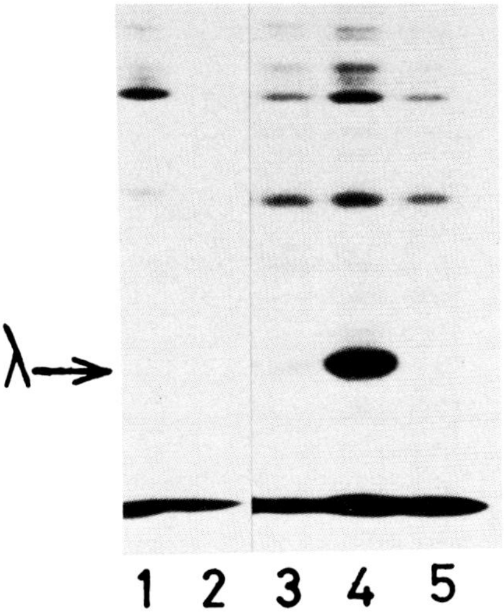

FIGURE 2. Analysis of lambda gene expression in *E.coli*. HB101 cells containing plasmids were grown under inducing conditions and pulse labeled with [^{35}S] methionine. Cells were lysed and samples analysed by immunoprecipitation and sodium dodecyl sulfate - polyacrylamide gel electrophoresis. Lanes 1 and 2, respectively, extracts from pCT54 19-1 and pNP3, immunoprecipitated with normal rabbit serum; lanes 3, 4 and 5 were extracts immunoprecipitated with rabbit anti-lambda serum and represent pNP4 (lane 3), pNP3 (lane 4) and pCT54 19-1 (lane 5). The position of unlabeled lambda protein from MOPC104E is indicated on the left.

sequence AAGG and the ATG so that the distance between these sequence elements could be varied. As most *E.coli* mRNAs have 6-11 nucleotides between the SD sequence and the AUG (10), the distance in pCT54 19-1 was reduced by modification at the Cla1 site. pCT54 19-1 was cut with Cla1 and incubated with S_1 nuclease. The amount of S_1 nuclease was adjusted so that some DNA molecules would lose 1-2 extra base pairs as a result of "nibbling" by the enzyme. This DNA on religation with T4 DNA ligase and transformation into *E.coli* strain HB101 gave rise to a number of plasmids which had lost the Cla1 site. The nucleotide sequence across the modified region of two of these plasmids was determined (Figure 1). pNP4 and pNP3 were shorter than pCT54 19-1 by 5 and 4 nucleotides respectively giving SD-ATG distances of

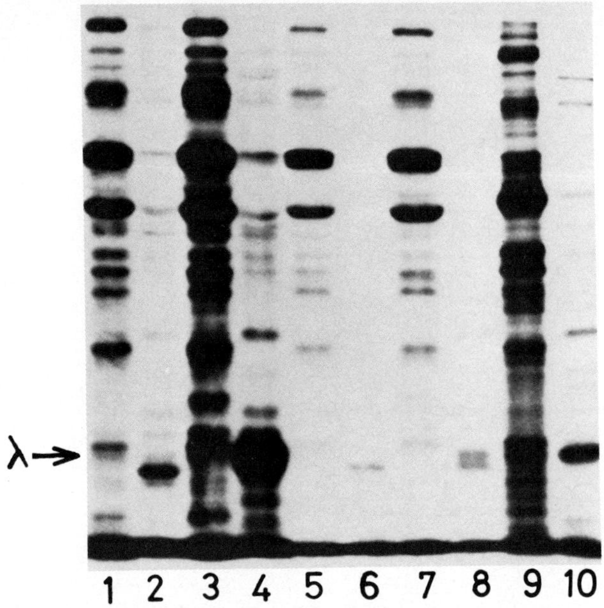

FIGURE 3. Distribution of recombinant lambda light chains. HB101 cells containing pNP3 or pNP4 were grown under inducing conditions, pulse labeled with [^{35}S] methionine, separated into soluble and insoluble fractions and analysed by sodium dodecyl sulfate - polyacrylamide gel electrophoresis. Lanes 1 and 3, respectively, soluble fraction from HB101-pNP3 immunoprecipitated with normal rabbit serum or rabbit anti-lambda serum; lanes 2 and 4, respectively, insoluble fraction from HB101-pNP3 immunoprecipitated with normal rabbit serum or rabbit anti-lambda serum; lanes 5 and 7, respectively, soluble fraction from HB101-pNP4 immunoprecipitated with normal rabbit serum or rabbit anti-lambda serum; lanes 6 and 8, respectively, insoluble fraction from HB101-pNP4 immunoprecipitated with normal rabbit serum or rabbit anti-lambda serum; lane 9, soluble fraction from HB101-pNP3; lane 10, insoluble fraction from HB101-pNP3. The position of unlabeled lambda protein from MOPC104E is indicated on the left.

9 and 10 nucleotides.

HB101 cells containing such plasmids were grown under inducing conditions to an OD_{600}, pulsed with [^{35}S] methionine

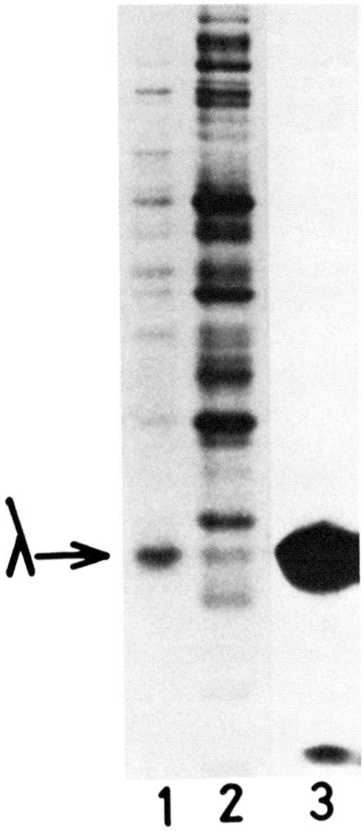

FIGURE 4. Purification of recombinant lambda light chain polypeptide. HB101 cells containing pNP3 were grown under inducing conditions and lysed. The insoluble fraction was dissolved in 8M urea and dialysed into 1M acetic acid or Tris containing buffer (data not shown). The mixture was centrifuged and the fractions analysed by sodium dodecyl sulfate - polyacrylamide gel electrophoresis. Lane 1, soluble fraction following dialysis; lane 2, insoluble fraction following dialysis; lane 3, soluble fraction following dialysis immunoprecipitated with rabbit anti-lambda serum. The position of unlabeled lambda protein from MOPC104E is indicated on the left.

for 20 minutes and the labeled proteins examined by immunoprecipitation. pNP3 showed the presence of a band at 25K daltons upon immunoprecipitation with rabbit anti-lambda serum (Figure 2, lane 4), which comigrated with authentic lambda light chain from the mouse myeloma MOPC104E. Such a 25Kd protein was not evident in immunoprecipitations of pNP3 with normal rabbit serum (Figure 2, lane 2) nor was it present in extracts from pCT54 19-1 (Figure 2, lanes 1 and 5). Interestingly pNP4 which had a SD-ATG distance one base pair shorter than that of pNP3 expressed only a very low level of lambda light chain (Figure 2, lane 3). *In vitro* coupled transcription and translation (11) of pNP3 confirmed that this plasmid coded for a protein which comigrated with authentic lambda light chain and is expressed *in vitro* at about equal rates as the β-lactamase gene carried on the same plasmid (Alan D. Bennett, M.A.B. and J.S.E. unpublished observations).

After cell lysis and centrifugation of HB101 containing pNP3, lambda light chain was detected in the insoluble (Figure 3, lane 4) but not in the soluble fraction (Figure 3, lane 3). The specificity of the immunoprecipitation was confirmed using normal rabbit serum controls (Figure 3, lanes 1 and 2). Extract from pNP4 when subjected to the same analysis showed that the small amount of lambda light chain synthesised was also present exclusively in the insoluble fraction (Figure 3, lanes 7 and 8). This finding suggests that the apparent insolubility of normally soluble eucaryotic proteins, which is not an unusual finding when eucaryotic genes are expressed in *E.coli* (1), is not simply a property of expressing foreign genes to a high level. In the absence of immunoprecipitation a major band of 25Kd was evident in the insoluble (Figure 3, lane 10) but not the soluble (Figure 3, lane 9) fraction from HB101 containing pNP3. The presence of lambda light chain in the pellet fraction was a useful purification step since it both concentrated the protein and separated it from the bulk of *E.coli* soluble proteins.

For further purification the lambda light chain was dissolved in 8M urea, which was subsequently removed by dialysis. After dialysis the lambda light chain was present in the soluble (Figure 4, lane 1) rather than in the insoluble fraction (Figure 4, lane 2). After dialysis, immunoprecipitation with anti-lambda serum confirmed the presence of lambda protein in the soluble fraction (Figure 4, lane 3). It is evident that the solubilisation procedure (Figure 4, lane 1) also removed some contaminating *E.coli* proteins. The recombinant light chain is now amenable to further purification. Preliminary results indicate that a high level of purity can be achieved using an anti-lambda immunoaffinity column.

ACKNOWLEDGEMENTS

We gratefully thank Drs A. Bothwell and D. Baltimore for giving us plasmid pABλ1-15. We thank members of the Chemistry Department for synthesis of oligodeoxyribonucleotides; Tim Harris and Paul Thomas for nucleotide sequencing; and Pete Lowe for advice on protein purification.

REFERENCES

1. Harris TJR (1983). Expression of eukaryotic genes in *E.coli*. In Williamson R (ed): "Genetic Engineering 4", Academic Press, p 127.
2. Amster O, Salomon D, Zemel O, Zamier A, Zeelon EP, Zantor F, Schechter I (1980). Synthesis of part of a mouse immunoglobulin light chain in a bacterial clone. Nucleic Acids Res 8:2055.
3. Kemp DJ, Cowman AF (1981). Direct immunoassay for detecting *Escherichia coli* colonies that contain polypeptides encoded by cloned DNA segments. Proc Natl Acad Sci 78:4520.
4. Bothwell ALM, Paskind M, Reth M, Imanishi-Kari T, Rajewsky K, Baltimore D (1982). Somatic variants of murine immunoglobulin lambda light chains. Nature 298:380.
5. Maniatis T, Fritsch EF, Sambrook J (1982). "Molecular Cloning (a laboratory manual)" Cold Spring Harbor Laboratory.
6. Patel TP, Millican TA, Bose CC, Titmas RC, Mock GA, Eaton MAW (1982). Improvements to solid phase phosphotriester synthesis of deoxyoligonucleotides. Nucleic Acids Res 10:5605.
7. Laemmli UK (1970). Cleavage of structural proteins during the assembly of the head of bacteriology T4. Nature 227:680.
8. Emtage JS, Angal S, Doel MT, Harris TJR, Jenkins B, Lilley G, Lowe PA (1983). Synthesis of calf prochymosin (prorennin) in *Escherichia coli*. Proc Natl Acad Sci In press.
9. Dunn JJ, Studier FW (1980). The transcription termination site at the end of the early region of bacteriophage T7 DNA. Nucleic Acids Res 8:2119.
10. Gold L, Pribnow D, Schneider T, Shinedling S, Singer BS, Stormo G (1981). Translational initiation in prokaryotes. Ann Rev Microbial 35:365.

Gene Expression, pages 523-541
© 1983 Alan R. Liss, Inc., 150 Fifth Avenue, New York, NY 10011

EXPRESSION VECTORS FOR FUSIONS OF THE E. COLI GALK GENE TO YEAST CYC1 GENE REGULATORY SEQUENCES

Richard S. Zitomer,* Brian C. Rymond,* Daniel Schumperli,+ and Martin J. Rosenberg+

*Department of Biological Sciences, State University of New York at Albany, Albany, New York 12222
+Laboratories of Biochemistry and Molecular Biology National Cancer Institute, Bethesda, Maryland 20205

ABSTRACT A family of fusions were constructed between the regulatory sequences of the yeast CYC1 gene, encoding the iso-1-cytochrome c protein, and the coding sequence of the E. coli galactokinase gene, galK. These fusions were capable of complementing the galactokinase deficiency of gal1 yeast mutants. Two types of fusions were characterized. The first, YCpR1, led to the production of a fused protein; the CYC1 5' sequences through the first four codons were fused to the bulk of the galK coding sequence lacking the first five codons. Galactokinase was expressed at high levels from this fusion and was regulated by both glucose and oxygen in the same fashion as the CYC1 gene. The other type of fusion was between the CYC1 5' sequences including the first four codons and the galK leader sequence plus the coding region. Galactokinase expression was poor from these fusions due to the presence of the CYC1 AUG initiation codon preceding the galK initiation codon on the mRNA. Deletion of this first AUG greatly increased galactokinase expression.

The usefulness of these fusions was demonstrated by the isolation of oxygen constituitive mutants selected for their ability to express the CYC1/galK fusion in the absence of oxygen. Two of these mutants were partially characterized. The general usefulness of this system for the study of the expression of yeast genes is discussed.

INTRODUCTION

Studies of the expression of many eukaryotic genes are hampered by a difficult assay for the protein and/or by the inability to select regulatory mutants. Even with a simple, genetically well characterized eukaryote like yeast, there are many genes for which there is no specific selective phenotype for regulatory mutants. Prokaryotic geneticists, when faced with these problems, have routinely turned to gene fusions as a solution. In such fusions, the region controlling the expression of the gene of interest is joined to the protein coding segment of a gene whose product can be easily assayed and for which a selectable phenotype exists. Recently such approaches have been used for the study of yeast genes, in particular using the same gene, the E. coli ß-galactosidase gene, that prokaryotic workers have favored (1,2). Fusions using this gene have many advantages, but unfortunately suffer from the inability to select for expression of ß-galactosidase activity because yeast cells, even with a functional ß-galactosidase enzyme, cannot utilize lactose for growth. As a result, when confronted with the problems described above in our studies of the yeast CYC1 gene, encoding the iso-1-cytochrome c protein, we decided to construct a fusion between this gene and the galK gene of E. coli encoding galactokinase.

Wildtype yeast can utilize galactose as an energy source and catabolize it in the same manner as E. coli. The yeast galactokinase enzyme is encoded by the GAL1 gene (3), and a successful CYC1/galK fusion would be able to complement a gal1 yeast mutation. Regulatory mutants could be selected by their ability to utilize galactose in conditions under which the expression of the CYC1 gene would normally be repressed. In addition, a simple and sensitive assay exists for the quantitation of galactokinase activity in crude cell extracts (4). Thus the galK system appeared to fulfill all our requirements for an ideal fusion system.

Here we describe such fusions and show that they can indeed be used to study CYC1 expression. We present data that shows that both protein coding fusions and mRNA leader sequence fusions can be used providing that, in the latter case, the requirements of the yeast translational machinery are satisfied. Finally we present our preliminary findings on the isolation of oxygen constituitive mutants which demonstrate the potential of using galK fusion genes contained on plasmids for the isolation of regulatory mutants.

METHODS

Strains

A galK⁻ Escherichia coli C600 strain obtained from Ursula Schmeisser was used for the initial experiments,but HB101, also galK⁻ (5) proved to be more useful because of its m⁻r⁻ genotype and was used for the latter experiments described here. E. coli transformations were carried out as described (6).

The haploid Saccharomyces cerevisiae strain aBR10 (a, gal1, trp1-1, his4-519, ade) was used for all yeast transformations which were carried out as described (7) using the enzyme lyticase (8) to make yeast sphereoplasts.

Plasmids

The plasmid YRp CYC1(2.4) was constructed by Lowry et al. (9). It consists of pBR322 into which the TRP1-ARS1 1.4 kb EcoR1 fragment was inserted (10) at the pBR322 EcoR1 site, but with the EcoR1 sites destroyed in the process. This plasmid also contains a 2.4 kb HindIII-BamH1 fragment containing the CYC1 coding sequence plus 1.8 kb of 5' and 300 bp of 3' sequences inserted into the pBR322 HindIII and BamH1 sites. The plasmid YCp CYC1(2.4) is identical to YRp CYC1(2.4) but with the addition of the 2 kb BglII-BamH1 CEN3 sequence into the BamH1 site (9).

The plasmid pYe(CEN3)41 contains the centromeric sequence of yeast chromosome III on a 2 kb BglII-BamH1 fragment (11).

The plasmid pDSΔ3.03 containing the E. coli galK gene was obtained from Martin Rosenberg and Daniel Schumperli.

CYC1/galK Fusion Constructions

To join the bulk of the galK protein coding sequence to the regulatory and coding sequence of the CYC1 gene, first a galK gene with one EcoR1 site within the beginning of its coding sequence and a second 3' to the coding sequence was constructed, then this EcoR1 fragment was inserted into the EcoR1 site comprising the third and fourth codons of the CYC1 gene. The first step was achieved by digesting pDSΔ3.03 with HindIII which cleaves the plasmid 40bp 5' to the galK ATG initiation codon. The linearized plasmid was then subjected to digestion with the exonuclease Bal31 then recircularized with T4 ligase in the presence of synthetic EcoR1 linkers. E. coli cells were transformed and a pool of plasmids containing the Bal31 deletions was prepared. An EcoR1 site was added 3' to the galK gene by digesting the plasmids in this pool with HpaI which cleaves 44 bp 3' to the coding sequence and religating in the presence of EcoR1 linkers. These plasmids were then amplified by transformation into E. coli. The EcoR1 fragments containing the galK gene were then inserted into the plasmid YRp CYC1(2.4) at the EcoR1 site. After amplification in E. coli, this plasmid pool was transformed into yeast cells and gal^+ transformants selected.

The CEN3 containing BglII-BamH1 fragment was inserted into the unique BamH1 site of the YRp fusion plasmids using T4 ligase.

The deletions of the CYC1 ATG codon were constructed by digestion of EcoR1 cleaved YCp CYC1(2.4) with Bal31 followed by ligation in the presence of EcoR1 linkers. E. coli cells were then transformed, and plasmids containing appropriate sized deletions were identified by preparing DNA from individual clones by the modified alkaline mini-prep procedure (12). Sizes of these deletions were determined initially by polyacrylamide gel electrophoresis and later by sequence analysis. The galK EcoR1 fragments from the plasmids YCpR3, YCpR6, and YCpR72 were then inserted into the EcoR1 site of chosen deletion plasmids.

Growth of Yeast Transformants

Galactose selective media consisted of, per liter, 6.7 g yeast nitrogen base without amino acids (Difco), 40 mg adenine, 40 mg uracil, 20 mg each of leucine, histidine, phenylalanine, tyrosine, threonine, isoleucine, tryptophan, arginine, and methionine, and 20 g of galactose. For plates, 15 g of agar was added. Tryptophan was omitted when it was desirable to select for the plasmid using the TRP1 marker. For galactokinase assays cells were grown in this media with galactose and tryptophan omitted and 20 g of either glucose or raffinose added.

Galactokinase Assay

For the preparation of extracts, 5 ml of cells were grown to a density of 2.5 x 10^7 per ml, chilled quickly to 0°, harvested by centrifugation, and resuspended in 250 µl of lysis buffer (20 mM-HEPES, 1 mM-dithiothreitol, 300 µg/ml-bovine serum albumin, pH adjusted to 7.5 with KOH). The cells were broken by rigorous mixing with an equal volume of glass beads (0.45 - 0.5 mm diameter). The extract was removed from the glass beads and the beads were washed with 250 µl of lysis buffer which was added to the extract. The extract was then clarified by centrifugation for 2 min in a microfuge and frozen at -20°.

The galactokinase assay was adapted from one previously described (13). The reaction mix was assembled on ice and consisted of 80 µl of 125 mM Tris · HCl, pH 7.9, 5 mM-$MgCl_2$, 1.25 mM-dithiothreitol, 4 mM-NaF, 2 mM-ATP, and ^{14}C-galactose. The specific activity of the ^{14}C-galactose was varied depending upon the sensitivity required, but was typically between 0.5 and 10 µCi/µmole with a final concentration of 1 to 10 mM. This reaction mix was preincubated at 32° for 30 sec, then 20 µl of cell extract was added. Time points were taken by transferring 25 µl to a tube containing 2 µl of 250 mM-EDTA and 0.5 M-galactose at 0°. The levels of ^{14}C-galactose-1-phosphate present in these time points were determined by spotting 20 µl onto Whatman DE81 filters and washing these filters extensively against several changes of a 0.1 % galactose solution at 0°. Radioactivity bound to the filter was determined in a liquid scintillation counter. The calculations of µmoles ^{14}C-galactose-1-phosphate formed per min were performed by determining the initial slopes of the curves plotted for cpm bound per filter versus reaction time.

DNA Sequence Analysis

The fusion points between the yeast *CYC1* gene and the *galK* gene were determined by subcloning the EcoR1 fragments containing the *galK* gene into M13 mp8 using the cloning procedures described (14), and then sequencing the insert by the dideoxynucleotide chain termination method (15). Only the *galK* side of the fusion was sequenced because the *CYC1* side was not altered in the construction procedure.

Nonetheless, this was confirmed by reconstructing each fusion by inserting the galK EcoR1 fragment of each sequenced fusion into the EcoR1 site of YCp CYC1(2.4) and ensuring that transformants containing the reconstructed fusions had the same phenotype as the original transformants.

The extent of the CYC1 leader sequence deletions in the plasmids shown in figure 3 were determined by subcloning the EcoR1-BamH1 CYC1 fragment into mp8 followed by DNA sequencing. The primer used for the sequencing reactions was purchased from Collaborative Research.

Enzyme Reactions

All reactions involving restriction enzymes, ligase, and Bal31 were carried out according to the recommendations of the vendors.

Genetic Analysis

Mutants constituitive for galactokinase expression with respect to oxygen were induced by u.v. mutagenesis. 10^6 cells were spread on galactose minus tryptophan plates and irradiated with u.v. light to 20 % survival. The plates were incubated for two hours at 30° then placed in an anaerobic chamber for five days at 30°.

To isolate cells that had lost the plasmids, cells were grown on non-selective medium (1 % Difco Yeast-Extract, 2 % Difco Bacto-Peptone, 2 % glucose) in two successive 2 ml overnight cultures, then plated on non-selective medium. Colonies of trp^- cells were identified by replica plating.

Complementation analysis was performed by mating aBR10 transformants with BRSc1 (trp1-1, gal1, ura) and scoring the diploids, selected on glucose medium lacking tryptophan, adenine, and uracil, for anaerobic growth on galactose.

RESULTS

Construction of CYC1/galK fusions

To construct a fusion which would place the expression of galactokinase enzyme activity under the control of the CYC1 regulatory sequences, we designed a scheme to join the bulk of the E. coli galK coding sequence to the CYC1 5' sequences plus the first four codons, thereby maintaining the yeast gene transcriptional and translational signals intact. This scheme is outlined in the Methods section and resulted in the family of yeast transforming plasmids diagramed in figure 1. The junction point of the two coding sequences was heterogeneous because it was necessary to digest the 5' end of the galK gene with the exonuclease Bal31 to delete the galK leader sequence (which contained an in-frame termination signal) and to ensure that, in at least some of the fusions, the galK coding sequence would be in-frame with the first four CYC1 codons.

Fusions capable of expressing galactokinase activity were identified by transformation of the fusion plasmids into gal1 (galactokinase deficient) yeast and selection for transformants capable of growth on galactose medium. A number of such transformants were identified and showed a wide range of growth rates. In all cases, the galactokinase activity was dependent on the presence of the plasmid within the cell, and when the plasmid was extracted from one yeast clone and transformed back into gal1 yeast, the new transformants were gal^{+} and grew at a rate characteristic of the original yeast clone from which the plasmid was extracted.

The junctures of some of the successful fusions were sequenced. One fusion YRpR1 which gave high levels of galK expression proved to be the type of protein coding fusion desired with the fusion point between the fourth codon of CYC1 and the fifth codon of galK as shown in figure 2A. Three other fusions, all of which gave varying degrees of poor galK expression, were characterized and proved to contain fusions between the galK leader sequence and the CYC1 coding sequence as indicated in figure 2B. In the cases of YRpR3 and YRpR72, the CYC1 initiation codon is out-of-frame with the galK coding sequence, and, therefore, the CYC1 ATG cannot be used for galK translation. Similarly, although the CYC1 initiation codon is in-frame with the

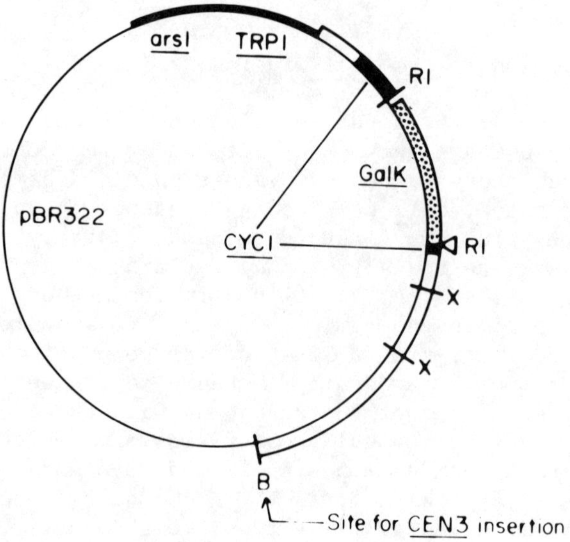

Figure 1. The construction of the fusion plasmids is described in the Methods section. The CYC1 gene is split between the fourth and fifth codons by the insertion of the 1.2 kb galK gene. The direction of transcription of the fused gene is counterclockwise. The heterogeneity of the fusions at the 5' end of the CYC1/galK fusion is represented by the △R1 symbol. The symbols represent: B, BamH1 restriction sites; X, XhoI restriction sites; and R1, EcoR1 restriction sites. The diagram represents the YRp plasmids and the site of CEN3 insertion in the YCp plasmids is indicated.

galK coding sequence in YCpR6, there is an in-frame termination codon between the two and again the CYC1 initiation codon cannot be used. Thus the sequence data suggest that the low level of galK expression observed in these fusions results from initiation of protein synthesis at an internal ATG of the mRNA.

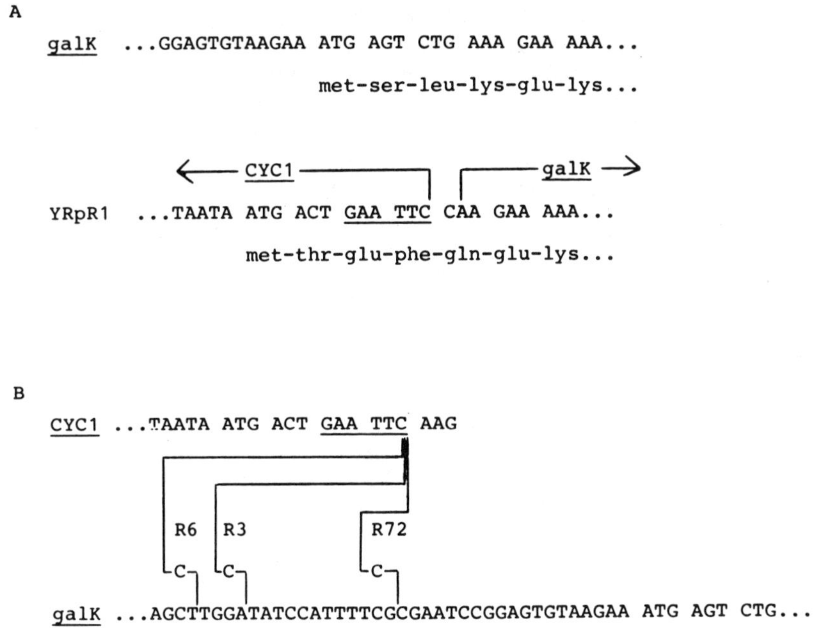

Figure 2. The DNA sequences read 5' to 3' from left to right.
A. The fusion point in the plasmid YRpR1 is indicated. The underlined portion is the EcoR1 site. An additional C residue immediately following this site results from the use of EcoR1 linkers having the sequence GGAATTCC.
B. The fusion points in the plasmids YRpR3(R3), YRpR6(R6), and YRpR72(R72) are represented. The lines connecting the upper CYC1 sequence with the lower galK sequence indicate the points of fusion. Again an extra C residue arises from the use of the EcoR1 linkers.

Expression and Regulation of the CYC1/galK Fusion

For the initial detection of successful fusions, we employed YRp plasmids containing the ARS1 yeast replication origin because such plasmids are maintained in multiple copies (10) and would presumably raise the sensitivity of the selection. However, these plasmids are also unstable and this feature plus their variation in copy number makes both the quantitation of galactokinase activity and the selection of mutants difficult. Therefore, once fusions capable of galactokinase expression were identified, CEN3, the centromeric sequences from yeast chromosomeIII, was added (see figure 1). CEN3 renders the plasmids much more stable and maintains the copy number at one per cell (11). Table 1 gives the galactokinase activity present in transformants carrying YRp and YCp fusion plasmids.

TABLE 1
GALACTOKINASE ACTIVITY IN
YEAST TRANSFORMANTS

Plasmid	Enzyme Activity		Percent of YCpR1
	Derepressed	Repressed	
YRpR1	3414	1844	
YRpR3	35.9		
YRpR6	29.4		
YRpR72	1282		
+CEN3			
YCpR1	163	18	
YCpR3	0.96		0.6
YCpR6	0.86		0.5
YCpR72	2.7		1.7

Enzyme activity is expressed as pmoles galactose-1-phosphate formed per min per 1.25 x 10^6 cells. Derepressed extracts were prepared from cells grown on 2 % raffinose; repressed extracts were prepared from cells grown on 2 % glucose.

As can be seen the introduction of CEN3 reduces enzyme levels as expected. Interestingly, however, plasmids containing YCpR1 grow faster in galactose containing medium than do cells containing the YRpR1 plasmid even though populations of YRpR1 transformants contain higher galactokinase levels. The explanation for this discrepancy probably lies in the high rate of loss of YRp plasmids so that many cells in the population do not divide. On the otherhand, the presence of CEN3 in the YCpR3, YCpR6, and YCpR72 plasmids decreases levels of galactokinase sufficiently so that only pinpoint colonies can form on galactose containing plates.

The data presented in Table 1 also demonstrate that the expression of the fused gene in YRpR1 and YCpR1 is subject to the same catabolite repression as is characteristic of CYC1 expression. In cells grown in glucose or other quickly fermented sugars, the rate of CYC1 transcription is repressed 8 to 10-fold as compared to that in cells grown in non-fermentable energy sources or in slowly fermented sugars such as, in this case, raffinose (16). The repression in YRpR1 transformants is less than that in YCpR1 transformants probably due to the higher copy number; catabolite repression of the intact CYC1 gene on YRp plasmids is also less extensive than that of the gene on YCp plasmids or in the chromosome (C.V. Lowry and R.S. Zitomer unpublished results). The transcription of the CYC1 gene is also regulated by oxygen levels; no CYC1 mRNA is detectable under anaerobic growth (9). Similarly, no galactokinase enzyme activity is apparent in anaerobic cells. Thus the expression of galactokinase activity from the YCpR1 plasmid is regulated in an identical fashion as that of CYC1 gene expression.

The mRNA produced from the YCpR1 fusion plasmid appears to initiate and terminate at the respective CYC1 sites both of which are present in the plasmid (see figure 1). Northern analysis (data not shown) indicated that the length of the mRNA, 1850 nucleotides, is sufficient to start transcription at the CYC1 transcriptional initiation site (61 nucleotides 5' from the ATG codon, 17), include the galK coding sequence, and extend through the CYC1 coding and trailing sequences to end 175 nucleotides 3' from the CYC1 coding region at the normal CYC1 termination site (17). In addition, hybridizational analysis confirmed that the CYC1 coding region 3' to the galK coding sequence is present in the fusion mRNA.

Deletions Causing Increased galK Expression

The poor expression of galactokinase activity from the fusion plasmids YCpR3, YCpR6, and YCpR72 was presumed due to the inability of ribosomes to translate the galK gene from its own ATG initiation codon. This might arise either because this initiation codon is not recognizable by the yeast translational machinery or because the CYC1 initiation codon precedes it on the mRNA and is preferentially used while precluding the use of any subsequent ATG codons. To distinguish between these possibilities, the CYC1 initiation codon was deleted from the wildtype CYC1 gene by Bal31 digestion and the fusion plasmids YCpR3, YCpR6, and YCpR72 were reconstructed using these altered CYC1 genes. Figure 3 shows the deletions created and the resulting new CYC1/galK junctions.

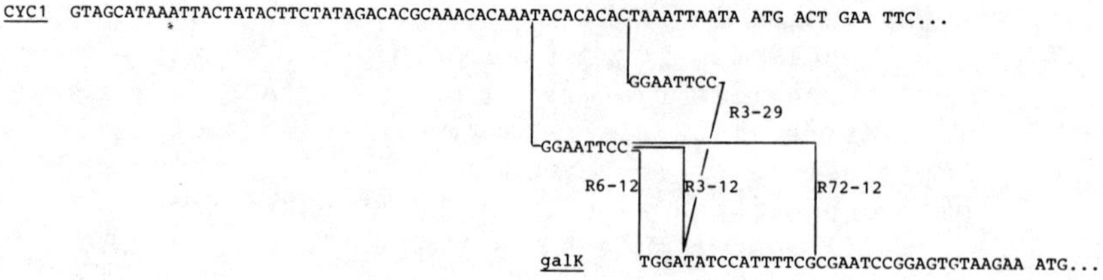

Figure 3. The point of the fusions are represented for a variety of plasmids by the lines connecting the upper CYC1 sequence with that for galK. In each case an EcoR1 linker has been added. The fusion plasmids are designated:YCpR3-12,R3-12; YCpR3-29,R3-29; YCpR6-12,R6-12; and YCpR72-12,R72-12. The CYC1 sequence begins at -61, the beginning of the mRNA. The fusion points within the galK region are unchanged.

These new fusions joined the CYC1 mRNA leader sequence to that of galK. As can be seen in Table 3 deletion of the CYC1 ATG and the sequences immediately surrounding it dramatically increases galK expression, raising it 100-fold to more than half that of the protein coding fusion. Thus we can conclude that the galK translational initiation codon can be used almost as efficiently in yeast as the CYC1 initiation codon, and the sequences immediately 5' and 3' to the yeast initiation codon are not essential to translation.

TABLE 2
GALACTOKINASE ACTIVITY IN
TRANSFORMANTS CONTAINING DELETION PLASMIDS

Plasmid	Galactokinase Activity	Percent of YCpR1
YCpR1	163	
YCpR3-12	99	61
YCpR3-29	99	61
YCpR6-12	127	76
YCpR72-12	99	61

Enzyme activity is expressed as pmoles galactose-1-phosphate formed per min per 1.25 x 10^6 cells.

Selection of CYC1 Regulatory Mutations

The original purpose for the construction of these CYC1/galK fusions plasmids was to enable us to select mutations affecting CYC1 expression and regulation. For such selections the plasmid phenotype must be stable; the plasmid must not be subject to variations in copy number or internal rearrangements and deletions at frequencies higher than inducible mutation rates. To test the suitability of these plasmids for mutational analysis we isolated the mutants described below and our results clearly demonstrate that these plasmids are quite useful for our purposes.

Wildtype yeast cells are capable of growth under anaerobic conditions with galactose as an energy source. However, gal1 cells transformed with YCpR2 (a plasmid with the same fusion as YCpR1, see figure 4) are incapable of such anaerobic growth because the CYC1/galK fusion gene is not expressed in the absence of oxygen. YCpR2 transformants were subjected to u.v. mutagenesis on galactose plates and incubated anaerobically. Colonies appeared after several days at a frequency of 10^{-4} of the survivors. Cells from two particularly large colonies were characterized as containing plasmid dependent genomic mutations. When cured of the plasmid, these mutants reverted to the gal^- phenotype and when retransformed with unmutagenized YCpR2 regained the ability to grow anaerobically on galactose. One mutant proved to be dominant while the other is recessive. Both mutants express not only galactokinase under anaerobic conditions, but also accumulate CYC1 mRNA transcribed from the wildtype chromosomal gene as determined by northern analysis (data not shown) indicating that these cells harbor authentic CYC1 constituitive mutations. Interestingly, although constituitive for the oxygen dependent regulation of galactokinase activity, galactokinase levels in both mutants are still subject to catabolite repression as indicated in Table 3. Thus it appears that separate pathways are involved in oxygen and catabolite regulation of CYC1. While these and other mutations are being

TABLE 3
GALACTOKINASE ACTIVITY IN OXYGEN CONSTITUITIVE MUTANTS

Mutant	Enzyme Activity		Ratio
	Derepressed	Repressed	Derepressed/Repressed
aBR10-2	79	20	4
aBR10-7	53	22	2.4

Enzyme activity is expressed as pmoles galactose-1-phosphate formed per min per $1.25x10^6$ cells. Derepressed extracts were prepared from cells grown on 2 % raffinose; repressed extracts were prepared from cells grown on 2 % glucose.

characterized further, these preliminary data clearly indicate that CYC1 regulatory mutants can be selected using these plasmids.

DISCUSSION

We have described here fusions between the regulatory sequences of the yeast CYC1 gene and the coding sequence of the *E. coli* galactokinase gene. These fusions when transformed into galactokinase deficient yeast cells produce a galactokinase enzyme capable of complementing the galactokinase deficiency. We have shown that the protein coding fusion YCpR1 is subject to the same catabolite repression and oxygen regulation as is the CYC1 gene product, and that the transcription of the mRNA appears to both begin and end at the normal CYC1 initiation and termination sequence. In all respects, the expression of the CYC1/galK fusion gene parallels that of the wildtype chromosomal CYC1 gene. Thus with this fusion, the expression of the CYC1 gene can be quantitated using the accurate, sensitive, and simple galactokinase assay rather than the laborious radioimmune assay and hybridization analysis we have previously used (16,18).

The more fruitful applications of these fusions may well be the isolation of regulatory mutations affecting the CYC1 gene. We have already demonstrated the usefulness of the YCpR1 fusion for the isolation of oxygen constituitive mutants and have had preliminary success in selecting mutants with increased levels of galactokinase expression starting with the poorly expressing YCpR3 fusion and selecting mutants that form normal sized colonies on galactose. In addition, we anticipate the possibility of selecting mutants with decreased levels of galactokinase expression by exploiting the fact that in gal7 or gal10 mutants the toxic galactose-1-phosphate product of galactokinase accumulates (3). Mutants which decrease expression of the CYC1/galK fusion gene are expected to survive exposure to galactose. We hope to isolate both genomic mutations and, by random *in vitro* mutagenesis of plasmids, mutations within the CYC1 regulatory regions.

We believe that the galactokinase fusion system is applicable to the study of other yeast genes and the plasmids described here represent a convenient starting points for such studies. We envision three types of fusions that can be successfully used. The first involves protein coding fusions of the YCpR1 type. For such fusions the question remains as to how much protein can be placed at the amino terminus of galactokinase without loss of enzyme function. In the case of YCpR1, the first four amino acids were replaced with five new residues which did not appear to affect protein function drastically. Also we found that galactokinase expression from YCpR6 could be greatly increased by the addition of an ochre suppressor gene to transformants (unpublished results). This allows ribosomes that initiate translation at the CYC1 ATG codon to read through the ochre codon preceding the galK coding sequence and thus synthesize the entire galactokinase enzyme with an additional sixteen amino terminal residues (see the sequence in figure 2B). This enzyme is clearly functional and to date we do not know the limits of this type of fusion. For convenience in the construction of other protein coding fusions, we have constructed two derivatives of YCpR1 shown in figure 4A.

The second type of fusion involves joining the mRNA leader sequences as described for the YCpR3, YCpR6, and YCpR72 deletion derivatives. Here the requirement is clearly that the galK ATG be the first ATG on the mRNA.

Finally, a third type of fusion involves the substitution of the upstream sequences of a yeast gene for those of CYC1. For example in YCpR1 any number of sites indicated in figure 4B 5' to the start of transcription might be used to insert the upstream sequences of interest. Although we have not tested this possibility with the galK plasmids, we have shown that the upstream sequences of CYC1 can be replaced by those of another gene and cause cytochrome c to be regulated according to the foreign sequences (9). The same principle has been demonstrated by others (19).

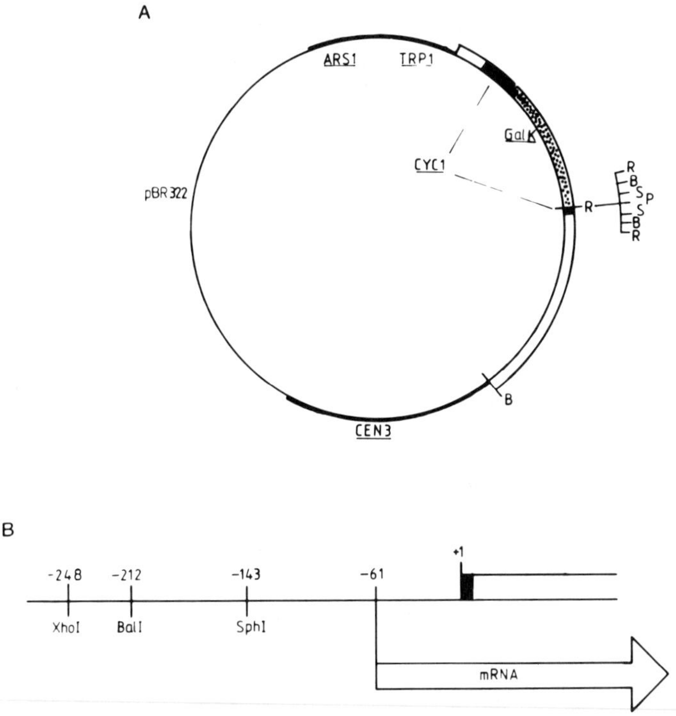

Figure 4. A. The YCpR1 plasmid is represented with two modifications. In both YCpR2 and YCpR7, the EcoR1 site 3' to the galK coding region has been deleted. Also in YCpR7 the 42 bp sequence from M13mp7(14) which contains the designated restriction site has been added. The restriction sites shown are: B, BamH1; P, PstI; R , EcoR1; S, SalI.
B. The diagram represents a 250 bp region 5' to the CYC1 coding sequence plus the initial part of the coding sequence in YCpR1. Base pairs are numbered counting the A in the ATG codon as 1 and using negative integers for base pairs 5' from that point.

ACKNOWLEDGMENTS

This work was supported by a grant from the NSF. R.S.Z. is supported by an RCDA from NIH and, for part of this work, by a fellowship from the Alexander von Humboldt Foundation. We would like to thank Dr. Cornelis Hollenberg in whose laboratory some of these experiments were carried out.

REFERENCES

1. Rose M, Casadaban MJ, Botstein D (1981). Yeast genes fused to ß-galactosidase in Escherichia coli can be expressed normally in yeast. Natl Acad Sci USA 78:2460.
2. Guarente L, Ptashne M (1981). Fusion of Escherichia coli LacZ to the cytochrome c gene of Saccharomyces cerevisiae. Proc Natl Acad Sci USA 78:2199.
3. Douglas HC, Hawthorne DC (1964). Enzymatic expression and genetic linkage of genes controlling galactose utilization in Saccharomyces. Genetics 49:837.
4. Wilson DB, Hogness DS (1966). Galactokinase and uridine diphosphogalactose 4-epimerase from Escherichia coli. Methods Enzymol 8:229.
5. Boyer HW, Roulland-Dussoix D (1969). A complementation analysis of the restriction and modification of DNA in Escherichia coli. J Mol Biol 41:459.
6. Cohen SN, Chang ACY, Hsu CI (1972). Nonchromosomal antibiotic resistance in bacteria: genetic transformation of Escherichia coli by R-Factor-DNA. Proc Natl Acad Sci USA 69:2110.
7. Hinnen A, Hicks JB, Fink GR (1978). Transformation of yeast. Proc Natl Acad Sci USA 75:1929.
8. Scott JH, Schekman R (1980). Lyticase: endoglucanase and protease activities that act together in yeast cell lysis. J Bacteriol 142:414.
9. Lowry CV, Weiss JL, Walthall DA, Zitomer RS (1983). Modulator sequences mediate oxygen regulation of CYC1 and a neighboring gene in yeast. Proc Natl Acad Sci USA 80:151.
10. Stinchcomb DT, Struhl K, Davis RW (1979). Isolation and characterization of a yeast chromosomal replicator. Nature 282:39.

11. Clarke L, Carbon J (1980). Isolation of a yeast centromere and construction of functional small circular chromosomes. Nature 287:504.
12. Maniatis T, Fritsch EF, Sambrook J (1982). In Molecular Cloning. Cold Spring Harbor Laboratory p 368.
13. McKenney K, Shimatake H, Court D, Schmeissner U, Brady C, Rosenberg M (1981). In Gene Amplification and Analysis. Structural Analysis of Nucleic Acids Vol 2 Elsevier North Holland Press, New York, p 383
14. Messing J, Crea R, Seeburg PH (1981). A system for shotgun DNA sequencing. Nucl Acid Res 9:309.
15. Sanger F, Nicklen S, Coulson AR (1977). DNA sequencing with chain-terminating inhibitors. Proc Natl Acad Sci USA 74:5463.
16. Zitomer RS, Montgomery DL, Nichols DL, Hall BD (1979). Transcriptional regulation of the yeast cytochrome c gene. Proc Natl Acad Sci USA 76:3627
17. Boss JM, Gillam S, Zitomer RS, Smith M (1981). Sequence of the yeast iso-1-cytochrome c mRNA. J Biol Chem 256: 12958.
18. Zitomer RS, Hall BD (1976). Yeast cytochrome c messenger RNA. J Biol Chem 251:6320.
19. Guarente L, Yocum RR, Gifford P (1982). A GAL10-CYC1 hybrid yeast promoter identifies the GAL4 regulatory region as an upstream site. Proc Natl Acad Sci USA 79: 7410.

Gene Expression, pages 543-554
© 1983 Alan R. Liss, Inc., 150 Fifth Avenue, New York, NY 10011

EXPRESSION OF HEPATITIS B VIRUS SURFACE ANTIGEN BY INFECTIOUS VACCINIA VIRUS RECOMBINANTS

Geoffrey L. Smith, Michael Mackett and Bernard Moss

Laboratory of Biology of Viruses, National Institute of Allergy and Infectious Diseases, Bethesda, MD 20205

ABSTRACT The coding sequence for hepatitis B virus surface antigen (HBsAg) has been inserted into the genome of vaccinia virus under control of vaccinia early promoters. Recombinants, selected by loss of thymidine kinase expression, are stable and retain infectivity. Cells infected with recombinant virus synthesize and excrete particles of HBsAg indistinguishable by antigenicity, polypeptide composition, buoyant density, sedimentation rate and size from particles present in the serum of humans infected with hepatitis B virus. Rabbits vaccinated with the recombinant virus rapidly produce antibodies against HBsAg, suggesting a potential use as a live vaccine in man.

INTRODUCTION

Vaccinia, the virus widely used for smallpox vaccination, is the best characterized member of the poxvirus family (1, 2). It possesses a large double stranded DNA genome of about 187 kb that encodes over 150 polypeptides including some involved in RNA and DNA synthesis. These enzymes account for the ability of poxviruses to replicate in the cytoplasm of infected cells. Vaccinia virus also appears to have evolved unique transcriptional regulatory sequences (3-5) recognized by the virus RNA polymerase and or other regulatory proteins.

Recent pilot studies indicated the feasibility of developing vaccinia virus as a general selectable eukaryotic cloning and expression vector (6, 7). This has been accomplished by constructing plasmid vectors containing vaccinia virus transcriptional regulatory sequences, restriction endonuclease sites for insertion of a foreign gene, and flanking vaccinia virus DNA containing an interrupted thymidine kinase (TK) gene (8). After insertion of the foreign protein coding sequence to form a chimeric gene, the plasmid was used to transfect cells that had been infected with wild-type vaccinia virus. Homologous recombination resulted in the insertion of the chimeric gene into the vaccinia virus genome at the TK locus. The resulting TK^- phenotype of recombinants facilitated their isolation. This combination of in vitro and in vivo methods provided a facile method for constructing chimeric molecules and overcoming problems associated with the size and lack of infectivity of vaccinia virus DNA. The procedure has recently resulted in expresison of a number of foreign genes including HBsAg (9). In this communication, we provide additional information regarding synthesis of HBsAg by a vaccinia virus vector and the potential use of such recombinants as live vaccines.

RESULTS

Insertion of HBsAg gene into vaccinia virus.

The DNA genome of HBV has been cloned and sequenced and the region encoding the surface antigen identified (10-13). A 1,350 bp BamHI DNA fragment from HBV strain adw that has previously been shown to express HBsAg in a SV40 recombinant (14) was chosen for insertion into vaccinia virus. Fig. 1 outlines the cloning of the HBsAg gene into a plasmid (insertion vector) so that the HBsAg gene is under the control of a vaccinia promoter. This promoter was taken from an early vaccinia gene coding for a 7.5 kilodalton protein (3) and inserted at the EcoRI site of the vaccinia thymidine kinase gene (5). A unique BamHI site engineered 30 bp downstream from the transcriptional start site of the translocated 7.5K promoter was used for the insertion of the HBsAg gene. The resultant plasmid called a recombination vector was used to transfect vaccinia-infected cells. Similar recombination vectors in which the HBsAg gene is incorrectly orientated with respect to the translocated promoter or in which the promoter and HBsAg gene run in the opposite direction to the TK gene have also been constructed (Fig. 2). These plasmids were all used to generate recombinant viruses as follows:

CV-1 cells were infected with wild type vaccinia virus at 0.05 pfu/cell and transfected 2 hr later with mixtures of plasmid, WT vaccinia and calf thymus DNA. After two days, progeny virus from infected cell lysates was screened for TK^- recombinants by plaque assay on TK^- cells with bromodeoxyuridine (BUdR) selection. Routinely, between 30-60% of TK^- isolates were the desired TK^- recombinants, the remainder being spontaneous TK^- mutants. A DNA:DNA hybridiza-

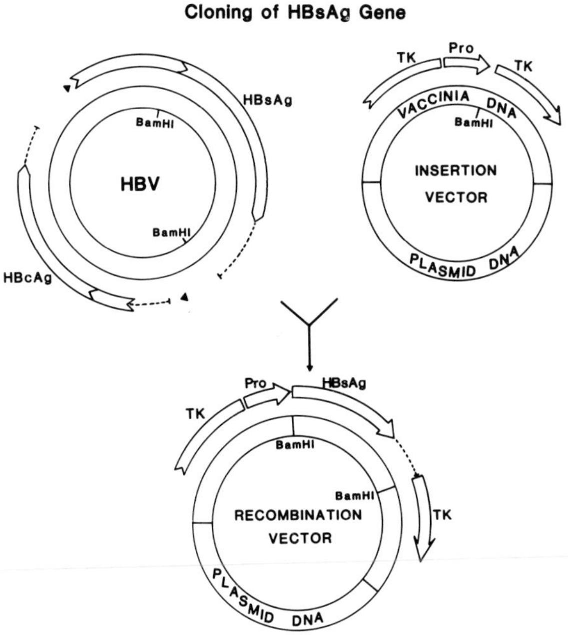

Figure 1. Construction of plasmid vectors for transfection of vaccinia virus infected cells. A BamHI fragment containing the entire mature form of HBsAg was ligated to BamHI cleaved insertion vector. The HBV genome map also shows the position of putative pre-HBsAg coding sequences and transcriptional start site (filled triangle). The putative 3'untranslated segments of HBsAg mRNA is indicated by a dashed line. The genome location of the core antigen (HBcAg) is also indicated in a similar manner. The vaccinia promoter (pro) in the insertion vector is a 275 bp fragment containing the transcriptional regulatory sequence of an mRNA encoding a 7.5 polypeptide.

tion technique, using nick translated ^{32}P-labeled hepatitis DNA as a probe, served to identify the true recombinants. TK^- isolates were grown in small cultures of TK^- cells with BUdR selection and individual lysates subsequently transferred to nitrocellulose sheets. After denaturation, neutralization and fixing of the DNA, the filter was probed with ^{32}P-nick translated hepatitis DNA. Recombinant viruses selected in this way were plaque purified twice before amplification and use in subsequent experiments.

Analysis of vaccinia virus recombinants.

The purity of the vaccinia virus recombinants was established in several ways. Plaquing of recombinant viruses in TK^- cells with and without BUdR selection gave similar

Virus Recombinant	Structure of Vaccinia - Hepatitis Surface Antigen Chimeric Genes	Yield of HBsAg ng/5x10^{6}cells cell / medium	Yield of Virus pfu x10^{-8} cell / medium
Wild Type		0 / 0	7.8 / 0.77
vHBs1	HBsAg → 7.5K	11 / 20	8.3 / 1.02
vHBs2	HBsAg ← 7.5K	835 / 1700	7.9 / 0.99
vHBs3	7.5K ← HBsAg	14 / 25	9.1 / 0.90
vHBs4	7.5K → HBsAg	930 / 1700	8.8 / 0.98
vHBs5	TK → HBsAg	35 / 80	10.3 / 0.96

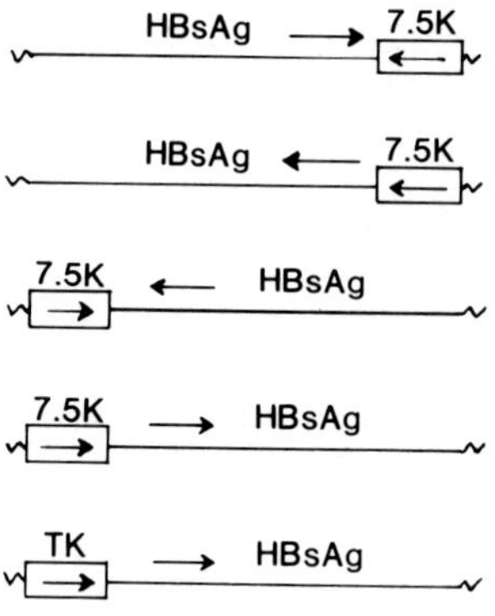

Figure 2. Structure of chimeric genes and expression of HBsAg. The yield of HBsAg was determined by radioimmunoassay at 24 hr after infection. Vaccinia virus titers were determined by plaque assay.

virus titers indicating the absence of contamination with TK^+ virus. Additionally, all the TK^- plaques contained hepatitis DNA. This was established by using in situ plaque hybridization (15) with ^{32}P-hepatitis DNA as probe. Southern blots of restricted DNA extracted from recombinant viruses also indicated the absence of any submolar bands or wild type TK-containing HindIII J fragment (9). These blotting experiments also showed that the entire 1.35 kb hepatitis DNA fragment was inserted into the HindIII J fragment of the virus genome as predicted. Significantly, no other genomic alterations were present indicating that the recombinant genome was stable.

Expression of the HBsAg gene.

Fig. 2 shows the structure of the chimeric vaccinia-HBsAg genes present in five different recombinants (9). Each virus was used to infect CV-1 cells and 24 hr later cell extracts and culture media were tested for HBsAg by radioimmunoassay and for infectious virus by plaque assay. Recombinants vHBs2 and vHBs4, which have correctly positioned vaccinia promoters, produced significant amounts of HBsAg, while recombinants vHBs1 and vHBs3 which have incorrectly positioned promoters did not. vHBs5, which has the TK gene promoter correctly positioned relative to the HBsAg gene, produced lower amounts of HBsAg. In all cases, approximately 65% or more of the detectable HBsAg was present in the culture fluid of infected cells. This was not a result of cell lysis caused by virus infection since 80%-90% of infectious virus remained cell associated. Excretion of HBsAg is also characteristic of some hepatoma cell lines and large amounts of HBsAg can be found in sera of humans chronically infected with hepatitis B virus. The yield of infectious virus from cells infected with similar amounts of wild-type or recombinant viruses were equivalent, indicating that neither the insertion of the HBsAg gene nor expression of it impaired virus growth and infectivity.

Immunoprecipitation of excreted HBsAg.

To analyze the polypeptide composition of HBsAg produced from cells infected with vaccinia recombinant vHBs4, infected cells were pulse-labeled with ^{35}S-methionine and cell extracts and culture medium immunoprecipitated with antibodies against HBsAg. Fig. 3 illustrates the composition of HBsAg immunoprecipitated from the culture medium

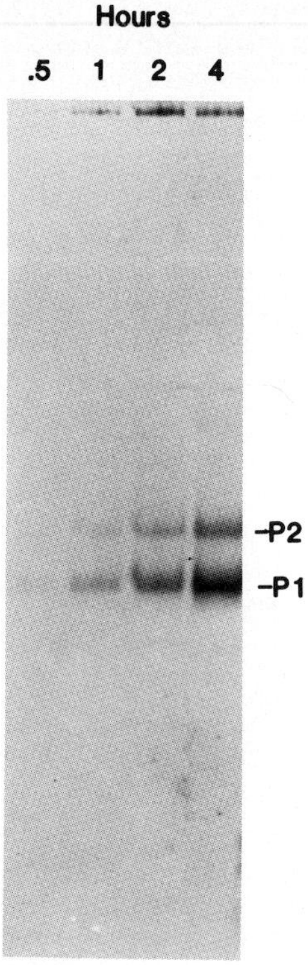

Figure 3. Immunoprecipitation of HBsAg released from vaccinia-infected cells. CV-1 cells were infected with recombinant vHBs4 at 30 pfu/cell. Two hr after infection, cells were pulse-labeled for 20 min with ^{35}S-methionine and then washed and chased in medium containing 1 mM methionine. Culture media taken at indicated times after chase was incubated with guinea-pig pre-immune serum followed by formalin-treated staphylococcal A suspension. After centrifugation, the supernatant was incubated with guinea-pig anti-HBsAg serum and then precipitated with staphylococcal A suspension. Precipitated proteins were eluted from staph A particles and analyzed by electrophoresis through a 15% polyacrylamide gel. P1 and P2 have molecular weights estimated to be 23,000 and 25,400 and correspond to non-glycosylated and glycosylated forms of HBsAg.

of infected cells at various times after pulse-labeling and subsequent chase with excess unlabeled methionine. Two polypeptides (P1 and P2) were identified. These had electrophoretic mobilities similar to HBsAg polypeptides released from pulse-labeled hepatoma cell line PLC/PRF/5 (not shown) and represent non-glycosylated (P1) and glycosylated (P2) forms of HBsAg. Analysis of extracts of cells infected with recombinant vHBs4 also demonstrated the presence of both forms of HBsAg (9). However, the P2/P1 ratio of cell-associated HBsAg was lower than that of excreted HBsAg. Fig. 3 also shows that HBsAg was excreted within 30 min after pulse-labeling and that excretion continued for at least 4 hr.

Electron microscopy of HBsAg particles.

HBsAg present in the serum of infected humans is present as characteristic 22 nm particles. To examine if the

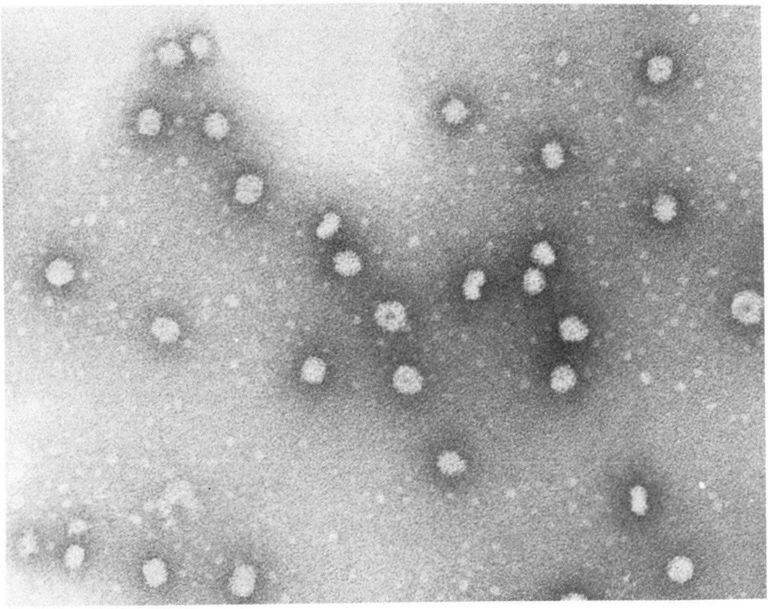

Figure 4. Electron microscopy of HBsAg particles. HBsAg particles were purified from the supernatant of cells infected with recombinant vHBs4. Cell media was clarified by low speed centrifugation 48 hr after infection and made to 10% (w/v) with polyethylene glycol. Precipitated protein was pelleted 30 min, 10,000 X g, resuspended in 10 mM Tris-HCl (pH 7.5), 1 mM EDTA (TE) and adjusted to 1.2 g/cm^3 with CsCl. After centrifugation to equilibrium, the gradient fractions were tested for HBsAg by radioimmunoassay and their density determined by refractometry. Peak fractions of HBsAg were recentrifuged in a second CsCl gradient and gradient fractions assayed as above. Peak fractions were dialyzed against TE buffer and then centrifuged on a 5-30% (w/v) sucrose gradient (Beckman SW 25.2, 10 hr, 24,000 rpm). Gradient fractions were screened for HBsAg and peak fractions pooled, dialysed against TE, adjusted to 1.2 g/cm^3 with CsCl and centrifuged to equilibrium. Peak gradient fractions containing HBsAg were dialzyed against TE and used for electron microscopy. Particles were adsorbed to carbon coated copper grids, stained with 1% phosphotungstic acid and observed X 50,000. Fifty particles were sized and had a mean diameter of 21.4 nm.

HBsAg released from vaccinia infected cells had similar characteristics, the particles were purified and visualized by electron microscopy. HBsAg was purified from culture medium of CV-1 cells 48 hr after infection with vaccinia recombinant vHBs4. Purification followed a procedure similar to that used for isolation of particles from human serum. Particles were precipitated by addition of 10% polyethylene glycol and then centrifuged sequentially in two CsCl gradients, one sucrose gradient and a third CsCl gradient. HBsAg particles from vaccinia-infected cells had the same buoyant density in CsCl (1.2 g/cm^3) and the same sedimentation rate in sucrose as particles from hepatoma cells (9). Electron microscopy identified HBsAg particles released from vaccinia-infected cells that had an average diameter of 21.4 nm (Fig. 4).

Vaccination of rabbits with vHBs4.

To determine if a recombinant vaccinia virus expressing HBsAg could induce the production of antibodies against HBsAg in animals, rabbits were vaccinated with either recombinant vHBs4 or wild-type virus. All rabbits exhibited typical local skin lesions between 5 and 10 days post vaccination which subsequently healed. No HBsAg or vaccinia virus was detectable in serum up to 14 days post vaccination. However, Fig. 5 shows that rabbits vaccinated with recombinant vHBs4 rapidly produced antibodies against HBsAg. An initial rapid response reached a peak of 240 mIU/ml in one animal and 365 mIU/ml in another at 9 days post vaccination. This response was presumed to correspond to IgM and declined after day 9. Starting between days 20 and 48, both rabbits developed a further antibody response, presumably IgG, that reached higher levels and persisted for at least three months. A control rabbit vaccinated with wild-type vaccinia produced no antibody against HBsAg.

Purified HBsAg from vaccinia infected cells was subtyped (16) and shown to contain both the cross-reactive "a" and type-specific "d" determinants. This was consistent with the source of HBV DNA. Rabbit antisera positive for antibodies against HBsAg was also subtyped and shown to recognize the cross-reactive "a" determinant. This was important since the presence of antibodies aginst the "a" determinant has been shown to protect against different HBV subtypes (17).

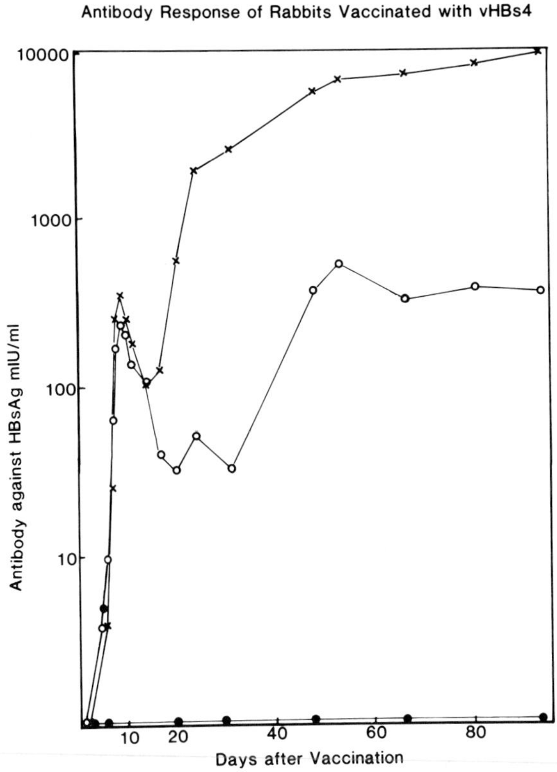

Figure 5. Antibody response of rabbits vaccinated with vHBs4. Rabbits were innoculated by intradermal injection of wild-type vaccinia (●-●) or recombinant vHBs4 (X-X and O-O). Rabbits ●-● and O-O were innoculated intradermally with 10^8 pfu of virus at each of four shaved sites of the rabbits' back. Rabbit X-X received a single inoculation of 10^8 pfu of virus on its back. Rabbit sera was taken and tested for antibodies against HBsAg by radioimmunoassay (AUSAB-Abbot Laboratories). Samples between days 1 and 14 were also screened for HBsAg by radioimmunoassay (AUSRIA II-Abbot) and for infectious virus by plaque assay in CV-1 cells. The level of antibody against HBsAg was standardized using positive human immunoglobulin. Background values obtained with pre-immune rabbit sera were subtracted.

DISCUSSION

A stable infectious vaccinia virus recombinant that expresses HBsAg has been produced. We estimate that approximately 1.4 X 10^8 molecules of HBsAg per cell are made during a lytic cycle in cells infected with the vaccinia recombinant. Moreover, these cells excrete HBsAg as particles approximately 22 nm in size. These particles contain antigenic determinants "a" and "d" consistent with the source of HBV DNA. Rabbits vaccinated with the virus recombinant developed transient local skin lesions and subsequently antibodies against HBsAg. These antibodies reached high levels which persisted for at least three months after vaccination and which recognized the cross-reactive "a" antigenic determinant of HBsAg. Similar levels of antibodies in man would confer protection against HBV infection. Longer term experiments to evaluate whether the recombinant vaccinia virus can be used to protect susceptible chimpanzees against HBV infection have been initiated.

These experiments demonstrate two important features of the use of vaccinia virus as a eukaryotic cloning vector. First, that virus infectivity is not affected by insertion and expression of foreign DNA in contrast to other eukaryotic viral vectors. Second, recombinant vaccinia viruses have potential as live vaccines against pathogenic organisms of man or animals. Vaccinia itself has already proved to be a highly effective live vaccine in the eradication of smallpox. Additionally, the large size of the vaccinia genome (187 kb) enables large pieces of foreign DNA to be accommodated without affecting virus stability or infectivity. This has been demonstrated by the insertion of 25 kb of bacteriophage λ DNA into vaccinia (G. L. Smith, unpublished). Consequently, it may be possible to construct polyvalent vaccines which may confer simultaneous protection against several pathogenic agents of man or animals.

ACKNOWLEDGEMENTS

We thank B. H. Hoyer and J. L. Gerin for providing cloned HBV DNA, Research Resources Section of the National Institute of Allergy and Infectious Diseases for HBsAg antiserum, J. Shih for human anti-HBsAg immunoglobulin, D. Djurickovic for vaccinating rabbits and collecting sera, J. L. Gerin and J. Ford for electron microscopy, R. Engel for subtyping of antigen and antibody, N. Cooper for technical assistance and J. Carolan for typing the manuscript.

REFERENCES

1. Moss B (1978). Poxviruses. In Nayak DP (ed): "The Molecular Biology of Animal Viruses," Marcel Dekker, Inc. 2: p 849.
2. Dales S, Pogo BGT (1981). Biology of poxviruses. Virology Monographs 18:1.
3. Venkatesan S, Baroudy BM, Moss B (1981). Distinctive nucleotide sequences adjacent to multiple initiation and termination sites of an early vaccinia virus gene. Cell 25:805.
4. Venkatesan S, Gershowitz A, Moss B (1982). Complete nucleotide sequence of two adjacent early vaccinia virus genes located within the inverted terminal repetition. J. Virol. 44:637.
5. Weir JP, Moss B (1983). Nucleotide sequence of the vaccinia virus thymidine kinase gene and the nature of spontaneous frameshift mutants. J. Virol., in press.
6. Panicali D, Paoletti E (1982). Construction of poxviruses as cloning vectors: insertion of the thymidine kinase gene from herpes simplex virus into the DNA of infectious vaccinia virus. Proc. Natl. Acad. Sci. USA 79:4927.
7. Mackett M, Smith GL, Moss B (1982). Vaccinia virus: a selectable eukaryotic cloning and expression vector. Proc. Natl. Acad. Sci. USA 79:7415.
8. Moss B, Smith GL, Mackett M (1983). Use of vaccinia virus as an infectious molecular cloning and expression vector. In Rosenberg M, Papas T (eds): "Eukaryotic Viral Vectors." Elsevier North Holland, in press.
9. Smith GL, Mackett M, Moss B (1983). Infectious vaccinia virus recombinants expressing hepatitis B surface antigen. Nature 302:490.
10. Charnay P, Pourcel C, Louise A, Fritsh A, Tiollas P (1979). Cloning in _Escherichia coli_ and physical structure of hepatitis B virion DNA. Proc. Natl. Acad. Sci. USA 76:2222.
11. Valenzuela P, Gray P, Quiroga M, Zaldivar J, Goodman HM, Rutter WG (1979). Nucleotide sequence of the gene coding for the major protein of hepatitis B virus surface antigen. Nature 280:815.
12. Pasek M, Gato F, Gilbert W, Zink B, Schaller H, Mackay P, Leadbetter G, Murray K (1979). Hepatis B virus genes and their expression in _E. coli_. Nature 282:575.

13. Mackay D, Pacek M, Magazin M, Kovacic RT, Allet B, Stahl S, Gilbert W, Schaller H, Bruce SA, Murray K (1981). Production of immunologically active surface antigens of hepatitis B virus by Escherichia coli. Proc. Natl. Acad. Sci. USA 78:4510.
14. Moriarity AM, Hoyer BH, Shih JW-K, Gerin JL, Hamer DH (1981). Expression of the hepatitis B virus surface antigen gene in cell culture by using a simian virus 40 vector. Proc. Natl. Acad. Sci. USA 78:2606.
15. Villareal LP, Berg P (1977). Hybridization in situ of SV40 plaques: detection of recombinant SV40 virus carrying specific sequences of non-viral DNA. Science 196:183.
16. Hoofnagle JH, Gerety RJ, Smallwood LA, Baker LF (1979). Subtyping of hepatitis B surface antigen and antibody by radioimmunoassay. Gastroenterology 72:290.
17. Szmuness W, Stevens CE, Harley EJ, Zang EA, Alter HJ, Taylor PE, DeVera A, Chen GTS, Kellner A, Dialysis Vaccine Trial Study Group (1982). Hepatitis B vaccine in medical staff of hemodialysis units: efficacy and subtype cross-protection. New Engl. J. Med. 307:1481.

Gene Expression, pages 555-563
© 1983 Alan R. Liss, Inc., 150 Fifth Avenue, New York, NY 10011

GENE TRANSFER USING RETROVIRAL VECTORS: INFECTIOUS VIRUS CONTAINING A FUNCTIONAL HUMAN HPRT GENE[1]

A. Dusty Miller*, Inder M. Verma*, Theodore Friedmann+, and Douglas J. Jolly+

*Molecular Biology and Virology Laboratory
The Salk Institute, P.O. Box 85800
San Diego, California 92138

+Department of Pediatrics
University of California at San Diego
La Jolla, California 92093

ABSTRACT A cDNA corresponding to the human gene for hypoxanthine phosphoribosyltransferase (HPRT) was ligated into a murine retroviral vector such that it is under the transcriptional control of the viral long terminal repeats. This replication-defective virus could be rescued using a variety of helper viruses, and was capable of transmitting the $HPRT^+$ phenotype to rodent or human $HPRT^-$ cells by infection. Cells infected with the HPRT-virus contained human HPRT enzyme activity at levels similar to that in normal $HPRT^+$ cells.

INTRODUCTION

Our laboratories have been interested in development of retroviral vector systems for studies of the potential role of gene transfer in the modification of human genetic disease. Retroviruses have many properties which make them

[1]This work was supported by grants from the Gould Foundation, Tamarac, Florida, NIH grants GM28223 and CA24288 (T. F.) and ACS and NIH grants (I. M. V.). A. D. M. is a Leukemia Society Fellow.

particularly suitable as vehicles for such studies: i) the viral genome (RNA) is efficiently transmitted to recipient cells and integrates into chromosomes as DNA, ii) integration is specific with respect to the viral genome, iii) the viral genome allows packaging of inserts of up to at least 7 kb, iv) retroviruses have a wide host range and can infect a variety of cell types, v) viral long terminal repeats (LTRs) provide efficient signals for initiation and termination of transcription, and vi) viral proteins required for the replication of the viral genome can be supplied in trans.

Previous studies have shown that retroviruses can be used to transfer a viral gene (herpes simplex thymidine kinase) (1-3) and a bacterial gene (E. Coli xanthine-guanine phosphoribosyltransferase) (4). We chose human HPRT as a model gene for our experiments because HPRT provides a convenient selectable marker, and because a defect in the gene is responsible for the crippling Lesch-Nyhan syndrome (5). Unlike previous experiments involving retroviruses for gene transfer, only one gene (HPRT), in the absence of other viral or transforming genes, can be expressed by the recombinant construct described here.

RESULTS

The HPRT-containing viral vector is based on the Moloney murine leukemia virus (Mo-MLV) genome (Fig. 1). The human HPRT cDNA coding sequence was obtained from the SV-40 based cDNA expression vector in which it was initially isolated (6), and was inserted between Moloney murine virus LTRs such that the HPRT translational start codon was in approximately the same position as the gag gene start codon in the parental virus. The 5'-LTR and adjacent sequences were actually derived from a clone of Moloney murine sarcoma virus (Mo-MSV), but these sequences are functionally analogous and similar in sequence to Mo-MLV. The 3' end of the HPRT cDNA was linked to sequences in the viral envelope gene. This HPRT-virus DNA construct has the feature that all known viral coding sequences have been either deleted entirely, or truncated, so that no viral proteins can be expressed. In initial experiments, we found that two 5'-AATAAA-3' sequences near the 3' end of the original HPRT cDNA functioned as polyadenylation signals, and reduced the number of full

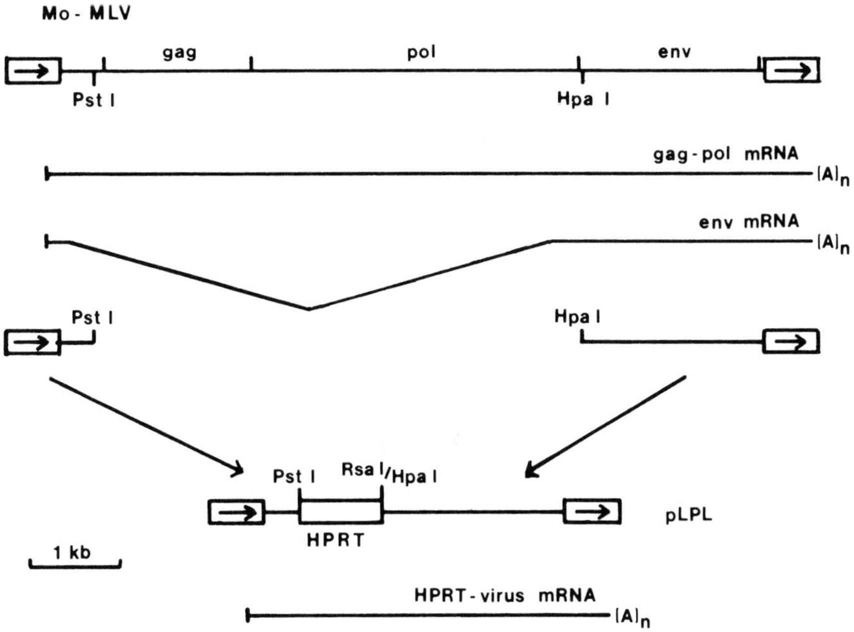

FIGURE 1. Derivation of the HPRT Virus. The integrated DNA form of the Mo-MLV genome, with its coding regions and major mRNA transcripts, is shown at the top of the figure. LTRs are denoted by boxes with arrows, indicating the direction of transcription. These LTRs are flanked by cellular DNA in the integrated proviral Mo-MLV genome. There are two *PstI* sites and two *HpaI* sites in Mo-MLV, but only those sites used for insertion of the HPRT cDNA are shown. A *RsaI* site at the 3' end of the HPRT cDNA was joined to the *HpaI* site of Mo-MLV, destroying both endonuclease sites.

length viral transcripts. These non-coding sequences were deleted in the construct (pLPL) shown in Fig. 1.

The HPRT-virus DNA was transfected into a variety of $HPRT^-$ cell lines using the calcium phosphate precipitation technique (7). Cells expressing HPRT were selected in HAT (hypoxanthine, amethopterin, and thymidine) medium (8). The HAT^r phenotype could be transferred to all cell types tested, including rat, mouse, and human cells. The retroviral construct (pLPL) is at least as effective in transferring the HAT^r phenotype to $HPRT^-$ 208F rat cells (9)

TABLE 1

TRANSFECTION EFFICIENCIES OF HPRT CONSTRUCTS[a]

DNA	Colonies/ug/10^6 cells
carrier	>5
p4aA8	400
pLPL	650

[a] HPRT⁻ rat cells were plated at 5×10^5 cells per 5 cm dish on day one. On day two, 0.5 ug of the indicated plasmid DNA with 8 ug carrier DNA (from NIH 3T3 cells) was introduced into cells using the calcium phosphate precipitation procedure. On day three, the cells were divided 1:5 into dishes containing HAT medium. The medium was replaced with HAT medium on day six, and duplicate dishes were stained and colonies counted on day nine.

as the original cDNA expression vector (p4aA8), in which the HPRT cDNA was isolated (Table 1).

HPRT-virus transfected cells did not release virus because the HPRT-virus was replication defective. However, we were able to rescue virus capable of transmitting HAT resistance using several replication-competent helper viruses, including the ecotropic viruses Mo-MLV and FBJ-MLV, and the amphotropic virus 1504A.

Cell culture medium from such transfected cells superinfected with helper virus, containing a mixture of rescued HPRT-virus and helper virus, was used to infect $HPRT^-$ rat 208F cells, and clonal cell lines were isolated which contained HPRT-virus in the absence of helper virus. These non-producer lines were found to contain 1-3 integrated copies of HPRT-virus DNA, as estimated by Southern analysis (10) using HPRT and LTR probes (11). The integrated proviruses are stable since cells grown in the absence of selection for one month retained the HAT^r phenotype. Conversely, transfected cells are less stable, since only 70% of transfected cells grown for one month in the absence of selection remained HAT^r.

HPRT and LTR probes hybridize to a single species of RNA from non-producer cells of a size corresponding to the

TABLE 2
RESCUE OF HPRT-VIRUS[a]

Infected Cells	HPRT cfu/ml / pfu/ml		
	Mo-MLV	FBJ-MLV	1504A
Rat	<10 / $10^{4.4}$	<10 / $10^{3.8}$	<10 / <10
Rat-LPL	$10^{3.3}$ / $10^{4.8}$	$10^{3.0}$ / $10^{4.1}$	$10^{3.5}$ / <10
Mouse-LPL	$10^{5.3}$ / $10^{6.3}$	--	--

[a] HPRT-virus was rescued from rat non-producer cell lines after superinfection with various helper viruses. The $HPRT^-$ murine cell line 2TGOR (13) was infected with virus from rat cells producing HPRT-virus and Mo-MLV helper virus, selected in HAT medium and the resultant colonies pooled for assay. Results are expressed as $HPRT^+$ colony forming units (cfu) and helper virus XC plaque forming units (pfu). The amphotropic virus 1504A does not form plaques in the XC assay and was not quantitated.

expected genome length RNA. Conversely, transfected cells display many additional RNA species which hybridize to these probes (11).

In consideration of these results, it is not surprizing that the highest titers of HPRT-virus were obtained from helper virus superinfected, non-producer cell lines when compared with helper virus superinfected, HPRT-virus transfected cells. Table 2 shows titers of HPRT-virus and helper virus produced by several cell lines after rescue with various helper viruses. Titers of 2 x 10^5 HPRT-virus per ml of cell supernatant were obtained from mouse cells,

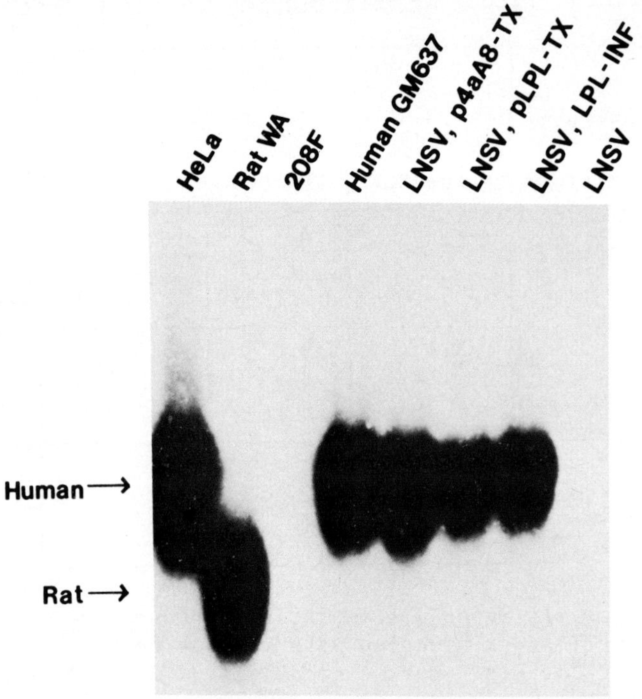

FIGURE 2. HPRT Enzyme Activity Analysis. Enzyme activity was measured in cell extracts after electrophoresis (14). TX indicates cells transfected with the indicated plasmid and selected for HAT resistance, and INF indicates cells infected with the indicated virus. Rat WA cells are $HPRT^+$ rat cells, GM637 cells are SV-40 transformed $HPRT^+$ normal human fibroblasts, and LNSV cells are SV-40 transformed $HPRT^-$ fibroblasts from a Lesch-Nyhan patient.

and the ratio of HPRT-virus to helper virus (titered using the XC plaque assay (12)) was about 10%.

Amphotropic retroviruses are capable of infecting human cells, and HPRT-virus rescued with amphotropic virus was capable of transmitting HAT resistance to $HPRT^-$ human cells, including diploid Lesch-Nyhan lymphoblasts and fibroblasts (15). HPRT enzyme activity was measured in such genetically transformed cells (Fig. 2) by isoelectric focusing in polyacrylamide gels. Control Lesch-Nyhan fibroblasts contained undetectable activity, whereas the same cells infected with the HPRT-virus and selected in HAT medium display HPRT activity which co-migrates with normal human fibroblast HRPT. Total HPRT activity in the infected cells was similar to that of normal human fibroblasts. Normal rat HPRT migrates differently from human HPRT, so it is unlikely that the transmitted HPRT activity in the human cells derived from the rat cells used to grow the virus for human cell infection. Cells transfected with either pLPL or the SV-40 based expression vector p4aA8 also display human HPRT activity.

DISCUSSION

We have constructed a transmissible retroviral vector that allows the expression of human HPRT in enzyme-deficient cells, including cells derived from patients with Lesch-Nyhan disease. Levels of enzyme activity in recipient cells are similar to those in normal human cells. HPRT is constitutively expressed from the normal cellular gene at relatively constant levels in all human cells, and is not under major control of cell cycle or other metabolic events, thus the control of virally expressed HPRT cannot be studied at this level. The availability of lymphoblasts infected with the HPRT-virus, however, now makes possible the evaluation of purine pools and the rate of de-novo purine biosynthesis as parameters of the fidelity of the control of virally expressed HPRT in relation to the cell's overall biosynthetic needs.

The vector described in this study does not encode any viral gene products, and therefore, in the absence of helper virus, no viral proteins can be expressed. Cells infected with the HPRT-virus alone will therefore not be antigenic in recipient animals, and such cells would not be subjected to immunological rejection. Infection of animals with HPRT-virus in the presence of helper virus may aid in the spread of HPRT-virus, because of the presence of helper virus, but may lead to an antigenic response or other unanticipated illness. HPRT virus might be isolated from helper virus by physical methods due to the difference in genome size. Of greater interest is the use of cells synthesizing viral replication proteins but not producing infectious virus, and preliminary results indicate that the HPRT-virus can be produced in the absence of helper virus in such cells.

REFERENCES

1. Wei C, Gibson M, Spear PG, Scolnick EM (1981). Construction and isolation of a transmissible retrovirus containing the src gene of Harvey murine sarcoma virus and the thymidine kinase gene of herpes simplex virus type 1. J Virol 39:935.
2. Shimotohno K, Temin HM (1981). Formation of infectious progeny virus after insertion of herpes simplex thymidine kinase gene into DNA of an avian retrovirus. Cell 26:67.
3. Tabin CJ, Hoffmann JW, Goff SP, Weinberg RA (1982). Adaptation of a retrovirus as a eukaryotic vector transmitting the herpes simplex virus thymidine kinase gene. Mol Cell Biol 2:426.
4. Mulligan RC (1982). Development of new mammalian transducing vectors. In Gluzman Y (ed): "Eukaryotic viral vectors," New York: Cold Spring Harbor Laboratory, p. 133.
5. Caskey CT, Kruh GD (1979). The HPRT locus. Cell 16:1.
6. Jolly DJ, Okayama H, Berg P, Esty AC, Filpula D, Bohlen P, Johnson GG, Shively JE, Hunkapillar T, Friedmann T (1983). Isolation and characterization of full length expressible cDNA for human hypoxanthine phosphoribosyl-transferase. Proc Natl Acad Sci USA 80:477.
7. Corsaro CM, Pearson ML (1981). Enhancing the efficiency of DNA-mediated gene transfer in mammalian cells. Somatic Cell Genetics 7:603.
8. Hakala MT, Taylor E (1959). The ability of purine and thymidine derivatives and of glycine to support the growth of mammalian cells in culture. J Biol Chem 50:126.
9. Quade K (1979). Transformation of mammalian cells by avian myelocytomatosis virus and avian erythroblastosis virus. Virol 98:461.
10. Southern EM (1975). Detection of specific sequences among DNA fragments separated by gel electrophoresis. J Mol Biol 98:503.
11. Miller AD, Jolly DJ, Friedmann T, Verma IM. submitted Proc Natl Acad Sci USA.
12. Rowe WP, Pugh WE, Hartley JW (1970). Plaque assay technique for murine leukemia viruses. Virol 42:1136.
13. Jha KK, Siniscalco M, Ozer HL (1980). Temperature-sensitive mutants of Balb/3T3 cells. Somatic Cell Genetics 6:603.

14. Johnson GG, Eisenberg LR, Migeon BR (1974). Human and mouse hypoxanthine-guanine phosphoribosyltransferase: dimers and tetramers. Science 203:174.
15. Bakay B, Nyhan WL, Croce CM, Koprowski H (1975). Reversion in expression of hypoxanthine-guanine phosphoribosyltransferase following cell hybridization. J. Cell Sci. 17:567.

WORKSHOP SUMMARIES

Gene Expression, pages 567-570
© 1983 Alan R. Liss, Inc., 150 Fifth Avenue, New York, NY 10011

WORKSHOP SUMMARY: POST-INITIATION CONTROL MECHANISMS

Terry Platt

Molecular Biophysics & Biochemistry
Yale University
333 Cedar Street
New Haven, Connecticut 06510

This workshop covered a variety of regulatory mechanisms that govern levels of active gene expression, with the sole constraint that the targeted events temporally follow initiation of transcription.

K. McKenney (1) isolated and analyzed numerous base mutations in the rho independent λ oop terminator site using genetic selections or directed by synthetic oligonucleotides. *In vivo* function was quantitated using a *galK* gene fusion vector. Mutations affecting termination occured in the symmetric region just preceding the termination site, and have the potential to change the RNA hairpin stability and the size or base composition of the loop. Single mutations recombined *in vitro* to construct double mutations were quantitated *in vivo*, and gave expected results. Sequences within 8bp distal to the termination site appear sufficient for normal function.

J. McSwiggen (2) used synthetic ribopolymers to define the mechanism and site of rho interaction with RNA. Electron micrographs confirm that rho is hexameric in solution and upon binding poly(C); some cooperativity is also seen. Ethenoadenosine (εA), a fluorescent analog of A, can be used to monitor binding of rho to the random copolymers, εA:C or εA:U (1:4), in either the presence or absence of competing poly(C) or poly(U). Rho interacts with, and excludes from further binding, about 12 nucleotides per monomer. Rho binding is moderately cooperative ($\omega \simeq 200$), and favors poly(C) ~100x over poly(U). RNase protection of poly(C) yields six RNA size classes with lengths in multiples of 12, hence each hexamer must bind ~72 nucleotides. Trypsin digestion of rho yields 2 major fragments, of 31 kD (f1) and 15 kD (f2). Rho monomer is 46 kD; sequencing shows that f1 is an N-terminal fragment, and that f2 begins 136 residues from the C-terminus. UV-activated crosslinking of azido-ATP to rho labels both f1 and f2; since only f2 labelling is competed with cold ATP, f2 may contain the ATP binding site.

J. Greenblatt (3) summarized the participation of *E. coli nusA* protein (L-factor) in antitermination. As a component of the elongation complex with core polymerase, nusA can protect it from mild tryptic digestion; mutant nusA1 protein has a strong effect on this sensitivity. Though nusA elicits long pauses at some termination sites in vitro, the polymerase allele rif501 conveys resistence to this pausing. Termination specified by some random fragments can be relieved in the absence of nusA, and λ N-protein (required to antiterminate for lytic infection) is not necessary when intracellular levels of nusA are limited by ts suppression of an amber mutation. Thus the fidelity of reading termination signals appears dependent on nusA. The λ antitermination response also may require species specific recognition by nusA of the sequence CGCTCCTA: a 1bp difference in this "BoxA" of S. typhimurium prevents heterologous function with the respective proteins (Friedman & Olson, in press).

D. Gallwitz (4) showed that chimaeric genes carrying intron-containing DNA fragments from higher eukaryotes in the coding region of the split actin gene of S. cerevisiae exhibit selective splicing: a yeast but not a foreign intervening sequence (IVS) is correctly excised from such hybrid transcripts. A second yeast intron in the active coding region can be correctly removed, thus more than one splicing event can occur during mRNA processing. Deletions into the actin intron identify short regions near both 5' and 3' ends of the IVS that are required for splicing to occur. Inserting an intron fragment containing the 3' sequence TACTAAC into the coding region of the actin gene resulted in the use of cryptic splice sites and the excision of RNA fragments of different lengths. This signal sequence of the yeast intron may be recognized by some factor(s) of the splicing machinery as an anchor site from which the transcript can be scanned for potential splice sites.

S. Gerrard (5) reported experiments on the X. laevis 5S RNA genes with polymerase III. If transcripts from gene fragments are hybridized to the complete 5S gene, and non-hybridized RNA digested with RNase, a discrete fragment of RNA results (if termination at its 3' end was correct). Though 5S RNA was correctly transcribed by Xenopus germinal vesicle extracts, purified polymerase III initiated randomly and made no correct 5S transcripts. Nevertheless, the same discrete fragment of correctly terminated RNA was obtained in both cases. Thus none of the factors required for transcription in vitro of 5S RNA genes are required for correct termination of transcription: purified X. laevis RNA polymerase III is sufficient.

R. Kaufman's studies (6) on dihydrofolate reductase (DHFR) synthesis utilized genes constructed from the adenovirus major late promoter, a DHFR cDNA segment, and fragments of SV40 (for polyadenylation signals), introduced into Chinese hamster ovary cells. DHFR mRNAs are identical at their 5' ends but differ at their 3' ends in different transformants, because different sites are utilized for polyadenylation. Those utilizing either DHFR polyadenylation signals or the SV40 late polyadenylation signal exhibit growth dependent DHFR synthesis (mRNA levels in growing cells are 10x those in stationary cells), resulting from post-transcriptional events. Three transformants using the SV40 early polyadenylation signal and one using a cellular polyadenylation signal are not growth dependent in DHFR synthesis, thus polyadenylation specificity must be the governing factor. Since growing and stationary cells differ in the fraction of mRNA polyadenylated at different sites in a tandem array, the metabolic state is important in determining either the efficiency of polyadenylation at various sites or the stability of the corresponding mRNAs.

C. Cole (7) used the herpes thymidine kinase (tk) gene, resected from its 3' end to eliminate the normal AAUAAA signals, to examine signal requirements for processing and polyadenylation. After transfer to an SV40-origin-containing plasmid, for replication and analysis of transcription in cos-1 cells, processing/polyadenylation signals from SV40 and polyoma DNA fragments were inserted at the 3' end of the resected gene. Very large differences in levels of tk mRNA are found in cos-1 cells transfected by different gene constructs, due to differential utilization of the signal, stability of the tk mRNAs, or both. Insertion of an 88 bp fragment of SV40 DNA containing AAUAAA (but normally not involved in polyadenylation), or fragments of African green monkey DNA, restored tk expression to a low level and was always associated with the production of polyA+ mRNA. The high frequency of restored gene expression suggests a simple signal is necessary for processing/polyadenylation, though additional features are clearly important.

Y. Aloni (8) proposed an attenuation mechanism for SV40 late transcription. RNA polymerase II can terminate _in vitro_ 94 nucleotides downstream from its startpoint, in a region typical of prokaryotic termination signals. Termination to produce this aborted RNA, or production of the full primary transcript, may depend on alternative conformations at the 5' end of the RNA. The 16S mRNA encodes the leader agnoprotein and the capsid protein VP1. In one of the two RNA conformations (A), the initiating AUG of agnoprotein is sequestered, leading to (cytoplasmic) synthesis of VP1; in conformation B agnoprotein is synthesized. In the nucleus, in conformation A, RNA transcription is attenuated, in conformation B, a full length transcript is synthesized. These fundamental elements may also regulate the production of the other two capsid proteins, VP2 and VP3.

M. Bendig (9) described the dramatic switch from tadpole to adult hemoglobin production in Xenopus. Though injected adult globin genes are not transcribed in oocytes, both α and β adult globin genes are transcribed if injected into fertilized eggs. These genes remain unmethylated, and detectable of transcripts with authentic 5' ends exist by gastrula stage and persist through hatching and early swimming tadpole stages. In vitro methylation of the adult genes before injection reduces but does not eliminate their expression during development, thus the injected genes are not in the same totally repressed state as the endogenous adult genes.

D. Peabody (10) reported on SV40-derived vectors that produce polycistronic mRNA's in animal cells, to examine factors influencing ribosome recognition of AUG's other than those nearest the 5' end. A dicistronic transcription unit (mouse DHFR followed by E. coli xanthine-guanine phosphoribosyl transferase, XGPT) transferred into monkey cells produces only the expected dicistronic mRNA, and both enzymes. When the translation terminator of DHFR is deleted, initiation at the downstream XGPT sequence is abolished, suggesting that initiation at an internal AUG occurs by a termination-reinitiation mechanism, though additional factors may also be involved. When several ATG's are placed upstream and in frame with XGPT, a series of electrophoretic variants of the enzyme is produced, each apparently corresponding to initiation at one of the upstream initiators. Thus some AUG's are inefficiently recognized, and ribosomes can scan past them and initiate translation at downstream AUG's.

Abstracts of speakers and their coworkers appear in the Journal of Cellular Biochemistry, Supplement 7A, 1983. (1) McKenney (MRC, Cambridge). (2) McSwiggen, Bear, Feinstein, Singer, Morgan, Platt, & von Hippel (U. of Oregon). (3) Greenblatt (U. of Toronto). (4) Gallwitz, Nellen, & Langford (Marburg). (5) Gerrard & Cozzarelli (UC Berkeley). (6) Kaufman, Lo & Sharp (MIT). (7) Cole & Santangelo (Yale). (8) Aloni, Hay, & Skolnik-David (Weizmann). (9) Bendig, Mahbubani, & Williams (ICRF, London). (10) Peabody & Berg (Stanford).

Gene Expression, pages 571-575
© 1983 Alan R. Liss, Inc., 150 Fifth Avenue, New York, NY 10011

DNA MODIFICATIONS AND GENE EXPRESSION: WORKSHOP SUMMARY

Suzanne Bourgeois

The Salk Institute, San Diego, California 92138

Two types of DNA "modifications" were considered: methylation and gene amplification. In prokaryotes a DNA adenine methylase (dam) methylates the adenine residue in the sequence GATC, and a DNA cytosine methylase (dcm) methylates the internal cytosine in the sequence CCA/TGG. In addition, prokaryotes contain a variety of restriction modification methylases that methylate a cytosine or adenine at specific sites recognized by endonucleases. Thus, a major role of DNA methylases in bacteria is to protect the cellular DNA against endonucleases designed to attack invading foreign DNA and, in particular, phage DNA. However, recent evidence indicates that some bacterial methylases can also act as positive control elements activating the expression of specific genes. S. Yanofsky (Univ. of Cal., San Francisco) presented data showing positive control of EcoRI endonuclease expression by the EcoRI methylase. Since the presence of EcoRI endonuclease in cells lacking the cognate methylase is lethal, a mutant endonuclease was used that allows the cells to survive when the methylase is deleted. Deletion of the methylase results in drastically reduced endonuclease production, but near normal levels can be restored by the presence of a methylase provided in trans on a plasmid. The question of the mechanism by which this activation might occur was raised: obvious possibilities include methylation of a site or interaction with a site near the endonuclease promoter. The existence of a methylase deletion, which destroys the methylating activity while retaining stimulation of endonuclease production, makes it unlikely that activation is simply the result of methylation. Recently, Hattman (1) observed dam-dependent transcription of the phage Mu mom gene and obtained results favoring the idea that the dam enzyme may act by virtue of its ability to methylate a specific sequence. It will be interesting to see whether *E. coli* RNA polymerase can distinguish unmethylated from methylated promoters and whether the expression of other prokaryote genes is regulated by methylases.

In higher eukaryotes the only modified base found in DNA is methylcytosine (m^5C), mainly present in CG sequences. The observation that levels of m^5C vary widely in different organisms implies that the functions served by m^5C are not universal or are mediated by other mechanisms in some species. The workshop focused on the hypothesis that m^5C is involved in regulation of gene expression and differentiation. Several molecular mechanisms are known which could account for a role of m^5C in gene expression, e.g. effects on protein-DNA interactions and on the transition of B DNA to left-handed forms. Moreover, methylation patterns are tissue-specific and conserved throughout cell divisions. Although the correlation between DNA undermethylation and gene expression is not perfect, several striking examples were presented at the workshop.

How can methylation patterns be changed? Since no demethylase has been found it is assumed that methyl groups are lost upon DNA replication in the absence of methylation. This may account for the spectacular effects of 5-azacytidine, widely used to activate silent genes, although such effects should be interpreted with caution since they could possibly result from inhibition of processes other than DNA methylation. The fundamental question is that of the mechanism(s) by which methyl groups might be removed at specific sites, during the normal process of differentiation, in response to signals such as hormones.

M. McGeady (NCI, Bethesda) presented data showing that *in vitro* methylation of cloned Molony sarcoma virus (MSV) proviral DNA results in a drastic decrease in its capacity to transform NIH 3T3 cells (2). This effect is reversed by treatment of the transfected cells with 5-azacytidine. The cells that were transformed by methylated DNA contained only copies of MSV DNA which had lost methyl groups. These results suggest that undermethylation is necessary for expression of *v-mos*, the oncogene of MSV. However, the identification of the specific sites of demethylation is still incomplete. P. Yen (Harbor-UCLA Medical Center, Torrance) addressed the controversial issue of the role of methylation in X-chromosome inactivation and presented data demonstrating a distinctive methylation pattern of the hypoxanthine-guanine phosphoribosyl transferase (HPRT) gene associated with the inactive human and murine X-chromosomes.

A Wilks (Friedrich Miescher Inst., Basel) presented data showing estrogen-induced demethylation at the 5' end region of the chicken vitellogenin gene (3). Since the expression of this gene is estrogen dependent, undermethylation appears to result from the activation of transcription of that gene. The demethylated site is located 611 base pairs (bp) from the <u>cap</u> site of the gene, a region which appears to bind estrogen receptor. A similar observation was made by Mermod et al. (4) who obtained evidence of glucocorticoid-induced demethylation near the promoter of mouse mammary tumor proviral DNA. These results support a model in which steroid receptor-mediated initiation of transcription can lead to site-specific demethylation, a mechanism that could conceivably play a role in hormone-induced differentiation.

The expression of the human β-like globin genes cluster, 5'ε-Gγ-Aγ-δ-β 3', switches during development from the early embryonic ε gene, to the fetal γ genes and finally to the adult δ and β genes. The only changes observed in the globin gene cluster upon switching occur in the DNA methylation patterns. F. Grosveld (Natl. Inst. Med. Res., London) presented results obtained by transfecting into L cells a cloned γ-globin gene hemimethylated <u>in vitro</u> in specific regions. Methylation of a 600 bp region at the 5' end of the gene inhibits expression, whereas methylation at any other region of the γ-globin DNA has no effect. This system will allow the identification of the expression-sensitive methylated sites within this 600 bp region, which still contains 10 CG sites.

A. Nienhuis (NIH, Bethesda) reported on the clinical use of 5-azacytidine to activate the expression of the fetal γ-globin synthesis in patients with severe β^+-thalassemia and sickle-cell disease. Clinical effects were striking: a 4 to 6 fold increase in γ-globin synthesis resulted in partial correction of these anemias (5). Surprisingly, 5-azacytidine selectively increased γ-globin gene expression, despite global demethylation within the β-globin genes cluster and near other genes such as the insulin gene. These results imply that, even if the effect of 5-azacytidine is a consequence of DNA demethylation, other regulatory mechanisms determine the final pattern of globin gene expression. Although the treatment was well tolerated, the long-term risks (including the carcinogenic potential) of 5-azacytidine need to be assessed before extending its therapeutic use. Another 5-azacytidine-sensitive system with potential clinical applications was recently described by Gasson et al. (6) who demonstrated in vitro that a glucocorticoid-resistant mouse T-lymphoma line was made glucocorticoid-sensitive by brief treatment with low doses of 5-azacytidine. This result suggests that some glucocorticoid-resistant T-cell malignancies may be sensitive to combination therapies including a glucocorticoid and 5-azacytidine. Again, demethylation of the T-lymphoma DNA by 5-azacytidine was global, involving close to 50% of the methylated DNA sites, without change in cell morphology or doubling time. This indicates that many m^5C do not play a role in gene expression and/or that demethylation often is not a sufficient condition for gene expression.

The last three presentations at the workshop dealt with gene amplification, a mechanism widely used by cells to overcome drug toxicity by overproducing a specific protein. This phenomenon can be exploited to facilitate the cloning of specific genes. This was illustrated by Y-F.C. Lau (Univ. of Cal., San Francisco) who described a vector in which a human α-globin gene was covalently linked to the dihydrofolate reductase (DHFR) gene. After transfection into CHO cells and initial selection to retain the DHFR and linked sequences, the cells were cultured in increasing concentrations of methotrexate. Both the α-globin and DHFR genes, integrated together in the CHO genome, were amplified 500 to 1,000 fold and their transcriptional activity increased.

The expression of the mammalian metallothionein genes (MTI and MTII) is affected by DNA methylation and gene amplification, as well as regulated by metals and glucocorticoid hormones. C. Hildebrand (Los Alamos Natl. Lab.) presented evidence that the switch of CHO cells from a Cd^{2+} sensitive (Cd^s) to a Cd^{2+} resistant (Cd^r) phenotype correlates with inducibility of both MTs, and is controlled by DNA methylation in the region of the MT genes. Increased Cd^{2+} resistance, developed during exposure of

MT inducible Cd^r cells to increasing concentrations of Cd^{2+}, correlates with overproduction of both MTs and coordinate amplification of both MT genes. J. Koropatnick (Univ. of Calgary) presented results indicating that, in mouse liver, the number of MTI gene copies is increased not only following treatment with toxic levels of Cd^{2+}, but also during liver development: compared to the adult MTI gene concentration, 19-day fetal livers have a 2 to 3 fold higher number of MTI genes, rising to 4 to 6 fold higher 6 days after birth and falling to adult levels by 30 days after birth. The role of this short-term amplification, resulting in increased MTI production during development, is intriguing and still unclear.

REFERENCES

(1) Hattman S (1982) DNA methyltransferase-dependent transcription of the phage Mu mom gene. Proc Natl Acad Sci USA 79:5518.

(2) McGeady ML, Jhappan C, Ascione R, and Vande Woude GF (1983). In vitro methylation of specific regions of the cloned Moloney sarcoma virus genome inhibits its transforming activity. Mol and Cell Biol 3:305.

(3) Wilks AF, Cozens PJ, Mattaj IW, and Jost J-P (1982). Estrogen induces a demethylation at the 5' end region of the chicken vitellogenin gene. Proc Natl Acad Sci USA 79:4252.

(4) Mermod J-J, Bourgeois S, Defer N, and Crépin M (1983). Demethylation and expression of murine mammary tumor proviruses in mouse thymoma cell lines. Proc Natl Acad Sci USA 80:110.

(5) Ley TJ, DeSimone J, Anagnou NP, Keller GH, Humphries RK, Turner PH, Young NS, Heller P, and Nienhuis AW(1982). 5-Azacytidine selectively increases γ-globin synthesis in a patient with β^+ thalassemia. The New England Journal of Medicine 307:1469.

(6) Gasson JC, Ryden T, and Bourgeois S (1983). Role of de novo DNA methylation in the glucocorticoid resistance of a T-lymphoid cell line. Nature 302:621.

Index

Eukaryotic cloning vector, 543